M. Cresti · G. Cai · A. Moscatelli (Eds.)

Fertilization in Higher Plants

Springer
Berlin
Heidelberg
New York
Barcelona
Hong Kong
London
Milan
Paris
Singapore
Tokyo

M. Cresti · G. Cai · A. Moscatelli (Eds.)

Fertilization in Higher Plants

Molecular and Cytological Aspects

With 159 Figures, Including 6 Color Plates

Springer

Prof. Dr. MAURO CRESTI
Dr. GIAMPIERO CAI
Dr. ALESSANDRA MOSCATELLI

Universita degli Studi di Siena
Dipartimento di Biologia Ambientale
Via P.A. Mattioli 4
53100 Siena
Italia

ISBN 3-540-64879-8 Springer-Verlag Berlin Heidelberg New York

Library of Congress Cataloging-in-Publication Data

Fertilization in higher plants : molecular and cytological aspects / [edited by] Mauro Cresti, Giampiero Cai, Alessandra Moscatelli. p. cm. Includes bibliographical references and index. ISBN 3-540-64879-8 (alk. paper). 1. Fertilization of plants. 2. Angiosperms – Molecular aspects. 3. Angiosperms – Cytology. I. Cresti, M. (Mauro) II. Cai, Giampiero, 1963– . III. Moscatelli, Alessandra, 1960– .
571.8'6422 – dc21

Production: PRO EDIT GmbH, D-69126 Heidelberg
Cover Design: design & production GmbH, D-69121 Heidelberg
Cover Illustration: See Chap. 8 Molecular Approach to
Female Meiosis in *Petunia Hybrida*,
Porceddu et al.

Typesetting: Zechnersche Buchdruckerei, D-67346 Speyer

SPIN: 10654542 31/3137-5 4 3 2 1 0 – Printed on acid-free paper

Preface

Fertilization requires an intimate partner relationship. It takes place at the cellular level: the egg cell interacts with the sperm cell.

The peculiar situation in Higher Plants is the Double Fertilization. Each syngamic process in Angiosperms consists of two events: the fusion of the egg cell with one sperm cell resulting in the diploid zygote, and the fusion of embryosac nuclei with the other sperm cell, leading to the production of the triploid endosperm. A prerequisite for the double fertilization of Angiosperms is the synchronuous transport of two sperm cells by the pollen tube.

The fertilization in higher plants is not only preceded by the pollination process, with its complicated interactions and with the assistance of pollinating agents (wind, water, animals), but also by a long lasting interaction period between the diploid pistil and the haploid pollen tube, as the carrier of the male material. This interaction phase can be described as the progamic phase, which starts with the landing of the pollen grain on the stigmatic surface, and ends with the syngamy, the fusion of the sexual cells.

All these coordinated events are well known for a limited number of plant species and have been treated for a long time as a section of plant science called embryology. During the last 40 years it became evident that the physiology of the events, which precede the final fusion of the sexual cells in flowering plants, is of great importance for understanding the fertilization processes and their genetic barriers (incompatibility and incongruity).

The fertilization of flowering plants results in the transformation of the zygote into seeds and fruits, the final products of fertilization. Fruit and seed formation are basic processes for the world food supply. Crops, like cereals (e.g. rice, corn, wheat), legums (e.g. soya, peas, beans), fruits (e.g. apples, oranges, almonds, chestnuts), grapes are essential staples to human nutrition, not to mention spices (e.g. pepper, nutmeg). Also, a greater part of the mankind provisions are fats and oils (e.g. olive, corn, oil palm, rape, peanut), semi-luxurious items (e.g. coffee) and fibre delivering plants (e.g. cotton, capoc). The pigments and many natural remedies are the final results of the successful fertilization processes in higher plants. Fundamental understanding and the increased ability to manipulate the gene-controlled stages of fertilization are essential for the survival of mankind.

The papers compiled in this book are the written and extended versions of lectures from the European Union Experts which have been given during a Euro Advanced Course.

This Euro Course entitled "Sexual Plant Reproduction and Biotechnological Applications: Recent Advances by Molecular Biology, Biochemistry and Morphology" has been supported by the European Commission, and was organized by the undersigned.

Lectures were given in the Conference Center of the University of Siena, the Certosa di Pontignano, whereas the practical courses and demonstrations took place in the Laboratories of the Dipartimento di Biologia Ambientale – Università degli Studi di Siena.

This compilation is a snapshot of this distinctive activity in a field of great importance. The Authors have full responsibility for their papers, which have not been reviewed or edited. We thank all the Speakers and the Instructors of the Course for their contributions, the Participants for their enthousiastic approach, the Technical Staff for their devoted assistance and the Publisher for the fast production of these proceedings.

Our hope is that this picture of the present state of the art demonstrates the fast progress made in this field of research, especially when compared with the earlier proceedings of the 1990 workshop entitled "Sexual Plant Reproduction" (Springer, Berlin, Heidelberg, New York, 1991).

In addition, it became evident that pollen, pollen tubes and egg cells are excellent examples for plant cell research, especially suitable for investigations on cell tip growth and polarization, signal transduction, channel and ion flux activity, gene expression, cytoskeleton and cell wall structure, biosynthesis and accumulation of specific substances.

This basic research and its results are important for future steps toward manipulation of the fertilization processes in higher plants. The manipulation will not only concern the testing of inhibiting drugs, but also the effects of pesticides used in agriculture, horticulture, viniculture and arboriculture. The future application of basic information gained on the steps of fertilization process will contribute to the promotion of genetic engineering in plants, in order to transform and improve crop plants.

Cell culture and suspension culture are already methods, where basic science and biotechnology have met. Pistil activation is the next step to be handled for the improvement of important economic fruit production. The role of special genes and mutations will offer insight into the regulation of early zygotic embryogenesis in higher plants, and also into the effects of externally applicated phytohormones and their influence on the control of maturation of zygotic embryos. Basic knowledge of the induction processes, local control and patterning of cell division, especially the meiotic processes, which are essential for the formation of the sexual cells, are necessary for the progress of biotechnology in sexual plant reproduction.

Siena, Spring 1998

Mauro Cresti

Contents

CHAPTER 1

Polyamines and Gene Expression of Biosynthetic Enzymes in Sexual Plant Reproduction

N. Bagni* · A. Tassoni · M. Franceschetti

* Department of Biology and Interdipartmental Center for Biotechnology, University of Bologna, Via Irnerio 42, 40126 Bologna, Italy
e-mail: bagninel@kaiser.alma.unibo.it
telephone: 0039-51-351280
fax: 0039-51-242576

Abstract. Content, biosynthesis and mechanism of action of aliphatic polyamines in higher plants are reviewed, pointing out especially their role in plant growth and development. Pollen formation and germination and sexual reproduction have been correlated with polyamines metabolism and gene expression in particular in two different model systems: apple and tobacco. The role of polyamines in the regulation of sexual reproduction has been emphasized.

1.1 Polyamines in Higher Plants: An Overview

In old books of plant biology, polyamines were considered secondary metabolites and/or precursors of alkaloid synthesis, but over the last three decades an impressive body of evidence has suggested that aliphatic polyamines play an essential role in the growth of plants, animals and microorganisms (Bagni, 1989). The roles of polyamines in plants, in decreasing order of importance are the following:

Plant growth regulators
Plant hormones
Second messengers
Precursors of alkaloids
Secondary metabolites
Organic polycations
Nitrogen source

The most common aliphatic polyamines, the diamine putrescine (1,4-diaminobutane), the triamine spermidine (1,8-diamino-4-azaoctane) and the tetramine spermine (1,12-diamino-4,9-oliazaoloole-cane), can be considered ubiquitous compounds synthesized in microorganisms, animals and plants (Figure 1.1). The biosynthesis of polyamines starts mainly from basic aminoacids ornithine and/or arginine through direct decarboxylation. Agmatine, a guanidine-amine decarboxylation product of arginine, occurs widely throughout the plant kingdom and was found for the first time in ragweed pollen (*Ambrosia artemisifolia*). Figure 1.1 shows the polyamine biosynthesis pathway, the interconversion between arginine and ornithine, the possible utilization of polyamines for alkaloid synthesis through putrescine-methyltransferase, and the pivotal role of S-adenosylmethionine (SAM) as precursor of spermidine and spermine. SAM can also be utilized, via 1-aminocyclopropane-1-carboxylic acid, for the synthesis of ethylene, hormone responsible of senescence processes in plants.

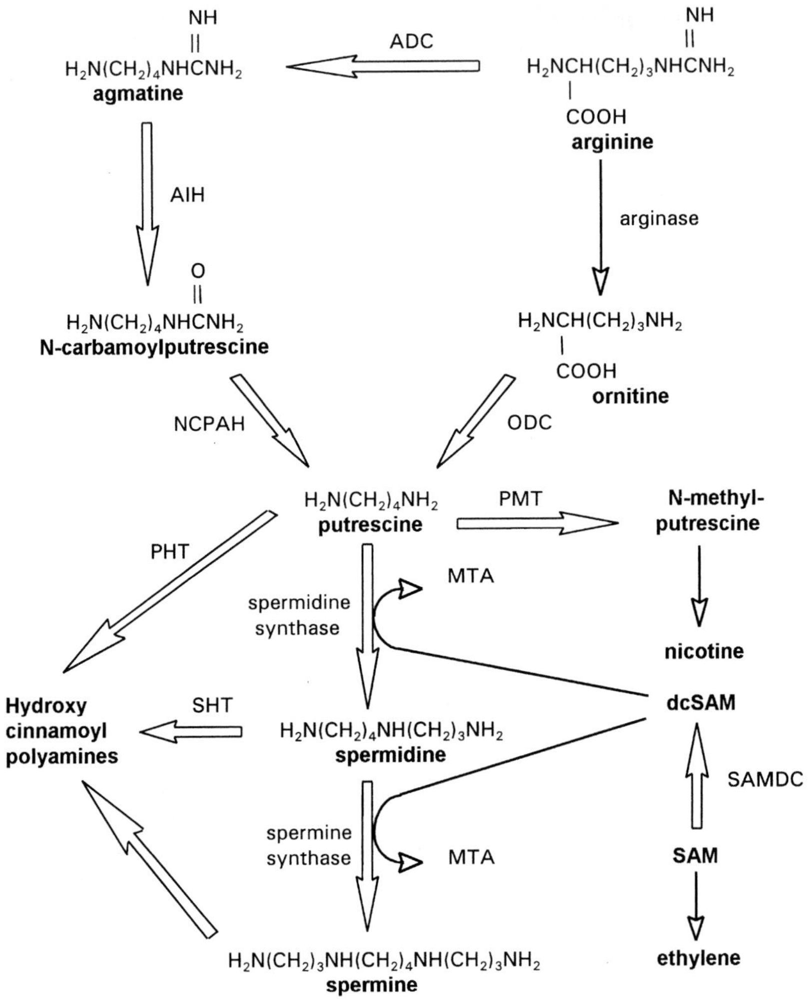

Fig. 1.1. Metabolic pathway for the biosynthesis of polyamines in higher plants and interaction with ethylen and alkaloid synthesis. *ADC*, arginine decarboxylase; *AIH*, agmatine iminohydrolase; *NCPAH*, N-carbamoylputrescine aminohydrolase; *ODC*, ornithine decarboxylase; *MTA*, 5′methyltioadenosine; *SHT*, spermidine hydroxylcinnamoyl transferase; *SAMDC*, S-adenosylmethionine decarboxylase; *dcSAM*, decarboxylated S-adenosylmethionine; *SAM*, S-adenosylmethionine

A high homologue of putrescine, the diamine cadaverine (1,5-diaminopentane), derived through decarboxylation of lysine, is frequently found, especially in members of the Leguminosae. Homologues of spermidine and spermine are quite frequent in lower plants and unusual polyamines are also found in members of Leguminosae, notably *Canavalia gladiata* and *Vicia sativa* (Matsuzaki et al., 1990; Hamana et al., 1991). Polyamines act, like other plant hormones or plant growth regulators, in the free forms (Bagni et al., 1994). By virtue of their cationic nature at physiological pH, they can interact both specifically and aspecifically with several cell molecules, such as nucleic acids (Feuerstein and Marton, 1989), acidic phospholipids (Tadolini et al., 1985) and pectic substances (D'Orazi and Bagni, 1987), thus affecting their structure and function. Polyamines can also be conjugated with monomers and dimers of hydroxycinnamic acids (TCA-soluble fraction) (Figure 1.2) or with more complex structures (TCA-insoluble fraction), as trisubstituted hydroxycinnamic acid spermidines, recently found in *Quercus dentata* pollen (Bokern et al., 1995). The presence of these conjugates is of particular importance both for the regulation of polyamine concentration inside the cell (Bagni and Pistocchi, 1990) and for their hypothetic interaction with cell wall components in mature pollen. In fact hydroxycinnamic acids bridge, through ester-ether linkages, different cell wall polymers, essentially hemicelluloses and lignin (Markwalder and Neukom, 1976; Hartley and Jones, 1976; Lam et al., 1992).

Polyamines can also be conjugated with proteins by transglutaminases (TGases), a family of enzymes that catalyse the incorporation of primary amines into endogluta-

Fig. 1.2. Amides of hydroxycinnamic acids formed with polyamines which can be found in different higher plants

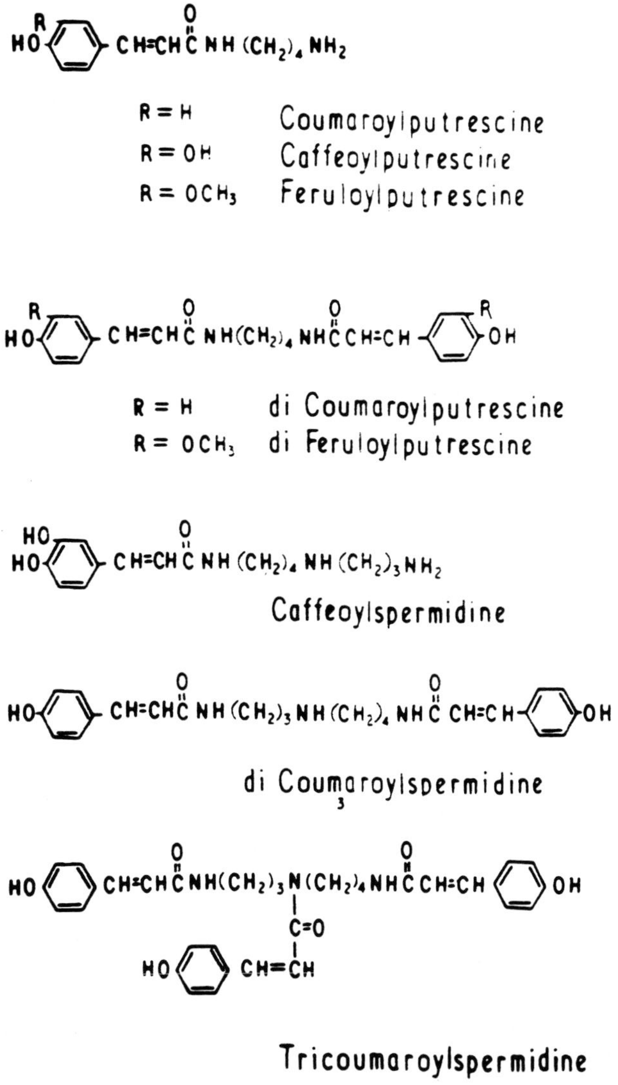

mine residues of specific proteins whose importance in pollen germination will be discussed later (Del Duca et al., 1997).

Polyamines have been correlated to a variety of growth and developmental processes in plants, including cell division, that can be considered their main effect, inhibition of xylem differentiation, embryo formation in tissue culture, root initiation, adventitious shoot formation, flower initiation and development and control of fruit ripening and senescence. In addition, changes in endogenous concentrations and ratio between polyamines and other hormones or plant growth regulators, modify morphogenetic processes. However, polyamines alone are able to induce and substain cell division (see review Bagni et al., 1993), but seem not to play a specific role in morphogenesis.

Studies on polyamine functions can be carried out using mutants, tissues temporarily deficient of polyamines or inhibitors of their biosynthesis.

Mutants or variants with altered polyamine biosynthesis or action, have been successfully isolated in microorganisms including *E. coli*, yeasts and *Neurospora*. Tabor and Tabor (1989) clearly showed that in *E. coli* a mutation inactivating the ornithine decarboxylase (ODC) locus prevents growth, while a mutation at the spermidine synthase locus permits slow growth, but not sporulation. This clearly demonstrates the essential and specific role of particular polyamines. Because higher plants have arginine decarboxylase (ADC) pathway in addition to the ornithine decarboxylase one, it is more difficult, using traditional methods of mutational analysis, to get definitive results with single mutants as can be done on the contrary with *E. coli* and yeast. Consequently, at the present time, polyamine action was demonstrated only in tissues temporarily deficient in polyamines, such as *Helianthus tuberosus* tuber during dormancy, or by using inhibitors of polyamine synthesis that reduce endogenous concentrations (Bagni, 1989).

Recently, molecular biology techniques have been employed to try to achieve unambiguous results. The mutant characterization, the production of transgenic plants with antisense messages that damp out polyamine production, or the overexpression of biosynthetic genes, should help us to understand the effect of variation in gene expression and consequently in polyamine content.

Why polyamines and reproduction?

The answer to this question goes back three hundred years, when in 1677 Anthonii van Leeuwenhock, the discoverer of the microscope, sent a letter to the Royal Academy of London with a description of crystals found in human seminal fluid. These crystals were identified, two hundred years later in 1888, by Schreiner, as an organic base of phosphate named spermine.

In different animal tissues, the distribution of polyamines, particularly spermidine and spermine, and of their biosynthetic enzymes S-adenosylmethionine decarboxylase (SAMDC) and spermidine and spermine synthases, suggest again a role of these compounds in normal and tumor growth processes. In particular, spermidine seems to be involved, in rat seminal plasma, in the modulation of immunocompatibility of spermatozoa (Raina et al., 1978; Porta et al., 1988). Moreover, in human seminal fluid, the high spermine level plays an important role as antioxidant, thus preventing the potential risk of oxidative destruction of genetic material (Løvaas, 1994).

As regard plant sexual reproduction, pollen contains a really high polyamine content, especially of spermidine which was detected for the first time in wheat pollen (Bagni et al., 1967). The increase of nucleic acid synthesis, induced by polyamines, in germinating pollen of *Petunia* (Linskens et al., 1968) suggests a specific role of these molecules in influencing growth and development of male gametophyte.

1.2 The Role of Polyamines in Pollen

The role of polyamines in pollen was previously reviewed by Bagni (1986).

In this chapter we describe two model systems, extensively investigated in our laboratory, which are representative of two different categories of pollens, one entomophyl (apple) and one self pollinated (tobacco).

1.2.1 Polyamines During Pollen Germination and Fertilization Processes in Apple

In apple mature pollen, polyamines are present in free forms and bound to soluble peptides at low molecular weight (~ 5 kD); spermidine is the most abundant, while putrescine and spermine are present in low amount. After 30 minutes of hydration no changes of polyamines are detectable, while a general decrease, particularly of spermidine, is evident during tube emergence and progression of tube elongation, probably due to their release in the medium. A spermidine increase is noticeable only after 2 h of germination. Polyamines bound to peptides show a similar trend (Bagni et al., 1981). As regards polyamine biosynthesis, arginine, present in high amounts (about 4.2 mM), represents the main putrescine precursor. By using labelled arginine, high arginase activity, responsible for ornithine formation as well as arginine decarboxylase activity and putrescine formation, throughout agmatine and N-carbamylputrescine, were detected. An exogenous supply of agmatine in the medium inhibited ADC and enhanced the production of N-carbamylputrescine, putrescine and spermidine (Bagni et al., 1981; Bagni et al., 1986). During germination, contemporarily to polyamine synthesis, there is an increasing release of free polyamines in the medium, higher for spermine, the minor polyamine (Speranza e Calzoni, 1980) which is slowly synthetized throughout germination, suggesting the involvement of an active mechanism of secretion. In plant cells, only energy-dependent polyamine uptake systems have been identified at the present time (Bagni and Pistocchi, 1990). Nevertheless, the presence, at plasmalemma level, of ionic channels utilized by polyamines for efflux from cell, was pointed out (Colombo et al., 1992).

The results concerning content and biosynthesis of RNAs in pollen are particularly interesting because of their relation with polyamines. As regard RNAs biosynthesis, polyA RNA is rapidly synthetized within 15 minutes of germination, followed by the formation of 5S RNA, rRNA and later tRNA, suggesting the activation of genes, that code for rRNA, tRNA and mRNA, before germination. It is interesting to underline that polyamine and RNA synthesis occur at the same time. Among total RNAs, polyamines and proteins, only the last ones display a decrease during pollen hydration, while all those compounds show a similar trend and generally decrease during germination (Bagni et al., 1981). Moreover, peaks of ribonuclease and protease activities exactly correspond to the decrease of total RNA and proteins (Speranza et al., 1984). Therefore, the interaction polyamine-RNAse could be a key factor in regulating the life of RNAs and the possibility of its synthesis during germination.

Recently, the polyamines covalently linked to proteins by TGase, were investigated in apple pollen. This enzyme is active in hydrated, as well as in germinating pollen where it doubled activity after 90 minutes germination. TGase catalyses the incorporation of polyamines mainly into proteins of 43 kD and 52–58 kD in both ungerminated and germinating pollen. Autoradiography of SDS-PAGE, clearly showed that both actin and tubulin are substrates of TGase. Thus, the pollen TGase may be involved, during pollen rehydration, in the rapid cytoskeletal rearrangement which is essential during pollen tube apical growth (Del Duca et al., 1997). In this rapid cell expansion, the supplied polyamines, by virtue of their interactions with pollen membrane constituents, are involved and can partly substitute the cation bridges necessary for membrane integrity in absence of Ca^{2+}. This might be true for the entire pollen membrane system, and/or in particular for the new membrane-forming zone of the tube tip, where polyamines themselves might allow growth through control of secretion (Speranza et al., 1983). Moreover, also specific bindings of polyamines to plasmalemma tube tip proteins cannot be excluded, as recently demonstrated for zucchini hypocotyls (Tassoni et al., 1996).

In conclusion, biosynthesis of RNAs and polyamines and TGase activity precede and are involved in tube emergence, also occurring a polyamine mediated control of RNAse activity (Bagni et al., 1986; Speranza et al., 1984). It is possible to suggest a role of polyamines in the progamic phase of the fertilization processes, this hypothesis being also supported by additional results on the effect of polyamines on field fertilization. Aliphatic polyamines when sprayed after full bloom, on open and artificially self pollinated flowers of apple, increase fruit-set and yield per tree (Costa and Bagni, 1983) when environmental conditions, particularly temperature, reduce or block natural endogenous polyamine synthesis in flowers. Polyamines generally increased fruit growth rate during stage 1 (cell division), but not thereafter (cell enlargement). Flower bud formation was also increased the following year. A good correlation between polyamine content, biosynthetic enzymes activities (ODC and ADC) and fruit growth was observed before and after anthesis and during initial stages of fruit development (Biasi et al., 1988; 1991).

1.2.2
Polyamines and Gene Expression of Polyamine Synthesis During Microsporogenesis, Pollen Maturation and Germination in Tobacco

As polyamines are present in high amount (10–14 mM) in different pollens such as apple (Bagni et al., 1981) and kiwi (Dicotiledons) (Franceschetti et al., unpublished data), but not in *Pinus parolinii* (Conifers) (Bagni and Bordoni, 1980) in which the concentration is near 350 μM, it was interesting to understand the importance of polyamine biosynthesis during the different stages of microsporogenesis as well as the expression of genes involved. To this purpose, tobacco, a plant in which the polyamine content and the rate and gradient of biosynthesis both *in planta* and *in vitro* culture were well known (Altamura et al., 1993; Bagni et al., 1993), was utilized.

Putrescine, spermidine and spermine were present both in free and TCA-soluble form in the mid-binucleate pollen, characterized by an intermediate level of starch, and increase throughout pollen maturation. Spermidine was the most abundant polyamine in all the stages, followed by putrescine and spermine. During germination polyamines decrease (Chibi et al., 1993) in a similar pattern to that described for apple pollen (Bagni et al., 1981).

Polyamine biosynthetic inhibitors, α-difluoromethylariginine (DFMA) and α-difluoromethylornithine (DFMO), irreversible inhibitors of ADC and ODC respectively, and cyclohexylamine, competitive inhibitor of spermidine synthase, were used to study maturation and germination processes. Although both enzymes ODC and ADC appeared to be involved in germination processes, putrescine biosynthesis via ODC is certainly required for pollen maturation. The inihibition of germination induced by DFMO and DFMA was reverted by the addition of putrescine (Chibi et al., 1993). Plants regenerated from cultures resistant to methylglyoxal-bis(guanylhydrazone), competitive inhibitor of SAMDC, show alteration in floral morphology (Malmberg and McIndoo, 1983). Their characterization, together with the high levels of SAMDC activity in tobacco flowers and the general high level of spermidine in pollen (Malmberg et al., 1985) prompted the analysis of the expression of SAMDC and spermidine synthase in greater detail. These determinations were carried out in tobacco flowers at different developmental stages and in separated ovaries, anthers, petals, and mature pollen. The data were correlated with polyamine content and enzyme activities.

A SAMDC transcript, of approximately 2.0 Kb, displayed a relatively high expression in early stages of ovary and anther development and in mature pollen. Pollen contained more transcript than the pre-anthesis whole anthers. Spermidine synthase showed a similar trend but with a lower expression.

The increase in SAMDC activity did not correlate with the pattern of SAMDC transcript accumulation in ovaries, while this correlation existed for anther development. In spite of a relative high accumulation of SAMDC transcript, no enzyme activity was detected in mature pollen (Torrigiani et al., 1998). This lack of activity could signify, according to Sari-Gorla (1992), that ungerminated tobacco pollen contains a significant amount of stored mRNA translated upon germination, leading to the hypothesis of a post-transcriptional control of SAMDC activity

1.3 Concluding Remarks

Polyamine synthesis is present during microsporogenesis and pollen maturation, but also throughout germination. Polyamine levels (free and conjugated with hydroxycinnamic acids) are very high and spermidine is the most abundant one. Pollen maturation and tube emergence are blocked by the inhibition of polyamine synthesis. Free poly-

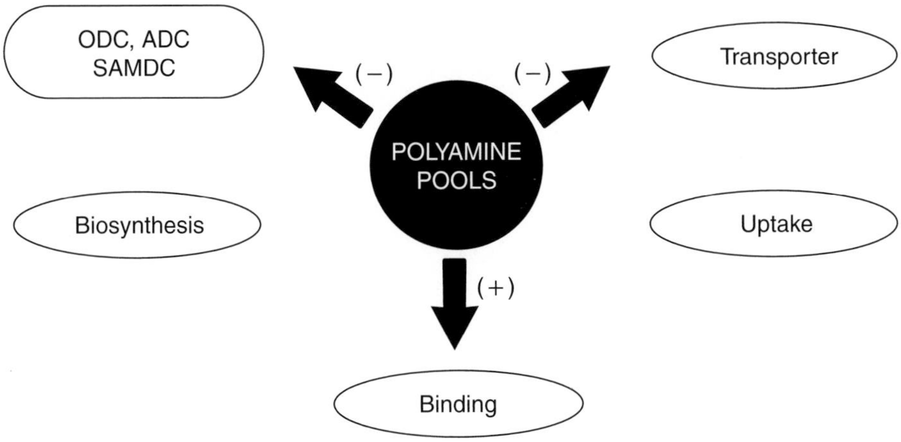

Fig. 1.3. Polyamine homeostasis in plant cells. The three effectors of polyamine pools are: (1) biosynthetic enzymes (*ADC*, *ODC* and *SAMDC*), (2) polyamine transporters, and (3) binding. The first two are negatively regulated by polyamine pool, while the third one is positively regulated as well as catabolic enzymes not shown in the figure

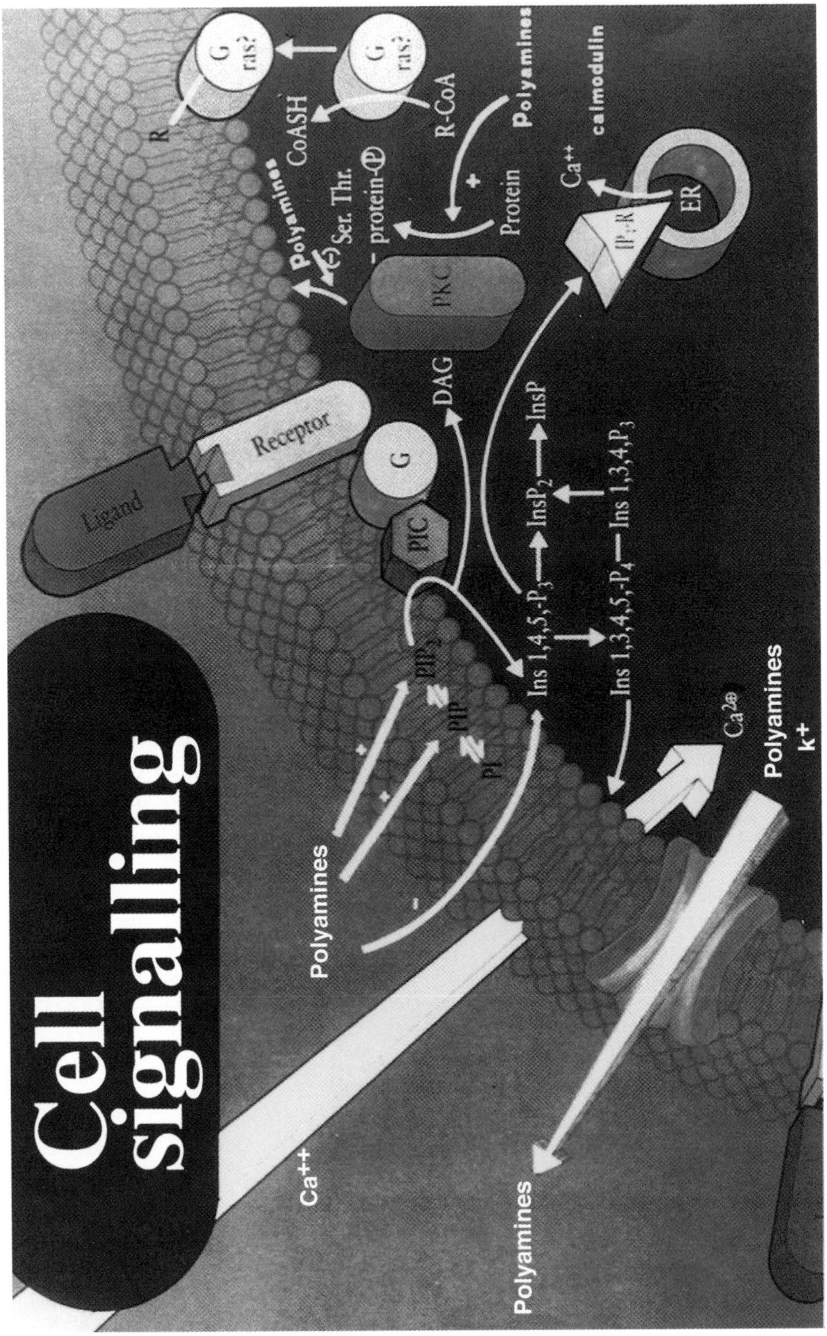

Fig. 1.4. Intracellular signalling transduction affected by polyamines

amines are involved in pollen development, while polyamines conjugated to hydroxycinnamic acids could be important in modifying cell wall components of mature pollen as evidenced in *Quercus dentata* and other pollens of Betulaceae, Fagaceae and Juglandaceae (Bokern et al., 1995). Polyamines linkages with other cell wall components, such as hemicelluloses and lignin, and in addition ionic interactions with pectic substances (Berta et al., 1997) can be responsible for the modification of the rigidity and/or flexibility of cell wall structure. Moreover, in plants, their intracellular concentration is also highly regulated (Pegg, 1986) by processes that control the turnover (Figure 1.3). The majority of tissues and cells, show basal low levels of ODC, ADC and SAMDC, but various stimuli induce their rapid turnover and activity changes (Bagni, 1989). Polyamine metabolism can also be regulated by transport and degradation systems.

Even if polyamines are present in high contents, according to transport and compartmentation studies (Bagni and Pistocchi, 1990), most polyamines are sequestrated in vacuol and cell walls, therefore the cytosolic availability is between two and three order of magnitude lower in respect to total endogenous concentration of the cell. Consequently, a role of polyamines in the transduction of extracellular messages has been proposed (Figure 1.4). Polyamines might act as intracellular signals by stimulating or inhibiting different steps of the phosphoinositols cycle, by stimulating protein kinases activities and blocking their attachment to plasmalemma, by interfering with Ca^{2+}-calmoduline interaction, with specific polyamine uptake Ca^{2+} dependent and Ca^{2+} influx (Bagni and Pistocchi, 1990; Antognoni et al., 1994). Therefore, as a working hypothesis, transduction signals induced by polyamines can be separated into fast responses, responsible for the activation of the systems mentioned above, and slow responses, that require or influence gene expression. Both signals are strictly linked, as also demonstrated in pollen, by polyamine uptake and release.

References

Altamura MM, Torrigiani P, Falasca G, Rossini P, Bagni N (1993) Morpho-functional gradients in superficial and deep tissues along tobacco stem: polyamine levels, biosynthesis and oxidation, and organogenesis in vitro. J Plant Physiol 142:543–551

Antognoni F, Casali P, Pistocchi R, Bagni N (1994) Kinetics and calcium-specificity of polyamine uptake in carrot protoplasts. Amino Acids 6:301–309

Bagni N (1986) The role of polyamines in pollen germination. In: Cresti M, Dallai R (eds) Biology of Reproduction and Cell Motility in Plants and Animals, University of Siena Press, Siena, pp 113–118

Bagni N (1989) Polyamines in plant growth and development. In: Bachrach U, Heimer YM (eds) Physiology of Polyamines, vol 2. CRC Press, Boca Raton Florida, pp 107–120

Bagni N, Adamo P, Serafini-Fracassini D, Villanueva VR (1981) RNA, proteins and polyamines during tube growth in germinating apple pollen. Plant Physiol 68:727–730

Bagni N, Altamura MM, Biondi S, Mengoli M, Torrigiani P (1993) Polyamines and morphogenesis in normal and transgenic plant cultures. In: Roubelakis-Angelakis KA, Tran Thanh Van K (eds) Morphogenesis in Plants, Plenum Press, New York, pp 89–111

Bagni N, Bordoni A (1980) Aspetti biochimici nel processo di invecchiamento del polline. Inf Bot Ital 12:393–398

Bagni N, Caldarera CM, Moruzzi G (1967) Spermine and spermidine distribution during wheat growth. Experientia 23:139–140

Bagni N, Pistocchi R (1990) Binding, transport and subcellular compartmentation of polyamines in plants. In: Flores HE, Arteca RN, Shannon JC (eds) Polyamines and Ethylene: Biochemistry, Physiology and Interactions, Am Soc Plant Physiol, Rockville, Maryland, pp 62–72

Bagni N, Scaramagli S, Bueno M, Della Mea M, Torrigiani P (1994) Which is the active form of polyamines in plants? In: Caldarera CM, Clò C, Moruzzi MS (eds) Polyamines: Biological and Clinical Aspects, CLUEB, Bologna, pp 131–138

Bagni N, Serafini-Fracassini D, Torrigiani P, Villanueva VR (1986) Polyamine biosynthesis in germinating apple pollen. In: Mulcahy D, Bergamini D, Ottaviano E (eds) Biotechnology and Ecology of Pollen, Springer-Verlag, New York, pp 363–368
Berta G, Altamura MM, Fusconi A, Cerruti F, Capitani F, Bagni N (1997) The plant cell wall is altered by inhibition of polyamine biosynthesis. New Phytol 137:569–577
Biasi R, Costa G, Bagni N (1988) Endogenous polyamines in apple and their relationship to fruit set and fruit growth. Physiol Plant 73:201–205
Biasi R, Costa G, Bagni N (1991) Polyamine metabolism as related to fruit set and growth. Plant Physiol Biochem 29:497–506
Bokern M, Witte L, Wray V, Nimtz M, Meurer-Grimes B (1995) Trisubstituted hydroxycinnamic acid spermidine from *Quercus dentata* pollen. Phytochemistry 39:1371–1375
Chibi F, Angosto T, Garrido D, Matilla A (1993) Requirement of polyamines for in vitro maturation of the mid-binucleate pollen of *Nicotiana tabacum*. J Plant Physiol 142:452–456
Colombo R, Cerana R, Bagni N (1992) Evidence for polyamine channels in protoplasts and vacuoles of *Arabidopsis thaliana* cells. Biochem Biophys Res Comm 182:1187–1192
Costa G, Bagni N (1983) Effect of polyamines on fruit-set of apple. HortSci 18:59–61
D'Orazi D, Bagni N (1987) In vitro interactions between polyamines and pectic substances. Biochem Biophys Res Comm 148:1259–1263
Del Duca S, Bregoli AM, Bergamini C, Serafini-Fracassini D (1997) Transglutaminase-catalyzed modification of cytoskeletal proteins by polyamines during the germination of *Malus domestica* pollen. Sex Plant Reprod 10:89–95
Feuerstein BG, Marton LJ (1989) Specificity and binding in polyamine/nucleic acid interaction. In: Bachrach U, Heimer YM (eds) The Physiology of Polyamines, vol 1. CRC Press, Boca Raton, Florida, pp 109–124
Hamana K, Niitsu M, Samejima K, Matsuzaki S (1991) Linear and branched pentaamines, hexaamines and heptaamines in seeds of *Vicia sativa*. Phytochemistry 30:3319–3322
Hartley RD, Jones EC (1976) Diferulic acid as a component of cell walls of *Lolium multiflorum*. Phytochemistry 15:1157–1160
Lam TBT, Iiyama K, Stone BA (1992) Cinnamic acid bridges between cell wall polymers in wheat and *Phalaris* internodes. Phytochemistry 31:1179–1183
Linskens HF, Kuchuyt ASL, So A (1968) Regulation der nucleisäuren-synthase durch polyamine in keimenden pollen von *Petunia*. Planta 82:111–122
Løvaas E (1994) Antiinflammatory and metal chelating effect of polyamines. In: Caldarera CM, Clò C, Moruzzi MS (eds) Polyamines Biological and Clinical Aspects, CLUEB, Bologna, pp 161–167
Malmberg RL, McIndoo J (1983) Abnormal floral development of a tobacco mutant with elevated polyamine levels. Nature 305:623–625
Malmberg RL, McIndoo J, Hiatt AC, Lowe BA (1985) Genetics of polyamine synthesis in tobacco: developmental switches in the flower. Cold Spring Harbor Symp Quant Biol 50:475–482
Markwalder HV, Neukom H (1976) Diferulic acid as a possible crosslink in hemicelluloses from wheat germ. Phytochemistry 15:836–837
Matsuzaki S, Hamana K, Okada M, Niitsu M, Samejima K (1990) Aliphatic pentaamines found in *Canavalia gladiata*. Phytochemistry 29:1311–1312
Pegg AE (1986) Recent advances in the biochemistry of polyamines in eukaryotes. Biochem J 234: 249–262
Porta R, Esposito C, Metafora S, Malorni A, Pucci P, Marino G (1988) Purification and structural characterization of in vitro synthesized (γ-glutamyl) spermidine conjugates as a major protein secreted from the rat seminal vesicles. In: Zappia V, Pegg AE (eds) Progress in Polyamine Research. Novel, Biochemical, Pharmacological and Clinical Aspects, Plenum Press, New York, pp 403–409
Raina A, Pajula R, Eloranta T, Tuomi K (1978) Synthesis of polyamines and S-adenosylmethionine in rat tissues and tumor cells: effect of DL-α-hydroxino-δ-aminovaleric acid on cell proliferation. In: Campbell RA, Morris DR, Bartos D, Doyle Daves G, Bartos F (eds) Advances in Polyamine Research, vol 1. Raven Press, New York, pp 75–82
Sari-Gorla M (1992) Gene expression during the male gametophytic phase. Giorn Bot Ital 126:99–109
Speranza A, Calzoni GL (1980) Compounds released from incompatible apple pollen during in vitro germination. Z Pflanzenphysiol 97:95–102
Speranza A, Calzoni GL, Bagni N (1983) Effect of exogenous polyamines on in vitro germination of apple pollen. In: Mulchay I, Ottaviano E (eds) Pollen: Biology and Implications for Plant Breeding. Elsevier Publ Co, New York, pp 21–27
Speranza A, Calzoni GL, Bagni N (1984) Evidence for a polyamine-mediated control of ribonuclease activity in germinating apple pollen. Physiol Veg 22:323–331
Tabor CW, Tabor H (1989) Microbial mutants deficient in polyamine synthesis. In: Bachrach U, Heimer YM (eds) The Physiology of Polyamines, vol 2. CRC Press, Boca Raton, Florida, pp 63–72
Tadolini B, Cabrini L, Varani E, Landi L, Pasquali P, Sechi AM (1985) Polyamine interactions with phospholipid vesicles. In: Selmeci L, Brosnan ME, Seiler N (eds) Recent Progress in Polyamine Research, Akademiai Kiado, Budapest, pp 217–227

Tassoni A, Antognoni F, Bagni N (1996) Polyamines binding to plasma membrane vesicles isolated from zucchini hypocotyls. Plant Physiol 110:817–824
Torrigiani P, Scaramagli S, Franceschetti M, Michael A, Bagni N (1998) S-adenosylmethionine decarboxylase (SAMDC) activity regulation in tobacco. In: Bardocz S, White A, Tiburcio AF (eds) Cost 917 "Biogenically active amines in food", vol 1, Biogenically active amines in transgenic plants, European Communities, Luxembourg, pp 37–42

Carotenoid Biosynthesis in Plant Reproductive Organs: Regulation and Possible Functions

G. Giuliano

ENEA, Innovation Dept., Biotechnology and Agriculture Division, Casaccia Res. Ctr., PO Box 2400, Roma 00100 AD, Italy
e-mail: giulianog@casaccia.enea.it
telephone: +39 6 30483192
fax: +39 6 30483215

Abstract. Reproductive organs of many plants accumulate carotenoids in specialized plastids called chromoplasts. In this chapter the regulation of carotenogenesis in reproductive organs as well as the possible functions of these pigments during plant reproduction is discussed.

2.1 Biosynthesis

Carotenoids are terpenoid pigments containing a C40 backbone. They are synthesized from the precursor GGPP, which is an intermediate in the biosynthesis of many plant terpenoids (Chappell, 1995). The pathway giving rise to the GGPP precursor pool is debated, although recent data favour a mevalonate-independent pathway (Arigoni et al., 1997; Lichtenthaler et al., 1997). Evidence exists in plants for both intraplastidic and extraplastidic GGPP pools and for multiple GGPP synthases (GGPS) (Bartley and Scolnik, 1995). The major GGPS which is induced during pepper fruit ripening is a plastidic enzyme encoded by the nucleus (Kuntz et al., 1992). The subsequent enzymes in the pathway are also encoded by the nucleus and are localized in the plastid (Bartley et al., 1994; Bartley and Scolnik, 1995). The first dedicated reaction in the pathway is the condensation of two molecules of GGPP (C20) to form a molecule of phytoene, the first C40 compound, through the action of phytoene synthase (PSY) (Fig. 2.1). Phytoene contains only three conjugated double bonds in the center of the molecule. This short system of conjugated double bonds, called the *chromophore*, absorbs light in the UV region of the spectrum, resulting in a colourless molecule. Subsequent desaturation reactions by the enzymes phytoene desaturase (PDS) and zeta-carotene desaturase (ZDS) increase the number of double bonds in the chromophore to 5 (phytofluene, colourless), 7 (zeta-carotene, pale yellow), 9 (neurosporene, yellow-orange) and finally 11 (lycopene, red). None of these compounds is usually accumulated in normal plant tissues, with the notable exception of lycopene, which is the major carotenoid in ripe tomato fruits.

Lycopene is the substrate of two competing cyclases, beta-LCY and epsilon-LCY (Fig. 2.1), which introduce beta- and epsilon- ionone rings at the ends of the molecule. Beta-cyclization is very important for human and animal nutrition, because animals, which are unable to make carotenoids *de novo*, utilize beta-carotenoids as nutritional precursors of all retinoids, including vitamin A (Fig. 2.2).

Beta- and epsilon-LCY are similar in primary structure, suggesting that they may have arisen from gene duplication followed by functional divergence (Cunningham et al., 1996). This is not an uncommon event in the evolution of carotenoid cyclases: red varieties of pepper (*Capsicum annuum*) fruits contain two unusual carotenoids, called

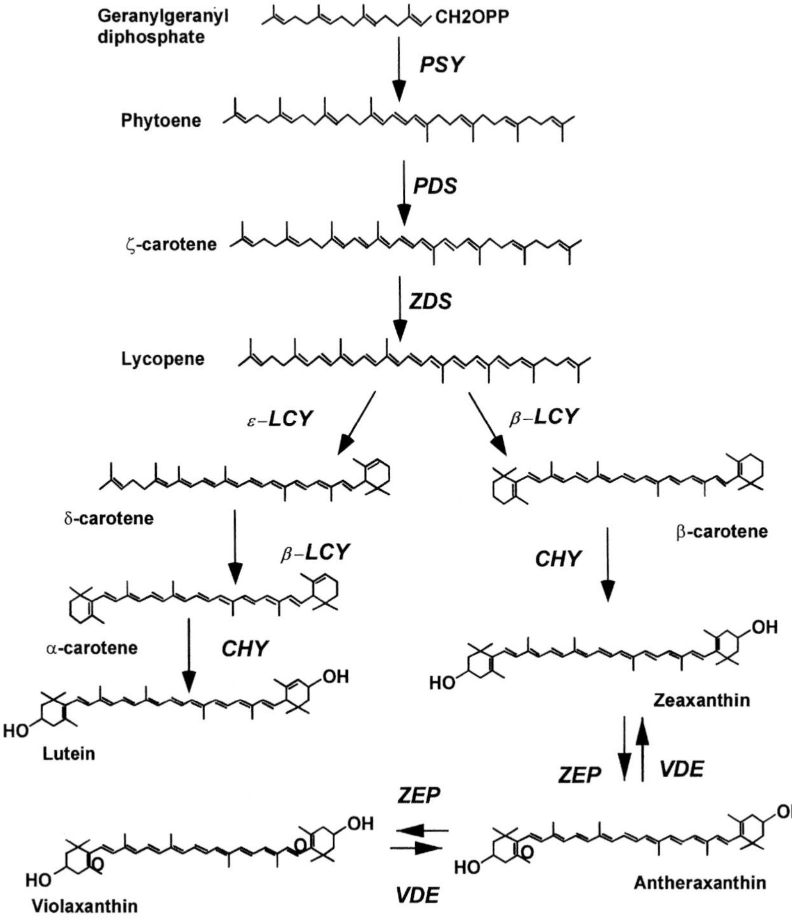

Fig. 2.1. Schematic pathway of carotenoid biosynthesis in plants, showing the key intermediates and the shortened names of the enzymes involved

capsanthin and capsorubin (also found in the flowers of some Liliaceae). These carotenoids arise from antheraxanthin and violaxanthin respectively (Fig. 2.3a), through the action of an enzyme, capsanthin-capsorubin synthase, (CCS) which is in fact a specialized cyclase, also known as kappa-cyclase. The similarity of the primary sequence of pepper CCS/kappa-cyclase and beta-cyclase (Fig. 2.3b) and the fact that CCS/kappa-cyclase carries a residual beta-cyclase activity (Bouvier et al., 1994; Hugueney et al., 1995) suggest that the former may have arisen from the latter by gene duplication.

Late reactions in carotenoid biosynthesis introduce oxygen atoms in the molecule, resulting in the production of xanthophylls. Carotene hydroxylase (CHY) is responsible for the production of zeaxanthin and lutein (the major carotenoids in maize endosperm and leaves, respectively). The competing actions of zeaxanthin epoxidase (ZEP) and violaxanthin de-epoxidase (VDE) give rise to the xanthophyll cycle found in leaves (Fig. 2.1), which has been suggested as playing a major role in preserving photosynthetic efficiency under varying light intensities (Demmig-Adams et al., 1996). In spite of this

Fig. 2.2. Retinoid metabolism in mammals

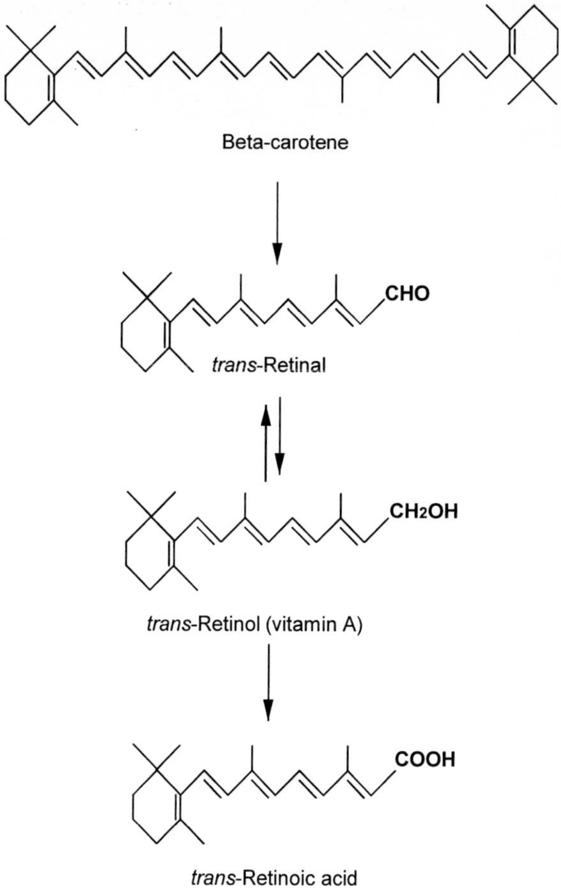

suggestion, mutations that alter the relative amounts of leaf xanthophylls have, surprisingly, only minor effects on photosynthesis (Hurry et al., 1997; Pogson et al., 1996).

Most carotenoids have a central axis of symmetry, dividing the molecule into two identical parts. Correspondingly, most enzymes are *versatile* (Cerdà-Olmedo, 1994) i.e. able to perform the same reaction twice, on the two similar moieties of the molecule. In the case of CCS, versatility is extended to performing different reactions on different substrates. The combinatorial action of several versatile enzymes can give rise to a plethora of biosynthetic intermediates (Misawa et al., 1995) and, together with gene duplication followed by functional divergence of enzymatic activities, is probably at the basis of the incredible variety of carotenoid compounds found in nature.

2.2 Regulation

In plants, carotenoids are mainly accumulated in the thylakoid membranes of leaf chloroplasts, and in the chromoplasts of many fruits and flowers. Very low levels of caroten-

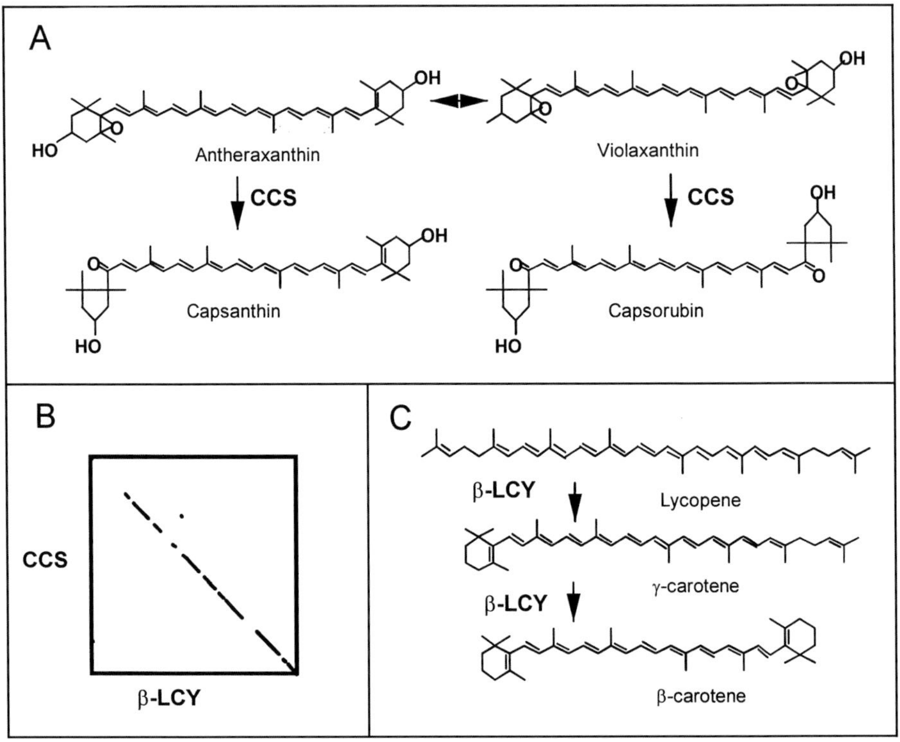

Fig 2.3. (A) Reactions catalyzed by CCS/kappa-cyclase; **(B)** protein similarity between CCS and beta-cyclase; **(C)** reactions catalyzed by beta-cyclase

oids are also found in roots, where they probably act as abscisic acid precursors (Parry and Horgan, 1992).

Chromoplasts arise from chloroplasts through the combined action of gene silencing (Piechulla et al., 1986) gene activation (Giuliano et al., 1993) degradation of photosynthetic complexes, and the formation of specialized carotenoid-bearing structures. The latter structures can be of different nature: fibrillar, as in the pepper fruit, crystalloid, as in the tomato fruit, or membranaceous, as in daffodil petals (for a detailed review on chromoplast differentiation see Camara et al., 1995). In some cases, carotenoid-binding proteins, induced during chromoplast differentiation, have been described and the relative genes have been cloned (Deruere et al., 1994; Vishnevetsky et al., 1996). In the case of pepper fruit, a detailed model has been proposed for the assembly of fibrils, which are composed of a fibrillin layer surrounding a xanthophyll core. Although pepper fibrils mostly contain esterified xanthophylls, it is not known if esterification is a prerequisite for their assembly.

During chromoplast differentiation in fruits and flowers, both the activity and the levels of most early enzymes in the pathway increase (Kuntz et al., 1992; Hugueney et al., 1992; Fraser et al., 1994; AlBabili et al., 1996; Schledz et al., 1996). This can be accompanied by spectacular increases in relative abundance of the corresponding mRNA, as in the case of pepper *GGPS* (Kuntz et al., 1992), tomato or daffodil *PSY* (Giuliano et al., 1993; Schledz et al., 1996) or much more modest mRNA increases, as in the case of pepper, daffodil or corn *PDS* (Hugueney et al., 1992; AlBabili at al, 1996; Li et al., 1996). In the case of pepper or daffodil *PDS*, protein levels increase much faster than mRNA levels during chromoplast differentiation, suggesting a possible post-transcriptional regulation (Hugueney et al., 1992; AlBabili et al., 1996). Tomato *PDS* mRNA shows a dual regulation, being induced only 4-fold in ripening fruits, and about 10-fold in opening flowers (Giuliano et al., 1993). The promoter of this gene is strongly inducible in both fruits and flowers, and a region of the 5′-untranslated leader is essential for this inducibility (Corona et al., 1996).On the other hand, very little DNA upstream of the TATA and CAAT boxes is necessary for *PDS* promoter function (unpublished data).

PSY, the first dedicated gene in the pathway, appears to be a key regulatory step in tomato. Firstly, *PSY* mRNA is highly induced during tomato fruit ripening (the corresponding clone, *pTOM5*, was initially isolated as a ripening-inducible mRNA (Slater et al., 1985)). Secondly, in tomato two *PSY* genes exist. The gene corresponding to *pTOM5* maps on chromosome 3 and is responsible for the pigmentation of the fruit and, to a lesser extent, of the flower, as shown by the phenotype of mutants and cosuppressing or antisense transgenes affecting its expression (Bartley et al., 1992; Bramley et al., 1992; Fray et al., 1993). The other gene maps on chromosome 2 and is probably responsible for the pigmentation of chloroplast-containing tissues, as suggested by its pattern of expression (Bartley and Scolnik, 1993). The existence of duplicated *PSY* genes, one chromoplast- and one chloroplast-associated, whose protein products have undistinguishable functions, strongly suggests that regulation of this step is highly different in reproductive and non-reproductive tissues. A similar situation possibly exists in maize, where the gene *Y1*, affecting pigmentation of endosperm and, to a much lesser extent, of leaves, encodes PSY. The absence of alleles of *Y1* that completely inactivate carotenoid biosynthesis in leaves has prompted the authors to postulate the existence of a second *PSY* gene in maize (Buckner et al., 1996). In any case, *PSY* gene duplication must have occurred independently in different species after the evolutionary radiation of Angiosperms, because *Arabidopsis*, that does not develop appreciable amounts of chromoplasts in petals or fruits, contains only one *PSY* gene, whose regulation is similar to the leaf-specific gene of tomato (VonLintig et al., 1997). The single *PSY* gene of *Arabidopsis* is under positive control by phytochrome (both phyA and a light-stable phytochrome different from phyB), while at least one *GGPS* and the single *PDS* gene are phytochrome-independent, suggesting that *PSY* is also a key regulatory step in phytochrome-regulated leaf carotenoid biosynthesis (VonLintig et al., 1997).

Little information is still available on the regulation of later steps of the pathway, with the exception of pepper CCS, which is highly induced at both the mRNA and protein levels during ripening (Bouvier et al., 1994), and tomato beta-LCY, which is instead inhibited during the same process (Pecker et al., 1996). Regulatory mutants, affecting this pattern of regulation, exist in both species (Bouvier et al., 1994; Tomes et al., 1956). In red pepper fruits, that, contrary to tomato, synthesize high amounts of cyclic xanthophylls in the fruit, beta-cyclase is not significantly induced. This apparent paradox can be explained by postulating that most of the beta-cyclization in pepper is performed by the ancillary beta-cyclase activity carried by CCS.

The decision to make (or not make) chromoplasts in a given reproductive organ differs in closely related plant species (the yellow petals of tomato contain chromoplasts, while the pink or white petals of tobacco or potato do not; similarly, ripe fruits of tomatoes or red peppers contain chromoplasts, while those of tobacco or green peppers do not). This suggests that carotenoids do not have a fundamental function in the development of these organs (see below). As a consequence, carotenoid biosynthesis must respond to *organelle-specific*, rather than *organ-specific* regulatory signals. In other terms, the decision to differentiate chromoplasts, and accumulate carotenoids, in a given organ, is not made solely through the action of general regulatory proteins governing the shape of this organ (Meyerowitz, 1997). Other, species-specific or even variety-specific regulatory factors must play a role in chromoplast-associated gene expression. In accordance with this model, the *PDS* promoter of tomato, which is highly induced in tomato petals and fruits, is silent in the corresponding organs of tobacco, which are devoid of chromoplasts.On the other hand, this promoter is highly expressed in anthers of both species, suggesting that chromoplast differentiation in this floral organ may serve some general function (Corona et al., 1996).

Chromoplast development is under hormonal control: ethylene, in the case of ripening of climacteric fruits such as tomato or melon (Maunders et al., 1987; Aggelis et al., 1997) and gibberellins in the case of flower development (Vainstein et al., 1994). Therefore, it is not surprising that carotenoid biosynthesis genes in climacteric fruits respond to ethylene (Maunders et al., 1987; Aggelis et al., 1997). Unpublished data obtained in our laboratory indicate that ethylene induction of carotenoid genes in tomato is modulated differently in different organs, suggesting that this hormone acts indirectly *via* activation of a developmental process (ripening), rather than directly *via* carotenoid gene activation (unpublished data).

2.3 Possible Functions

As discussed above, the species-specific or even variety-specific differences in carotenoid accumulation found in flowers, fruits and seeds suggest that carotenoids do not play a major functional role in the development and physiology of these organs. A notable exception is seed dormancy. In maize, most mutants affected in carotenoid biosynthesis germinate precociously on the cob, giving rise to the so-called *viviparous* phenotype. This observation is at the basis of the contention that carotenoids, and more specifically violaxanthin, are the metabolic precursors of ABA, which is the hormone controlling, among other things, seed dormancy. This hypothesis is corroborated by the fact that ABA-deficient mutants in plants as different as Arabidopsis, tobacco and sunflower are also affected at some stage of carotenoid biosynthesis (Rock and Zeevaart, 1991; Fambrini et al., 1993; Marin et al., 1996). While the role of carotenoid-derived ABA in the establishment of seed dormancy is thus generally accepted, its primary site of synthesis remains much more unclear. The tomato carotenoid-deficient mutant *ghost* (Scolnik et al., 1987) does not show altered seed dormancy, suggesting that in this case enough ABA is provided by the surrounding maternal tissue. On the other hand, in the case of maize carotenoid-deficient mutants the viviparous phenotype is observed irrespective of the genetic composition of the cob, suggesting that in that case maternal ABA, provided by the cob, is insufficient to establish seed dormancy. In the maize kernel, endosperm contains by far the largest amounts of carotenoids. Nevertheless, mutants affecting carotenoid biosynthesis only in this tissue (such as the *yellow* mutants) have generally a very weak viviparous phenotype, suggesting that endosperm carotenoids are used only marginally to synthesize the ABA required for establishment of dormancy. In accordance with this idea, rice, in which the endosperm is totally devoid of carotenoids, undergoes normal seed dormancy. The relative insensitivity of cereal seed dormancy to the carotenoid levels found in the endosperm facilitates research efforts aimed at engineering the carotenoid levels in this tissue, a very important goal for overcoming nutritional vitamin A deficiency (Burkhardt et al., 1997).

Apart from their role as ABA precursors, no other functions for carotenoids in plant reproductive organs have been clearly demonstrated. Certainly, part of the vivid colours found in some flowers and fruits are a result of man-made selection, but in other cases carotenoids may play a role as attractants for pollinating insects and animals active in seed dispersal. Therefore, carotenoid diversity may constitute a fertile field of research for ecologists interested in animal-plant interactions. Furthermore, it should not be forgotten that plants possess a wide variety of blue light photoreceptors (cryptochromes, Short and Briggs, 1994). Cryptochromes regulate plant responses as diverse as plant growth, phototropism, flowering, and stomatal opening. Therefore, it is not unlikely that carotenoids may alter the amount of environmental blue light in dense plant communities, and that this alteration may be perceived by the cryptochromes of neighbouring plants and alter their development. If this is the case, carotenoids and cryptochromes may be a pivotal duo for the physiology of dense plant communities, as is already known to be for chlorophyll and phytochromes (Smith and Whitelam, 1997).

Acknowledgements

Work in the laboratory of GG is supported by grants from the European Commission (contracts FAIR CT96 1633 and BIO4 CT97 2077) and from the Italian Ministry of Agriculture, Biotechnology project.

Notes added in proof

A thorough review on carotenoid genes and enzymes has just appeared (Cunningham, F. X. and E. Gantt (1998). "Genes and enzymes of carotenoid biosynthesis in plants." Annu Rev Plant Physiol 49 : 557–583). The cloning of the maize vp5 gene has established an unequivocal relation between vivipary and lesion in a structural gene for carotenoid biosynthesis (Hable, W. E., K. K. Oishi, et al. (1998). "Viviparous-5 encodes phytoene desaturase, an enzyme essential for abscisic acid (ABA) accumulation and seed development in maize." Mol Gen Genet 258 (2) : 167–176). Several mutants altering carotenoid composition in the tomato fruit (*high-beta, delta, crimson*) have been positionally cloned and shown to map in structural genes (Prof. Joseph Hirschberg, personal communication).

References

Aggelis A, John I, Grierson D (1997) Analysis of physiological and molecular changes in melon (Cucumis melo L) varieties with different rates of ripening. J Exp Bot 48 : 769–778

AlBabili S, VonLintig J, Haubruck H, Beyer P (1996) A novel, soluble form of phytoene desaturase from Narcissus pseudonarcissus chromoplasts is Hsp70-complexed and competent for flavinylation, membrane association and enzymatic activation. Plant J 9 : 601–612

Arigoni D, Sagner S, Latzel C, Eisenreich W, Bacher A, Zenk MH (1997) Terpenoid biosynthesis from 1-deoxy-D-xylulose in higher plants by intramolecular skeletal rearrangement. Proc Natl Acad Sci USA 94 : 10600–10605

Bartley GE, Scolnik PA (1993) cDNA cloning, expression during development, and genome mapping of PSY2, a second tomato gene encoding phytoene synthase. J Biol Chem 268 : 25718–25721

Bartley GE, Scolnik PA (1995) Plant carotenoids: Pigments for photoprotection, visual attraction, and human health. Plant Cell 7 : 1027–1038

Bartley GE, Scolnik PA, Giuliano G (1994) Molecular biology of carotenoid biosynthesis in plants. Ann Rev Plant Physiol Plant Mol Biol 45 : 287–301

Bartley GE, Viitanen PV, Bacot KO, Scolnik PA (1992) A tomato gene expressed during fruit ripening encodes an enzyme of the carotenoid biosynthesis pathway. J Biol Chem 267 : 5036–5039

Bouvier F, Hugueney P, d'Harlingue A, Kuntz M, Camara B (1994) Xanthophyll biosynthesis in chromoplasts: isolation and molecular cloning of an enzyme catalyzing the conversion of 5,6-epoxycarotenoid into ketocarotenoid. Plant J 6:45–54
Bramley P, Teulieres C, Blain I, Bird C, Schuch W (1992) Biochemical characterization of transgenic tomato plants in which carotenoid synthesis has been inhibited through the expression of antisense RNA to pTOM5. The Plant Journal 2:343–349
Buckner B, Miguel PS, Janick-Buckner D, Bennetzen JL (1996) The Y1 gene of maize codes for phytoene synthase. Genetics 143:479–488
Burkhardt PK, Beyer P, Wunn J, Kloti A, Armstrong GA, Schledz M, VonLintig J, Potrykus I (1997) Transgenic rice (Oryza sativa) endosperm expressing daffodil (Narcissus pseudonarcissus) phytoene synthase accumulates phytoene, a key intermediate of provitamin A biosynthesis. Plant J 11:1071–1078
Camara B, Hugueney P, Bouvier F, Kuntz M, Moneger R (1995) Biochemistry and molecular biology of chromoplast development. Int Rev Cytol 163:175–247
Cerdà-Olmedo E (1994) The genetics of chemical diversity. Crit. Rev. Microbiol. 20:151–160
Chappell J (1995) Biochemistry and molecular biology of the isoprenoid biosynthetic pathway in plants. Annu Rev Plant Physiol Plant Mol Biol 46:521–547
Corona V, Aracri B, Kosturkova G, Bartley GE, Pitto L, Giorgetti L, Scolnik PA, Giuliano G (1996) Regulation of a carotenoid biosynthesis gene promoter during plant development. Plant J 9:505–512
Cunningham FX, Pogson B, Sun ZR, Mcdonald, KA, DellaPenna D, Gantt E (1996) Functional analysis of the beta and epsilon lycopene cyclase enzymes of Arabidopsis reveals a mechanism for control of cyclic carotenoid formation. Plant Cell 8:1613–1626
Demmig-Adams B, Gilmore AM, Adams WW (1996) Carotenoids. 3. In vivo functions of carotenoids in higher plants. FASEB J 10:403–412
Deruere J, Romer S, d'Harlingue A, Backhaus RA, Kuntz M, Camara B (1994) Fibril assembly and carotenoid overaccumulation in chromoplasts: a model for supramolecular lipoprotein structures. Plant Cell 6:119–133
Fambrini M, Pugliesi G, Vernieri P, Giuliano G, Baroncelli S (1993) Characterization of a sunflower (*Helianthus annuus* L.) mutant deficient in carotenoid synthesis and abscisic acid content induced by in vitro tissue culture. Theor Appl Genet 87:65–67
Fraser PD, Truesdale MR, Bird CR, Schuch W, Bramley PM (1994) Carotenoid biosynthesis during tomato fruit development. Plant Physiol 105:405–413
Fray RG, Grierson D (1993) Identification and genetic analysis of normal and mutant phytoene synthase genes of tomato by sequencing, complementation and co-suppression. Plant Mol Biol 22:589–602
Giuliano G, Bartley GE, Scolnik PA (1993) Regulation of carotenoid biosynthesis during tomato development. Plant Cell 5:379–387
Hugueney P, Badillo A, Chen HC, Klein A, Hirschberg J, Camara B, Kuntz M (1995) Metabolism of cyclic carotenoids: A model for the alteration of this biosynthetic pathway in *Capsicum annuum* chromoplasts. Plant J 8:417–424
Hugueney P, Romer S, Kuntz M, Camara B (1992) Characterization and molecular cloning of a flavoprotein catalyzing the synthesis of phytofluene and z-carotene in *Capsicum* chromoplasts. Eur J Biochem 209:399–407
Hurry V, Anderson JM, Chow WS, Osmond CB (1997) Accumulation of zeaxanthin in abscisic acid-deficient mutants of Arabidopsis does not affect chlorophyll fluorescence quenching or sensitivity to photoinhibition in vivo. Plant Physiol 113:639–648
Kuntz M, Romer S, Suire C, Hugueney P, Weil JH, Schantz R, Camara B (1992) Identification of a cDNA for the plastid-located geranylgeranyl pyrophosphate synthase from *Capsicum annuum*: correlative increase in enzyme activity and transcript level during fruit ripening. Plant J 2:25–34
Li ZH, Matthews PD, Burr B, Wurtzel ET (1996) Cloning and characterization of a maize cDNA encoding phytoene desaturase, an enzyme of the carotenoid biosynthetic pathway. Plant Mol Biol 30: 269–279
Lichtenthaler HK, Schwender J, Disch A, Rohmer M (1997) Biosynthesis of isoprenoids in higher plant chloroplasts proceeds via a mevalonate-independent pathway. FEBS Lett 400:271–274
Marin E, Nussaume L, Quesada A, Gonneau M, Sotta B, Hugueney P, Frey A, MarionPoll A (1996) Molecular identification of zeaxanthin epoxidase of Nicotiana plumbaginifolia, a gene involved in abscisic acid biosynthesis and corresponding to the ABA locus of Arabidopsis thaliana. EMBO J 15: 2331–2342
Maunders MJ, Holdsworth MJ, Slater A, Knapp JE, Bird CR, Schuch W, Grierson D (1987) Ethylene stimulates the accumulation of ripening-related mRNAs in tomatoes. Plant Cell Environ 10:177–184
Meyerowitz EM (1997) Genetic control of cell division patterns in developing plants. Cell 88:299–308
Misawa N, Satomi Y, Kondo K, Yokoyama A, Kajiwara S, Saito T, Ohtani T, Miki W (1995) Structure and functional analysis of a marine bacterial carotenoid biosynthesis gene cluster and astaxanthin biosynthetic pathway proposed at the gene level. J Bacteriol 177:6575–6584
Parry AD, Horgan R (1992) Abscisic acid biosynthesis in roots I. The identification of potential abscisic acid precursors, and other carotenoids. Planta 187:185–191

Pecker I, Gabbay R, Cunningham FX, Hirschberg J (1996) Cloning and characterization of the cDNA for lycopene beta-cyclase from tomato reveals decrease in its expression during fruit ripening. Plant Mol Biol 30:807–819

Piechulla B, Pichersky E, Cashmore AR, Gruissem W (1986) Expression of nuclear and plastid genes for photosynthesis-related proteins during tomato fruit development and ripening. Plant Mol Biol 7:367–376

Pogson B, Mcdonald KA, Truong M, Britton G, Dellapenna D (1996) Arabidopsis carotenoid mutants demonstrate that lutein is not essential for photosynthesis in higher plants. Plant Cell 8:1627–1639

Rock CD, Zeevaart JAD (1991) The *aba* mutant of *Arabidopsis thaliana* is impaired in epoxy-carotenoid biosynthesis. Proc Natl Acad Sci USA 88:7496–7499

Schledz M, AlBabili S, VonLintig J, Haubruck H, Rabbani S, Kleinig H, Beyer P (1996) Phytoene synthase from Narcissus pseudonarcissus: Functional expression, galactolipid requirement, topological distribution in chromoplasts and induction during flowering. Plant J 10:781–792

Scolnik PA, Hinton P, Greenblatt IM, Giuliano G, Delanoy MR, Spector DL, Pollock D (1987) Somatic instability of carotenoid biosynthesis in the tomato *ghost* mutant and its effect on plastid development. Planta 171:11–18

Short TW, Briggs WR (1994) The transduction of blue light signals in higher plants. Ann Rev Plant Physiol Plant Mol Biol 45:143–171

Slater A, Maunders MJ, Edwards K, Schuch W, Grierson D (1985) Isolation and characterization of cDNA clones for tomato polygalacturonase and other ripening-related proteins. Plant Mol Biol 5:137–147

Smith H, Whitelam GC (1997) The shade avoidance syndrome: Multiple responses mediated by multiple phytochromes. Plant Cell Environ 20:840–844

Tomes ML, Quackenbush FW, Kargl TE (1956) Action of the *B* gene in the biosynthesis of carotenes in the tomato. Bot Gaz 117:248–253

Vainstein A, Halevy AH, Smirra I, Vishnevetsky M (1994) Chromoplast biogenesis in *Cucumis sativus* corollas. Plant Physiol 104:321–326

Vishnevetsky M, Ovadis M, Itzhaki H, Levy M, LibalWeksler Y, Adam Z, Vainstein A (1996) Molecular cloning of a carotenoid-associated protein from Cucumis sativus corollas: Homologous genes involved in carotenoid sequestration in chromoplasts. Plant J 10:1111–1118

VonLintig J, Welsch R, Giuliano G, Batschauer A, Kleinig H (1997) Light-dependent regulation of carotenoid biosynthesis occurs at the level of phytoene synthase expression and is mediated by phytochrome in *Sinapis alba* L. and *Arabidopsis thaliana* L. seedlings. Plant J 12:625–634

CHAPTER 3

Lipid Accumulation and Related Gene Expression in Gametophytic and Sporophytic Anther Tissues

P. Piffanelli · D. J. Murphy

John Innes Centre, Norwich Research Park, Colney Lane, Norwich NR4 7UH, United Kingdom
e-mail: denis.murphy@bbsrc.ac.uk
telephone: +44 1603 452571
fax: +44 1603 259882

Abstract. In recent years the ontogenesis and development of the male gametophyte in angiosperms has attracted an increasing interest among plant scientists. The availability of molecular techniques has contributed significantly to unveiling how the expression and regulation of specific classes of sporophytic (tapetal) and gametophytic genes contribute to the formation of a viable pollen grain. Almost a quarter of a pollen grain is made of lipids, which are structural components of both intracellular and extracellular pollen domains. It is not surprising, therefore, that the biogenesis and function of the different lipidic and lipid-derived components has become one of the main areas of interest in sexual plant reproduction. In addition to their structural roles it is becoming clear that lipid-derived molecules are also strongly involved in the pollen-stigma signalling and recognition events. Moreover, lipidic volatiles are released from the pollen coat and act as attractants for pollinators. Whilst the majority of extracellular pollen exine and pollen coat lipid components are synthesised in tapetal cells and relocated to the pollen surface, the majority of pollen intracellular membrane and storage lipids are synthesised within the pollen vegetative cell under the control of the gametophytic genome. The purpose of this Chapter is to review, in the light of a wealth of recent and exciting experimental data, the biogenesis of pollen lipidic structures in tapetal and gametophytic cells and their concerted deposition and function in the assembly of viable pollen grains.

3.1 Introduction

Lipids are structural components of all plant cells (Harwood, 1989). In addition to making up the membrane lipid bilayer, lipids are also the milieu within which such central processes as photosynthesis and respiration occur (Browse and Somerville, 1991). Many plant lipids or their metabolic derivatives have important biological activities in the regulation of plant development and of specific cellular events (e.g. phosphoinositides). Moreover, lipids are structural constituents of the surface layers (e.g. epicuticular wax, cutin, suberin) which serve as a vital barrier between the plant tissues and the surrounding environment. These layers prevent water loss and, conversely, the entry of pathogenic micro-organisms.

Only in recent years have we begun to unveil how the chemical properties and molecular shapes of lipid molecules affect the biogenesis and function of the various membranes of the plant cell (e.g. thylakoids) and of specialised cellular structures (i.e. exine).

Lipids are also accumulated as storage products within the cells of certain plant tissues (Töpfer et al., 1995) and are an important source of energy not only as edible substances, but, increasingly, as raw materials for industry. The two major sites of lipid accumulation in plants are the seed and the anther. Within the anther, sporophytic tapetal cells become filled with cytoplasmic lipid bodies and elaioplasts which, following the developmentally-regulated tapetal lysis, are transferred to the surface of the developing pollen grains where they constitute the major structural components of the pollen coat. The developing pollen grains of numerous entomophilous species also synthesise and accumulate, under the control of the gametophytic genome, storage lipids in the form of triacylglycerols (TAGs), which function as a supply of carbon skeleton and energy to support the rapid growth of the pollen tube from germinated pollen grains.

Recent data, mainly from mutant analysis, has revealed that lipids are not only important structural components of the male gametophyte unit, but are also important precursors of signalling molecules involved in the regulation of anther development (i.e. jasmonic acid; Mc Conn and Browse, 1996) pollen-stigma interaction (i.e. long-chain fatty acids; Preuss et al., 1993) and attraction of pollinators (i.e. lipidic volatiles, Dobson et al., 1996).

In this chapter the biosynthesis and accumulation of membrane and storage lipids in tapetal and pollen cells and their interaction to form a viable pollen grain will be analysed together with their proposed functions for the propagation and survival of angiosperm plants.

3.2
Lipid Biosynthetic Pathways in Anther Tapetal and Pollen Cells

Lipid biogenesis in plant cells involves two main cellular components: the plastid and the endoplasmic reticulum (Stumpf, 1981). In the stroma of all the different types of plastids, long-chain fatty acids [C16 to C18] are synthesised *de novo* from acetyl-CoA monomers by the fatty acid synthetase enzymatic complex (Harwood, 1996; Somerville and Browse, 1991). Fatty acid chains are then cleaved from the acyl carrier protein (ACP) by highly specific thioesterases (Jones et al., 1995) (with high affinity for a determined chain length and saturation level of the newly synthesised fatty acid chain) and re-esterified to a CoA molecule, before being exported to the cytoplasm and directed towards specialised domains within the endoplasmic reticulum (ER) to be further modified (Harwood, 1989). The main modification is the introduction of additional double bonds at specific positions to the monounsaturate oleic acid (C18:1), which leads to the formation of polyunsaturate fatty acid chains (mainly linoleic, C18:2, and linolenic acid, C18:3) by the enzymes oleoyl-desaturase and linoleoyl-desaturase, respectively (Browse and Somerville, 1991; Töpfer et al., 1995).

In all plant cells, the acyl CoAs are esterified to a glycerol backbone by specialised (e.g. substrate-specific) acyltransferases to form phospholipids or galactolipids, which are structural components of all intracellular and cell membranes (Browse and Somerville, 1994). This extra-plastidial lipid pathway is also known as the “eukaryotic pathway” (Browse and Somerville, 1994) and is the main lipid biosynthetic route in non-photosynthesising tissues (e.g. in roots, developing pollen grains). In the ER the diacylglycerol acyltransferase (DAGAT) is responsible for the esterification of fatty acid chains to the third (or *sn3*) position of the glycerol backbone (Murphy, 1993; Ohlrogge et al., 1991), leading to the synthesis of triacylglycerols, which are the most abundant form of storage lipids in seed (Murphy and Cummins, 1989) and anther tissues (Murphy, 1993; Evans et al., 1992; Piffanelli et al., 1997; Wu et al., 1997). This enzyme has been suggested as controlling the flux through the TAG biosynthetic pathway in seed tissues (Weselake et al., 1993) and it is likely that the same protein or isoforms of it are expressed in both tapetal and male gametophytic cells, where high levels of TAGs are synthesised (Evans et al., 1992; Wu et al., 1997).

In both tapetal cells and male gametophyte vegetative cells, TAGs are accumulated and stored in subcellular organelles named cytoplasmic lipid bodies and pollen oil bodies (Murphy and Ross, 1998; Wu et al., 1997; Piffanelli et al., 1997), respectively.

Each of the three TAG-storing organelles (seed oil bodies, pollen intracellular oil bodies and tapetal cytoplasmic lipid bodies) and the non-TAG storing tapetal elaioplasts have unique and distinct morphological characteristics as discussed further in this Chapter and illustrated in Fig. 3.1.

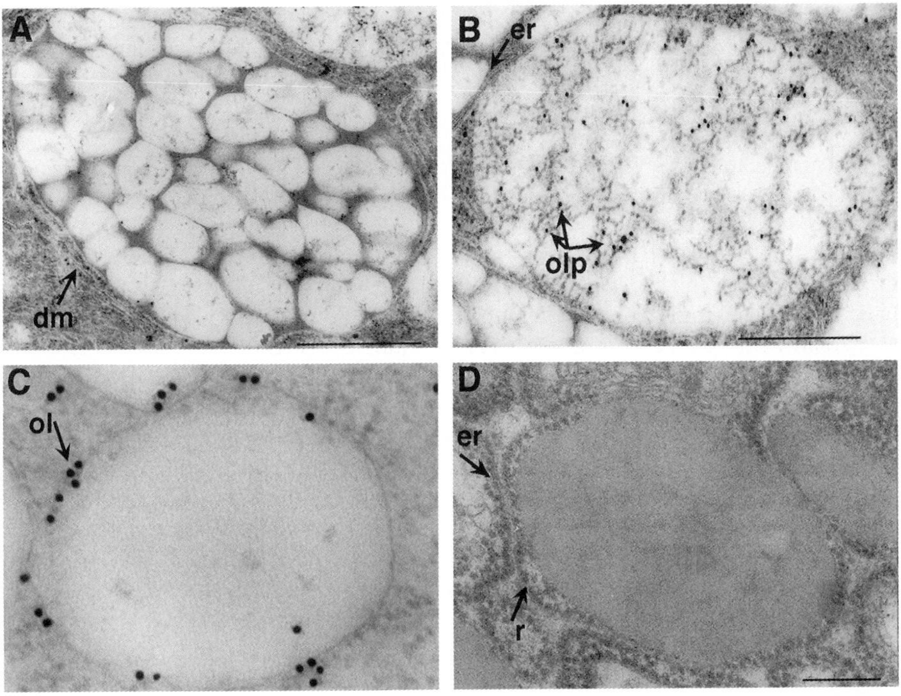

Fig. 3.1A–D. Electron micrographs of the four different types of lipid-storing organelles in *Brassica napus* anther tapetum, pollen grain and seed cells. **A** tapetal elaioplast showing the surrounding double plastidial membrane (dm) and the densely packed lipid droplets (also called plastoglobuli). Scale bar 1 µm. **B** tapetal cytoplasmic lipid body which is characterised by a regularly organised internal proteinaceous network. Immunolocalisation using an antibody raised against the tapetal-specific oleosin-like proteins (olp) revealed their association with the internal fibrous structures. The tapetal cytoplasmic lipid bodies are also found in close contact with the ER. Scale bar 1 µm. **C** seed oil body characterised by a homogeneous TAG-filled space that is bounded by an annulus of oleosin proteins (ol), as shown by the immunolocalisation using a polyclonal antibody raised against seed-specific *Brassica napus* oleosins. Bar 0.2 µm. **D** pollen intracellular oil body which is also homogeneosly packed with TAGs and is encircled by ER cisternae, as shown by the ribosomes (r) that are oriented towards the cytoplasmic face. Scale bar 0.2 µm. lu, lumen; r, ribosomes; er, endoplasmic reticulum; dm, double plastidial membrane; olp, tapetal-specific oleosin-like proteins; ol seed-specific oleosins
Legend: *dm*, double membrane; *er*, endoplasmic reticulum; *ol*, oleosin; *olp*, oleosin-like protein; *r*, ribosome

A series of elongations of long-chain fatty acids by the addition of C2 units from malonyl-CoA, carried out by microsomal enzymes, leads to the biosynthesis of very long-chain fatty acids (Post-Beittenmiller, 1996). This involves the same four basic reactions of condensation, reduction, dehydration and second reduction, as carried out by the plastidial FAS complex (Post-Beittenmiller, 1996). These four extraplastidial membrane-associated (Whitfield et al., 1993) enzyme activities, collectively termed elongases (von Wettstein-Knowles, 1982), were shown to require ATP and NADPH for their activities (Kollatukudy, 1980; Evenson and Post-Beittenmiller, 1995). The elongase complexes have a central role in wax production (von Wettstein-Knowles, 1995) in stem and anther tissues (Aarts et al., 1995; Hannoufa et al., 1993) and biosynthesis of very long-chain fatty acids (i.e. erucic acid, C22 : 1) esterified to storage lipids accumulated in the seeds of Brassicaceae (Murphy and Cummins, 1989) (as TAGs) and of jojoba (as liquid wax esters) (Metz and Lassner, 1996).

The elongated fatty acid chains may enter different reaction pathways and can be released as free fatty acids, they can be reduced to aldehydes, decarboxylated to alkanes or esterified with alcohols and give rise to a variety of cellular and extracellular structures (e.g. suberin, cutin, exine).

Recently, genes encoding not only seed elongase components, but also enzymes involved in wax esters biosynthesis in epidermal and tapetal cells were cloned and characterised in plants (Metz and Lassner, 1996; James et al., 1995; Aarts et al., 1995; Hannoufa et al., 1996). The characterisation of these enzymes, to which only a putative function was assigned, will clarify the exact pathway leading to the biosynthesis of these important lipid structures (i.e. very long-chain acyl esters) for pollen viability (see next paragraph) and stem or leaf epicuticular waxes biosynthesis.

Thus, long-chain fatty acids produced in the plastid from synthesis *de novo* are utilised by at least three extra-plastidial biosynthetic pathways that lead to the production of phospholipids, waxes (also comprising cutin in epidermal cells and suberin in root and damaged tissues) and storage triacylglycerols. Presently, the mechanisms that regulate this partitioning are not well understood but are likely to be accomplished: 1) by enzyme specificities (i.e. acyltranferases), 2) by substrate availabilities (i.e. compartmentalisation and/or metabolic channelling), 3) by the supply of specific fatty acids (i.e. saturated or unsaturated fatty acid chains) in individual cell types (i.e. tapetal cells, epidermal cells, zygotic embryo cells; Post-Beittenmiller, 1996).

A variable proportion, depending upon tissue-types, of the *de novo* synthesised fatty acid chains are also retained as acyl-ACP esters in the plastids and are esterified to plastid-specific lipids in the plastid envelope to constitute the structural components (i.e. galactolipids, sulfolipids) of the thylakoid membranes (Browse and Somerville, 1991). This plastidial lipid biosynthetic pathway, also known as the "prokaryotic pathway", includes a set of nuclear-encoded and plastid-targeted lipid-linked desaturase enzymes (e.g. *fad7, fad8*) which lead to the biosynthesis of highly polyunsaturated lipids (i.e. galactolipids, sulfolipids) (Somerville and Browse, 1991). Moreover, in specialised plastids, termed elaioplasts, medium (i.e. C14 : 0, myristic acid) and long-chain fatty acids mainly saturated (C16:0, palmitic acid and C18 : 0, stearic acid), are esterified to various backbones (e.g. sterols, terpenes) to constitute a class of lipid structures known as neutral or non-polar esters. To date, little is known about the exact structures of this class of lipids and of their biosynthetic pathways in both tapetal and pollen gametophytic cells. The isolation of tapetal elaioplasts (Wu et al., 1997) and the detailed chemical analysis of neutral esters before and after their transfer to the pollen will shed some light onto their synthesis, accumulation and relocation to the pollen coat.

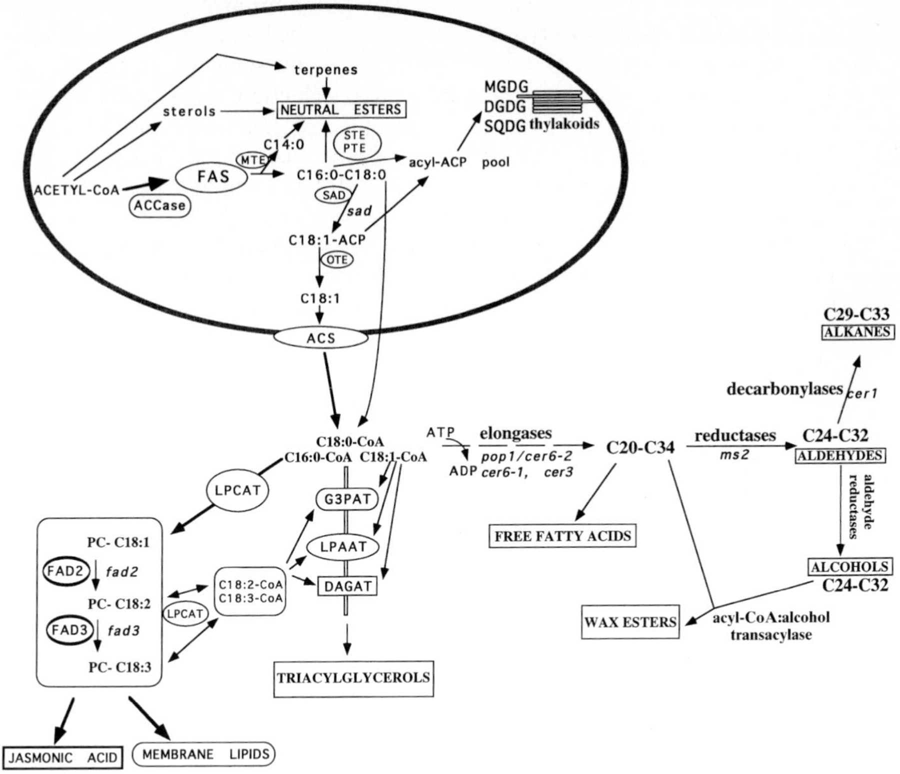

Fig. 3.2. Schematic outline of the main lipid biosynthetic pathways in anther sporophytic and gametophytic cells (tapetal cells and vegetative pollen cell). Some pathways (e.g. wax ester, alkane biosynthesis) are only active in tapetal cells and the expression levels of some enzymes dramatically change in gametophytic and sporophytic cells (e.g. acyl-ACP thioesterases) leading to dramatic differences in the fluxes through the various lipid pathways and to the biosynthesis of distinct fatty acid chains esterified to storage and membrane lipids.
De novo fatty acid biosynthesis from acetyl-CoA occurs exclusively in the plastid of all cells. The newly synthesised medium/long fatty acid chains (C14–C18) can follow two distinct parallel routes:
1) can be exported to the cytoplasm as acyl-CoAs , directed to ER membranes and there become the substrates for a number of further modifications (e.g. desaturation, elongation, reduction) and and of esterification to various backbones, giving rise to membrane, storage and signalling lipids (e.g. TAGs, wax esters, jasmonic acid, alkanes, phospholipids).
2) can be retained within the plastid (as acyl-ACPs) and there modified (e.g. desaturation) and esterified to a variety of backbones leading to the biosynthesis of glycerolipids and of neutral esters.
Some of the genes involved in fatty acid and lipid biosynthesis shown to be expressed in anther and pollen tissues are indicated in italics (see text for details).
Legend: *ACCase*, acetyl-CoA carboxylase; *ACS*, acyl-CoA synthetase; *FAS*, fatty acid synthase complex; *OTE*, oleoyl-ACP thioesterase; *STE*, stearoyl-ACP thioesterase; *PTE*, palmitoyl-ACP thioesterase; *MTE*, myristoyl-ACP thioesterase; *LPCAT*, lysophosphatydilcholine acyltransferase; *G3PAT*, glycerol-3-phosphate acetyltransferase; *LPAAT*, lysophosphatydil acetyltransferase; *DAGAT*, diacylacylglycerol acyltransferase; *TAG*, triacylglycerols; *SAD*, stearoyl-ACP desaturase; *MGDG*, monogalactosyldiacylglycerols; *DGDG*, digalactosyldiacylglycerols; *SQDG*, sulfoquinovosyldiacylglycerols

The biosynthesis of highly differentiated tissue-specific lipids and at specific stages of anther development (as described in the two following paragraphs) provides an indication of the complexity of the lipid biosynthetic pathways and of their numerous branching points.

3.3 Lipid Accumulation in Anther Tapetal Cells

The tapetal cell layer that surrounds the developing male gametophytes not only has a central role in the nourishment (i.e. amino acids and sugars) of developing meiocytes/microspores (Echlin, 1971), but also contributes to the biosynthesis of most or all of sporopollenin and pollen coat structural components (Shaw, 1971; Hesse and Hess, 1993; Wu et al., 1997; Dickinson and Lewis, 1973; Schrauwen et al., 1996).

Recent molecular data in relation to the temporal expression of lipid biosynthetic genes in developing anthers (Piffanelli et al., 1997; Aarts et al., 1995; 1997), strongly support the idea that tapetal cells actively synthesise long/very long-chain fatty acids and wax esters during the immediate post-meiotic phases of male gametophyte development (Fig. 3.3). The temporally restricted (from microspore release to first pollen mitosis) tapetal expression pattern of some lipid biosynthetic genes (e.g. *ms2*) and of a number of other proteins (e.g. lipid transfer proteins, Scott et al., 1991a; Foster et al., 1992; glycine-rich proteins, Koltunow et al., 1990; SATAP proteins of unknown function, Staiger, 1994) suggests the accumulation within and release from the tapetal cells of various lipid precursors involved in the synthesis of sporopollenin, the complex polymer constituting the exine or pollen wall (Schrauwen et al., 1996).

Following pollen meiosis, numerous lipidic structures synthesised in the tapetal cells have been shown to be secreted into the anther locule as osmiophilic droplets, which become structural components of the sporopollenin (Heslop-Harrison, 1962; 1968). At the early stages of exine formation (before microspore release), tapetally-derived lipid bodies must transverse the callose wall in order to reach the microspore surface. Electron micrograph studies support the idea that the callose wall may not be an impermeable barrier to the diffusion of such materials (Dunbar, 1973; Mascarenhas, 1975; Gabarayeva, 1992; 1995).

Ultrastructural studies have also revealed that, following microspore release, spherical osmiophilic bodies, termed "orbicules" or "Ubisch bodies" are released from the tapetal cells, into the locule, concurrently with sporopollenin deposition and are accumulated at the surface of the developing microspores (Echlin, 1971; Echlin and Godwin, 1968; Wiermann and Gubatz, 1992; Suarez-Cervera et al., 1995). These lipidic bodies which are likely to arise from the ER as pro-Ubisch bodies in the tapetal cytoplasm can be released without further modifications (Staiger et al., 1994) or they can be modified into mature Ubisch bodies by acquiring an electron-dense coating that resembles sporopollenin in its staining properties (Suarez-Cervera et al., 1995; Wiermann and Gubatz, 1992; Echlin, 1971). The use of gentle degradation methods and of sophisticated analytical techniques, such as ^{13}C NMR, have revealed that sporopollenin is made up of a series of related polymers derived from long-chain fatty acids, oxygenated aromatic rings (in much lower amounts) and phenolic acids (Guilford et al., 1988; Wehling et al., 1989; Wiermann and Gubatz, 1992; Gubatz and Wiermann, 1992; 1993; Wilmesmeier et al., 1993). Moreover, the use of thiocarbonates, a class of herbicides which inhibits the activity of long-chain fatty acid elongases, has confirmed that very long-chain fatty acids (C20–C40) are structural components of the exine wall (Wilmesmeier and Wiermann, 1995).

Furthermore, the characterisation of the *ms2* gene, encoding a putative fatty acid elongase/reductase, is in agreement with its involvement in the biosynthesis of long-chain wax esters. The *Arabidopsis ms2* gene (Aarts et al., 1997) was shown to be expressed solely in tapetal cells and to be activated around the time of microspore release, a stage when hardly any pollen wall has been formed, and its message rapidly disappears before the occurrence of the first pollen mitosis, when most of the exine has already been deposited.

The analysis of the *ms2 Arabidopsis* mutant not only revealed the complete absence of the pollen exine layer in developing male gametophytes, as shown by the sensitivity to the acetolysis treatment (Aarts et al., 1997), but also the arrest of pollen development at the microspore stage. Similar phenotypes were also reported in other species, like the *ms9* mutant of tomato (Rick, 1948) or the *ms2* mutant of maize (Albertsen and Philips, 1981), both of which are male-sterile and severely disturbed in the formation of the pollen wall. Also in *Oenothera* species two mutations affecting anther lipid metabolism were described: the mutant termed *fr* failed to develop any sporopollenin while the other, termed *ster*, had a highly abnormal sporopollenin appearance (Noher de Halac and Harte, 1995). Despite the fact that the exact chemical composition and mechanisms for the assembly and polymerisation of the sporopollenin are still largely unknown, the absence of exine wall deposition in *ms2* developing microspores strongly supports the hypothesis that the reduction/elongation of fatty acids to fatty alcohols is one of the central steps in the formation of sporopollenin and that fatty alcohols may be among the major precursors from which sporopollenin polymerisation proceeds (Southworth, 1990; Scott, 1994).

A number of tapetal-specific genes encoding lipid-transfer proteins (Koltunow et al., 1990; Foster et al., 1992; Crossley at al., 1995) were also shown to be expressed in tapetal cells during the exine formation phases and to contain an N-terminal secretory signal sequence. Similar lipid-transfer proteins were found in epidermal cells where they are believed to be involved in the trafficking of long-chain fatty acids contributing to the biosynthesis of epicuticular waxes (Sterk et al., 1991; Thoma et al., 1993; 1994). It is possible that these anther-specific lipid-transfer proteins are secreted and participate in the transport of fatty acids and/or other sporopollenin lipid precursors from the tapetum to the developing microspores during exine deposition.

Tapetal secretion into the anther locule of fibro-granular proteinaceous material continues after completion of the exine formation (Dickinson and Lewis, 1973; Owen and Makaroff, 1995). This fibrillar material accumulates around the exine and may anchor the lipidic mass of the pollen coat material released after tapetal cell lysis and in such a way control the sites where the pollen coat will be deposited. However, in mature pollen grains the pollen coat is distributed to a relatively even depth between the baculae, covering the entire sculpted surface of the pollen grain, apart from the colpi region of the exine. This observation implies that there is some kind of attractive mechanism, possibly a capillary action caused by the pressure built up by the expanding pollen grains (Keijzer, 1987; Keijzer and Cresti, 1987), which contributes to the distribution of the pollen coat materials around the pollen grains. During the final phases of anther development (dehydration phase) the pollen coat is subject to a number of modifications (also referred to as pollen coat condensation; Dickinson and Lewis, 1973), which only recently started to be analysed in detail (Murphy and Ross, 1998, Ruiter et al., 1997a; 1997b; Dickinson and Elleman, 1985).

Concomitantly with the final phases of exine deposition, the tapetal plastids undergo a major developmental change. These plastids, which in earlier anther developmental phases (meiocyte/microspore stages of male gametophyte development) resemble chloroplasts, with thylakoid membranes and starch granules (Dickinson, 1973; Pacini and Casadoro, 1981; Zavada et al., 1984), gradually start to synthesise and accumulate osmiophilic droplets. As more of these small lipidic droplets accumulate in the stroma of the tapetal plastids, these organelles develop into elaioplasts, as described in numerous genera (e.g. *Brassica*, Polowick and Sawhney, 1990; *Arabidopsis*, Owen and Makaroff, 1995; *Cucurbita*, Ciampolini et al., 1993; *Ledebouria*, Hess and Hesse, 1994; *Olea*, Pacini and Casadoro, 1981; and *Geranium*, Weber, 1996). These elaioplasts were shown to synthesise and accumulate two distinct classes of neutral esters (Wu et al., 1997) esterified to mainly saturated medium and long-chain fatty acids. The mechanism by which these organelles are released from the ruptured tapetum cells and transferred onto the surface of the maturing pollen remains an intriguing open question.

The chemical analysis of washed *Brassica napus* pollen coat fractions revealed that 41% of its lipids are neutral esters, predominantly esterified to myristic acid (C14:0), palmitic acid (C16:0) and stearic acid (C18:0) (Piffanelli et al., 1997). It is possible that specific classes of these neutral esters are degraded within the tapetal cells or during the packing phases of the pollen coat before anther anthesis (Dickinson and Lewis, 1973), as only the more-hydrophobic neutral ester fraction accumulated in the tapetal elaioplasts was detected in the ether-washed *Arabidopsis* pollen coat (Wu et al., 1997). Similar neutral lipids have also been reported from other species including sterol esters in *Typha latifolia* pollen (Caffrey, 1987) and triterpene esters in maize pollen grains (Bianchi et al., 1990). Both lipid classes were also found to be predominantly esterified to medium and long-chain saturated fatty acid chains, as seen for the *Brassica napus* pollen coat neutral esters.

One possible function of these relatively saturated neutral esters, which appear to make up the bulk of the pollen coat lipids, could be to maintain the pollen coat in a semi-solid state, in order to enclose and hold important classes of proteins (e.g. the glycoproteins involved in self-incompatibility and the oleosin-like proteins) and other substances (e.g. carotenoids, phenylpropanoids and volatiles) which are embedded within it. Experimental evidence for this is found in *Brassica napus*, where it was reported that pollen coat fractions appear as a semi-solid waxy translucent substance at 20°C, which melts at 30°–40°C to form a translucent liquid (Murphy and Ross, 1998). Also the pollen coat of *Typha latifolia* contains a lipid structure, probably a saturated sterol ester, that is present in a rigid gel phase at room temperature and undergoes a melting transition at temperatures above 55°C (Caffrey et al., 1987).

The analysis of pollen lipids has also shown the presence of relatively small but significant amounts of long-chain wax esters (Scott and Strohl, 1962; Bianchi et al., 1990; Preuss et al., 1993). Moreover, the characterisation of mutants affecting wax biosynthesis or wax composition in *Arabidopsis thaliana* revealed a striking phenotype: the production of non-fertile pollen grains (*cer1* and *cer6-2* or *pop1* mutants; Aarts et al., 1995; Preuss et al., 1993). The molecular characterisation of the *cer1* gene expression profile, which was predicted to encode a novel integral membrane protein involved in the conversion of long aldehydes to alkanes (a putative aldehyde decarbonylase enzyme), revealed high levels of *cer1* mRNA, not only in stem and siliques (whose epidermal cells accumulate high levels of epicuticular waxes) but also in flower buds, most probably in tapetal cells. In fact, the pollen coat of *cer1* mutants showed a more granular appearance, smaller lipid droplets and more inclusions, even if the total amount of lipid in the pollen coat was similar to that in the wild type pollen coat (Aarts et al., 1995).

A secondary effect of the altered pollen coat lipid environment in *cer* mutants is the accumulation of callose in stigma papillae upon their interaction with *cer1* or *cer6* pollen grains (Aarts et al., 1995). This observation does not appear to be a general mecha-

Table 3.1. Some male sterile mutants known to be caused by lesions in genes regulating the biosynthesis and accumulation of lipid structures in the tapetum and/or developing pollen grains

Mutant	Species	Phenotype	Reference
ms2	*A. thaliana*	absence of exine layer, block of pollen development at microspore stage, reduced fatty acid alcohols in pollen grains	Aarts et al, 1997
ms2	*T. aestivum*	abnormal pollen wall formation	Albertsen and Philips, 1981
cer1	*A. thaliana*	abnormal pollen coat with smaller lipid droplets, absence of alkanes in pollen grains stem and fruits, failure to successfully interact with stigmatic papillae	Aarts et al., 1995 Hülskamp et al., 1995
cer3	*A. thaliana*	normal pollen coat appearance, pollen fails to hydrate in contact with the stigmatic papillae, deficient in C29–C30 fatty acids in pollen, stem and leaf rescued by pollination at low temperatures (18°C)	Hülskamp et al., 1995 Hannoufa et al., 1996
cer6-1	*A. thaliana*	smaller lipid droplets in the pollen coat, impaired pollen-stigma interaction, partial male sterility rescued by pollination at low temperatures (18°C)	Hülskamp et al., 1995
pop1 (cer6-2)	*A. thaliana*	lack of pollen coat, pollen fails to hydrate, deficient in very long-chain lipids (C29–C30) in pollen and stem	Preuss et al., 1993
triple fad [fad3, fad7 fad8]	*A. thaliana*	absence of jasmonic acid biosynthesis, impaired pollen maturation, lack of anther dehiscence	Mc Conn and Browse, 1996
ster	*O. hookeri*	abnormal sporopollenin deposition microspores collapse before first pollen mitosis	Noher de Halac and Harte, 1995
fr	*O. hookeri*	abnormalities in tapetal lipid accumulation, lack of sporopollenin deposition, dissolution of developing microspores	Noher de Halac and Harte, 1995

nism to prevent pollination by other species, as it was not observed when wild-type *Arabidopsis* stigma was pollinated with non-specific petunia or *Brassica* pollen (Aarts at al., 1995).

The analysis of the *pop1* mutant (also referred to as *cer 6-2*), instead, revealed clear abnormalities in the pollen coat morphology and the pollen lipid showed strongly reduced levels of C29–C30 very long-chain fatty acids and, correspondingly, higher levels of C24–C27 fatty acid chains and unaffected levels of C16–C18 fatty acids, suggesting that the *pop1* gene may encode a specific fatty acid elongase (Preuss et al., 1993). The sterility of the *pop1* mutant was shown to result from the inability of the pollen grain to hydrate on a suitable stigmatic surface, although pollen was shown to be viable and able to germinate *in vitro* or in conditions of high humidity. All available data support the hypothesis, proposed by Preuss et al. (1993), that very long-chain lipid molecules, in particular alkanes, are needed for proper pollen-pistil signalling. In fact, within a few minutes of contact between pollen and the stigmatic papillae, the tryphine spreads and creates a region of continuity to form a foot-like adhesion zone (Elleman et al., 1992; Kandasamy et al., 1994). In this region the so called "coat conversion" process, which involves the appearance of numerous membranous inclusions, takes place, enabling the pollen coat to act as a conduit for the hydration of the pollen grain (Dickinson and Elleman, 1985; Elleman and Dickinson, 1986) It remains to be elucidated whether the very long-chain lipids are themselves the signalling molecules or whether they stabilise another signalling ligand within the pollen coat. The importance of very long-chain lipids in such cell to cell signalling events has recently been emphasised by the characterisation of the *Arabidopsis fiddlehead-1* mutant, which shows abnormal pollen-epidermis interactions and specific alterations in epidermal extracellular long-chain lipids (Lolle et al., 1997).

In addition to elaioplasts, another type of lipid-containing organelle has been described in the tapetal cytoplasm of a wide range of plant families (e.g. *Liliaceae*, Reznickova and Willemse, 1980; *Tiliaceae*, Hesse, 1993 and *Lamiaceae*, Ubera Jimenez et al., 1996). The structure and lipid composition of these sucbellular organelles have been particularly characterised and analysed in *Brassica* and *Arabidopsis* species (Murphy and Ross, 1998; Wang et al., 1997; Wu et al., 1997; Ferreira et al., 1997).

The tapetal cytoplasmic lipid bodies differ from the oil bodies found in seeds of Brassicaceae (Murphy, 1993) in respect to their size, which, on average, is approximately five times larger than the latter (approximately 3 µm in diameter), and their internal organisation (Murphy and Ross, 1998). The tapetal cytoplasmic lipid bodies, also termed "tapetosomes" (Wu et al., 1997), are characterised by a regularly organised internal fibrous proteinaceous network (Owen and Makaroff, 1995; Murphy and Ross, 1998). Recent immunocytochemical studies have demonstrated the association between the internal hexagonal fibrous structures and tapetal-specific oleosin-like proteins (OLPs) (Murphy and Ross, 1998). In contrast, seed oil bodies are bounded by an external oleosin-phospholipid membrane and are internally homogeneous (Murphy, 1993; Huang, 1996).

The tapetal-specific oleosin-like proteins in *Brassica* species are encoded by a large family of genes whose expression was shown to start at the unicellular microspore stage, to reach a maximum at the bicellular pollen stage and to decrease after the second pollen mitosis (Ross and Murphy, 1996; Ruiter et al., 1997b). The tapetal-specific oleosin-like proteins were shown to be characterised by a highly hydrophobic domain (characteristic and in common with all plant oleosins described to date), located near the N-terminus, followed by a highly polar C-terminal domain, characterised by Gly-rich and/or repetitive Pro/Ala and Lys/Gly-rich motifs (Ross and Murphy, 1996, Ruiter et al., 1997b; Robert et al., 1994). Cluster analysis based upon the amino acid sequence of the highly conserved central hydrophobic domain of all available oleosin sequences showed that the seed- and anther-expressed oleosins fall into two major distinct clusters (Ross and Murphy, 1996) suggesting their divergence and a different function in seed and anther tissues, respectively. The *Brassica oleracea* tapetal oleosin-like genes were also shown to be post-transcriptionally regulated, partly by mechanisms of alternative splicing, leading to the biosynthesis of truncated versions of the same protein in tapetal cells (Ruiter et al., 1997b). The significance of this post-transcriptional processing and the association of OLPs with specific stages or physiological states (e.g. dehy-

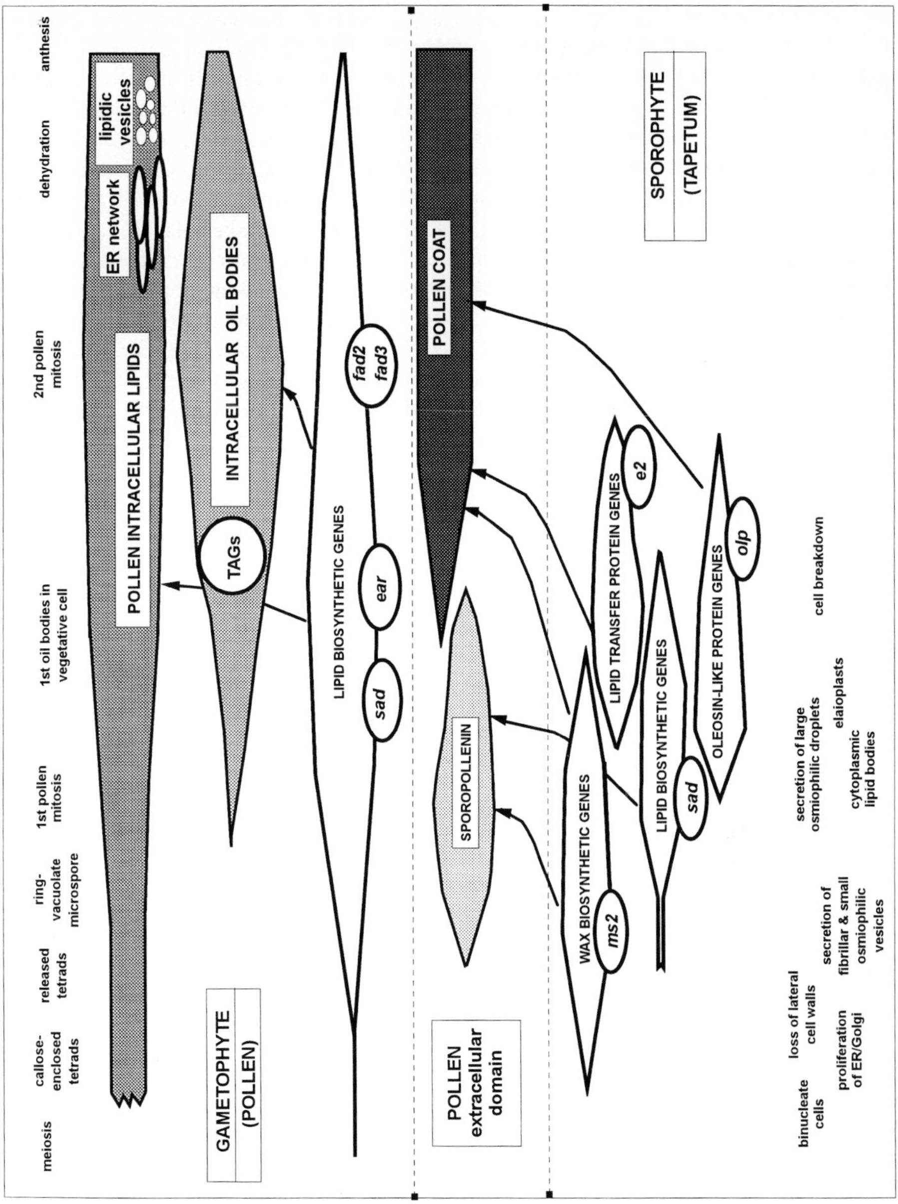

Fig. 3.3. Diagramatic representation of temporal regulation of gene expression of major classes of lipid biosynthetic enzymes, lipid transfer proteins and oleosin-like oil-body associated proteins in the tapetum and in pollen vegetative cells of Brassicaceae (based upon data from *Brassica napus*, *Brassica oleracea* and *Arabidopsis thaliana*). The main physiological and anatomical changes occurring in developing microspore/pollen grains and associated sporophytic tapetal cells, are indicated at the top and bottom of the two panels, respectively. Within elliptical circles are indicated (in italics) some of the recently characterised genes upon whose patterns of gene expression this diagram is based. The filled boxes indicate the four major lipidic structural components of the mature pollen grain (i.e. intracellular pollen lipids, pollen intracellular oil bodies, sporopollenin and pollen coat). Legend *ear*, enoyl-ACP reductase; *sad*, stearoyl-ACP desaturase (Piffanelli et al., 1997); *fad2*, oleoyl-desaturase (Piffanelli et al., 1997); *fad3* linoleoyl-desaturase (Piffanelli et al., 1997); *olp* tapetal- specific oleosin-like proteins (Ross and Murphy, 1996); *e2*, lipid transfer protein from *Brassica napus* (Foster et al., 1992); *ms2*, putative fatty acyl reductase/elongase (Aarts et al., 1997)

dration) of anther development awaits further investigations. The oleosin-like proteins appear to be co-translationally targeted to the ER (Murphy and Ross, 1998), which is consistent with the likely biogenesis of the tapetal-specific cytoplasmic lipid bodies from ER membranes. In fact, during the active stage of lipid-particle formation not only the close association between the "tapetosomes" and ER but also the penetration of ER filaments into the cytoplasmic lipid bodies were observed (Wang et al., 1997).

The tapetal cytoplasmic lipid bodies were shown to accumulate TAGs and to contain higher amounts of polar lipids (mainly phosphatidylcholine and phosphatidylethanolamine) than seed oil bodies (Wu et al., 1997). It remains unclear as to what the destiny of all the TAGs stored in the tapetal cytoplasmic bodies is, as only trace amounts of TAGs were found to be present in the pollen coat extracts from *Brassica napus* and *Arabidopsis* (Piffanelli et al., 1997; Wu et al., 1997). It was suggested that the tapetal-derived TAGs can become a source of polyunsaturated lipids, which are precursors of essential hormones (i.e. jasmonic acid) for the final stages of pollen maturation and anther dehiscence (McConn and Browse, 1996).

Detailed ultrastructural studies have clearly shown that, upon tapetal lysis, the cytoplasmic lipid bodies are released intact into the locule (Murgia et al., 1991b; Dickinson and Lewis, 1973). Then they are deposited and attached to the exine walls and only there are they subject to a complete rupture (Dickinson and Lewis, 1973; Murphy and Ross, 1998). At this developmental stage it was also shown that the oleosin-like proteins associated with the cytoplasmic lipid bodies are cleaved close to the junction between the hydrophobic domain and the highly polar C-terminal domain (Ross and Murphy, 1996; Murphy and Ross, 1998). The resulting C-terminal domains then constitute the major proteinaceous component of the pollen coat in *Brassica* and *Arabidopsis* species (Ross and Murphy, 1996; Wang et al., 1997; Preuss and Davis, 1995; Ruiter et al., 1997b; Murphy and Ross, 1998). The oleosin-like proteins may play a role in stabilising the various pollen coat components on the exine during anther dehydration and in facilitating pollen rehydration and germination following a successful compatibility response (Ross and Murphy, 1996; Ruiter et al., 1997b). The combination of a predicted random-coiled secondary structure and the high occurrence of glycine residues in pollen coat-associated oleosin-like proteins were suggested as features to bind and guide water from the stigmatic papillae to the pollen grains (Ruiter et al., 1997b).

The oleosin-like N-terminal hydrophobic (which is cleaved off during the assembly of the pollen coat lipidic components) may be responsible for the targeting of OLPs to the large cytoplasmic lipid bodies in the tapetum (Murphy and Ross, 1998; Wu et al., 1997). It is also possible to speculate that the cleavage of the hydrophobic oleosin domain of the OLPs enables the coalescence of the associated lipid droplets and their adherence to other pollen coat components, as it was shown for seed oil bodies devoid of oleosins (Cummins et al., 1993).

The tapetally-derived pollen coat has also been reported to carry various lipidic volatile compounds (Henning and Teuber, 1992; Robertson et al., 1993; Dobson et al., 1996), which can act as either attractants or deterrents to pollen-seeking flower visitors such as honeybees, moths and hummingbirds (Dobson, 1988). Only recently has the chemical structure of these pollen coat volatile components begun to be analysed (Robertson et al., 1993, Dobson et al., 1996, Murphy, 1997) and it is expected that in the next few years the biosynthetic pathways and the mechanism of action of these compounds will be clarified, possibly enabling researchers to improve the pollination process through the attraction of the most suitable pollinators and by protection from pathogen attack.

3.4 Intracellular Lipid Accumulation in Developing Pollen Grains

The internal or cytoplasmic domains of pollen grains accumulate nutrient reserves (i.e. starch and storage lipids) for the metabolic support of the male germ unit during the pollen germination phase on the stigmatic surface and during the pollen tube growth through the style and the ovary structures.

Numerous pollen grains, predominantly entomophilous (Baker and Baker, 1979), synthesise and accumulate large amounts of storage lipids, normally TAGs, as oil bodies (Evans et al., 1991: Wetzel and Jensen, 1992; Murgia et al., 1991a; Piffanelli et al., 1997). The chemical analysis of lipids in the internal cytoplasmic domain of *Brassica napus* mature pollen grains revealed that the polar lipids and neutral TAGs each constitute approximately 40% of the total lipid content, whilst the remaining 20% is mainly neutral esters (approx. 5%), free fatty acids (approx. 5%) and other non-identified lipid classes. The synthesis of storage lipids and the accumulation of the oil bodies is restricted to the vegetative cell of maturing pollen grains (Charzynska et al., 1989; Van Aelst et al., 1993) as in most species the generative cell does not contain plastids and, as a consequence, is not able to synthesise fatty acid chains, which are the precursors of both membrane and storage lipids.

The intracellular pollen oil bodies are spherical organelles with a diameter ranging from 0.5 to 2.0 μm (Evans et al., 1992; Piffanelli et al., 1997; Wang et al., 1997), which is similar to that of oil bodies found in most temperate oil-storing seeds (Huang, 1996; Murphy, 1993). Both seed and pollen oil bodies are homogeneous organelles filled up with triacylglycerols and contain about 2–3% phospholipids, which are likely to form a monolayer around the surface of the oil body. Despite repeated efforts to isolate oil body-associated proteins from intracellular pollen domains, no oleosin or oleosin-like proteins were found to be associated with the intracellular pollen oil bodies (Ross, 1996; Murphy and Ross, 1998). In oil-storing seed tissues of temperate species, oleosin proteins encircle the oil bodies and are mainly deposited during the mid-late phases of seed maturation (Murphy, 1993: Huang, 1996; Hills et al., 1993). Oleosin protein function has been related to the stabilisation of oil body structures during seed dehydration and the subsequent rehydration during germination (Leprince et al., 1998).

The triacylglycerols synthesised and accumulated in *Brassica napus* pollen oil bodies (Evans et al., 1992; Piffanelli et al., 1997) were shown to be esterified mainly to palmitic acid (27%) and linolenic acid (63%). This differs markedly from the composition of triacylglycerols accumulated in zygotic embryos of the same species, which are predominantly esterified to oleic acid (65% in zero erucic acid cultivars) or erucic acid (40% in high erucic acid cultivars) (Evans et al., 1987; Piffanelli et al., 1997). It is interesting to note that in pollen grains from freshly dehisced anthers most intracellular oil bodies appear to be enfolded in the extensive network of ER membranes (Fischer et al., 1968; Jensen et al., 1968; Ross, 1996). In *Gossypium hirsutum* and *Impatiens walleriana*, pockets of ER are formed around oil bodies during the final phases of pollen maturation. It is possible to hypothesise that pollen oil bodies are protected from coalescence during dehydration/rehydration phases by the close association with enfolding pockets of ER membranes (Wetzel and Jensen, 1992; Piffanelli et al., 1997).

The biosynthesis and accumulation of lipid structures in the internal or cytoplasmic domain of the developing pollen grains was shown to be strictly developmentally regulated by the gametophytic genome (Piffanelli et al., 1997). A basal level of expression of genes encoding lipid biosynthetic enzymes ensures the synthesis of membrane structures to support the growth of the developing meiocytes and microspores. Following the first pollen mitosis, a concerted transcriptional activation of genes encoding both plastidial- and ER-localised lipid biosynthetic enzymes takes place in the newly formed vegetative cell, leading to a sharp increase in total pollen lipid content (Piffanelli et al., 1997; Evans et al., 1992). In entomophilous pollen grains, like *Brassica napus*, an extensive network of ER cisternae and oil-body droplets starts to accumulate just after the first pollen mitosis and, by the time of the second pollen mitosis, these have filled up most of the vegetative cell (Evans et al., 1992; Owen and Makaroff, 1995; van Aelst, 1993).

In fact, the amount of pollen membrane lipids (also referred as polar lipids) doubles between the first and the second mitoses. The majority of these lipids form the ER network, although a significant proportion is required to form the many vesicles and organelles including plastids, mitochondria and dictyosomes found in the vegetative cell (Murgia et al., 1991a; Charzynska et al., 1989). Compositional analysis has shown that intracellular pollen membranes are mainly made up of phospholipids, namely phosphatidylcholine, phosphatidylethanolammine and phosphatidylinositol (Evans et al., 1990). Fatty acid analysis has also revealed that the intracellular pollen membrane lipids change their composition significantly during pollen development, with particular increases in palmitic and linolenic acids (which at pollen maturity represent 65% and 20% of the fatty acid chains esterified to polar lipids, respectively) and corresponding declines in the amount of oleic and linoleic acids (Piffanelli et al., 1997). The similarities in fatty acid composition of the pollen intracellular membrane and storage lipids (both TAGs and neutral esters) may well reflect their shared fatty acid biosynthetic pathways (see Fig. 3.2) and their common regulation by the gametophytic genome.

The important role of linolenic acid in pollen viability and anther dehiscence was recently demonstrated by the characterisation of an *Arabidopsis thaliana* triple mutant, defective in all three isoforms (one ER-localised and two plastid-localised) of the enzyme linoleate desaturase, which is responsible for linolenic acid biosynthesis. The only obvious phenotypes of this triple mutant at optimal growth conditions were pollen infertility and the lack of anther dehiscence, caused by the inability to produce jasmonic acid (a derivative of linolenic acid) in anther tissues (Mc Conn and Browse, 1996). Both phenotypes could be rescued by spraying jasmonic acid a few hours before anther dehiscence or by supplying exogenous linolenic acid, which suggest that jasmonic acid is an essential signalling molecule for correct anther maturation and dehiscence. This idea is also supported by the characterisation of the *Arabidopsis thaliana coi* methyl-jasmonate insensitive mutant, which was shown to be male sterile due to aborted pollen and lack of anther dehiscence (Feys et al., 1994).

During the final phases of pollen maturation (from the second pollen mitosis to anther dehiscence) a reproducible decrease of about 18% of the total pollen lipid content was observed in *Brassica napus* (Piffanelli et al., 1997). Quantitative analysis of lipid fractions throughout pollen development revealed that only TAGs and neutral esters decreased, whilst the polar lipid content remained unaltered. The presence of glyoxysome-like microbodies in mature pollen grains (Pais and Fejo, 1987; Charzynska et al., 1989) and the expression of genes encoding glyoxylate cycle enzymes during pollen maturation phases (Zhang et al., 1994) are consistent with a controlled mobilisation of part of the storage lipids. This controlled mobilisation of storage lipids may represent a developmentally regulated mechanism by which polyunsaturated fatty acid chains are released and become substrates for enzymatic modifications leading to the biosynthesis of jasmonic acid, which was shown to be an essential signalling molecule for pollen maturation and anther dehiscence (McConn and Browse, 1996).

Very few studies investigated the mobilisation and catabolism of storage TAGs during pollen germination and pollen tube growth. A detailed ultrastructural study of germinating pollen and pollen tubes of *Tradescantia reflexa* revealed that oil bodies that are encircled by pockets of ER become associated with thin electron-dense vesicles which then develop into larger vesicles and vacuoles (Noguchi, 1990). This is, to date, the most direct demonstration of the conversion of pollen oil bodies into the membranous vesicles that serve as precursors for the growth of plasma membrane in the pollen tubes. As described in castor bean seed endosperm tissues, it is possible to envisage the existence of a specific transacylase (which would catalyse the transfer of a fatty acid chain from TAGs to monoacylglycerol with the consequent formation of diacylglycerol, DAG) and the rapid transfer of fatty acid chains directly from DAG to the membrane lipid, phosphatidylcholine (Stobart et al., 1997). This hypothesis is in agreement with the failure to observe high levels of expression of genes involved in fatty acid catabolism (e.g. malate synthase, isocitrate lyase) during either germination or pollen tube growth in *Brassica napus* (Zhang et al., 1994).

In addition to the extensive network of ER cisternae, pollen grains of several species were also shown to contain numerous vesicles lying immediately beneath the intine surface, where they are available to fuse with the plasma membrane and are likely to facilitate its rapid expansion immediately following pollen germination (Dorne et al.,1988; Cresti and Tiezzi, 1990; Hess, 1995). It is also relevant to note that several enzymes and cofactors involved in fatty acid biosynthesis have been detected in intracellular fractions of mature dehydrated *Brassica napus* pollen grains (Evans et al., 1992; Piffanelli et al., 1997), suggesting that the rapid onset of membrane synthesis to ensure pollen tube growth immediately after pollen germination may be, at least partially, due to a set of lipid biosynthetic enzymes already present in the mature pollen grain (Whipple and Mascarenhas, 1978).

A detailed molecular analysis of the genes encoding stearoyl-ACP desaturase (the plastidial-targeted enzyme responsible for the biosynthesis of oleic acid by the introduction of a double bond to the C18 saturated fatty acid chain of stearic acid) in both gametophytic and sporophytic anther and seed tissues of *Brassica napus* conclusively showed that the same messages were expressed in all tissues analysed (Piffanelli et al., 1997). These results were also supported by GUS expression data from transgenic tobacco and rapeseed plants transformed with a *Brassica napus* stearoyl-ACP desaturase promoter fused to the GUS reporter gene (Slocombe et al., 1994, Piffanelli, unpublished data). A variety of genetic, molecular and biochemical data suggest that, like stearoyl-ACP desaturase, other lipid biosynthetic genes, such as linoleoyl desaturase, acyl carrier protein, and enoyl-ACP reductase, also have a haplo-diploid pattern of expression (Baerson and Lamppa, 1993; Evans et al., 1988; Jourdren et al., 1996; McConn and Browse, 1996; Aarts et al., 1995; 1997; Preuss et al., 1993). This haplo-diploid pattern of expression of lipid biosynthetic genes may have relevant implications in applying the concept of "pollen selection" to alter the composition of storage lipids accumulated in seed tissues (Evans et al., 1988; Jourdren et al., 1996). It is important to stress that, due to their non-tissue-specific expression, the use of the promoter regions of such genes will not be useful to drive expression of novel enzymatic functions aimed at the specific alteration of seed lipid profiles.

It also appears that some lipid biosynthetic genes, like the enoyl-ACP reductase, are constitutively expressed at high levels, while others, as the stearoyl-ACP desaturase, are finely regulated, mainly at the transcriptional level, in sporophytic and gametophytic tissues. It was suggested that the stearoyl-ACP desaturase expression levels in gametophytic pollen and zygotic embryo sporophytic tissues may represent a central regulatory point leading to the biosynthesis and esterification of monounsaturated versus polyunsaturated fatty acid chains in TAGs (Piffanelli, 1997).

3.5 Conclusions and Future Prospects

The isolation and characterisation of a significant number of male-sterile mutants which are affected in the biosynthesis or accumulation of various lipidic structures (see Table 3.1) suggests that lipidic and lipid-derived structures are very important components of viable pollen grains. It is becoming clear that in tapetal and gametophytic cells there is a concerted activation of a large number of lipid biosynthetic and lipid-related pathways. Much experimental evidence from mutant analysis and from gene expression studies points to the importance of the temporal and spatial co-ordination between biosynthesis and processing of lipids in the tapetum and the developing male gametophyte. We are just beginning to dissect and understand the pathways for assembly of the lipid components of complex structures like the sporopollenin polymer and the pollen coat matrix.

What until recently appeared to be a disorganised rupture and lysis of the tapetal cells is now coming to be recognised as a series of programmed and controlled events which lead to the transfer of specialised lipid structures into the interstices of the pollen wall. The disappearance/appearance of specific classes of tapetally-synthesised lipids and the processing of lipid-associated proteins clearly underlies a tightly regulated deposition and packing of tapetal-derived pollen coat, which has been shown to be essential for the success of pollen-stigma interactions and pollen rehydration. In that respect, the precise function of the tapetal-specific "tapetosomes" and of the associated oleosin-like proteins remains to be determined. It is hoped that an antisense/cosuppression strategy to down-regulate the genes encoding the oleosin-like proteins will soon elucidate the events leading to the relocation of lipid-related material to the pollen surface and to their modifications to form the pollen coat that lines the exine of viable pollen grains.

It has also become clear that the gametophyte is able autonomously to synthesise and accumulate high levels of membrane and storage lipids. The same lipid biosynthetic genes which are expressed in the tapetal cells, under the control of the sporophytic diploid genome, are also active in the vegetative cell but here are regulated in a distinct temporal and spatial pattern by the haploid gametophytic genome.

Based upon experimental data it may be reasonable to hypothesise a form of cross-talk between the tapetum and the developing pollen grain, which ensures the biosynthesis and release of structural components in both gametophytic and sporophytic tissues.

In the future, progress towards elucidating the structure, function, biosynthesis and processing of the various lipidic components of the sporopollenin polymer and pollen coat matrix, is likely to come from a combination of molecular genetic and biochemical/cell biological studies of male-sterile mutants and mutants affected in specific steps of lipid biosynthetic pathways. Reverse genetics may also prove informative to down-regulate specific classes of genes encoding lipid biosynthetic enzymes or lipid-associated proteins (e.g. fatty acid desaturases, fatty acid thioesterases, lipid-transfer proteins) using tapetal or gametophytic-specific promoters, as already shown for enzymes and proteins involved in other pathways (van der Meer et al., 1992; Matsuda et al., 1996; Xu et al., 1995).

Acknowledgements
We would like to thank Dr Joanne Ross for the electron micrograph pictures of the different forms of oil bodies and of the tapetal elaioplasts shown in Figure 3.1.

Abbreviations
DAGAT, diacylglycerol; **SAD**, stearoyl-ACP desaturase (protein); ***sad***, stearoyl-ACP desaturase (gene), **TAGs**, triacylglycerols; **ER**, endoplasmic reticulum; **OLPs** tapetal oleosin-like proteins.

References

Aarts MGM, Hodge R, Kalantidis K, Florak D, Wilson Z, Mulligan BJ, Stiekema WJ, Scott R, Pereira A (1997) The Arabidopsis MALE STERILITY 2 protein shares similarity with reductases in elongation/condensation complexes. Plant J 12:615–623
Aarts MGM, Keijzer CJ, Stiekma WJ, Pereira A (1995) Molecular characterisation of the *CER1* gene of *Arabidopsis* involved in epicuticular wax biosynthesis and pollen fertility. Plant Cell 7:2115–2127
Albertsen MC, Philips RL (1981) Developmental cytology of 13 genetic male sterile loci in maize. Can J Genet Cytol 23:195–208
Baerson SR and Lamppa GK (1993) Developmental regulation of an acyl carrier protein gene promoter in vegetative and reproductive tissues. Plant Mol Biol 22:255–267
Baker HG, Baker I (1979) Starch in angiosperm pollen grains and its evolutionary significance. Amer J Bot 66:591–600
Bianchi G Plant waxes (1987) in The Metabolism, Structure and Function of Plant Lipids pp 553–594 edited by Stumpf PK, Mudd JP, Nes WD Plenum Press New York
Bianchi G, Murelli C, Ottaviano E (1990) Maize pollen lipids. Phytochem 29:739–744
Browse J, Somerville C (1991a) Glycerolipid synthesis: biochemistry and regulation. Annu Rev Plant Physiol Plant Mol Biol 42:467–506
Browse J, Somerville C (1994) Glycerolipids. In: Arabidopsis, Cold Spring Harbor Laboratory Press, pp 881–912
Caffrey M, Werner BG, Priestley DA (1987) A crystalline lipid phase in a dry biological system: evidence from X-ray diffraction analysis of *Typha latifolia* pollen. Biochim Biophys Acta 921:124–134
Charzynska M, Murgia M, Cresti M (1989) Ultrastructure of the vegetative cell of *Brassica napus* pollen with particular reference to microbodies. Protoplasma 152:22–28
Ciampolini F, Nepi M, Pacini E (1993) Tapetum development in *Cucurbita pepo* (Cucurbitaceae). Plant Syst Evol 7:13–22
Cresti M, Tiezzi A (1990) Germination and pollen tube formation. In: Blackmore, S and Knox, RB (eds) Microspores: Evolution and Ontogeny. Academic Press, London pp 239–263
Crossley SJ, Greenland AJ, Dickinson HG (1995) The characterisation of tapetum-specific cDNAs isolated from a *Lilium henryi* L. meiocyte subtractive cDNA library. Planta 196:523–529
Cummins I, Hills MJ, Ross JHE, Hobbs DH, Watson MD and Murphy DJ (1993) Differential temporal and spatial expression of genes involved in storage oil and oleosin accumulation in developing rapeseed embryos: implications for the role of oleosins and the mechanisms of oil-body formation. Plant Mol Biol 23:1015–1027
Dickinson HG (1973) The role of plastids in the formation of pollen grain coatings. Cytobios 8:25–40
Dickinson HG, Elleman CJ (1985) Structural changes in the pollen grain of *Brassica oleracea* during dehydration in the anther and development on the stigma as revealed by anhydrous fixation techniques. Micron Microsc Acta 16:255–270
Dickinson HG, Lewis D (1973) The formation of the tryphine coating the pollen grains of *Raphanus*, and its properties relating to the self-incompatibility system. Proc Roy Soc Lond (B) 184:149–165
Dobson HEM (1988) Survey of pollen and pollenkitt lipids – chemical cues to flower visitors? American Journal of Botany 75:170–182
Dobson HE, Groth I, Bergstrom G (1996) Pollen advertisement: chemical contrasts between whole-flower and pollen odors. Amer J Bot 83:877–885
Dorne A-J, Kappler R, Kristen U, Heinz E (1988) Lipid metabolism during germination of tobacco pollen. Phytochem 27:2027–2031

Dunbar, A (1973) Pollen development in the *Eleocharis palustris* group (Cyperaceae). 1. Ultrastructure and ontogeny. Bot Not 126:197–254

Echlin P (1971) The role of the tapetum during microsporogenesis of angiosperms. In: Heslop-Harrison J (ed) Pollen: Development and Physiology. Butterworths, London, pp 41–61

Echlin P, Godwin H (1968) The ultrastructure and ontogeny of pollen in *Helleborus foetidus* L. J Cell Sci 3:161–174

Elleman CJ, Dickinson HG (1986) Pollen stigma interactions in *Brassica*. 4. Structural reorganisation in the pollen grains during hydration. J Cell Sci 80:141–157

Elleman CJ, Franklin-Tong Vand Dickinson HG (1992) Pollination in species with dry stigmas: the nature of the early stigmatic response and the pathway taken by pollen tubes. New Phytol 121:413–24

Evans DE, Sang JP, Cominos X, Rothnie NE, Knox RB (1990) A study of phospholipids and galactolipids in pollen of two lines of *Brassica napus* L. (rapeseed) with different ratios of linoleic to linolenic acid. Plant Physiol 92:418–424

Evans DE, Taylor PE, Singh MB, Knox RB (1991) Quantitative analysis of lipids and protein from the pollen of *Brassica napus* L. Plant Sci 73:117–126

Evans DE, Taylor PE, Singh MB, Knox RB (1992) The interrelationship between the accumulation of lipids, protein and the level of acyl carrier protein during the development of *Brassica napus* L. pollen. Planta 186:343–354

Evenson KJ and Post Betteinmiller D (1995) Fatty acid elongating activity in rapidly-expanding leek epidermis. Plant Physiol 109:707–16

Ferreira MA, de Almeida Engler J, Miguens FC, Van Montagu M, Engler G, de Oliveira DC (1997) Oleosin gene expression in *Arabidopsis thaliana* tapetum coincides with accumulation of lipids in plastids and cytoplasmic bodies. Plant Physiol Biochem 35:729–739

Feys BJF, Benedetti CE, Penfold CN and Turner JG (1994) Arabidopsis mutants selected for resistance to the phytotoxin coronatine are male sterile, insensitive to methyl jasminate, and resistant to a bacterial pathogen. Plant Cell 6:751–59

Fisher DB, Jensen WA, Ashton MA (1968) Histochemical studies of pollen: storage pockets in the endoplasmic reticulum (ER). Histochem 13:169–182

Foster GD, Robinson SW, Blundell RP, Roberts MR, Hodge R, DraperJ, Scott R (1992) A *Brassica napus* mRNA encoding a protein homologous to phospholipid transfer proteins is expressed specifically in the tapetum and developing microspores. Plant Sci 84:187–192

Gabarayeva NI (1992) Sporoderm development in *Asimina triloba* (Annonaceae). I The developmental events before callose dissolution. Grana 31:213–222

Gabarayeva NI (1995) Pollen wall and tapetum development in *Anaxagorea brevipes* (Annonaceae): sporoderm substructure, cytoskeleton, sporopollenin precursor particles, and the endexine problem. Rev Palaeobot Palynol 85:123–152

Gubatz S, Wiermann R (1992) Studies on sporopollenin biosynthesis in *Tulipa* anthers. Bot Acta 105: 407–413

Gubatz S, Wiermann R (1993) Studies on sporopollenin biosynthesis in *Cucurbita maxima* I. The substantial labelling of sporopollenin from *Curcurbita maxima* after application of ^{14}C-phenylalanine. Z Naturforsch 48:910–915

Guilford WJ, Schneider DM, Labovitz J, Opella SJ (1988) High resolution solid state ^{13}C NMR spectroscopy of sporopollenins from different plant taxa. Plant Physiol 86:134–136

Hannoufa A, McNevin JP and Lemieux B (1993) Epicuticular waxes of ecefiferum mutants in Arabidopsis thaliana. Phytochemistry 33:851–55

Hannoufa A, Negruk V, Eisner G and Lemieux B (1996) The CER3 gene of Arabidopsis thaliana is expressed in leaves, roots, flowrs and apical meristems. Plant J 10:459–67

Harwood JL (1989) Lipid metabolism in plants. Crit Rev Plant Sci 8:1–44

Harwood JL (1996) Recent advances in the biosynthesis of plant fatty acids. Biochem. Biophys. Acta 1301:7–56

Henning JA, Teuber LR (1992) Identification of pollenkitt variation among alfalfa germplasm sources. Crop Sci 32:653–656

Heslop-Harrison J (1962) Origin of exine. Nature 195:1069–1071

Heslop-Harrison J (1968) Tapetal origin of pollen coat substances in *Lilium*. New Phytol 67 : 779–786

Hess MW (1995) High pressure freeze fixation reveals novel features during ontogensis of the vegetative cell in *Ledebouria* pollen: an ultrastructural and cytochemical study. Biochem Cell Biol 73:1–10

Hess MW, Hesse M (1994) Ultrastructural observations on anther tapetum development of freeze-fixed *Ledebouria socialis* Roth (Hyancinthaceae). Planta 192:421–430

Hesse M (1993) Pollenkitt development and composition in *Tilia platyphyloos* (Tiliaceae) analysed by conventional and energy filtering TEM. Plant Syst Evol 7:39–52

Hesse M, Hess MW (1993) Recent trends in tapetum research. A cytological and methodological review. Plant Syst Evol 7:127–145

Hills MJ, Watson MD, Murphy DJ (1993) Targeting of oleosins to the oil bodies of oilseed rape (*Brassica napus* L.). Planta 189:24–29

Huang, AHC (1996) Oleosins and oil bodies in seeds and other organs. Plant Physiol 110:1055–1061

Hulskamp M, Kopczak SD, Horejsi TF, Kihl BK, Pruitt RE (1995) Identification of genes required for pollen-stigma recognition in *Arabidopsis thaliana*. Plant J 8:703–714

James DW, Lim E, Keller J, Plooy I, Ralston E, and Dooner HK (1995) Directed tagging of the Arabidopsis FATTY ACID ELONGATION (FAE1) gene with the maize transposon activator. Plant Cell 7: 309–319

Jensen W, Fisher DB, Ashton, ME (1968) Cotton embryogenesis: the pollen cytoplasm. Planta 81:206–228

Jones A, Davies HM and Voelker TA (1995) Palmitoyl-acyl carrier protein (ACP) thioesterase and the evolutionary origin of the plant acyl-ACP thioesterases. Plant Cell 7:359–371

Jourdren C, Simmoneaux D and Renard M (1996) Selection of pollen for linolenic acid content in rapessed, *Brassica napus* L. Plant Breeding 115:1–5

Kandasamy MK, Nasrallah JB, Nasrallah ME (1994) Pollen-pistil interactions and developmental regulation of pollen tube growth in *Arabidopsis*. Development 120:3405–3418

Keijzer CJ (1987) The processes of anther dehiscence and pollen dispersal. II. The formation and the transfer mechanism of pollenkitt, cell-wall development of the loculus tissues and a function of orbicules in pollen dispersal. New Phytol 105:499–507

Keijzer CJ, Cresti M (1987) A comparison of anther tissue development in male sterile *Aloe vera* and male fertile *Aloe cilaris*. Ann Bot 59:533–542

Kollatukudy PE (1980) Cutin, suberin and waxes. In: Stumpf PK, Conn EE (eds) The Biochemistry of Plants. New York, Academic Press 4:571–645

Koltunow AM, Truettner J, Cox KH, Wallroth M, Goldberg RB (1990) Different temporal and spatial gene expression patterns occur during anther development. Plant Cell 2:1201–1224

Lemieux B (1996) Molecular genetics of epicuticular wax biosynthesis. Trends Plant Sci 1:312–318

Leprince O, van Aelst AC, Pritchard HW, Murphy DJ (1998) A novel role for oleosins in preventing oil-body coalescence during seed imbibition as suggested by a low-temperature scanning electron microscope study in desiccation-tolerant and -sensitive oilseeds. Planta 204:109–119

Lolle SJ, Berlyn GP, Engstrom EM, Krolikowski KA, Reiter W-D, Pruitt RE (1997) Developmental regulation of cell interactions in the *Arabidopsis fiddlehead-1* mutant: A role for the epidermal cell wall and cuticle. Dev Biol 189:311–321

Mascharenhas JP (1975) The biochemistry of angiosperm pollen development. Bot Rev 41:259–314

Matsuda N, Tsuchiya T, Kishitani S, Tanaka Y, Toriyama K (1996) Partial male sterility in transgenic tobacco carrying antisense and sense PAL cDNA under the control of a tapetum-specific promoter. Plant Cell Physiol 37:215–222

McConn M, Browse J (1996) The critical requirement for linolenic acid is pollen development, not photosynthesis, in an *Arabidopsis* mutant. Plant Cell 8:403–416

Metz J and Lassner M (1996) Reprogramming of oil synthesis in rapessed: industrial applications. Ann N Y Acad Sci 792:82–90

Murgia M, Detchepare S, van Went JL, Cresti M (1991a) *Brassica napus* pollen development during generative cell and sperm cell formation. Sex Plant Reprod 4:176–181

Murgia M, Detchepare S, Van Went JL and Cresti M (1991b) Secretory tapetum of Brassica Oleracea L.: polarity and ultrastructural features. Sex Plant Reprod 4:28–35

Murphy DJ (1993) Structure, function and biogenesis of storage lipid bodies and oleosins in plants. Prog Lipid Res 32:247–280

Murphy DJ (1998) Why is rapeseed pollen such a sticky subject? In: Reproductive Biology 1996. Kew, London, in press

Murphy DJ, Cummins I (1989) Biosynthesis of storage products during embryogenesis in rapeseed. J Plant Physiol 135:63–69

Murphy DJ, Ross JHE (1998) Biosynthesis and processing of the major pollen coat proteins of *Brassica napus*. Plant J 13:1–16

Negruk V, Yang P, Subramanian M, McNevin JP, Lemieux B (1996) Molecular cloning and characterization of the *CER2* gene of *Arabidopsis thaliana*. Plant J 9:137–145

Noguchi T (1990) Consumption of lipid granules and formation of vacuoles in the pollen tube of *Tradescantia reflexa*. Protoplasma 156:19–28

Noher de Halac I, Fama G, Cismondi IA (1992) Changes in lipids and polysaccharides during pollen ontogeny in *Oenothera* anthers. Sex Plant Reprod 5:110–116

Noher de Halac I, Harte C (1995) Genetics and development of morphological and physiological characters of male sterility in *Oenothera*. Protoplasma 187:22–30

Ohlrogge JB, Browse J, Somerville C (1994) The genetics of plant lipids Biochim. Biophys. Acta 1082: 1–26

Owen HA, Makaroff CA (1995) Ultrastructure of microsporogenesis and microgametogenesis in *Arabidopsis thaliana* (L) Heynh ecotype Wassilewskija (*Brassicaceae*). Protoplasma 185:7–21

Pacini E, Casadoro G (1981) Tapetum plastids of *Olea europaea* L. Protoplasma 106:289–296

Pais MS, Feijo J (1987) Microbody proliferation during the microsporogenesis of *Ophrys lutea* Cav. (Orchidaceae). Protoplasma 138:149–155

Piffanelli P (1997) Molecular and biochemical studies of stearoyl-ACP desaturase in gametophytic and sporophytic generations of plants. PhD Thesis, University of East Anglia, UK

Piffanelli P, Ross JHE, Murphy DJ (1997) Intra and extracellular lipid composition and associated gene expression patterns during pollen development in *Brassica napus*. Plant J 11:549–652
Polowick PL, Sawhney K (1990) Microsporogenesis in a normal line and in the *ogu* cytoplasmic male-sterile lines of *Brassica napus*. I. The influence of high temperatures. Sex Plant Reprod 3:263–276
Post Beittenmiler D (1996) Biochemistry and molecular biology of wax production in plants. Annu Rev Plant Phys Plant Mol Biol 47:405–30
Preuss D, Davis R (1995) Cell signalling during fertilization: Interactions between the pollen coating and the stigma. J Cell Biochem 19A:140
Preuss D, Lemieux B, Yen G, Davis RW (1993) A conditional sterile mutation eliminates surface components from *Arabidopsis* pollen and disrupts cell signalling during fertilization. Genes Dev 7:974–985
Reznickova SA, Willemse MTM (1980) The formation of pollen in the anther of *Lilium*. II The function of the surrounding tissues in the formation of pollen and pollen wall. Acta Bot Neerl 29:141–156
Rick CM (1948) Genetics and development of nine male-sterile tomato mutants. Hilgardia 18:599–633
Robert LS, Gerster J, Allard S, Cass L, Simmonds J (1994) Molecular characterization of two *Brassica napus* genes related to oleosins which are highly expressed in the tapetum. Plant J 6:927–933
Robertson GW, Griffiths DW, MacFarlane Smith W, Butcher RD (1993) The application of thermal desorption – gas chromatography-mass spectrometry to the analyses of flower volatiles from five varieties of oilseed rape (*Brassica napus* spp *oleifera*). Phytochem Anal 4:152–157
Ross JHE (1996) Oleosin-like genes and proteins in the tapetum and pollen coat of *Brassica napus*. PhD thesis, University of East Anglia, UK
Ross JHE, Murphy DJ (1996) Characterisation of anther-expressed genes encoding a major class of extracellular oleosin-like proteins in the pollen coat of Brassicaceae. Plant J 9:625–637
Ruiter RK, Mettenmeyer T, van Laarhoven D, van Eldick GJ, Doughty J, van Herpen MMA, Schrauwen JAM, Dickinson HG, Wullems GJ (1997a) Proteins of the pollen coat of *Brassica oleracea*. J Plant Physiol 150:105–113
Ruiter RK, van Eldik, GJ, van Herpen MMA, Schrauwen JAM (1997b) Characterization of oleosins in the pollen coat of *Brassica oleracea*. Plant Cell 9:1621–1631
Schrauwen JAM, Mettenmeyer T, Croes AF and Wullems GJ (1996) Tapetum-specific genes: what role do they play in male gametophyte development? Acta Biol Neerl 45:1–15
Scott RJ (1994) Pollen exine – the sporopollenin enigma and the physics of pattern. In: Scott RJ, Stead AD (eds) Molecular and cellular aspects of plant reproduction. Cambridge University Press, Cambridge, pp 49–81
Scott R, Dagless E, Hodge R, Wyatt P, Soufleri I, Draper J (1991) Patterns of gene expression in developing anthers of *Brassica napus*. Plant Mol Biol 17:195–207
Scott RW, Strohl MJ (1962) Extraction and identification of lipids from loblolly pine pollen. Phytochemistry 1:189–193
Shaw G (1971) The chemistry of sporopollenin. In: Brooks J, Grant PR, Muir P, van Gijzel P, Shaw G (eds) Sporopollenin. Academic Press, London, pp 305–348
Somerville C and Browse J (1991) Plant lipids: Metabolism, Mutants and Membranes. Science 252:80–87
Southworth D (1990) Exine Biochemistry. In: Blackmore S, Knox RB (eds) Microspores. Evolution and Ontogeny. Academic Press, London, pp 193–212
Staiger D, Kappeler S, Muller M, Apel K (1994) The proteins encoded by two tapetum-specific transcripts, Satap35 and Satap-44 from *Sinapis alba* L. are localized in the exine wall layer of developing microspores. Planta 192:221–231
Sterk P, Booij H, Schellekens GA, van Kammen A, de Vries SC (1991) Cell-specific expression of the carrot EP2 lipid transfer protein gene. Plant Cell 3:907–921
Stobart K, Mancha M, Lenman M, Dahlqvist A and Stymne S (1997) Triacylglycerols are synthesised and utilized by transacylation reactions in microsomal preparations of developing safflower (Carthamus tinctorius L.) seeds. Planta 203:58–66
Stumpf PK (1981) Plants, fatty acids, compartments. TIBS 6:173–176
Suarez-Cervera M, Marquez J, Seoane-Camba J (1995) Pollen grain and Ubisch body development in *Platanus acerifolia*. Rev Palaeobot Palynol 85:63–84
Thoma S, Kaneko Y, Somerville C (1993) A non-specific lipid transfer protein from *Arabidopsis* is a cell wall protein. Plant J 3:427–436
Thoma S, Hecht U, Kippers A, Botella J, De Vries S, Somerville C (1994) Tissue-specific expression of a gene encoding a cell wall-localized lipid transfer protein from *Arabidopsis*. Plant Physiol 105:35–45
Töpfer R, Martini N and Schell J (1995) Modification of Plant Lipid Synthesis. Science 268:681–85
Ubera Jimenez JL, Fernandez PH, Schlag MG, Hesse M (1996) Pollen and tapetum development in male fertile *Rosmarinus officinalis* L. (Lamiaceae). Grana 34:305–316
van Aelst AC, Pierson ES, Van Went JL and cresti M (1993) Ultrastructural changes of Arabidopsis thaliana pollen during final maturation and rehydration. Zygote 1:173–79
van der Meer IM, Stam ME, van Tunen AJ, Mol JNM, Stuitje AR (1992) Inhibition of flavonoid biosynthesis in *Petunia* anthers by antisense RNA: a novel way to engineer nuclear male sterility. In: Ottaviano E, Mulcahy DL, Sari Gorla M, Bergamini Mulcahy G (eds) Angiosperm pollen and ovules. Springer Verlag, New York, pp 22–27

von Wettstein-Knowles PM (1982) Elongases and epicuticular wax biosynthesis. Physiol Veg 20:797–809
von Wettstein-Knowles PM (1995) Biosynthesis and Genetics of Waxes. In: Hamilton RJ (ed) Waxes: Chemistry, Molecular Biology and Functions. Allowry, Ayr, Scotland: Oily Press 6:91–130
Wang T-W, Balsamo RA, Ratnayake C, Platt KA, Ting JTL, Huang AHC (1997) Identification, subcellular localisation and developmental studies of oleosins in the anther of *Brassica napus*. Plant J 11: 475–487
Weber M (1996) The existence of a special exine coating in *Geranium robertianum* pollen. Int J Plant Sci 157:195–202
Wehling K, Niester C, Boon JJ, Willemse MTM, Wiermann R (1989) p-Coumaric acid – a monomer in the sporopollenin skeleton. Planta 179:376–380
Weselake RJ, Pomeroy MK, Furukawa TL, Golden JL, Little DB and Laroche A (1993) developmental profile of diacylglycerol acyltransferase in maturing seeds of oilseed rape and safflower and microspore-derived cultures of oilseed rape. J Plant Physiol 102:565–571
Wetzel CLR, Jensen WA (1992) Studies of pollen maturation in cotton: the storage reserve accumulation phase. Sex Plant Reprod 5:117–127
Whipple AP, Mascarenhas JP (1978) Lipid synthesis in germinating *Tradescantia* pollen. Phytochem 17:1273–1274
Whitfield HV, Murphy DJ and Hills MJ (1993) Sub-cellular localization of fatty-acid elongase in developing seeds of Lunaria annua and Brassica napus. Phytochemistry 32:255–58
Wiermann R, Gubatz S (1992) Pollen wall and sporopollenin. Int Rev Cytol 140:35–72
Wilmesmeier S, Steuernagel S, Wiermann R (1993) Comparative FTIR and ^{13}C CP/MAS NMR spectroscopic investigations on sporopollenin of different systematic origins. Z Naturforsch 48c:697–701
Wilmesmeier S, Wiermann R (1995) Influence of EPTC (S-Ethyl-Dipropyl-Thiocarbamate) on the composition of surface waxes and sporopollenin structure in *Zea mays*. J Plant Physiol 146:22–28
Wu SSH, Platt KA, Ratnayake C, Wang T-W, Ting JTL and Huang AHC (1997) Isolation and characterisation of neutral-lipid-containing organelles and globuli-filled plastids from *Brassica napus* tapetum. Proc Natl Acad Sci 94:12711–12716
Xu H, Knox BR, Taylor PE and Singh MB (1995) Bcp1, a gene required for male fertility in *Arabidopsis*. Proc Natl Acad Sci 92:2106–2110
Ylstra B (1995) Molecular control of fertilisation in plants. PhD Thesis, Free University of Amsterdam.
Zavada MS (1984) Pollen wall development of *Austrobaileya maculata*. Bot Gaz 145:11–21
Zhang JZ, Laudencia-Chingcuanco DL, Comai L, Li M, Harada JJ (1994) Isocitrate lyase and malate synthase genes from *Brassica napus* L. are active in pollen. Plant Physiol 104:857–864

CHAPTER 4

Sex Determination or Sexual Dimorphism? On Facts and Terminology

A. Lardon · C. Delichère · F. Monéger · I. Negrutiu

Ecole Normale Supérieure de Lyon, Reproduction et Développement des Plantes, E.N.S./LR5,
46, Allée d'Italie, 69364 LYON Cedex 07
e-mail: Ioan.Negrutiu@ens-lyon.fr
telephone: 33-4-72 72 86 12
fax: 33-4-72 72 86 00

4.1 Introduction

Most Angiosperms bear morphologically and functionally hermaphroditic flowers. Unisexuality is, however, a widespread condition in the plant kingdom (Renner and Ricklefs, 1995 and ref. therein). Selective arrest in development or non-development of reproductive organs is the key feature flower plants have evolved to establish unisexuality. It is generally accepted that the hermaphroditic form is the ancestral condition from which unisexuality has emerged independently many times (Charlesworth, 1991; Dellaporta and Calderon-Urea, 1993).

As an almost general rule, mono- and dioecious species have bipotential floral buds in which, according to the species, the development of reproductive organs of one sex is arrested or undergoes degenerative processes at widely different developmental stages. Thus, the cellular mechanisms involved in sexual dimorphism vary among plant species (Dellaporta and Calderon-Urea, 1993), each bearing either developmental (early dimorphism) or sexual (late dimorphism) mutations. The idea that unisexual development implies mutations leading to male and female sterility or consists in a "simple" breakdown of reproductive development (Charlesworth, 1991), is most likely an over-simplification of the situation in most unisexual species.

The acquisition of the unisexual condition is considered to be the typical plant equivalent of "sex determination and differentiation", as termed in other eukaryotic systems (Nöthiger and Steinmann-Zwicky, 1985; Sala and Negrutiu, 1991; Irish, 1996). We need to recall here the general criteria of sex determination, namely:

- a genetic and molecular switch which, when mutated, results in sex reversal; this implies that every individual has the complete genetic information to differentiate both sexes
- a hierarchically built pathway in which a primary signal or a key gene act to set the activity of a cascade of regulatory genes which finally determine the mutual exclusive activity of the male or female differentiation genes
- a sex control system frequently incorporated into sex chromosomes
- a phylogenetic conservation of the basic scheme of sex determination.

The resulting sexual dimorphism and derived genetic alterations responsible for sex reversal do reflect the listed characteristics.

The question then is whether in Angiosperms the processes underlying unisexuality meet these general criteria of sex determination and whether the observed range of sexual dimorphic states reflects a variety of alterations in developmental controls affecting reproductive organ formation.

The object of this report is to undertake a clarification of terminology and suggest that, in plants, "sexual dimorphism" is a better term for processes which, so far, have been coined *en masse* under the term "sex determination". To illustrate this point, we refer to two experimental model species, the monoecious maize (Gramineae) and the dioecious white campion (Caryophyllaceae). Through these two examples, we support the view that unisexuality in flowering plants has polyphyletic origins, which makes it very unlikely that homologous or similar genes have been more or less systematically recruited to control sexual dimorphism.

4.2 Sexual Dimorphism in Maize – A Summary

Maize develops separate male and female flowers at defined and distinct locations on the same plant. No sex chromosomes are involved in the control of sexual dimorphism in maize. Reviews on the regulation of "sex determination" in maize have regularly been published (Irish and Nelson, 1989; Dellaporta and Calderon-Urea, 1994; Irish, 1996). Here we will only recall the main trends. Sexual dimorphism in the species has complex genetic and environmental bases, with a set of (partially) redundant control systems identified in a large collection of mutants (Irish, 1996):

- masculinizing mutants, part of which result from defects in the biosynthesis or reception of gibberellin. Several such mutants exhibit perfect (bisexual) flower phenotypes.
- feminizing mutations, among which approximately ten *ts* (*tasselseed*) loci are known and correspond to at least two independent genetic pathways that do not act via gibberellins. Many of the *ts* mutants result in sex reversal.

Some of the above results are interpreted as stamen and carpel development being under the control of distinct pathways and monoecy being ensured independently of "sex determination". Furthermore, many of the identified genes that control sexual development also regulate the architecture of the plant, via both meristem activities and patterns of organogenesis in the inflorescence.

Finally, a key masculinizing gene, *ts2*, which corresponds to a sterol dehydrogenase controlling apoptotic processes in pistil tissues in the tassel, has been the first plant gene involved in maize "sex determination" to be cloned (DeLong et al., 1993). The corresponding gene in the dioecious white campion has been identified (Lebel-Hardenack et al., 1997) and has no equivalent role in the sexual development of the species.

4.3 Sexual Dimorphism in White Campion

Among the species that have evolved unisexuality in Angiosperms, white campion is one of the rare cases in which the sexual dimorphism, on one hand, is controlled by an active Y chromosome in a heteromorphic XY sex chromosome system (Westergaard, 1958; van Nigteveght, 1966; Ye, 1991) and, on the other hand, is achieved by alterations of reproductive organ formation at early stages of development (Farbos et al., 1997).

We can consider that white campion is a natural double "mutant", affected in very early processes of reproductive organ development, with the following flower formulae:

wt♂ (XY) S,P,A,Fil	⚥ → whorl 4 specific arrest of C
wt♀ (XX) S,P,a,C	♂ → early arrest of A differentiation

where S – sepal ; P – petal; A – stamen; a – arrested stamen; C – carpel; Fil – filamentous structure; ⚥ – hermaphroditic flower.

Experimentally induced mutations demonstrate the critical role the Y chromosome playsin the establishment of sexual dimorphism in this species. Thus, gamma-ray (**artificial**) – induced mutations in the two key functions of sexual dimorphism result in the following flower formulae:

(XYd1)	S,P,A,C	♂ → ⚥
(XYd2)	S,P,a,Fil	♂ → **O** cumulates wt (♂ + ♀) blocks (determinate growth)

where d1 – deletion of Gynoecium Suppressing Function on the Y chromosome; d2 – deletion of the Stamen Promoting Function on the Y chromosome; ⚥ – hermaphroditic flower; O – asexual flower.

Our work challenges the long rooted idea that, as in mammals, the Y chromosome is primarily the male determiner (Mittwoch, 1967, Westergaard, 1958). As a matter of fact, we consider that the earliest and most intriguing event controlled through the Y chromosome is the arrest of female development. The following sections summarise our knowledge on how sexual dimorphism is achieved in this species.

4.3.1 The female flower

In the absence of the "Stamen Promoting Function" (SPF) located on the Y chromosome, the male developmental arrest in the female flower (XX constitution) results from the lack of parietal initials and the degeneration of sporogenous cell initials during anther differentiation (flower stage 6; Farbos et al., 1997; Table 4.1). Thus, the female plant contains the genetic information necessary to initiate anther development up to the early sporogenous stage. Subsequently, anther development is arrested because the gene(s) required to proceed beyond that stage are not expressed or are missing. The slow regression of sporogenous cells into parenchymatous cells within the anther, without apparent cell lysis, tends to support the view that the SPF function, present in male plants (XY constitution), could be missing in the female plants (XX constitution). This assumption is supported by the asexual phenotype of mutant plants exhibiting both pistil and stamen arrest and bearing a deletion on the Y chromosome (Veuskens et al., 1995 and Table 4.2). Thus the earliest male differentiation function on the Y chromosome is most likely involved in initiating the parietal/sporogenous "dichotomy".

Functions that control the initiation of sporogenous/parietal differentiation should be conserved among plants as part of the male genetic pathway. Genes of this kind have

Table 4.1. Developmental time at which processes generating sexual dimorphism operate to arrest male or female development in individuals of opposite sex in white campion

	Carpel development		Stamen development	
Flower stage	in ♂ flower	in ♀ flower	in ♂ flower	in ♀ flower
4	+	+	+	+
5	–	+	+	+
6	–	+	+	–

not been cloned in plants so far, nor have corresponding mutants been reported in species other than white campion.

4.3.2
The Male Flower

In the presence of the "Gynoecium Suppressing Function" (GSF) located on the Y chromosome, the female developmental arrest in the male flower (XY configuration) results from a block in carpel initiation (flower stage 5; Table 4.1). The arrest of carpel development in the male flower is the earliest event in the establishment of sexual dimorphism in white campion (Farbos et al., 1997; Grant et al., 1996 and Table 4.1). The male plant (XY constitution) contains all the genetic information to produce a functional hermaphroditic flower, while the Gynoecium Suppressing Function causes a sudden arrest of cell proliferation in whorl 4 of male flowers at the time of partitioning between whorls 3 and 4. As a result, a filamentous structure in the centre of the male flower is located in the position where 5 carpels normally appear in the female flower. This substitution organ has no structural similarity with the wild type pistil and does not result from degenerative processes. Rather, the viable filament is produced by cell elongation from a whorl 4 primordium in which cessation of cell proliferation and a 5-fold deficit in cell number can no longer produce carpellar structures.

Genes that regulate cell proliferation processes associated with initiation of carpel development in whorl 4 are primary candidates for the Gynoecium Suppressing Function (also see Table 4.2). This aspect is discussed in the following section.

Table 4.2. Sex transformation in white campion based on loss-of-function and gain-of-function modifications of the Y-located functions responsible for sexual dimorphism (also see Table 4.1)

	Stamen Promoting Function (SPF)	Gynoecium Suppressing Function (GSF)
in WT ♂	present	present
in WT ♀	absent	absent
	Loss-of-Function	
	SPF^-	GSF^-
in WT ♂	ASEXUAL (acting at stage 6)	HERMAPHRODITE (acting at stage 5)
in WT ♀	(not applicable)	
	Gain-of -Function	
	SPF^+	GSF^+
in WT ♂	(not applicable)	
in WT ♀	HERMAPHRODITE[a] (acting at stage 6)	ASEXUAL (acting at stage 5)

[a] *Mycrobotryum violacea* phenotype which results from infection of female plants by the fungus (Ruddat et al., 1991).

4.3.3 The Gynoecium Suppressing Function Operates at or just Downstream of the ABC Pathway in the Flower Meristem

This type of "natural" mutation is interesting because it tells us that as soon as whorl 4 is spatially delimited, a whorl 4 – specific control on cell proliferation is established. A new and tightly controlled cell division activity in whorl 4 is a condition for gynoecium formation. This is the starting point of carpel initiation, i.e. the first specific event in female development.

The modification(s) leading to female developmental arrest in whorl 4 can be explained by alterations in functions such as those controlling carpel number or cell proliferation in the centre of the flower meristem. Such functions are under genetic control in *Arabidopsis* (Crone and Lord, 1993) and normal carpel formation requires the activity within the Flower Meristem Centre of genes such as *agamous (ag), clavata (cla), superman (sup), shootmeristemless (stm)* etc (Meyerowitz, 1997).

If such genes were involved in the Gynoecium Suppressing Function, the question is whether *Arabidopsis* mutants in such genes could actually simulate an early sexual dimorphism like the one observed in white campion? Appropriate examples are shown below (see Bowman, 1995; Meyerowitz, 1997; Jacobsen and Meyerowitz, 1997).

Loss-of-function mutants in the Flower Meristem Centre regulatory network or in genes known to confer male or female identity to flower organ primordia in *Arabidopsis*:

WT	S,P,A,C	⚥
ap3/pi	S,S,C,C	⚥ → super ♀
sup	S,P,A,(A/Fil)	⚥ → super ♂
fon	S,P,A,(**A/C→C**)	⚥ → late super ⚥
clv	S,P,**A,C**	⚥ → early super ⚥
ag	(S,P,P)n	⚥ → "asexual" (indeterminate meristem)

where *ap3/pi* – *apetala/pistilata* mutants; **A/C** – supranumerary stamens and carpels, frequently chimerical; *fon* – flower organ number mutants.

It can be seen from the listed phenotypes that defined loss-of-function mutants in *Arabidopsis* can simulate an early dioecious condition. In general, homeotic, cadastral or cell proliferation (multifunction) genes can generate gross versions of unisexual configurations in which more than one whorl is affected and chimerical structures are systematically produced. The *sup* mutants are so far the closest to the situation that occurs in the male flower of white campion.

For comparison, mutations in the Gynoecium Suppressing Function and Stamen Promoting Function loci are presented in Table 4.2. It is obvious that the major difference between white campion and *Arabidopsis* is the less precise spatial control of modifications generated in all the listed *Arabidopsis* mutants. Several interpretations are possible, such as Gynoecium Suppressing Function acting just downstream of some of the known Flower Meristem Centre regulatory genes, or Gynoecium Suppressing Function being one of these functions whose activity has been specifically modified in whorl 4 in white campion.

4.4 Conclusions

We have shown here that in two model plants used to study sexual development, namely maize and white campion, widely different mechanisms have been used to generate sexual dimorphism. In addition, the general criteria for sex determination are not met, with the exception of a sex chromosome system in white campion. In the latter species, the above analysis shows that sexual dimorphism is controlled by two independent developmental pathways operating at two distinct developmental stages in male and female flowers (Table 4.1):

- a "female suppressing" function acting to specifically limit the proliferative (and subsequently the differentiation) activities in whorl 4 primordium. This novel or recruited function on the Y chromosome has become part of the flower meristem organisation and represents the main "innovation" of sexual dimorphism in the species;
- a gene in the anther differentiation pathway acting during early sporogenesis and possibly missing or inactive in XX females (see below).

The existence of independent pathways for male and female developmental arrest and the fact that sex reversal generates either hermaphroditic or asexual mutants (Table 4.2), strongly indicates that no switch mechanism is at work in the white campion system of sexual dimorphism.

The key genes within the two pathways are both located on the Y chromosome. The two functions are tentatively located on the same arm of the Y chromosome (Westergaard, 1958), although this awaits clear demonstration using genetical and molecular tools. The Gynoecium Suppressing Function is a negative regulator of carpel formation. The Stamen Promoting Function is an activating function of stamen differentiation. This tandem of dominant functions with opposite growth effects is inherited as a unit and generates a Y chromosome – dependent pattern of sex expression. Thus, according to prevailing models on evolution of separate sexes and incipient sex chromosomes in plants, the Gynoecium Suppressing Function represents the dominant female sterile (f^S) mutation and the Stamen Promoting Function represents the wild type male fertile (m^F) function (Charlesworth, 1996). The corresponding Y/X genetic formula is therefore ($f^S m^F / f^f m^s$). In reality and based on the hermaphroditic and asexual phenotypes described and their respective karyotypes, we postulate that the genetic formula is in fact ($f^S m^F$/—). This reflects the fact that the Y-located f^S and m^F loci seem to have no functional counterparts within the female genome.

What do we know about the X and Y chromosomes in white campion? Unfortunately, very little so far. Genetic data suggest that these chromosomes have all the main features (including dosage compensation by X inactivation) of heteromorphic sex chromosomes known in animal systems (Vyskot et al., 1993). Preliminary molecular evidence tends to suggest that X and Y chromosomes in white campion have a high degree of commonality in DNA sequence, at least for some major classes of satellite DNA motifs (Scutt et al., 1997).

Despite the numerous similarities between plants and animals at the level of sex chromosome organisation and evolution, we consider that the genetic systems and associated regulatory switches and cascades controlling sexual dimorphism are fundamentally different. These are valid enough reasons for considering white campion a particularly valuable tool for both developmental and evolutionary studies.

References

Bowman J (1995) Arabidopsis. An atlas of morphology and development. Springer Verlag, pp 180–273

Charlesworth B (1996) The evolution of sex determination and dosage compensation. Current Biology 6:149–162

Crone WN, Lord EM (1993) Flower development in the organ number mutant clavata-1 of *Arabidopsis thaliana*. Amer J Bot 80:1419–1426

Dellaporta SL, Calderon-Urrea A (1993) Sex determination in flowering plants. The Plant Cell 5: 1241–1251

DeLong A, Calderon-Urrea A, Dellaporta S (1993) Sex determination gene Tasselseed2 of maize encodes a short-chain alcohol dehydrogenase required for stage-specific floral organ abortion. Cell 74: 757–768

Farbos I, Oliveira M, Negrutiu I, Mouras A (1997) Sex organ determination and differentiation in the dioecious plant *Melandrium album*: a cytological and hystological analysis. Sex Plant Reprod 10: 155–167

Grant S, Hunkirchen B, Saedler H (1994) Developmental differences between male and female flowers in the dioecious plant *Silene latifolia*. The Plant Journal 6:471–480

Hardenack S, Ye D, Saedler H, Grant S (1994) Comparison of MADS box gene expression in developing male and female flowers of the dioecious plant white campion. The Plant Cell 6:1775–1787

Irish VV (1996) Regulation of sex determination in maize. BioEssay 18:363–369

Irish EE, Nelson T (1989) Sex determination in monoecious and dioecious plants. The Plant Cell 1: 737–744

Jacobsen SE, Meyerowitz E (1997) Hypermethylated *SUPERMAN* epigenetic alleles in *Arabidopsis*. Science 277:1100–1103

Meyerowitz E (1997) Genetic control of cell division patterns in developing plants. Cell 88:299–308

Mittwoch U (1967) Sex chromosomes. Academic Press, London.

Nigtevecht G van (1966) Genetic studies in dioecious *Melandrium*. I. Sex-linked and sex influenced inheritance in *M. album* and *M. dioicum*. Genetica 37:281–306

Nöthiger R, Steinmann-Zwicky M (1985) A single principle for sex determination in insects. Cold Spring Harbor Symposia on Quantitative Biology, vol L:615–621

Renner SS, Ricklefs RE (1995) Dioecy and its correlates in the flowering plants. Amer J Bot 82:596–606

Ruddat M, Kokontis J, Birch L, Garber ED, Chiang K-S, Campanella J, Dai H (1991) Interactions of *Mycrobotryum violacea* (*Ustilago violacea*) with its host plant *Silene alba*. Plant Science 80:157–165

Sala F, Negrutiu I (1991) Sexual development in plants: an open question of strategic importance. Plant Science 80:1–6

Scutt CP, Kamisugi Y, Sakai F, Gilmartin PM (1997) Laser isolation of plant sex chromosomes: studies on the DNA composition of the X and Y sex chromosomes of *Silene latifolia*. Genome 40:705–715

Veuskens J, Marie D, Brown SC, Jacobs M, Negrutiu I (1995) Flow sorting of the Y sex chromosome in the dioecious plant *Melandrium album*. Cytometry 21:363–373

Vyskot B, Araya A, Veuskens J, Negrutiu I, Mouras A (1993) DNA methylation of sex chromosomes in *Melandrium album*. Molec Gen Genet 239:219–224

Westergaard M (1958) The mechanism of sex determination in dioecious flowering plants. Adv Genet 9:217–281

CHAPTER 5

Meiosis

T. Schwarzacher

John Innes Centre, Norwich Research Park, Colney, Norwich NR4 7UH, U.K.
e-mail: Trude.Schwarzacher@bbsrc.ac.uk
telephone: +44 16 03 45 25 71
fax: +44 16 03 45 68 44

5.1 Introduction

Meiosis, the unique and essential part of the life cycle of sexually reproducing organisms, is the division process where a diploid cell of the sporophyte gives rise to haploid cells which further develop to the gametophyte and gametes (for review see e.g. John 1990). As most frequently reported (and described here), meiosis in plants involves a single diploid nucleus with its DNA in the replicated state dividing twice, without further complete DNA replication, to give four haploid nuclei. The process is accompanied by the recombination of chromosomes and genes so the haploid cells have different and new combinations from those of the plant's parents. Meiosis involves two nuclear divisions: the first, reductional division is preceded by a long interphase and characterized by a complex and lengthy prophase I, during which the two homologous chromosomes (one from each parent) pair and synapse, the synaptonemal complex (SC; see below) is formed and reciprocal recombination – crossing over – takes place. At metaphase I, the paired chromosomes have formed bivalents that are held together at the chiasmata (the points of cross overs), and disjoin at anaphase I. After a short interphase, the second, equational divisions segregate the sister chromatids in the nuclei arising from meiosis I, proceeding through the stages of prophase II, metaphase II, anaphase II and telophase II as in a mitotic division.

Numerous variations of these processes are known, and many large groups of species (particularly among insects) have never been studied. In mammals, the products of meiosis themselves are the gametes, but a number of further divisions occur on both the male and female sides in plants before gamete production. Within insects, no recombination occurs in male *Drosophila melanogaster*, and in other species the reductional division occurs after the equational division. Within plants, some meiotic stages may be present, and others missing, in certain apomictic species. In the relatively frequent situations where a plant has other than two fully homologous chromosome sets (e.g. aneuploids, polyploids, rearranged chromosomes or supernumerary chromosomes) unpaired chromosomes – univalents – or groups of three or more chromosomes paired with one another – mutivalents – occur at metaphase I and give unequal segregation of chromosomes. Complex systems of balanced inter-chromosomal translocations may ensure segregation without gene recombination e.g. in *Oenothera* species.

Developments in molecular biology, genomics and molecular cytogenetics have led to major advances in our understanding of meiosis in the 1990s. The technology involved means that substantial work has been confined to less than a dozen model species – human, mouse, fission and budding yeast (*Saccharomyces cereviseae*), *Drosophila melanogaster*, wheat, lily, maize, and *Arabidopsis thaliana* – which have shown both parallels and notable differences; we are some way from a generally applicable model. These organisms vary widely in genome size the number of base pairs in the unreplicated haploid genome – from some 14 Mbp in yeast, through 150 Mbp in *Arabidopsis*, 2,500 Mbp in maize, 3,600 Mbp in human, to more than 15,000 Mbp in wheat and some lilies. It is probable that some models based on smaller genomes, for example exhaustive homology searches, would become unworkable in the larger genomes.

Analysis of meiosis has found a wide range of applications in basic and applied research. Chromosome behaviour and organization, the mechanisms of chromosome recognition and pairing, meiotic recombination and chromosome disjunction have been investigated using a combination of molecular, cytological and immunohistochemistry techniques, covered in this chapter. In cases of reduced fertility, meiosis is analysed in order to check whether it is normal and, if not, where it is going wrong. To study the relation of genomes, the level of pairing and bivalent formation at meiotic metaphase I in hybrids and polyploids can give information about homology of genomes and chromosomes and an indication which species can be crossed successfully and are fertile.

5.2 Terminology

Figs. 5.1 and 5.2 see color plate at the end of the book.

Meiotic prophase I is characterized by unique and specific events, unlike mitotic prophase, and has been divided into several substages depending on chromosomal morphology and behaviour (for a detailed description of meiosis in wheat see Bennett et al. 1973). At **leptotene**, individual chromosomes become organized into the typical meiotic prophase chromosomes with long thin threads (Fig. 5.1c, d). At **zygotene**, synapsis starts, and paired and unpaired chromosome threads are visible (Fig. 5.1e). Synapsis is complete at **pachytene** (Figs. 5.1f, 5.2a) and chromosomes condense during late pachytene and **diplotene** (Fig. 5.2b, f) when homologues start to separate except at sites of chiasmata. Chromosomes condense further and bivalents become recognizable at **diakenesis** (Fig. 5.2c). At **metaphase I**, bivalents are arranged in the equatorial plate (Figs. 5.1g and 5.2d, g, h) and homologues segregate to opposite poles at **anaphase I** (Fig. 5.2i).

In the past, the term **pairing** has been used to refer to a variety of events leading to the final stage of homologous chromosomes being held in register by the fully formed tripartite SC. However, it is now clear that homologous chromosome pairing is a multistep process (Kleckner 1996; Schwarzacher 1997; see below) and it is hence advisable to distinguish different aspects of pairing. **Association** exists when chromosomes have some contact with each other; **alignment** is the homologous orientation of the chromosomes, possibly at a distance greater than the SC; and **synapsis** is the formation of the tripartite SC.

Key events of meiosis relating to pairing and recombination are now believed to be initiated before meiotic prophase, during the preceding lengthy interphase (see below); hence it has become inappropriate to call this stage premeiotic interphase and I use **interphase before leptotene**. The term meiotic interphase has been used previously to denote the short interphase between division I and II.

5.3 Synaptonemal Complex (SC)

The SC is a tripartite proteinaceous structure that forms between homologous chromosomes as they synapse during zygotene (for reviews see Gillies 1984; von Wettstein et al. 1984; Moens and Pearlman 1988). Before pairing, at leptotene, each single chromosome along its entire length develops a proteinaceous axial core, or axial element, to which the two sister chromatids are attached in a serious of loops. When homologous chromosomes synapse these axial elements come together to become the lateral elements of the SC. The gap between the lateral elements is traversed by perpendicular, thin transverse filaments. The width of the gap varies slightly between different species, usually being somewhere between 100–300 nm. Lying parallel between the lateral elements is the central element. This, together with the lateral elements, makes up the tripartite structure of the synaptonemal complex.

Initially, analysis of meiotic prophase nuclei and the synaptonemal complex involved time-consuming three-dimensional reconstructions of serial sections analysed in the electron microscope (for review see von Wettstein et al 1984). Alternatively, surface spreading allowed flattening of the entire nuclear content in a two dimensional plane both preserving and spreading the SC and the surrounding chromatin loops. The SC can be stained with silver nitrate and visualized by either light or electron microscope (Fig. 5.1a and e.g. Dresser and Moses 1979; Holm 1986; Schwarzacher 1986; Albini and Jones 1987; Jenkins and White 1990), allowing detailed analysis of the dynamics of this meiosis specific structure. Recently, two developments have advanced the studies of the SC and the associated chromatin: DNA : DNA *in situ* hybridization and immunostaining using specific antibodies to SC components and meiotic gene products usually with fluorescent detection methods. *In situ* hybridization allows a labelled DNA probe to hybridize to homologous target DNA sequences on chromosome preparations so their location can be determined. Similarly, antibodies hybridize to specific protein fragments (or occasionally other molecules) in the cell and enable accurate localization of the molecule.

5.3.1 DNA Sequences Associated with the SC

Genomic and sequencing projects are revealing details of the DNA sequence and organization within many species. Only a small proportion – often less than 10% – of the genome is represented by single-copy sequences including genes, while the rest consists of repetitive sequences with motifs from two to 10,000 base pairs long, including retroelements and tandemly repeated (or satellite) DNA sequences. We can now attempt to analyse the organization of chromatin and DNA sequences within the SC and the relationship of specific sequence motifs to homologous chromosome cognition and recombination. Moens and Pearlman (1988) have found that the average size of the chromatin loops attached to the SC are species-specific and have postulated that one mechanism for the regulation of the loop size is the existence of specialized DNA sequences that associate with the meiotic chromosome core. DNA sequences isolated from purified, DNAse-treated rat SCs did not contain sequences that are unique to chromosome cores but proved to be notably different from random genomic fragments and contained an excess of simple sequence repeats (mostly the dinucleotide motif GT or, on the complementary DNA strand, CA) and retroelement-related repetitive sequences (LINE and

SINE elements, long and short interspersed sequences; Pearlman et al. 1992). Fluorescent *in situ* hybridization to mouse SCs has shown that unique sequences, the mouse minor satellite DNA sequence and some other tandemly repeated sequences are mainly found in the chromatin loops, while the signal from the telomere sequence does not come from the loops (see Moens and Pearlman 1993). Similarly, in humans, telomeric sequences were seen tightly associated with the SCs while centromeric alpha-satellites and classical satellites were all found to form loops that are associated with the SC only at their base (Barlow and Hultén 1996).

In rye, two tandemly repeated sequences, the 18S–25S rDNA and a 120 bp repeat from rye heterochromatin are closely associated with the bivalent axes, corresponding to the SC, and also located in the surrounding chromatin loops (Albini and Schwarzacher 1992). The relative length of the axis covered with 18S–25S rDNA signal corresponds closely to the proportion of the sequences in the genome, but is less than found on somatic metaphase chromosomes; the positions of the rDNA sites correspond exactly between the two stages. The relative lengths of the bivalent axes covered with signal from the 120 bp repeat appears to be less than expected from somatic metaphases, supporting the speculation that heterochromatin is underrepresented in the SC (Stack 1984), although perhaps only with respect to somatic metaphase. In contrast to the tight packing and close association of these classes of long tandem repeats, preliminary studies using simple sequence repeats as probes for fluorescent *in situ* hybridization showed that very little signal is associated with the SC, but is mainly found in the chromatin loops (Cuadrado and Schwarzacher unpublished).

5.3.2 SC Proteins

Many meiotic specific proteins that are components of the SC, and antibodies to them, have now been identified in yeast and other organisms (for review see Moens 1994; Heyting 1996; Roeder 1997). In mouse, Dobson et al. (1994) have shown that the meiotic core protein Cor1 is present in early unpaired cores, in the lateral elements of the SC and in the chromosome cores when they separate. Cor1 was also present at metaphase I in association with pairs of sister centromeres. The extended presence of Cor1 suggests that it may have a role in chromosome disjunction by fastening chiasmata at metaphase I and by joining sister kinetochores, which ensures co-segregation at anaphase I. Similarly, in lily, Suzuki et al. (1997) have found a specific antiserum that does not stain centromeres during mitotic division in somatic cells of lily but stains centromeres during the meiotic division (male and female). The antiserum was detected when the centromeres of the homologous chromosomes associated and fused at zygotene-pachytene in prophase I, the disjunction of the homologous centromeres at diplotene, the doubling of each centromere at metaphase I and non-separation of the sister centromeres at anaphase I. The protein was also detected during the G2-phase before meiotic prophase. The authors therefore postulate that the meiosis specific centromere protein is associated with conversion of a mitotic to a meiotic chromosome, that meiosis is regulated by modification of the structure of chromosomes and particular centromeres, and that a meiosis-specific centromere protein is required for the meiosis-specific behaviour of the centromere. The isolation of interacting protein components of the transverse filaments of the central element alone shows the complexity of the SC structure and its assembly (Dobson et al. 1994; Schmekel et al. 1996).

5.4
Recombination – Yeast Model

A model of meiosis is available for budding yeast, based on extensive data using molecular, cytological, immunological and fluorescent *in situ* hybridization analysis (reviews: Roeder 1995, 1997; Kleckner 1996). Meiotic recombination is initiated by double strand breaks that are turned into gaps by exonulcease activity. Repair mechanisms, using the homologous chromosome and not the sister chromatid, mend the gap and branch migration results in two Holliday junctions. Depending on how the Holliday junctions are resolved, either reciprocal recombination (cross over) or gene conversion (non-cross over) result. Molecular cross overs are seen as cytological chiasmata along the chromosomes at diakenesis (see Fig. 5.2c).

Due to the studies in yeast there is new insight into the sequence of events during synapsis and crossing over. The long-accepted hypothesis held that when homologues found each other and synapsed with the formation of the SC, then genetic recombination between the chromosomes could take place. It is now clear that recombination events are initiated before synapsis and in mutants with defective SC formation. Although recombination may be initiated as part of the pairing process, SC formation is essential for the maturation of these initial events to actual cross overs and chiasmata by preventing sister chromatid exchanges (Zenvirth et al. 1997;). The SC therefore plays a vital role in the control of the frequency and distribution of exchanges and ensures the proper disjunction of homologues at anaphase I (Moens 1994; Maguire 1996).

5.5
Recombination Nodules

Early meiotic nodules (also called recombination nodules) are proteinaceaous structures about 100nm in diameter that are associated with forming SCs during early prophase I, while late recombination nodules appear on the central element of the SC during pachytene and are fewer in number than early nodules. There is circumstantial evidence that only some early nodules mature into late nodules whose distribution coincides with chiasmata and which are thought to be involved in cross overs (see Roeder 1997). The role of early nodules is less clear, but they may be involved in homology searching before synaptic initiation. Two potential components of early nodules are Rad51 and Dmc1 proteins that are important for meiotic recombination in eukaryotes by converting double strand breaks into hybrid joint molecules (Bishop 1994). Rad51 proteins were found to form discrete nuclear foci from early zygotene to pachytene, to collocalize with lateral element proteins in yeast, mouse, rat human and lily meiosis (Bishop 1994; Terasawa et al. 1995; Barlow et al. 1997). Anderson et al. (1997) have used electron microscopic immunogold localization to spread zygotene and early pachytene SCs from lily to show that Rad51 and Dmc1 proteins are components of early nodules. Similarly, in rat and mouse spermatocytes Moens et al. (1997) demonstrated 100nm nodules that are immunolabelled by Rad51 antibodies. These proteins are homologous to RecA, the major protein that catalyzes homologous pairing and DNA strand exchange in prokaryotes. The role of late nodules in meiotic cross overs has been substantiated by Baker and Plug et al (1996) who found that the mismatch repair gene Mlh1 associates with late recombination nodules.

5.6 Homologous Chromosome Pairing

Some of the important events of meiosis are the processes of homologous chromosome recognition and alignment leading to synapsis and SC formation (Bennett 1984; Loidl 1990; Moens 1994), their timing and how they relate to meiotic recombination (Kleckner 1996; Roeder 1997). One of the impediments in looking at chromosome pairing was that until recently chromosomes prior to and at early meiosis were not amenable to cytological studies. By the time chromosomes are visible as axes at zygotene using conventional DNA staining methods (see Figs 5.1 and 5.2), homologues have partially paired, although the significance of the events happening earlier has long been recognized (e.g. McClintock 1933). Fluorescent *in situ* hybridization allows direct observation of the behaviour of individual chromosomes during interphase (Lichter et al. 1988; Schwarzacher et al. 1989, 1992a) and the technology has recently been applied to studies of early meiotic prophase. In yeast, it has been observed that some homologues are paired, possibly via multiple interstitial interactions, in premeiotic and likely also in vegetative cells, and that homologues align prior to synapsis (see Scherthan et al. 1992; Loidl et al. 1994; Weiner and Kleckner 1994). Kleckner (1996) argues that some of the early interactions are lost during meiotic S phase, before homologues identify one another and recombine during the first part of prophase. Both premeiotic and meiotic pairing in yeast was suggested to involve unstable side-by-side joints between intact DNA duplexes and to include interactions at multiple sites along each chromosome pair (see Roeder 1995). While mitotic association of homologues in budding (see above) and fission yeast (Kohli 1994) have been shown and such associations seem to be a universal condition in somatic Diptera cells (e.g. Metz 1916; Hiraoka et al. 1993), three-dimensional reconstructions of human cells (see Leitch et al. 1994) and cereal plant species (Heslop-Harrison et al. 1988) were not able to find evidence for homologous association of chromosomes, even at the last mitotic division before meiosis (Bennett 1984; Schwarzacher et al. 1992b). In human and mouse, several studies have used *in situ* hybridization to follow individual chromosomes from the premeiotic mitotic metaphase through to the sperm heads (Goldman and Hultén 1992; Armstrong et al. 1994; Cheng and Gartler 1994, O'Keefe et al. 1997) showing that human spermatogonia (meiotic stem cells) reveal compact, largely mutually exclusive chromosome territories that did not show homologue association. Scherthan et al. (1996) postulated movements of centromeres and then telomeres to the nuclear envelope and subsequent bouquet formation as conserved features of the pairing process. At the onset of meiotic prophase, the compact and separate chromosome territories developed into long thin threads. Subsequently, telomeres moved towards the clustered site and produced numerous encounters among now elongated chromosomes that are suggested as contributing to homology testing at exposed pairing sites. They suggest that convergence of chromosome ends increases the efficiency of homologue search and leads to prealignment of bent homologues (Scherthan et al. 1996).

In maize, Dawe et al. (1994) demonstrated that the homologous chromosomes, similar to mouse and man, are apart when entering meiosis, but undergo a dramatic structural reorganization prior to synapsis at zygotene. The unique features of prezygotene chromosomes are a partial separation of sister chromatids, an elongation of knob heterochromatin and increased surface complexity and total chromosome volume. Telomeres are localized peripherally and pair first. They conclude that the specialized prezygotene chromosome morphology may facilitate homology recognition once homologues have been brought together.

5.6.1
Homologous Chromosome Pairing in Wheat

In contrast to maize and mammals are the findings in the larger cereal genomes, such as wheat, where homologous chromosomes associate during the interphase before leptotene. Total genomic DNA from rye, used as a probe for *in situ* hybridization identified the rye chromosome arm in a wheat-rye translocation line (T5AS·5RL) at meiotic prophase and the preceding interphase (Schwarzacher 1997, Fig. 5.1b–f). These plants grow normally and chromosomes pair as bivalents at metaphase I (Fig. 5.1g). Accurate staging of the development of the meiocytes was attained by parallel studies of chromatin morphology, nucleolar behaviour and synaptonemal complex formation in electron microscope thin sections and silver stained surface spreads (e.g. Fig. 5.1a). Three stages of pairing were identified for the large cereal genomes that are organized in a Rabl configuration (Schwarzacher 1997): first, cognition occurs during the interphase before leptotene bringing the homologous chromosome domains into close proximity (see Fig. 5.1b). The interphase before leptotene is of seven times longer duration than the minimum found in cycling root tip cells (Bennett et al. 1973; Bennett 1984). Such a period might allow many complex processes to occur including transcription, translation and accumulation of the RNAs, enzymes and precursors needed through meiosis, although the rate-limiting step is most probably the physical reorganization of the chromatin within the nucleus. Bivalent formation at metaphase I is known to be particularly sensitive to drugs and temperature shock during the interphase before leptotene (Loidl 1990; Bennett et al. 1979, Bennett 1984), emphasizing the criticality of this time interval. The diffuse chromatin and enlarged nuclear volume observed at the interphase before leptotene could provide the extra space needed for meiotic pairing interactions, including any involved proteins.

Once the homologous domains are associated, condensation patterns are aligned, suggesting that not only homologous chromosomes but also homologous chromosome regions were substantially collocalized by this stage (Schwarzacher 1997). Karpen et al. (1996) have analysed pairing of achiasmatic chromosomes in *Drosophila melanogaster* and proposed that DNA and protein structures inherent to heterochromatin could produce a self complementary chromosome 'landscape' that ensures partner recognition and alignment by 'best fit' mechanisms. Recently, Cook (1997) argued that each chromosome in a haploid set has a unique array of transcription units strung along its length and that therefore chromatin fibres will be folded into a unique array of loops and that only homologues share similar arrays. As chromosomes only pair when transcriptionally active, pairing can become the inevitable consequence of transcription of partially condensed chromosomes (Cook 1997). Apart from specific coiling patterns with apparent denser and weaker zones, presumably reflecting more or less condensed chromatin observed in the homologous chromosome domains (Fig. 5.1b, Schwarzacher 1997) that could promote pairing, we noted that the association of homologous domains apparently starts predominantly at the centromeres (Fig. 5.1b). Hence, cognition of homology could be mediated by proteins or sequences associated with the wheat centromeres (Schwarzacher 1997).

Aragón-Alcaide et al. (1997a) using confocal microscopy have also demonstrated that homologous chromosomes associate before leptotene and that centromeres are clustered at that stage. However, homologous chromosomes failed to associate at equivalent stages in lines with deletions of the homologous paring gene *Ph1*. Centromeres at pre-meiotic interphase through to pachytene and anaphase I have a more diffuse structure in hexaploid wheat exhibiting high homeologous pairing, absence of *Ph1*, compared to low homeologous pairing wheat, presence of *Ph1* and *Ph2* (Aragón-Alciade et al. 1997b). Further, early resolution of sister chromatid centromeres and consequent sister chromatid disjunction of univalent chromosomes was observed in homeologous pairing situations at anaphase I. These observations in wheat suggest either that centromeric behaviour is important for pairing control, or that abnormal chromosomal pairing affects centromeric structure.

The second step in the pairing process is the organization of the chromosomes into the meiotic chromosome "threads", as recognized in conventionally stained preparations, and the alignment of homologous sequences during leptotene (Fig. 5.1d) and early zygotene (Fig. 5.1e). Homologous chromosome domains were associated in the majority of meiocytes (Schwarzacher 1997); but, once the chromosomes have organized into thin threads, the distance between homologous chromosome regions is often increased and hence, further alignment of homologous regions during late leptotene – early zygotene occurs. At this stage, the process could become based mainly on testing of DNA homologies.

At the third step of pairing, homologous chromosomes synapse and the tripartite SC is formed during zygotene. Using chromosome painting, synapsed and unsynapsed homologous chromosome segments can be clearly identified (Fig. 5.1e). In many cases, the telomeric region was amongst the first to synapse, while the middle of the rye chromosome arm synapsed last, often with several small interstitial non-paired sites (Schwarzacher 1997). This has also been found in studies using SC surface spreading in rye (Gillies 1985), wheat (Jenkins 1983; Holm 1986) and other plants (e.g. Albini and Jones 1987; Jenkins and White 1990). Knowledge of the importance of telomeres for synapsis comes from studies that in the absence of homology in the distal regions of chromosome arms, very long homologous segments may remain unrecognized in meiosis (Lukaszewski 1997). A yeast telomere associated meiotic protein, Tam1, has been identified that functions in chromosome synapsis and cross over interference (Chua and Roeder 1997).

At pachytene, the homologous rye chromosome arms are fully synapsed with each other and formed a single long compact chromatin thread (Fig. 5.1f). Metaphase I (Fig. 5.1g) is characterized by complete bivalent formation and wheat-rye translocation chromosomes form a ring bivalent in most cases indicating that normal levels of chiasma formation occur in the 5RL chromosome arm.

5.7 Chiasma Distribution

In the large cereal genomes, we and others have shown that physical distances between genes and markers along chromosomes correlate poorly with genetic map distances (Heslop-Harrison 1991; Lukasewski 1992; Fukui 1995; Schwarzacher and Heslop-Harrison 1995; Schwarzacher 1996; Endo and Gill 1996). Physically, genes and RFLP markers are clustered near the ends of chromosome arms, while genetically they are far apart, indicating that genetic recombination is frequent near the telomeres, but rare towards centromeres. *In situ* hybridization of cloned probes to meiotic metaphase I preparations allows genetic length measurements to be made by counting the chiasmata which occur in different segments of the rye chromosome 1R, distally or proximal of the 18S-25S rDNA site (Schwarzacher 1996, see also Fig. 5.2c, d; Table 5.1). Comparisons of both the relative frequency and total number of chiasmata show close correlation with genetic map distances described by Devos et al. (1993). The discrepancy between the genetic

Table 5.1. Position of 18S–25S rDNA on the long arm of chromosome 1R of rye (*Secale cereale*) measured in percent from the centromere

Genetic map	8
Chiasma counts	16
Somatic metaphase	67
Somatic interphase	52
Pachytene	58
SC	71

map and the physical map showing increased recombination near the telomere correlates with the distribution of chiasmata. Similarly, in tomato chiasmata were found almost exclusively in more distal, rather subterminal chromosome segments in *Lycopersicum esculentum* X *L. peruvianum* back crosses using genomic *in situ* hybridization (Parakonny et al. 1997). While in the small yeast chromosomes although showing non-random distribution of double strand breaks, defining large (39–105kb) chromosomal domains both hot and cold (Baudat and Nicolas 1997); no preferred meiotic DSB sites near the telomeres were found (Klein et al. 1996).

We have found that the length of the SC distal to the rDNA in chromosome 1R is not significantly different from the somatic metaphase measurements (Albini and Schwarzacher 1992, and Table 5.1) indicating that the increased recombination and chiasma formation in the distal chromosome segment does not need a longer SC (Schwarzacher and Heslop-Harrison 1995). This is in contrast to findings in yeast, humans and insects where an inverse correlation between the density of chromatin packing (SC length) and the rate of meiotic crossing over has been found (e.g. Quevedo et al. 1997) and where within a single chromosome the chromatin loop size is two to three times smaller near the telomeres than it is interstitially (Heng et al. 1996). However, we have also noted that the subtelomeric heterochromatin in rye forms very tight loops and is probably under-represented by the SC (see above) and hence could possibly compensate for any extended SC lengths of the distal euchromatin. Comparisons of genetic and physical maps have important implications for understanding genome organization, gene isolation, chromosome walking and transformation.

5.8 Future Outlook

Many advances in understanding the molecular mechanisms of meiosis during the last ten years have come from studies in yeast (see Moens 1994; Kleckner 1996; Roeder 1997). The recent isolation of meiotic genes in mouse and man, analysis of their mutants and use of antibodies against their products have allowed the analysis of SC components, and functional, temporal and spatial analysis of the gene expression and gene product localization in these complex organisms. It is notable that some of the genes also have a role in cancer development, and such parallels give pointers to the evolution of the meiotic recombination processes being linked to mitotic DNA repair. In plants, meiotic mutants have been known and characterized by genetics for 50 years, in maize and other species, and now meiotic genes are being isolated. The study of plants offers many advantages unavailable to mammalian workers: the plentiful availability of synchronized meiocytes, the large anthers of lily and many cereals, the possibility of making numbers of hybrids and large-scale progeny analysis, and development of tagging systems. Early pairing events have been studied in detail in wheat (Schwarzacher 1997; see above) and Rad51/Dmc1 homologous genes have been isolated for lily (Kobayashi et al. 1994), *Arabidopsis* (Klimyuk and Jones 1997) and barley (Garkoucha, Klimyuk and

Schwarzacher unpublished). Several mutants, effecting synapsis and chromosome fragmentation during meiosis have been described for *Arabidopsis* and tomato (Bhatt et al. 1996; Havekes et al. 1997; Ross et al. 1997). Some of these mutants have been produced by the insertion of special DNA fragments that will allow rescuing of the DNA sequences of the effected genes and their cloning. Other approaches for isolation of meiotic genes are isolation of meiotic specific cDNAs, sequence homology to existing meiotic genes from other organisms (Klimuyk and Jones 1997) or functional cloning of meiotic proteins complementing yeast mutant (Hirayama et al. 1997).

The increased molecular understanding of plant genomes including *Arabidopsis,* and the availability of extensive genetic marker maps, in combination with easy transformation systems, are making new approaches possible to understanding the developmental biology of plant sexual reproduction. Meiosis is the source of most of the new variability gained through recombination of genes and there are many exciting new research opportunities in the area that will lead to discoveries of both fundamental and applied importance.

5.9 Cytological Methods for Looking at Meiotic Chromosomes

In the following basic cytological procedures (aceto-carmine, Feulgen and DAPI staining), preparation for meiotic material for *in situ* hybridization and surface spreading of SCs is described.

5.9.1 Aceto-Carmine-Staining of Fresh Material

(see e.g. Darlington and LaCour 1976)

i. Place flower buds or ears likely to contain required stages of meiosis in a Petri dish with moist filter paper. Under the dissecting microscope open bud or floret and remove anthers.
ii. Transfer anthers to a drop of aceto-carmine or 45% acetic acid. Leave for 1–3 mins.
iii. Gently squeeze out the columns of pollen mother cells. Cover with a cover slip, heat over a spirit lamp until just under boiling and squash gently between layers of filter paper.
iv. Analyse under the microscope. Aceto-carmine stains DNA red, but sometimes stain is only weak. If so or 45% acetic acid only was used, phase contrast becomes necessary.
v. Put slide on dry ice for 5 mins and remove cover slip by flicking off with a razor blade. Fix slides in 100% alcohol/glacial acetic acid 3:1, for 30–60 mins, rinse twice in 100% alcohol 10 mins each and air dry.

5.9.2 Feulgen Staining of Fixed Material

(Feulgen and Rössenbeck 1924)

i. Fix anthers containing meiotic stages as determined by above method (5.9.1) in 100% alcohol/acetic acid 3:1 for a minimum of 2 hours. Material can be stored in the freezer for several weeks.

ii. Transfer anthers to 70% ethanol for 5 mins and then distilled water until they sink.
iii. Hydrolyse in 1N HCl at 60°C for 10 mins.
iv. Stain in Feulgen's reagent for 30–120 mins in the hood.
v. Transfer to distilled water. Can be stored in the refrigerator for several days.
vi. Transfer to 45% acetic acid and squeeze out pollen mother cells. Cover with a cover slip and squash.
vii. Examine slide under the microscope.
viii. To make a permanent preparation. Pass slide through pure alcohol and allow to dry. Mount in D.P.X. mountant.

5.9.3 DAPI-Staining

(Schweizer 1976)

i. Use chromosome preparation made with 45% acetic acid (see protocol 5.9.1) or after enzyme digestion (5.9.5).
ii. Apply 200 µl of DAPI solution on the marked area of the slide. Cover with a plastic cover slip and incubate at room temperature for 10 mins avoiding bright light. Remove coverslip.
iii. Briefly wash slides in distilled water. Add two drops of antifade solution to the preparation and cover with a large, thin coverslip (e.g. 24 × 40 mm No. 0). Gently, but firmly, squeeze excess antifade from the slide with filter paper.
iv. Examine hybridization signal with an epifluorescent microscope equipped with UV excitation filters and lenses.

5.9.4 Surface-Spreading

(Based on Albini and Jones 1987 and Albini and Schwarzacher 1992)

i. Determine stage of meiosis by method 5.9.1. Select anthers at pachytene.
ii. Place 1–3 anthers on a slide with 20 µl SC-enzyme mixture and release pollen mother cells. Incubate for 2 mins at room temperature.
iii. Add 20 µl of 0.5% (v/v) Lipsol detergent in water. Mix well and leave for 2 mins.
iv. Add 40 µl of paraformaldehyde fixative, mix well and dry down at room temperature in the hood. Leave slides overnight.
v. Wash slide in 0.1% Photoflo 2 times 5 mins each and air dry.
vi. Stain with 30–50% (w/v) silver nitrate in water under a nylon mesh at 60°C for 30–60 mins. Rinse slide and air dry.
vii. Analyse in the light or electron microscope.

5.9.5 Enzyme Digestion and Chromosome Preparation for in situ Hybridization

(Modified from Leitch et al. 1994 and Schwarzacher et al. 1994)

i. Clean slides with chromic acid
   a. Place slides into chromium trioxide solution in 80% (w/v) sulphuric acid for at least 3h at room temperature. Sulphuric acid is extremely corrosive.
   b. Wash slides in running water for 5 mins.

c. Rinse slides thoroughly in distilled water.
d. Air dry.
e. Place slides into 100% ethanol. Remove and dry slides immediately prior to use.

ii. Wash material 2 × 10 mins in enzyme buffer to remove fixative.
iii. Transfer to 2 × enzyme solution, incubate at 37°C for 1–2 hours (adjust time and concentration of enzyme treatment to suit the material, when using Onozuki cellulase use shorter times).
iv. Carefully transfer to enzyme buffer, leave for at least 15 min.
v. Place material, one at a time, in 45% aqueous acetic acid for a few minutes, replace by 60% acetic acid.
vi. Make chromosome spread on a clean slide by teasing the material to fragments with a fine needle. Gently squash the material in minimal liquid between glass slide and coverslip.
vii. Put slide on dry ice for 5–10 min, then flick off the coverslip with a razor blade, and air dry.
viii. Screen and select slides.

5.9.6 Solutions

Good laboratory practise is advised. Wearing of safety glasses and working in the hood where indicated is recommended.

Aceto-carmine. In the hood, reflux 1 g carmine in 45% acetic acid for 24 hours. Filter, and store in a dropping bottle for use over 5 years or more. Corrosive.

Feulgen Stain or Schiff's Reagent. Can be bought ready-made from Merck. Be careful with Feulgen. It is a potential carcinogen, and turns any organic material and benches purple. Alcohol might clean it. Work in the hood.

McIlvaine's buffer (pH 7.0). 18 ml of 0.2 M Na_2HPO_4 and 82 ml 0.1 M citric acid.

DAPI (4′,6-diamidino-2-phenylindole, Sigma). Prepare DAPI stock solution of 100 µg/ml in aqua dest. DAPI is a potential carcinogen. To avoid weighing out the powder, order small quantities and use the whole vial to make the stock solution. Aliquot and store at −20°C (it is stable for years). Prepare a working solution of 0.5 µg/ml by dilution in McIlvaine's buffer, aliquot and store at −20°C.

Antifade solution. Citifluor AF1 (Agaraids) or Vectashield mounting medium (Vector Laboratories).

Enzyme buffer. A: 0.1 M citric acid; B: 0.1 M tri-sodium-citrate; 10 × stock: 60 ml B + 40 ml A (pH = 4.8). Use: 1 : 9 diluted with aqua dest.

Enzyme solution: 1% cellulose (from *Aspergillus niger*, Calbiochem 21947) or a mixture of cellulase (0.9% Calbiochem, 0.1% Onozuki RS); 10% pectinase (from *Aspergillus niger*, solution in glycerol, Sigma P-9932) or 0.2% pectolyase (Sigma) and 1% cytohelicase (Sigma) in enzyme buffer.

SC-enzyme solution: 1% cytohelicase (Sigma), 1% bovine serum albumin, 10mM ethylene diamine tetraacetic acid (EDTA) in 6mM sodium phosphate buffer, pH 7.4.

Paraformaldehyde fixative: In the fume hood, add 4 g of paraformaldehyde (EM grade) and 1.5 g of sucrose to 80 ml water, heat to 60°C for about 10 mins, clear the solution with a few drops of concentrated NaOH, leave to cool down and adjust the final volume to 100 ml with water. Adjust pH to 9 with borate buffer.

Silver nitrate solution: Dissolve 1g in 1 ml aqua dest., Filter through a 25 µm filter. Silver nitrate is a clear solution, but after several hours stains skin, cloth and benches black and cannot be removed. Be extremely careful when using.

Acknowledgements

I would like to thank Shoabin Wu for technical help, Terry Miller for supplying the wheat translocation lines and Pat Heslop-Harrison for many fruitful discussions. The work is supported by a BBSRC senior fellowship.

References

Albini SM, Jones GH (1987) Synaptonemal complex spreading in *Allium cepa* and *A. fistulosum*. I. The initiation and sequence of pairing. Chromosoma 95:324–338

Albini SM, Schwarzacher T (1992) *In situ* localization of two repetitive DNA sequences to surface-spread pachytene chromosomes of rye. Genome 35:551–559

Anderson LK, Offenberg HH, Verkuijlen WMHC, Heyting C (1997) RecA-like proteins are components of early meiotic nodules in lily. Proc Natl Acad Sci USA 94:6868–6873

Aragón-Alcaide L, Reader S, Beven A, Shaw P, Miller T, Moore G (1997a) Association of homologous chromosomes during floral development. Curr Biol 7:905–908

Aragón-Alcaide L, Reader S, Miller T, Moore G (1997b) Centromere behaviour in wheat with high and low homeologous chromosome pairing. Chromosoma 106:327–333

Armstrong SJ, Kirkham AJ, Hultén MA (1994) XY Chromosome behaviour in the germ-line of the human male: a FISH analysis of spatial orientation, chromatin condensation and pairing. Chromosome Res 2:445–452

Baker SM, Plug AW, Prolla TA, Bronner CE, Harris AC, Yao X, Christie DM, Monell C, Arnheim N, Bradley A, Ashley T, Liskay RM (1996) Involvement of mouse *Mlh1* in DNA mismatch repair and meiotic crossing over. Nature Genet 13:336–342

Barlow AL, Benson FE, West SC, Hultén MA (1997) Distribution of the Rad51 recombinase in human and mouse spermatocytes. EMBO J 16:5207–5215

Barlow AL, Hultén MA (1996) Combined immunocytgenetic and molecular cytogenetic analysis of meiosis I human spematocytes. Chromosome Res 4:562–573

Baudat F, Nicolas A (1997) Clustering of meiotic double-strand breaks on yeast chromosome III. Proc Natl Acad Sci USA 94:5213–5218

Bennett MD (1984) Premeiotic events and meiotic chromosome pairing. Symp Soc Exp Biol 38:87–121

Bennett MD, Rao MK, Smith JB, Bayliss MW (1973) Cell development in the anther, the ovule and the young seed of *Triticum aestivum* L. var. Chinese Spring. Phil Trans R Soc Lond B 266:39–81

Bennett MD, Toledo LA, Stern H (1979) The effect of colchicine on meiosis in *Lilium speciosum* cv Rosemede. Chromosoma 72:175–189

Bhatt AM, Page T, Lawson EJR, Lister C, Dean C (1996) Use of Ac as an insertional mutagen in Arabidopsis. Plant J 9:935–945

Bishop DK (1994) *RecA* homologs *Dmc1* and *Rad51* interact to form multiple nuclear complexes prior to meiotic chromosome synapsis. Cell 79:1081–1092

Cheng EY and Gartler SM (1994) A fluorescent *in situ* hybridization analysis of X chromosome pairing in early human female meiosis. Hum Genet 94:389–394

Chua PR, Roeder GS (1997) Tam, a telomere-associated meiotic protein, functions in chromosome synapsis and cross over interference. Genes & Dev 11:1786–1800

Cook PR (1997) The transcriptional basis of chromosome pairing. J Cell Sci 110:1033–1040

Darlington CD, LaCour LF (1976) The handling of chromosomes. 6th edt. George Allen & Unwin, London

Dawe RK, Sedat JW, Agard DA, Cande WZ (1994) Meiotic chromosome pairing in maize is associated with a novel chromatin organization. Cell 76:901–912

Devos K, Gale M (1993) The genetic maps of wheat and their potential in plant breeding. Outlook on Agriculture 22:93–99

Dobson MJ, Pearlman RE, Karaiskakis A, Spyropoulos B, Moens PB (1994) Synaptonemal complex proteins: occurrence, epitope mapping and chromosome disjunction. J Cell Sci 107:2749–2760

Dresser ME, Moses MJ (1979) Silver staining of synaptonemal complexes in surface spreads for light and electron microscopy. Exp Cell Res 121:416–419

Endo TR, Gill BS (1996) The deletion stocks of common wheat. J.Heredity 87:295–307

Feulgen R, Rössenbeck H (1924) Mikroskopisch-chemischer Nachweis einer Nukleinsäure vom Typus der Thymonukleinsäure und die darauf beruhende selektive Färbung von Zellkernen in mikroskopischen Präparaten. Hoppe-Seylers Z Physiol Chem 135:203–248

Fukui K (1995) Quantitative chromosome maps as a basis of biosciences. In: Brandham PE, Bennett MD (eds) Kew Conference IV, Royal Botanic Gardens, Kew

Gillies CB (1984) The Synaptonemal complex. Ann Rev Plant Sci 2:81–116

Gillies CB (1985) An electron microscopic study of synaptonemal complex formation at zygotene in rye. Chromosoma 92:165–175

Goldman ASH, Hultén MA (1992) Chromosome *in situ* suppression hybridization in human male meiosis. J Med Genet 29:98–102
Havekes FWJ, de Jong JH, Heyting C, Ramanna MS (1994) Synapsis and chiasma formation in four meiotic mutants of tomato (*Lycopersicon esculentum*). Chromosome Res 2:315–325
Hawley RS and Arbel T (1993) Yeast genetics and the fall of the classical view of meiosis. Cell 72: 301–303
Heng HHQ, Tsui L-C, Moens PB (1994) Organization of heterologous DNA inserts on the mouse meiotic chromosome core. Chromosoma 103:401–407
Heslop-Harrison JS (1991) The molecular cytogenetics of plants. J Cell Sci 100:15–21
Heslop-Harrison JS, Smith JB, Bennett MD (1988) The absence of the somatic association of centromeres of homologous chromosomes in grass mitotic metaphases. Chromosoma 96:119–131
Heyting C (1996) Synaptonemal complexes – structure and function. Curr Opinion Cell Bio 8:389–396
Hiraoka Y, Dernburg AF, Parmelee SJ, Rykowski MC, Agard DA, Sedat JW (1993) The onset of homologous chromosome pairing during *Drosophila melanogaster* embryogenesis. J Cell Biol 120:591–600
Hirayama T, Ishida C, Kuromori T, Obata S, Shimoda C, Yamamoto M, Shinozaki K, Ohto C (1997) Functional cloning of a cDNA encoding Mei2-like protein from *Arabidopsis thaliana* using a fission yeast pheromone receptor deficient mutant. FEBS Letters 413:16–20
Holm PB (1986) chromosome-pairing and chiasma formation in allohexaploid wheat, *Triticum aestivum* analyzed by spreading of meiotic nuclei. Carlsberg Research Communications 51:239–294
Jenkins G (1983) Chromosome pairing in *Triticum aestivum* cv Chinese Spring. Carlsberg Res Commun. 48:255–283
Jenkins G, White J (1990) Elimination of synaptonemal complex irregularities in a *Lolium* hybrid. Heredity 64:45–53
John B (1990) Meiosis, Cambridge University Press, Cambridge
Karpen GH, Le M-H, Le H (1996) Centric heterochromatin and the efficiency of achiasmate disjunction in Drosophila female meiosis. Science 273:118–122
Kleckner N (1996) Meiosis: How could it work? Proc Natl Acad Sci USA 93:8167–8174
Klein S, Zenvirth D, Dror V, Barton AB, Kaback DB, Simchen G (1996) Patterns of meiotic double-strand breakage on native and artificial yeast chromosomes. Chromosoma 105:276–284
Klimyuk VI, Jones JDG (1997) *AtDMC1*, the *Arabidopsis* homologue of the yeast *DMC1* gene: characterization, transposon-induced allelic variation and meiosis-associated expression. Plant J 11:1–14
Kobayashi T, Kobayasgi E, Sato S, Hotta Y, Miyajima N, Tanaka A, Tabata S (1994) Characterization of cDNAs induced in meiotic prophase in lily microsporocytes. DNA Res 1:15–26
Kohli J (1994) Telomeres lead chromosome movement. Current Biol 4:724–727
Leitch AR, Brown JKM, Mosgöller W, Schwarzacher T, Heslop-Harrison JS (1994) The spatial localization of homologous chromosomes in human fibroblasts at mitosis. Hum Genet 93:275–280
Leitch AR, Schwarzacher T, Jackson D, Leitch IJ (1994) In situ Hybridization, a practical guide. RMS Microscopy Handbook 27. Bios Scientific Publishers, Oxford
Lichter P, Cremer T, Borden J, Manuelidis L, Ward DC (1988) Delineation of individual human chromosomes in metaphase and interphase cells by *in situ* suppression hybridization using recombinant DNA libraries. Hum Genet 80:224–234
Loidl J (1990) The initiation of meiotic chromosome pairing: the cytological view. Genome 33:759–778
Loidl J, Klein F, Scherthan H (1994) Homologous pairing is reduced but not abolished in asynaptic mutants of yeast. J Cell Biol 125:1191–1200
Lukaszewski AJ (1992) A comparison of physical distribution of recombination in chromosome 1R in diploid rye and in hexaploid triticale. Theor Appl Genet 83:1048–1053
Lukaszewski AJ (1997) The development and meiotic behavior of asymmetrical isochromosomes in wheat. Genetics 145:1155–1160
Maguire MP (1996) Is the synaptonemal complex a disjunction machine? J Heredity 86:330–340
McClintock B (1933) The association of non-homlgous parts of chromosomes in the midprophase of meiosis in *Zea mays*. Z Zellforsch Mikrosk Anat 19:191–237
Metz CW (1916) Chromosome studies on the Diptera II: The paired association of chromosomes in the Diptera and its significance J Exp Zool 21:213–279
Moens PB (1994) Molecular perspectives of chromosome pairing at meiosis. Bioessays 16:101–106
Moens PB, Chen DJ, Shen ZY, Kolas N, Tarsounas M, Heng HHQ, Spyropoulos B (1997) RAD51 immunocytology in rat and mouse spermatocytes and oocytes. Chromosoma 106:207–215
Moens PB, Pearlman RE (1988) Chromatin organization at meiosis. Bioessays 9:151–153
Moens PB, Pearlman RE (1993) Probing pachytene chromosomes. In: Heslop-Harrison JS, Flavell RB (eds) The Chromosome. Bios, Oxford, pp 183–192
O'Keefe C, Hultén MA, Tease C (1997) Analysis of proximal X chromosome pairing in early female mouse meiosis. Chromosoma 106:276–283
Parokonny AS, Marshal JA, Bennett MD, Cocking EC, Davey MR, Power JB (1997) Homeologous pairing and recombination backcross derivatives of tomato somatic hybrids (*Lycopersicon esculentum* x *L. peruvianum*). Theor Appl Genet 94:713–723
Pearlman RE, Tsao N, Moens PB (1992) Synaptonemal complexes of DNase-treated rat pachytene chromosomes contain (GT)n and LINE/SINE sequences. Genetics 130:865–872

Quevedo C, Del Cerro AL, Santos JL, Jones GH (1997) Correlated variation of chiasma frequency and synaptonemal complex length in *Locusta migratoria*. Heredity 78:515–519
Roeder GS (1995) Sex and the single cell: Meiosis in yeast. Proc Natl Acad Sci USA 92:10450–10456
Roeder GS (1997) Meiotic chromosomes: it takes two to tango. Genes Develop 11:2600–2621
Ross KJ, Fransz P, Armstrong SJ, Vizir I, Mulligan B, Franklin FCH, Jones GH (1997) Cytological characterization of four meiotic mutants of Arabidopsis isolated from T-DNA-transformed lines. Chromosome Res 5:551–559
Scherthan H, Loidl J, Schuster T, Schweizer D (1992) Meiotic chromosome condensation and pairing in *S. cerevisiae* studied by chromosome painting. Chromosoma 101:590–595
Scherthan H, Weich S, Schwegler H, Heyting C, Harle M, Cremer T (1996) Centromere and telomere movements during early meiotic prophase of mouse and man are associated with the onset of chromosome pairing. J Cell Biol 134:1109–1125
Schmekel K, Meuwissen RLJ, Dietrich AJJ, Vink ACG, Van Marle J, van Heen H, Heyting C (1996) Organization of Scp1 protein molecules within synaptonemal complexes of the rat. Exp Cell Res 226:20–30
Schwarzacher T (1986) Meiosis, SC-formation, and karyotype structure in diploid *Paeonia tenuifolia* and tetraploid *P. officinalis*. Plant Syst Evol 154:259–274
Schwarzacher, T (1996) The physical organization of Triticeae chromosomes. In: Heslop-Harrison JS (ed) Unifying Plant Genomes, Symposia of the Society for Experimental Biology, Number 50, The Company of Biologists Ltd, Cambridge, pp 71–75
Schwarzacher T (1997) Three stages of meiotic homologous chromosome pairing in wheat: cognition, alignment and synapsis. Sex Plant Rep 10:324–331
Schwarzacher T, Anamthawat-Jónsson K, Harrison GE, Islam AKMR, Jia JZ, King IP, Leitch AR, Miller TE, Reader SM, Rogers WJ, Shi M, Heslop-Harrison JS (1992a) Genomic *in situ* hybridization to identify alien chromosomes and chromosome segments in wheat. Theor Appl Genet 84:778–786
Schwarzacher T, Heslop-Harrison JS (1995) Molecular cytogenetics and investigations of meiosis. In: Brandham PE, Bennett MD (eds) Kew Conference IV, Royal Botanic Gardens, Kew, pp 407–416 + xl
Schwarzacher T, Heslop-Harrison JS, Anamthawat-Jónsson K, Finch RA, Bennett MD (1992b) Parental genome separation in reconstructions of somatic and premeiotic metaphases of *Hordeum vulgare* x *H. bulbosum*. J Cell Sci 101:13–24
Schwarzacher T, Leitch AR, Bennett MD, Heslop-Harrison JS (1989) *In situ* localization of parental genomes in a wide hybrid. Ann Bot 64:315–324
Schwarzacher T, Leitch AR, Heslop-Harrison JS (1994) DNA/DNA *in situ* hybridization – methods for light microscopy. In: Harris N, Oparka KJ (eds) Plant Cell Biology: a Practical Approach. Oxford University Press, Oxford, pp 127–155
Schweizer D (1976) DAPI fluorescence of plant chromosomes prestained with actinomycin D. Exp Cell Res 102:408–413
Stack SM (1984) Heterochromatin, the synaptonemal complex and crossing over. J Cell Sci 71:159–176
Suzuki T, Ide N, Tanaka I (1997) Immunocytochemical visualization of the centromeres during male and female meiosis in *Lilium longiflorum*. Chromosoma 106:434–445
Terasawa M, Shinohara A, Hotta Y, Ogawa H, Ogawa T (1995) Localization of RecA-like recombination proteins on chromosomes of the lily at various meiotic stages. Genes & Dev 9:925–934
von Wettstein D, Holm PB, Rasmussen SW (1984) The synaptonemal complex in genetic segregation. Ann Rev Genet 18:331–413
Weiner BM and Kleckner N (1994) Chromosome pairing via multiple interstitial interactions before and during meiosis in yeast. Cell 77:977–991
Zenvirth D, Loidl J, Klein A, Arbel A, Shemesh R, Simchen G (1997) Switching yeast from meiosis to mitosis: double-strand break repair, recombination and synaptonemal complex. Genes to Cells 2: 487–498

CHAPTER 6

Regulation of Gene Expression During Pollen Development

G. J. Wullems · J. A. M. Schrauwen

Department of Experimental Botany, Catholic University of Nijmegen, Toernooiveld 1, 6525 ED Nijmegen, The Netherlands
e-mail: Wullems@sci.kun.nl
Schrauw@sci.kun.nl
telephone: +31-24-3 65 27 61
fax: +31-24-3 65 34 50

6.1 Pollen Development as a Model System to Study Cell Differentiation

Pollen development has unique features since it offers the opportunity to study, in haploid cells, cell fate, division, signalling and origin of polarity. The molecular biology of pollen development as a model system for differentiation, can be studied at different developmental stages ranging from a single cell in the microspore up to a maximum of three cells (in tricellular pollen). Plant species are available that allow the collection of synchronously developing microgametophytes. A proof for correct pollen formation is its fertility in a bio-assay. By *in vitro* culture, pollen development can be directed to form mature pollen or somatic embryos. All these aspects make the developing pollen grain a unique model system to tackle major questions in developmental biology.

6.1.1 Pollen Development

In angiosperm plants, sexual reproduction requires the production of viable male and female gametophytes. The essential function of the male gametophyte, or pollen, in this process is to deliver two male gametes into the embryo sac as a prelude to double fertilisation (Heslop-Harrison, 1987; McCormick, 1993). An arrest in the development of one of the anther tissues, like stomium cells (Beals and Goldberg, 1997) effects pollen development and results in the production of sterile pollen. This means that for the development of pollen, occurring in the sporophytic anther, a coordinated gene expression is required in the gametophytic cells and the surrounding sporophytic tissues (Schrauwen et al., 1996).

The development of the gametophyte within the anther, starts with the formation of archesporial cells in the anther primordia. Divisions of the archesporial cells result in a primary parietal layer and an inner primary sporogenous layer. The cells of the primary sporogenous layer undergo mitotic divisions and diploid meiocytes are formed. These meiocytes or pollen mother cells enter meiosis. The onset of a synchronous development to form tetrads of haploid microspores is frequently observed in angiosperms (Esau, 1977; Knox, 1984).

6.1.2
Pollen Wall Formation

Pollen wall formation takes place during pollen development and starts as early as in the tetrad stage. The pollen wall of most species is composed of two layers, the inner intine and the outer exine. The intine is pectocellulosic and is of gametophytic origin. The exine is composed of sporopollenin (Zetsche, 1932), which is produced by the sporophytic tapetum (Heslop-Harrison, 1968). Both the exine and the intine surround the pollen grain, except for the apertures, where the exine is mostly absent.

The exine of pollen grains of most species is coated with the lipidic remaining material after tapetal breakdown, called pollenkit or tryphine (Dickinson 1973). The lipids that end up in the pollen coat are first accumulated in oil bodies and plastids in the tapetum, well before this layer degenerates. At tapetal rupture, lysis of these organelles occurs and lipids aggregate. This mass is released into the locule where it surrounds and coats the developing pollen grains (Dickinson, 1973; Murgia et al., 1991). Microscopical studies suggest deposition of materials on the exine also prior to tapetum degeneration (Murgia et al., 1991).

The lipidic coating is well-suited to protect the pollen protoplast, and its stickness is helpful in pollen transport. The pollen coat is also designed to play a role in pollen stigma interactions, such as adhesion to the stigma, hydration and recognition of the pollen, and early pollen tube growth (Dickinson, 1993; Doughty et al., 1993; Kandasamy et al., 1995; Preuss et al., 1993).

6.1.3
Definition of Microspore

The asymmetrical mitosis of the microspores forms the transition from the unicellular microspore to the bicellular pollen. However some researchers, mainly working on species with tricellular pollen, use the term microspore also for the bicellular or even for mature pollen grains to distinguish these structures from the female megaspore. We follow the definition of microspore as restricted to the unicellular structure (Esau, 1977; Mascarenhas, 1990a and Roeckel et al., 1990).

6.1.4
Influence of Tapetum on Early Pollen Development

Disruption of the tapetum during meiosis of the pollen mother cells causes male-sterility. A cytotoxic gene coupled to the tapetum-specific promoter *ta29* destroyed the tapetum and prevented pollen formation in transgenic tobacco and *Brassica napus* (Koltunow et al., 1990; Mariani et al, 1990). This illustrates the important role of the tapetum during early pollen development.

In petunia, a certain form of male-sterility is caused by untimely callase activity. This results in premature dissolution of the callose wall surrounding the microspores. This early tapetum function has been mimicked by introduction of a ß-1,3-glucanase coupled to a tapetum-specific promoter. The transformation resulted in microspores with an abnormally thin cell wall. The observed male-sterility is the result of bursting of aberrant microspores (Worrall et al., 1992). The correct timing of the dissolution of the callase of the tetrad is then essential for the correct development of microspores.

might bind them to oil bodies in the tapetum. Similar oleosins were found in the pollen coat of *B.alboglobra* and *B.napus* (Ruiter et al., 1997a, b).

The question is what the function of a highly glycine-rich pollen coat oleosin might be. The hydraulic continuity between the pollen grain and the stigmatic papilla is established by the conversion of pollen coat material into a substance allowing water passage, and is a prerequisite for pollen hydration (Dickinson 1995; Hulskamp et al., 1995; Kandasamy et al., 1995; Preuss, 1995). The total structure of these proteins may be functional in attracting water from the stigmatic papilla.

6.5 Expression of the Late Pollen Specific Gene ntp303 During Development and Pollen Tube Growth

The late pollen-specific gene *ntp303* (Weterings et al., 1992) was isolated by differential screening from a cDNA library prepared with RNA from mature pollen. Northern blot analysis confirmed that the *ntp303* RNA is only detectable in bicellular and germinating pollen. Besides, homologous RNAs are present in pollen of several other plant species, including mono and dicots. For instance the homology to *lat51* from *Lycopersicon esculentum* is 91% (McCormick, 1991) and to the genomic clone *bp10* from *Brassica napus* is 64% (Albani et al., 1992).

Analysis of the complete sequence of *ntpc303* showed a full length of 1961 bp, with an open reading frame of 1662 bp coding for a putative protein of 554 amino acids. The open reading frame predicted for a protein with molecular weight of 59 kD, a pI of 9.5 with six N-potential glycosylation sites.

In tobacco pollen transcription of the gene starts after pollen mitosis I and continues during pollen tube growth. The rise of the hybridisation signal of *ntp303* during the late pollen development is probably due to the overall accumulation of mRNA in maturing pollen grains (Tupy, 1982). *ntp303* is regulated concerted with a group of other genes in pollen of tobacco. Their transcripts appear after the first haploid mitosis and accumulate during the maturation of the microgametophyte (Schrauwen et al., 1990). Concluding from the fact that transcripts homologous to *ntp303* are present in pollen from other species it is asssumed that this regulatory mechanism is evolutionary conserved.

Gene expression of the *ntp303* gene initiates as the nursing tapetal cells have already decayed (Koltunow et al., 1990). This suggests that the expression of *ntp303* is regulated by the haploid gametophyte itself and that gene products of the surrounding diploid sporophytic tissue are not involved. In fact Northern blot analysis in our lab have proven that the pattern of *ntp303* gene expression during in vitro ripening of pollen is similar to the expression pattern of *ntp303* during in vivo maturation (van Herpen and de Groot, pers comm).

The sudden increase and accumulation of the amount of *ntp303* transcripts after the first haploid mitosis suggest a role for this RNA in germination and pollen tube growth rather than in development (Frankis et al., 1980). There already exists evidence that the correct development of pollen is determined during microsporogensis by coordinated gene expression of both the gametophyte and the sporophyte (Vasil, 1967; Koltunow et al., 1990). Therefore it is feasible that the *ntp303* is not involved in the development of the microgametophyte. It is most probable that the gene excerts its function during the germination process and during the growth of the pollen tube.

The association of *ntp303* with germination and/or pollen tube growth is additionally proven by analysis of *ntp303* expression during *in vitro* germination and pollen tube growth. The higher hybridisation signals of the Northern blot experiments suggest expression of the *ntp303* gene during imbibition and during the first hours of pollen tube growth *in vitro*.

The occurrence of transcription during pollen tube growth was confirmed by *in vivo* labeling of the RNA. Results from the pulse labeling assay show a signal originating from the hybrids of anti-sense *pntpc303* transcripts and ^{3}H-labeled newly formed *ntp303* RNA from the germinating pollen (Weterings et al., 1992). This proves unequivocally that the *ntp303* gene is transcribed during germination and tube growth in vitro. Now it is established that the *ntp303* actively contributes to the RNA synthesis in pollen tubes from tobacco (Suss et al., 1982). This active transcription of a pollen specific gene during germination is in contrast to previous reports on an anther specific clone from *Lycopersicon esculentum* (*lat52*), which suggest that no transcription occurs during the first hours of pollen tube growth (Twell et al,1989; Ursin et al., 1989).

The localisation of the *ntp303* transcripts at a high resolution level was obtained by the application of Confocal Laser Scanning Microscopy (CLSM) in combination with *in situ* hybridization. With the CLSM technique optical sections of the specimen could be built, that resulted in a 3-D image of the localisation of *ntp303* transcripts. These *in situ* localisation experiments proved that transcription starts at mid-bicellular pollen stage and was only detectable in the vegetative cell. *ntp303* transcripts are still present near the tip of pollen tubes 72 hours after pollination of the pistil (Reijnen et al., 1991; Weterings et al., 1995a). This data indicates a function of the gene during pollen tube growth or the process of fertilisation.

Southern blot analysis showed that the *ntpg303* was part of a gene family of at least five members.

Dissection of the *ntp303* promoter for the *cis*-regulotory elements was assayed by 5′deletion analysis and the microprojectile facilitated transient expression technique. These experiments showed that the minimal pollen-specific promoter of the gene is located between -103 and -51 nucleotides upstream of the transcriptional start site. In this region two hexameric sequences AAATGA, of which one was inverted, proved to have transcription activating properties (Weterings et al., 1995b). Using gel mobility shift assays, we demonstrated that this minimal functional promoter interacts with a leaf nuclear GT-1 binding activity. This data suggests that ubiquitous transcription factors are involved in the pollen-specific expression of the gene (Hochstenbach et al., 1996).

However all this data did not give much information on a function or even the location of the protein. In order to get information about the *ntp303* protein antisera were raised against recombinant proteins of *ntp303*. Preliminary results show that the protein is produced in the maturing pollen and in the pollen tubes. With immunolocalisation its location in association with the the plasma membrane could be established. More conclusive information on the function of the *ntp303* protein will be obtained with loss-of-function experiments with transgenic plants which are now in progress.

References

Albani D, Robert LS, Donaldson PA, Altosaar I, Arnison PG, Fabijanski SF (1990) Characterisation of a pollen-specific gene family from *Brassica napus* which is activated during early microspore development. Plant Mol Biol 15:605–622

Albani D, Sardana R, Robert LS, Altosaar I, Arnison PG, Fabijanski SF (1992) A *Brassica napus* gene family which shows sequence similarity to ascorbate oxidase is expressed in developing pollen. Molecular characterisation and analysis of promoter activity in transgenic tobacco plants. Plant J 2:31–342

Beals TP, Goldberg RB (1997) A novel cell ablation strategy blocks tobacco anther dehiscence. The Plant Cell 9: 1527–1545

Bedinger PA, Edgerton MD (1990) Developmental staging of maize micro spores reveals a transition in developing micro spore proteins. Plant Physiol 92: 474–479

Custers JBM, Oldenhof MT, Schrauwen JAM, Cordewener JHG, Wullems GJ, Van Lookeren Campagne M.M (1997) Analysis of microspore-specific promoters in transgenic tobacco. Plant Molecular Biology 35: 689–699

Dickinson HG (1973) The role of plastids in the formation of pollen coatings. Cytobios 8: 25040

Dickinson HG (1993) Pollen dressed for success. Nature 364: 573–574

Dickinson HG (1995) Dry stigmas, water and self-incompatibility in *Brassica*. Sex Plant Reprod 8: 1–10

Dickinson HG, Lewis D (1973a) Cytochemical and ultrastructural differences between intraspecific compatible and incompatible pollinations in *Raphanus*. Proc R Soc Lond B 183: 21–38

Doughty J, Hedderson F, McCubbin A, Dickinson HG (1993) Interaction between a coating-borne peptide of the *Brassica* pollen grain and stigmatic S (incompatibility)-locus-specific glycoproteins. Proc Natl Acad Sci USA 90: 467–471

Dumas C, Gaude T (1982) Stigma-pollen recognition and pollen hydration. Phytomorphology 31: 191–200

Elleman CJ, Dickinson HG (1990) The role of the exine coating in pollen-stigma interactions in *Brassica oleracea L.* New Phytol 114: 511–518

Esau K (1977) Anatomy of seed plants. John Wiley and Sons, New York

Frankis R, Mascarenhas JP (1980) Messenger RNA in the ungerminated pollen grain: a direct demonstration of its presence. Ann Bot 45: 595–599

Heslop-Harrison J. (1968) Pollen wall development. Science 161: 230–237

Heslop-Harrison J (1987) Pollen germination and pollen-tube growth. In: Giles KL, Prakash J (eds) Review of Cytology. 107. Pollen: Cytology and Development. Ac Press, Orlando, pp 1–78

Hochstenbach R, De Groot P, Jacobs J, Schrauwen J, Wullems G (1996) The promoter of a gene that is expressed only in pollen interacts with ubiquitos transcription factors. Sex Plant Reprod 9: 197–202

Holbrook LA, Van Rooijen GJH, Wilen RW, Moloney MM (1991) Oil body proteins in microspore-derived embryos of *Brassica napus*. Plant Physiol 97: 1051–1058

Hülskamp M, Kopczak SD, Horejsi TF, Kihl BK, Pruitt RE (1995) Identification of genes required for pollen-stigma recognition in *Arabidopsis thaliana*. Plant J 8: 703–714

Kandasamy MK, Nasrallah JB, Nasralla ME (1995) Pollen-pistil interactions and developmental regulation of pollen tube growth in *Arabidopsis*. Development 120: 3405–3418

Knox RB (1984) The pollen grain. In: Johri (ed) Embryology of Angiosperms. Springer Verlag, pp 197–261

Koltunow AM, Truettner J, Cox KH, Wallroth M, Goldberg RB (1990) Different temporal and spatial gene expression patterns occur during anther development. Plant Cell 2: 1201–1224

Mariani C, De Beuckeleer M, Truettner J, Leemans J, Goldberg RB (1990) Induction of male sterility in plants by a chimaeric ribonuclease gene. Nature 347: 737–741

Mascarenhas JP (1990a)Gene expression in the angiosperm male gametophyte. In: Blackmore S, Knox RB (eds) Microspores. Evolution and ontogeny. Academic Press, London, pp 265–280

Mascarenhas JP (1990b)Gene activity during pollen development. Ann Rev Plant Phys Mol Biol 41: 317–338

Mascarenhas JP (1992) Pollen gene expression: molecular evidence. Int Rev Cytol 140: 3–18

McCormick S (1991) Molecular analysis of male gametogenesis in plants. Trends Genet 7: 298–303

McCormick S (1993) Male gametophyte development. Plant Cell 5: 1265–1275

Murgia M, Charzynska M, Rougier M, Cresti M (1991) Secretory tapetum of *Brassica oleracea* L.: polarity and ultrastructural features. Sex Plant Reprod 4: 28–35

Oldenhof MT, de Groot PFM, Visser JH, Schrauwen JAM, Wullems GJ (1996) Isolation and characterisation of a microspore-specific gene from tobacco. Plant Mol Biol 31: 213–225

Preuss D, Lemieux B, Yen G, Davis RW (1993) A conditional sterile mutation eliminates surface components from *Arabidopsis* pollen and disrupts cell signaling during fertilisation. Genes Devel 7: 974–985

Preuss D (1995) Being fruitful: Genetics of reproduction in *Arabidopsis*. Trends Genet 11: 147–153

Reijnen WH, Vanherpen MMA, De Groot PFM, Olmedilla A, Schrauwen JAM, Weterings KAP, Wullems GJ (1991) Cellular localisation of a pollen-specific messenger RNA by in situ hybridisation and confocal laser scanning microscopy. Sex Plant Repr 4: 254–257

Robert LS, Gerster J, Allard S, Cass L, Simmonds J (1994) Molecular characterisation of two *Brassica napus* genes related to oleosins which are highly expressed in the tapetum. Plant J 6:927–933

Roberts MR, Boyes E, Scott RJ (1995) An investigation of the role of the anther tapetum during microspore development using genetic cell abletion. Sex Plant Reprod 8:299–307

Roeckel P, Chaboud A, Matthys-Rochon E, Russell S, Dumas C (1990) Sperm cell structure, development and organization. In: Blackmore S, Knox RB (eds) Microspores. Evolution and ontogeny. Academic Press, London, pp 281–309

Ross JHE, Murphy DJ (1996) Characterisation of anther-expressed genes encoding a major class of extracellular oleosin-like proteins in thepollen coat of *Brassicaceae*. Plant J 9:625–637

Ruiter RK, Mettenmeyer T, Van Laarhoven D, Van Eldik GJ, Doughty J, Van Herpen MMA, Schrauwen JAM, Dickinson HG, Wullems GJ (1997a) Proteins of the pollen coat of *Brassica oleracea*. J Plant Physiol 150:85–91

Ruiter RK, van Eldik GJ, van Herpen RMA, Schrauwen JAM, Wullems GJ (1997b) Characterization of oleosins in the pollen coat of Brassica oleracea. The Plant Cell 9:1621–1623

Schrauwen JAM, De Groot PFM, Vanherpen MMA, Vanderlee T, Reynen WH, Weterings KAP, Wullems GJ (1990) Stage-related expression of messenger-RNAs during pollen development in lily and tobacco. Planta 182:298–304

Schrauwen JAM, Mettenmeyer T, Croes AF, Wullems GJ (1996)Tapetum-specific genes: what role do they play in male gametophyte development? Acta Bot Neerl 45:1–15

Scott R, Hodge R, Paul W, Draper J (1991) The molecular biology of anther differentiation. Plant Science 80:167–191

Suss J, Tupy J (1982) Kinetics of uridine uptake and incorporation into RNA in tobacco pollen culture. Biol Plant 24:72–79

Tupy J, Rihova L, Zarsky V (1991) Production of tobacco pollen from microspores in suspension culture and its storage for *in situ* pollination. Sex Plant Reprod 4:284–287

Tupy J (1982) Alterations in polyadenylated RNA during pollen maturation and germination. Biol Plant 24:331–340

Twell D, Wing R, Yamaguchi J, McCormick S (1989) Isolation and expression of an anther-specific gene from tomato. Mol Gen Genet 217:240–245

Twell D, Patel S, Sorensen A, Roberts M, Scott R, Draper J, Foster G (1993) Activation and developmental regulation of an *Arabidopsis* anther-specific promoter in microspores and pollen of *Nicotiana tabacum*. Sex Plant Reprod 6:217–224

Twell D, Bate N, Sweetman J, Spurr C, Foster GD (1996) Transcriptional and translational control of pollen gene expression. Abstract, 14th International Congress of Sexual Reproduction, Lorne, Australia, p 57

Ursin VM, Yamaguchi J, McCormick S (1989) Gametophytic and sprorphytic expression of anther-specific genes in developing tomato anthers. Plant Cell 1:727–736

Vasil IK (1967) Physiology and cytology of anther development. A review Biol Rev 42:327–373

Van Rooijen GJH, Moloney MM (1995) Structural requirements of oleosin domains for subcellular targeting to the oil body. Plant Physiol 109:1353–1361

Van Rooijen GJH, Terning LI, Moloney MM (1992) Nucleotide sequence of an oleosin gene from *Arabidopsis thaliana*. Plant Mol Biol 18:1177–1179

Weterings K, Reijnen W, Vanaarssen R, Kortstee A, Spijkers J, Vanherpen M, Schrauwen J, Wullems G (1992) Characterisation of a pollen-specific cDNA clone from *Nicotiana tabacum* expressed during microgametogenesis and germination. Plant Mol Biol 18:1101–1111

Weterings K, Reijnen W, Wijn G, Vandeheuvel K, Appeldoorn N, Dekort G, Vanherpen M, Schrauwen J, Wullems G (1995a) Molecular characterisation of the pollen-specific genomic clone NTPg303 and in situ localisation of expression. Sex Plant Repr 8:11–17

Weterings K, Schrauwen J, Wullems G, Twell D (1995b) Functional analysis of the pollen-specific gene NTP302 reveals a novel pollen-specific, and conserved cis-regulatory element. Plant J 8:55–63

Worrall D, Hird DL, Hodge R, Paul W, Draper J, Scott R (1992) Premature dissolution of the microsporocyte callose wall causes male sterility in transgenic tobacco. Plant Cell 4:759–771

Zetsche H (1932) Sporopollenin. In: Klein G (ed) Handbuch de Pflanzenanalyze, vol. 3

CHAPTER 7

Double Fertilization in Flowering Plants: Origin, Mechanisms and New Information from *in vitro* Fertilization

J.-E. Faure

Ecole Normale Supérieure de Lyon, Laboratory of Plant Reproduction and Development,
CNRS-INRA-ENS, 46, allée d'Italie, 69364 Lyon Cedex 07, France
e-mail: jefaure@ens-lyon.fr
telephone: (33)-4-72-72-86-04
fax: (33)-4-72-72-86-00

7.1 Introduction

One hundred years ago, Sergius Nawaschin (1898), followed a few months later by Léon Guignard (1899) suggested for the first time that fertilization is actually double in *Lilium martagon* and *Fritillaria tenella*. Double fertilization is now believed to be ubiquitous among flowering plants. The pollen grain (male gametophyte) germinates on a pistil. It grows a tube in which two male gametes are moved toward an embryo sac (female gametophyte) that contains in particular two female gametes, the egg and the central cells, and two synergids. Once the pollen tube has reached the embryo sac, it releases the two male gametes into a degenerated synergid. One fuses with the egg cell to form a zygote that further develops into the embryo, and the other fuses with the central cell to form a polyploid (generally triploid) cell that develops into the endosperm. Thus, two fertilizations happen simultaneously. This sequence was well documented, especially in the sixties by William Jensen using transmission electron microscopy (Jensen 1964, Jensen and Fisher, 1968). However the underlying mechanisms are not well known, mainly because fertilization takes place inside the embryo sac itself embedded in sporophytic tissues. The isolation of both male and female gametes from the gametophytes (Cass, 1973; Hu et al., 1985; also see review: Theunis et al., 1991; Huang and Russell, 1992a) and their fusion under *in vitro* conditions (Kranz et al., 1991; Kranz and Lörz, 1993; Faure et al., 1994; also see review: Dumas and Faure, 1995; Kranz and Dresselhaus, 1996) now allows new experimental investigations. The interest in fertilization, the central step of sexual reproduction, has therefore been renewed over the last ten years. The purpose of this review is to give a general overview of double fertilization in flowering plants, to address the main questions we will now need to answer to understand how fertilization operates at the cellular and molecular level, and to discuss how fertilization done *in vitro* can provide us with new information.

7.2 Origin of Double Fertilization

There is a consensus to consider the gnetales (*Ephedra*, *Gnetum* and *Welwitschia*) as the most closely related seed plants to angiosperms (see review: Friedman and Carmichael, 1996). Studies of how fertilization operates in these species provided insights into the origin of double fertilization. A form of double fertilization (more likely to be termed 'double karyogamy') was observed in *Ephedra nevadensis* (Friedman, 1990), *Ephedra trifurca* (Friedman, 1992) and in *Gnetum gnemon* (Carmichael and Friedman, 1995,

1996). A binucleate sperm fuses with a binucleate egg cell in *Ephedra*, or with the multinucleate female gametophyte in *Gnetum*. Each sperm nucleus then fuses with a female nucleus. This double karyogamy leads to the formation of two viable zygote nuclei. In *Ephedra*, the two zygote nuclei divide twice to typically form eight nuclei. The structure then gets cellularized and results in eight cells that develop into eight proembryos. However, only one embryo ultimately survives. In *Gnetum* the two zygotic nuclei get surrounded by a cell wall and develop into proembryos. As in *Ephedra*, only one embryo ultimately survives. The biological significance of this double process is unknown. The supernumerary embryos might assist the normal embryo in its development. These observations suggest that the double process is likely to have evolved early, maybe before the divergence between gnetales and angiosperms. One possible model to explain the origin of double fertilization in flowering plants is (1) the appearance of 'gnetale-like' double karyogamy in a common ancestor, (2) then the appearance of a specific development for one of the supernumerary embryos (see review: Friedman and Carmichael, 1996). This second organism would have evolved into a nutritive structure, the endosperm.

7.3 Mechanisms of Fertilization From in Planta Observations

7.3.1 The Gametes

The two male gametes originating from the pollen grain have very special features when compared to other plant cells. They have no cell wall. These two protoplasts are embedded in a third cell, the vegetative cell that forms the pollen tube. Thus the plasma membrane of the two gametes is closely apposed to the inner plasma membrane of the vegetative cell (see review: Mogensen, 1992). The nucleus of the vegetative cell seems to be physically associated to this structure. This very close association was termed Male Germ Unit (Dumas et al., 1984), but its role is unresolved. The two male gametes are small spheroidal or elongated cells. In maize for example they are very elongated and sized about $25 \times 5 \times 2$ μm (McConchie et al., 1987). They have a highly condensed chromatin, few organelles (mitochondria and/or plastids, dictyosomes, endoplasmic reticulum) and a reduced cytoplasm. Mitochondria are occasionally observed in the nucleus as in tobacco (Yu and Russell, 1994). The gametes contain labile microtubules that depolymerize once the cells are isolated from pollen grains or tubes (see review: Palevitz and Tiezzi, 1992). In addition they might lack actin but this is uncertain. The two male gametes of flowering plants also have a very special feature when compared to gametes of animals or some other plants such as mosses and ferns: they do not have neither flagella nor cilia, and seem to lack independent motility. This is suggested in particular by the important role of the pollen tube cytoskeleton in moving the gametes toward the embryo sac (see review: Cai et al., 1997), and the absence of cell motility after isolation from pollen grains or tubes.

The egg and central cells are enclosed in the embryo sac. The embryo sac of *Polygonum* constitutes the most common pattern in flowering plants and also probably the ancestral one (see review: Huang and Russell, 1992a). It originates from only one of the four meiotic products and is therefore termed 'monosporic'. It contains three antipodal cells and two synergids in addition to the two female gametes. These last four cells are the four cells required to receive the pollen tube and undergo double fertilization. Authors have termed this functional unit as the Female Germ Unit (Dumas et al., 1984;

see also Russell, 1992). The two synergids and the egg cell are located in the micropylar end of the embryo sac where the pollen tube penetrates. They are commonly pear-shaped and of approximately the same size. As an example, the maize egg cell is about 60 µm long (Diboll, 1968; Diboll and Larson 1966) and its volume is about 300 times that of a male gamete (Faure et al., 1992). The egg cell is usually highly vacuolized as is the case in maize (Diboll, 1968; Diboll and Larson 1966; Faure et al., 1992). In addition, the egg cell is polarized: Its nucleus is central or located at one end of the cell and its cell wall is partial. The membrane section free of cell wall can directly interact with the male gamete plasma membrane at the time of fertilization. The central cell is contiguous to the egg cell and the two synergids and it occupies most of the embryo sac. In maize it is about 150 µm long. In the *Polygonum* type embryo sac, this big and highly vacuolized cell contains two haploid nuclei also termed polar nuclei (see review: Russell, 1992).

Many questions can be raised about the fate of these male and female gametes. (1) Gametes look like quiescent cells. So do they need to be capacitated prior to fertilization, like in mammals for example? There is no answer to this question to date. (2) How are these male and female gametes put into contact *in planta*? This is a striking question when comparing animal and plant fertilization because the male gametes of angiosperms seem to lack independent motility unlike animal gametes. (3) Then what is the microenvironment needed for the gamete adhesion and fusion and what are the molecules involved in these steps? Whereas some molecules for the adhesion and fusion of gametes have been identified in mammals (Blobel et al., 1992; Almeida et al., 1995) or algae such as *Fucus* sp. (Wright et al., 1995a,b), very little is known about the adhesion and fusion steps in angiosperms. (4) What components of the sperm cell are then integrated and how does karyogamy occur? (5) Finally how are the egg and the central cells activated to initiate development? We will now address some of these questions in a chronological order and review the information that *in planta* observations have provided.

7.3.2 Transport of the Male Gametes

How are the male gametes transported from the aperture of the pollen tube to the site of gamete fusion? The actin distribution is modified in the embryo sac before or at the time of pollen tube arrival. The use of rhodamine-phalloidin shows that actin bands appear along the presumed path of the male gametes. This was observed in *Nicotiana tabacum* (Russell, 1993; Huang et al., 1993a; Huang and Russell, 1994), *Plumbago zeylanica* (Huang et al., 1993b) and *Zea mays* (Huang and Sheridan, 1994). Scott Russell suggested an interaction of these bands with myosin at the surface of the male gametes (see review: Russell, 1996). Such an acto-myosin based mechanism to transport the gametes toward their site of fusion is very attractive. However neither the involvement of the actin in pulling the gametes nor the presence of myosin on the male gametes have been demonstrated to date. Therefore this model needs further examination. If this model is correct this would be, to our knowledge, the first case of an extracellular acto-myosin based system.

7.3.3 Microenvironment for Gamete Adhesion and Fusion

Once released in the embryo sac, the two male gametes are in a very complex microenvironment derived from a synergid that degenerated, some pollen tube components, and probably extracellular fluids. The osmolarity of this microenvironment surrounding the gametes is not known nor its pH and composition. It seems that the amount of the total calcium detected in the synergids by X-ray microanalysis is high, as well as the amount of loosely sequestered calcium detected with chlorotetracycline (CTC) or antimonate precipitates (Chaubal and Reger 1990, 1992a,b, 1993; Huang and Russell, 1992b). This might lead to a high level of free or loosely bound calcium once one synergid degenerates and participates in the microenvironment for the gamete fusion. However this is only a hypothesis and the amount of free calcium is unknown to date. To our opinion this is a critical question because the composition of the microenvironment is probably essential for the ability of the gametes to fuse. In mammals for example free calcium is required at a millimolar concentration (Yanagimachi, 1978; Evans et al., 1995). The knowledge of the microenvironment composition is also essential to improve the *in vitro* fertilization protocols that we will present later in this review.

7.3.4 Gamete Adhesion and Fusion

How do the gamete adhere and fuse? The only data available come from transmission electron microscopy observations mainly in *Plumbago zeylanica* where the two gamete fusions seem to happen fast (probably within a minute) and simultaneously (Russell, 1983). Understanding the molecular mechanism of fusion is critical in flowering plants because two male gametes fuse with two female gametes: what are the processes that lead to the fusion of only one male gamete with the egg cell, and the other male gamete with the central cell? We can only propose models and discuss their relevance.

The simplest model we can imagine would consist of two male gametes with the same surface components and the same morphology (Model 1, Figure 7.1a). Both could potentially fuse with the egg or the central cell. But after a first fertilization, a block to polyspermy would force the second male gamete to fuse with the other female gamete yet unfertilized. A second and more sophisticated model would consist of a segregation of the male gametes when released by the pollen tube into the embryo sac. The segregation could be due to a specific order of release from the tube and/or to a difference in morphology and therefore a difference of transport toward the female gametes. One male gamete could then specifically be presented first to the egg cell (as an example) and fuse preferentially with it (Model 2, Figure 7.1b). Such a model would again probably require the establishment of a block to polyspermy forcing the second male gamete to fuse with the second female gamete. Finally a third model would be the existence of two male gametes with different surface compounds, one set being complementary to egg cell compounds, the other to central cell compounds (Model 3, Figure 7.1c). In this ultimate model of preferential fertilization the male gametes could be either similar in morphology or dimorphic, and a block to polyspermy would be optional.

Which model is to be favored? There is no clear answer to date. Two plants however have provided some few information. In *Plumbago zeylanica* the two male gametes originating from the same pollen grain are dimorphic. The first gamete, adjacent to the vegetative nucleus, doesn't contain plastids. The second gamete contains plastids that are smaller and more contrasted in transmission electron microscopy than the egg cell plastids. Such paternal plastids were observed in about 94% of fertilized egg cells. This suggests a preferential fusion of the plastid-rich gamete with the egg cell (Russell, 1985). However such a preferential fertilization cannot be generalized, in particular because *Plumbago zeylanica* has an atypical embryo sac that lacks synergids. Similarly it seems

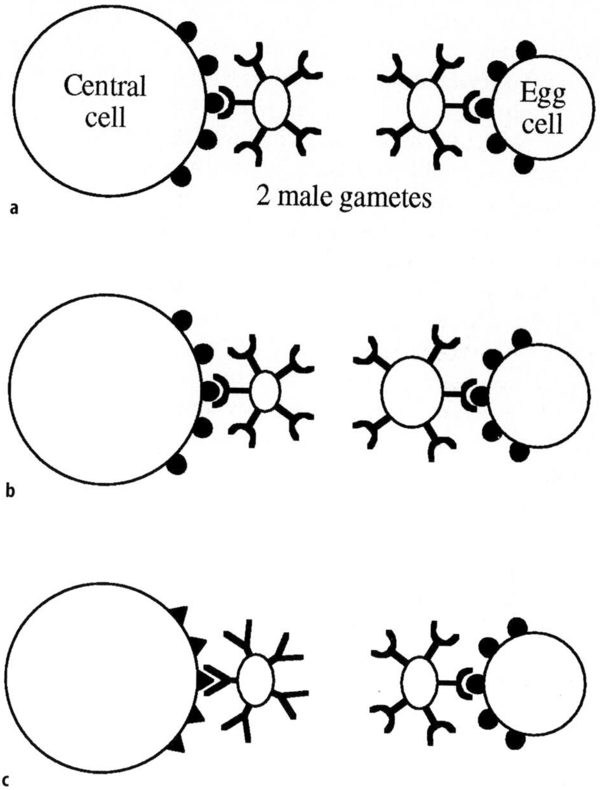

Fig. 7.1. Three prospective models for double fertilization in flowering plants. **(a)** No preferential fusion. One male gamete fuses either with the egg cell or the central cell. A block to polyspermy is established after this first fusion and forces the other male gamete to fuse with the other female gamete. **(b)** Preferential fusion. The male gametes are segregated by their order when released in the synergid and/or by a dimorphism as represented here. One specific male gamete is presented first to the egg, for example, and fuses with it. A block to polyspermy then forces the other male gamete to fuse with the other female gamete. **(c)** Targeted preferential fusion. The male gametes have different surface components for the adhesion and/or fusion. These components determine one specific male gamete to fuse with the egg cell, the other with the central cell. A male gamete dimorphism and a block to polyspermy are optional

that there is preferential fertilization in maize genotypes containing supernumerary chromosomes (B-chromosomes). In most of the pollen grains of such genotypes there is no disjunction of the B-chromatids during the mitosis that leads to the two male gametes (Roman, 1948; Shi et al., 1996). As a consequence one sperm receives all B-chromatids. It seems from genetic analysis that this male gamete fuses more frequently with the egg cell than the other male gamete that lacks B-chromosome (Roman, 1948; Carlson 1986). Again this cannot be generalized to other maize genotypes because this preferential fusion might be induced by the presence of B-chromosomes. In conclusion, data about the mechanisms of the double fusion is scarce and further investigations are needed.

7.3.5 Integration of the Male Components

The mode of inheritance of mitochondria and plastids is variable among flowering plants. However the majority of angiosperms seem to transmit plastids exclusively or predominantly from the female parent. The published data also suggests that uniparental mitochondria inheritance predominates (see review: Mogensen, 1996). Many different mechanisms have evolved to reduce the male cytoplasm, or reduce the number of male organelles or their DNA content, during pollen formation, at the time of plasmogamy, or after gamete fusion. As an example, enucleated bodies were observed in the degenerated synergid (with transmission electron microscopy), next to the fertilized egg and central cells in cotton and lily (Jensen and Fisher, 1968; Janson and Willemse, 1995), or next to the egg cell only, in barley (Mogensen 1988, 1990). It was suggested that these bodies are male cytoplasm remnants that were not incorporated at the time of gamete fusion. This might provide a mechanism to partly or entirely eliminate male mitochondria and plastids.

Obviously the other major male component to be integrated in the female cytoplasm is the male nucleus. It migrates toward the egg nucleus and fuses with it as seen in transmission electron microscopy (Jensen, 1964). However *in planta* observations didn't provide us with a very precise timing of these events. In maize for example, karyogamy is of premitotic type and occurs in the fertilized egg 14 to 20 hours after pollination (Mòl et al., 1994).

As a conclusion, *in planta* observations have provided us with very few answers to the questions we raised in this section, for two main reasons. First, the timing of fertilization is unprecise because it happens after pollination and pollen tube growth that are variable in time. Second, fertilization occurs deep in the maternal tissues. This strongly limits the experimentations. *In vitro* methods have therefore been proposed to bypass these difficulties. We will now review the *in vitro* protocols and the new information obtained on fertilization.

7.4 New Information From *in Vitro* Fertilization

7.4.1 The Methods

In vitro fertilization can now be performed using two main strategies. A first strategy consists of microinjecting male gametes or nuclei into isolated embryo sacs (Keijzer et al., 1988; Matthys-Rochon et al., 1994). However these *in vitro* fertilized structures have not been regenerated into plants. The injection of male gametes into female gametes completely isolated from the embryo sac has also not yet been achieved, unlike in animals (see Lanzendorf et al., 1988: Alikani et al., 1995).

A second strategy consists of fusing gametes completely isolated from the gametophytes. We will now focus this review on these methods of *in vitro* fertilization (*sensu stricto*). To date both male and female gametes can be isolated simultaneously in at least eight species: *Hordeum vulgare* (Cass, 1973; Holm et al., 1994), *Lolium perenne* (Van der Maas et al., 1993a, b), *Triticum aestivum* (Szakacs and Barnabas, 1989; Kovacs et al., 1994), *Zea mays* (Dupuis et al., 1987; Wagner et al., 1989), *Brassica napus* (Matthys-Rochon et al., 1988; Katoh et al., 1997), *Nicotiana tabacum* (Hu et al., 1985; Cao et al., 1996) *Plumbago zeylanica* (Russell, 1986; Cao and Russell, 1997) and *Torenia fournieri* (Mòl, 1986; Keijzer et al., 1988). The male gametes are released from pollen grains by an osmotic/pH shock (see review: Chaboud and Perez, 1992). If needed, large amounts of gametes can be then purified usually by Percoll or sucrose gradient centrifugations. About 10^5 cells can be isolated from a single isolation procedure. It is more difficult to isolate the female gametes, and this constitutes so far a limiting step. Indeed the isolation requires the manual microdissection of each ovule and embryo sac, with or without a preliminary enzymatic treatment of the tissues, depending on the species (Holm et al., 1994; see review: Huang and Russell, 1992a). The first *in vitro* fusions where obtained by Kranz et al (1991). Pairs of isolated male and female gametes are electrofused: they are first aligned by dielectrophoresis, then induced to fuse using one or more electric pulses. Fertile plants have been regenerated from these *in vitro* fusion products (Kranz and Lörz, 1993). This electrofusion procedure was also successfully applied to wheat (Kovacs et al., 1995). An alternative to electrofusion was developed in maize. It consists of bringing isolated gametes into contact by gently steering the medium (with a glass microneedle) that contains 1 to 10 mM calcium chloride at low pH (Faure et al., 1994), or 50 mM calcium chloride at high pH (Kranz and Lörz, 1994). This method should allow the experimental access to the adhesion and gamete fusion steps, bypassed by electric pulses during electrofusion. However the zygotes obtained after fusion in 5 mM calcium chloride and low pH have not yet been regenerated into plants. The number of species whose gametes can be isolated is increasing fast. We can therefore expect *in vitro* fertilization to be applied in many more species in the coming years.

7.4.2 *In Vitro* Observations

We will now mainly focus on the new information that we have started to obtain on fertilization using the calcium-mediated *in vitro* fusion assay (Faure et al., 1994). A major benefit from *in vitro* fertilization is that we can visualize and manipulate the gamete fusion. It appears that gametes apposed in a medium containing 5 mM calcium chloride adhere for a few minutes and fuse within a few seconds. The adhesion is difficult to obtain in the absence of calcium, and the fusion rate is lower at 1 mM calcium chloride. Thus, it seems that calcium or a divalent ion is important for the adhesion and/or fusion. Interestingly the calcium concentration in the microenvironment surrounding the gametes at the time of fertilization may also be high *in planta*, as already discussed in this review. In addition, interesting results come from the study of the effects of calcium on the viability of maize gametes isolated from pollen grains. Calcium at millimolar concentrations leads to a fast decrease of cell viability (Zhang et al., 1995). Altogether, *in vivo* and *in vitro* observations suggest a model for the possible role of calcium. A low calcium concentration in the pollen tube could prevent the fusion of the pollen tube cell and the two male gametes. Once released in the degenerated synergid and exposed to higher calcium concentrations, the male gamete membranes may be destabilized and their fusion with the female gametes favored (see review: Zhang and Cass, 1997). Such a possible change in calcium environment may have a capacitation effect on the gametes that we actually mimic under *in vitro* conditions.

In 5 mM calcium chloride, about 80% of the maize male gamete-egg cell pairs fuse. If we assume that the male gametes are picked randomly, this suggests that the two male gametes from the same pollen grain can fuse with the egg cells. This would favor the absence of preferential fusion of one of the male gametes with the egg cell (Figure 7.1a). However, this may highly depend on the fusion environment including the calcium concentration. So further investigations are needed. As an example, it would be of interest to perform *in vitro* fusions with B-chromosome containing lines, and observe if still 80% of the pairs fuse under the same conditions.

Several events occur after plasmogamy in maize. First, it seems that additional male gametes can no longer fuse. This suggests the presence of a block to polyspermy. It may be established as fast as during the first minute after fusion and it is effective at least during the next 45 minutes (Faure et al., 1994, and unpublished observations). To date we cannot be more precise because it took us at least 45 seconds after the first fusion to bring a new male gamete and make it adhere to the fertilized egg. A cell wall is regenerated within a few minutes after the gamete electrofusion (Kranz et al., 1995) as well as after calcium mediated fusion (Aldon D, Antoine A-F and Rougier M, personal communication). This could constitute at least one possible mechanism to block additional fusions. However, this hypothesis has to be tested, and it doesn't exclude some possible electrical block that would probably be established faster as in animals (Jaffe, 1976) or algae (Brawley, 1990). Electrophysiological studies are therefore needed.

Second, a transient increase of the cytosolic calcium concentration constitutes another event a few seconds after the fusion of a male gamete with an egg cell (Digonnet et al., 1997). This transient increase lasts several minutes and the calcium concentration is back to the resting level about 30 minutes after plasmogamy. Further investigations are required to understand if this constitutes a signal for the initiation of development, as in animals (see reviews: Schultz and Kopf, 1995; Jaffe, 1996). The topology of the increase (i.e. in a wave that propagates through the egg, versus a uniform elevation of the concentration), the origin of the calcium, and the possible targets for calcium will also have to be identified.

Third, the male nucleus migrates toward the egg nucleus during the calcium elevation. The *in vitro* fertilization provided a more precise picture and timing of this phenomenon than *in planta* observations. Unlike *in planta*, the fusion of the plasma membranes can again be visualized and samples can be fixed at very precise times after plasmogamy. The male nucleus reaches the egg nucleus about 30 minutes after the electrofusion (Faure et al., 1993) or the calcium-mediated fusion (Rougier M, Aldon D, Digonnet C, personal communication) as seen with transmission electron microscopy. The karyogamy occurs within the first hour after the gamete fusion. Further experiments could consist of using drugs on zygotes to test the role of the cytoskeleton in this nuclear migration and fusion.

7.5 Conclusions

In vitro fertilization has already provided some new information about fertilization such as: (1) a transient increase of the calcium concentration in the egg cell cytosol, (2) the fast formation of a cell wall around the egg cells, induced by fertilization, (3) the establishment of a block to polyspermy in fertilized egg cells. The functional relationship of these events has now to be established. In addition, *in vitro* procedures are now combined with molecular methods such as the construction of cDNA libraries from single cells or very small amounts of cells to identify fertilization-induced genes (Dresselhaus et al., 1994; Richter et al., 1996). The differential screening of libraries from electrofusion products (18 hours after plasmogamy) and unfertilized egg cells has already lead to the identification of few clones specifically expressed after fertilization, including a gene encoding calreticulin (Dresselhaus et al., 1996; Kranz and Dresselhaus, 1996). In the future, the combination of the *in vitro* methods with cellular and molecular analysis should allow us to continue to gain knowledge on the gamete adhesion and fusion, the activation and initiation of development, and the subsequent first division. A major challenge will be to not only focus on the egg cell fertilization but also to be able to investigate central cell fertilization which is essential for the successful reproduction of flowering plants.

Acknowledgements
The author is funded by the Centre National de la Recherche Scientifique. We thank Anne-Frédérique Antoine for helpful discussions and comments on the manuscript.

References

Alikani M, Cohen J, Palermo GD (1995) Enhancement of fertilization by micromanipulation. Curr Op Obstetrics and Gynecology 7:182–187
Almeida EAC, Huovila A-PJ, Sutherland AE, Stephens LE, Calarco PG, Shaw LM, Mercurio AM, Sonnenberg A, Primakoff P, Myles D, White JM (1995) Mouse egg integrin a6b1 functions as a sperm receptor. Cell 81:1095–1104
Blobel CP, Wolfsberg TG, Turk CW, Myles DG, Primakoff P, White JM (1992) A potential fusion peptide and integrin ligand domain in a protein active in sperm-egg fusion. Nature 356:248–252
Brawley SH (1990) The fast block against polyspermy in fucoid algae is an electrical block. Dev Biol 144:94–106
Cai G, Moscatelli A, Cresti M (1997) Cytoskeletal organization and pollen tube growth. Trends in Plant Sci 2:86–91
Cao Y, Russell SD (1997) Mechanical isolation and ultrastructural characterization of viable egg cells in *Plumbago zeylanica*. Sex Plant Reprod 10:368–373
Cao Y, Reece A, Russell SD (1996) Isolation of viable sperm cells from tobacco (*Nicotiana tabacum*). Zygote 4:81–84
Carlson WR (1986) The B chromosome of maize. Critical Review in Plant Science 3:201–226
Carmichael JS, Friedman WE (1995) Double fertilization in *Gnetum gnemon*: The relationship between the cell cycle and sexual reproduction. Plant Cell 7:1975–1988
Carmichael JS, Friedman WE (1996) Double fertilization *in Gnetum gnemon* (Gnetaceae): Its bearing on the evolution of sexual reproduction within the Gnetales and the anthophyte clade. Am J Bot 83: 767–780
Cass DD (1973) An ultrastructural and Nomarski-interference study of the sperms of barley. Can J Bot 51:601–605
Chaboud A, Perez R (1992) Generative cells and male gametes: Isolation, physiology, and biochemistry. Int Rev Cytol 140:205–232
Chaubal R, Reger BJ (1990) Relatively high calcium is localized in synergig cells of wheat ovaries. Sex Plant Reprod 3:98–102
Chaubal R, Reger BJ (1992a) Calcium in the synergid cells and other regions of pearl millet ovaries. Sex Plant Reprod 5:34–46
Chaubal R, Reger BJ (1992b) The dynamics of calcium distribution in the synergid cells of wheat after pollination. Sex Plant Reprod 5:206–213
Chaubal R, Reger BJ (1993) Prepollination degeneration in mature synergids of pearl millet: an examination using antimonate fixation to localize calcium. Sex Plant Reprod 6:225–238
Diboll AG (1968) Fine structural development of the megagametophyte of *Zea mays* following fertilization. Am J Bot 55:787–806
Diboll AG, Larson DA (1966) An electron microscopic study of the mature megagametophyte in *Zea mays*. Am J Bot 53:391–402
Digonnet C, Aldon D, Leduc N, Dumas C, Rougier M (1997) First evidence of a calcium transient in flowering plants at fertilization. Development 124:2867–2874
Dresselhaus T, Lörz H, Kranz E (1994) Representative cDNA libraries from few plant cells. Plant J 5: 605–610
Dresselhaus T, Hagel C, Lörz H, Kranz E (1996) Isolation of a full-length cDNA encoding calreticulin from a PCR library of *in vitro* zygotes in maize. Plant Mol Biol 31:23–34
Dumas C, Faure J-E (1995) Use of *in vitro* fertilization and zygote culture in crop improvement. Curr Op Biotech 6:183–188
Dumas C, Knox RB, McConchie CA, Russell SD (1984) Emerging physiological concepts in fertilization. What's new in Plant Physiology 15:17–20
Dupuis I, Roeckel P, Matthys-Rochon E, Dumas C (1987) Procedure to isolate viable sperm cells from corn (*Zea mays* L) pollen grains. Plant Physiol 85:876–878
Evans JP, Schultz RM, Kopf GS (1995) Mouse sperm-egg plasma membrane interactions: analysis of roles of egg integrins and the mouse sperm homologue of PH-30 (fertilin) β. J Cell Sci 108: 3267–3278
Faure J-E, Mogensen HL, Kranz E, Digonnet C, Dumas C (1992) Ultrastructural characterization and three-dimensional reconstruction of isolated maize (*Zea mays* L.) egg cell protoplasts. Protoplasma 171:97–103

Faure J-E, Mogensen HL, Dumas C, Kranz E, Lörz H (1993) Karyogamy after electrofusion of single egg and sperm cell protoplast from maize (*Zea mays* L.): Cytological evidences and time course. Plant Cell 5:747–755
Faure J-E, Digonnet C, Dumas C (1994) An *in vitro* system for adhesion and fusion of maize gametes. Science 263:1598–1600
Friedman WE (1990) Double fertilization in Ephedra, a nonflowering seed plant: its bearing on the origin of angiosperms. Science 247:951–954
Friedman WE (1992) Evidence of a pre-angiosperm origin of endosperm: implications for the evolution of flowering plants. Science 255:336–339
Friedman WE, Carmichael JS (1996) Double fertilization in Gnetales: Implications for understanding reproductive diversification among seed plants. Int J Plant Sci 157:S77–S94
Guignard L (1899) Sur les anthérozoïdes et la double copulation sexuelle chez les végétaux angiospermes. Rev Gén de Bot 11:129–135
Holm PB, Knudsen S, Mouritzen P, Negri D, Olsen FL, Roué C (1994) Regeneration of fertile barley plants from mechanically isolated protoplasts of the fertilized egg cell. Plant Cell 6:531–543
Hu S-Y, Li LG, Zhu C (1985) Isolation of viable embryo sacs and their protoplasts of *Nicotiana tabacum*. Acta Bot Sin 27:343–347
Huang B-Q, Russell SD (1992a) Female germ unit: Organization, isolation, and function. Int Rev Cytol 140:233–293
Huang B-Q, Russell SD (1992b) Synergid degeneration in *Nicotiana*: a quantitative, fluorochromatic and chlorotetracycline study. Sex Plant Reprod 5:151–155
Huang B-Q, Russell SD (1994) Fertilization in *Nicotiana tabacum*: cytoskeletal modifications in the embryo sac during synergid degeneration. Planta 194:200–214
Huang B-Q, Sheridan WF (1994) Female gametophyte development in maize: microtubular organization and embryo sac polarity. Plant Cell 6:845–861
Huang B-Q, Strout GW, Russell SD (1993a) Fertilization in *Nicotiana Tabacum* – Ultrastructural organization of Propane-jet-frozen embryo sacs *in vivo*. Planta 191:256–264
Huang B-Q, Pierson ES, Russell SD, Tiezzi A, Cresti M (1993b) Cytoskeletal organisation and modification during pollen tube arrival, gamete delivery and fertilisation in *Plumbago zeylanica*. Zygote 1: 143–154
Jaffe LA (1976) Fast block to polyspermy in sea urchin eggs is electrically mediated. Nature 261:68–71
Jaffe LA (1996) Egg membranes during fertilization. In: Schultz et al (eds) Molecular Biology of Membrane Transport Disorders. Plenum Press, New-York, pp 367–378
Janson J, Willemse MTM (1995) Pollen tube penetration and fertilization in *Lilium longiflorum* (Liliaceae). Am J Bot 82:186–196
Jensen WA (1964) Observations on the fusion of nuclei in plants. J Cell Biol 23:669–672
Jensen WA, Fisher DB (1968) Cotton embryogenesis: the entrance and discharge of the pollen tube in the embryo sac. Planta 78:158–183
Katoh N, Lörz H, Kranz E (1997) Isolation of viable egg cells of rape (*Brassica napus* L). Zygote: 5:31–33
Keijzer CJ, Reinders MC, Leferinkten Klooster HB (1988) A micromanipulation method for artificial fertilization in *Torenia*. In: Cresti M, Gori P, Pacini E (eds) Sexual Reproduction in Higher Plants, Springer-Verlag, Berlin pp 119–124
Kovacs M, Barnabas B, Kranz E (1994) The isolation of viable egg cells of wheat (*Triticum aestivum* L.). Sex Plant Reprod 7:311–312
Kovacs M, Barnabas B, Kranz E (1995) Electro-fused isolated wheat (*Triticum aestivum* L.) gametes develop into multicellular structures. Plant Cell Rep 15:178–180
Kranz E, Bautor J, Lörz H (1991) *In vitro* fertilization of single isolated gametes by electrofusion. Sex Plant Reprod 4:12–16
Kranz E, Lörz H (1993) *In vitro* fertilization with isolated, single gametes results in zygotic embryogenesis and fertile maize plants. Plant Cell 5:739–746
Kranz E, Lörz H (1994) *In vitro* fertilization of maize by single egg and sperm cell protoplast fusion mediated by high calcium and high pH. Zygote 2:125–128
Kranz E, Von Wiegen P, Lörz H (1995) Early cytological events after induction of cell division in egg cells and zygote development following *in vitro* fertilization with angiosperm gametes. Plant J 8:9–23
Kranz E, Dresselhaus T (1996) *In vitro* fertilization with isolated higher plant gametes. Trends in Plant Sci 1:82–89
Lanzendorf SE, Malony MK, Veeck LL, Slusser J, Hodgen GD, Rosenwaks Z (1988) A preclinical evaluation of pronuclear formation by microinjection of human spermatozoa into human oocytes. Fertility and Sterility 49:835–842
Matthys-Rochon E, Detchepare S, Wagner V, Roeckel P, Dumas C (1988) Isolation and characterization of viable sperm cells from tricellular pollen grains. In: Cresti M, Gori P, Pacini E (eds) Sexual Reproduction in Higher Plants. Springer-Verlag, Berlin, pp 245–250
Matthys-Rochon E, Mòl R, Heizmann P, Dumas C (1994) Isolation and microinjection of active sperm nuclei into egg cells and central cells of isolated maize embryo sac. Zygote 2:29–35

McConchie CA, Hough T, Knox RB (1987) Ultrastructural analysis of the sperm cells of mature pollen of maize, *Zea mays*. Protoplasma 139:9–19

Mogensen HL (1988) Exclusion of male mitochondria and plastids during syngamy in barley as a basis for maternal inheritance. Proc Natl Acad Sci USA 85:2594–2597

Mogensen HL (1990) Fertilization and early embryogenesis. In: Chapman CP (eds) Reproductive versatility in the grasses. New York: Cambridge University Press, pp 76–99

Mogensen HL (1992) The male germ unit: concept, composition, and significance. Int Rev Cytol 140: 129–147

Mogensen HL (1996) The hows and whys of cytoplasmic inheritance in seed plants. Am J Bot 83: 383–404

Mòl R (1986) Isolation of protoplasts from female gametophytes of *Torenia fournieri*. Plant Cell Rep 3: 202–206

Mòl R, Matthys-Rochon E, Dumas C (1994) The kinetics of cytological events during double fertilization in *Zea mays* L. Plant J 5:197–206

Nawaschin SG (1898) Resultate einer Revision der Befruchtungsvorgange bei *Lilium martagon* und *Fritillaria tenella*. Bul Acad Imp Sci St Petersburg 33:39–47

Palevitz BA, Tiezzi A (1992) Organization, composition, and function of the generative cell and sperm cytoskeleton. Int Rev Cytol 140:149–185

Richter J, Kranz E, Lörz H, Dresselhaus T (1996) A reverse transcriptase-polymerase chain reaction assay for gene expression studies at the single cell level. Plant Sci 114:93–99

Roman H (1948) Directed fertilization in maize. Proc Natl Acad Sci USA 34:36–42

Russell SD (1983) Fertilization in *Plumbago zeylanica*: Gametic fusion and fate of the male cytoplasm. Am J Bot 70:416–434

Russell SD (1985) Preferential fertilization in *Plumbago*: Ultrastructural evidence for gamete-level recognition in an angiosperm. Proc Natl Acad Sci USA 82:6129–6132

Russell SD (1986) Isolation of sperm cells from the pollen of *Plumbago zeylanica*. Plant Physiol 81: 317–319

Russell SD (1992) Double fertilization. Int Rev Cytol 140:357–388

Russell SD (1993) The egg cell: development and role in fertilization and early embryogenesis. Plant Cell 5:1349–1359

Russell SD (1996) Attraction and transport of male gametes for fertilization. Sex Plant Reprod 9: 337–342

Schultz RM, Kopf GS (1995) Molecular basis of mammalian egg activation. Curr Topics in Dev Biol 30: 21–62

Shi L, Zhu T, Mogensen HL, Keim P (1996) Sperm identification in maize by fluorescence *in situ* hybridization. Plant Cell 8:815–821

Szakacs E, Barnabas B (1989) Sperm cell isolation from wheat (*Triticum aestivum* L) pollen. In: Barnabas B, Liszt K (eds) Characterization of male transmission units in higher plants. MTA Copy, Budapest, pp 37–40

Theunis CH, Pierson ES, Cresti M (1991) Isolation of male and female gametes in higher plants. Sex Plant Reprod 4:145–154

Van Der Maas HM, Zaal MACM, De Jong ER, Van Went JL (1993a) Optimization of isolation and storage of sperm cells from the pollen of perennial ryegrass (*Lolium perenne* L.). Sex Plant Reprod 6:64–70

Van Der Maas HM, Zaal MACM, De Jong ER, Krens FA, Van Went JL (1993b) Isolation of viable egg cells of perennial ryegrass (*Lolium perenne* L.). Protoplasma 173:86–89

Wagner VT, Song YC, Matthys-Rochon E, Dumas C (1989) Observations on the isolated embryo sac of *Zea mays* L. Plant Sci 59:127–132

Wright PJ, Green JR, Callow JA (1995a) The *Fucus* (Phaeophyceae) sperm receptor for egg. I. Development and characteristics of a binding assay. J Phycol 31:584–591

Wright PJ, Callow JA, Green JR (1995b) The *Fucus* (Phaeophyceae) sperm receptor for egg. II. Isolation of a binding protein which partially activates eggs. J Phycol 31:592–600

Yanagimachi R (1978) Calcium requirement for sperm-egg fusion in mammals. Biol Reprod 19:949–958

Yu H-S, Russell SD (1994) Occurence of mitochondria in the nuclei of tobacco sperm cells. Plant Cell 6: 1477–1484

Zhang G, Williams CM, Campenot MK, McGann LE, Cutler AJ, Cass DD (1995) Effects of calcium, magnesium, potassium, potassium, and boron on sperm cells isolated from pollen of *Zea mays* L. Sex Plant Reprod 8:113–122

Zhang G, Cass DD (1997) Calcium signaling in sexual reproduction of flowering plants. Recent Res Devel in Plant Physiol 1:75–83

CHAPTER 8

Molecular Approach to Female Meiosis in *Petunia Hybrida*

A. Porceddu[1] · Ch. Moretti[1] · S. Sorbolini[1] · S. Guiderdone[1] · L. Lanfaloni[2]
F. Lorenzetti[1] · M. Pezzotti[1]

[1] Istituto Miglioramento Genetico Vegetale – Università degli Studi di Perugia, Italy
[2] Dipartimento di Biologia Cellulare e Molecolare – Università degli Studi di Perugia, Italy
e-mail: pezzotti@unipg.it
telephone: +39 75 5 85 62 13
fax: +39 75 58 56 22 24

8.1 Introduction

Because of the pivotal role the ovule plays in plant reproduction and agriculture, throughout the history of plant biology much attention has been focused on the angiosperm ovule. Ovule development has been proposed to occur in four distinct phases (Schneitz et al., 1995). The first phase involves the initiation of the ovule primordia from the carpel placenta. During the second phase, the identity of the ovule is specified. This is followed by the formation of spatially defined pattern elements within the developing ovule in the third phase. The final phase involves morphogenesis to form a mature ovule. In angiosperms, morphogenesis results in ovules that consists of a nucellus enveloped by one or two integuments and a supporting stalk, the funiculus, which attaches the ovule to the placenta (Bouman, 1984; Reiser and Fisher, 1993). The megasporocyte is produced within the nucellus and undergoes meiosis to produce four megaspores (megasporogenesis). A single surviving megaspore will undergo megagametogenesis to produce the female gametophyte (Willemse and Van Went, 1984; Mansfield et al., 1990; Reiser and Fischer, 1993).

Despite the wealth of descriptions of ovule anatomy and morphology, little is known about the molecular basis of ovule development and function. The relative inaccessibility of the ovule within the ovary and the difficulty in harvesting adequate amounts of tissue at known developmental stages have impeded progress toward understanding the molecular basis of ovule development and function. Genes that are involved in determining and regulating the identity and development of an ovule have been cloned in *Petunia* (Angenent et al., 1995) and several mutations affecting ovule and female gametophyte development have been detected (for review see Drews et al. 1998). Mutations that lead to defects in certain aspects of megasporogenesis have been identified; for example *megasporogenesis* (*msg*) gene in wheat (Joppa et al., 1987), *synaptic mutant-2* (*sy-2*) in *Solanum* (Parrott and Hanneman, 1988), and *female gametophyte factor* (*Gf*) in *Arabidopsis* (Redei, 1965).

Little progress has been made in identifying and analysing genes expressed specifically in the ovule during megasporogenesis. We chose *Petunia hybrida*, which, because of its flower size and the number of ovules present in each ovary can be considered a convenient experimental plant model, to explore the megasporogenesis pathway molecularly.

8.2 Callose Deposition During Megasporogenesis in Petunia Flowers

The megasporogenesis of angiosperms with monosporic or bisporic embryo sac development is marked by a well-defined pattern of callose deposition. A boundary of callose normally envelopes megaspore mother cells at meiotic prophase I and later callose accumulates in the transverse wall and around diads triads and tetrads (Figure 8.1a). Callose disappears at the functional megaspore stage or the beginning of embryo sac differentiation (Rodkiewicz 1970).

We started our study on the molecular control of megasporogenesis with a detailed fluorescence microscopic analysis to distinguish meiotic stages by visualising callose in the ovule. Five different ovary stages were defined and related to petal size (Figure 8.1b). They spanned from pre-meiosis to tetrad dissolution. Flower organs, including sepals, petals, anthers, stigma/styles and ovaries, were collected at each developmental stage and polyA$^+$RNA was extracted. Two ovary cDNA libraries were constructed; one from ovary mRNA from stages one to five, the other, specific of meiotic stages, from stage two to four.

8.3 Cell Division, Flower Development and Macrosporogenesis

Plant morphogenesis is mainly determined by the patterns in which cells are formed by cell division and by the directions in which they expand during the enlargement phase of their growth. Our understanding of the role of cell division in plant morphogenesis is based on analysis of the position and frequency of mitotic figures, or of cells which have incorporated labelled DNA precursors (Steeves and Sussex, 1989; Lindon, 1990). A highly conserved set of genes has been shown to control the timing of cell division in yeast, insects and vertebrates. Some of these genes have also been described in plants, including cdc2 and cyclins. The product of the cdc2 gene is a serine-threonine kinase which periodically associates with the product of different classes of cyclin genes to create successive waves of kinase activity which regulate mitotic activity. Structurally two classes of cdc2 related genes can be distinguished in plants based on the occurrence of the so called PSTAIRE hallmark. This is a motif of 16 aminoacids EGVPSTAIREISLLKE which is essential for cyclin binding. Genes with PSTAIRE binding motif, which belong to the first class, can often rescue yeast cdc2 mutants, whereas genes encoding proteins with only partially conserved PSTAIRE binding motif, belonging to the other class, share less similarity with cdc2 and generally lack the ability to rescue yeast cdc2 mutants.

At least three classes of cyclins can be distinguished in plants. Class A and B cyclins, often referred to as mitotic classes, are transcribed during late S-G2-M phases. Cyclins D, cloned by their ability to rescue *Saccharomyces cerevisae* mutants arrested in G1, are thought to play a role in G1-S transition.

Mitotic cyclin expression has been reported to correlate strictly with actual division activity of the cell in plant, whereas cdc2 expression also correlates with competence for division (Ferreira et al. 1994; Hemerly et al. 1993)

Despite the fact that the link between cell proliferation, morphogenesis and cell-cycle gene expression remains poorly understood, studies should be directed at investigating cell-cycle genes expression during plant development. We attempted to gain a better understanding of the role the cell cycle apparatus plays in regulating macrosporogenesis by studying cell cycle gene expression during *Petunia* flower development.

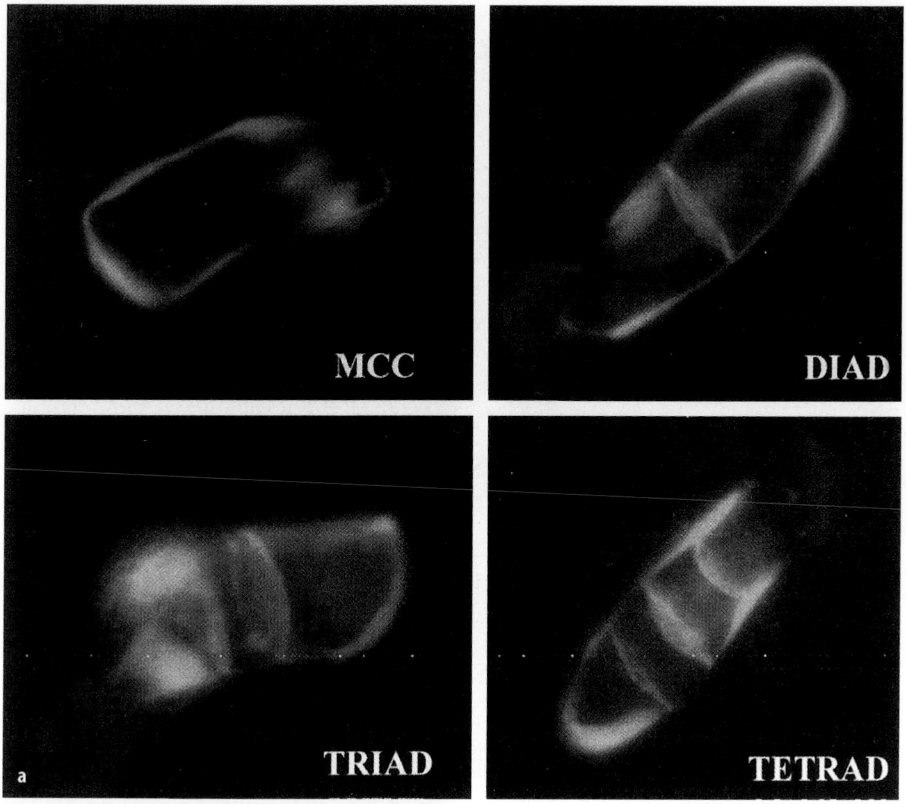

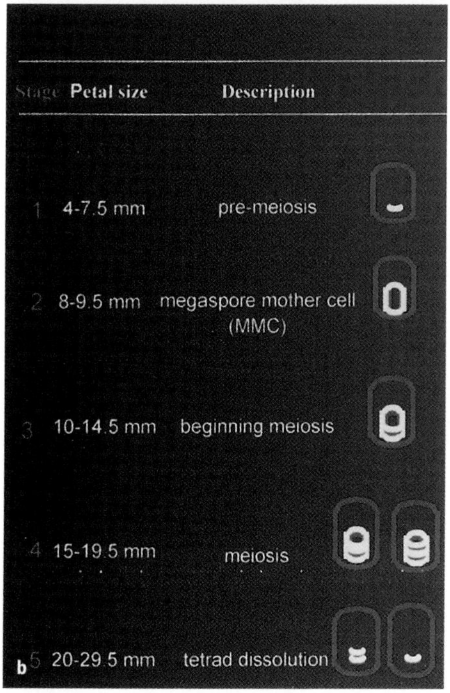

Fig. 8.1. (a) Callose deposition during megasporogenesis in *Petunia hybrida*. **(b)** Megasporogenesis and petal size stages in *Petunia hybrida*

The mitotic cyclins have two regions exhibiting a particularly high degree of conservation within their cyclin box (Minshull et al. 1989; Pines and Hunter, 1989). Based on these sequences two degenerated oligonucleotides were synthesised for use with PCR. The PCR that used *Petunia* ovary cDNA led to the amplification of a 186 bp cDNA fragment, which was used to screen our cDNA libraries. Several clones were isolated and one, called *Pet cyc 1*, has been further analysed.

The complete nucleotide sequence was determined. Comparison of predicted amino acid sequence revealed that *Pet cyc1* is very similar to other cyclins isolated in tobacco, alfalfa, *Arabidopsis*, soybean, *Antirrhinum* and other plants and is more closely related to A and B mitotic cyclins than to D cyclins (for a review see Renaudin et al. 1996). *Pet cyc1* contains several amino acid residues specific to the class B which allow it to be classified as B type cyclin.

In order to determine whether *Pet cyc1* was expressed in an organ-specific fashion, RNA was extracted from roots, leaves, sepals, petals, anthers, stigma/styles and ovaries. The results of the RNA gel blot analyses are reported in Figure 8.2a. The *Pet cyc1* transcript is absent in roots, leaves and sepals; weakly detectable in anthers; present in stigma/styles and ovaries; and abundant in petals. In addition, the transcript was also ab-

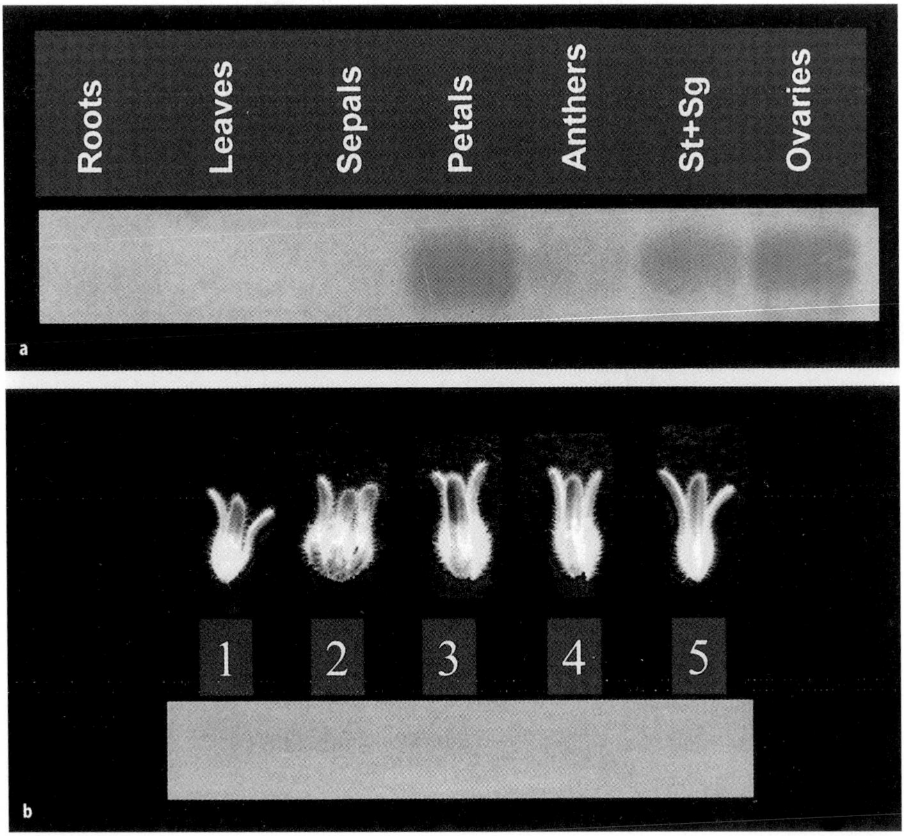

Fig. 8.2. (a) Accumulation of Pet cyc1 mRNA in *Petunia hybrida* organs. **(b)** Accumulation of Pet cyc1 mRNA during sepal development

sent in the apical, medial and distal part of roots, as well as in young and old leaves (data not shown). This experiment, therefore, demonstrated that the *Pet cyc1* expression is flower specific.

We have characterised the expression pattern of the *Pet cyc1* gene on gel blots with RNA isolated from sepals, petals, anthers, stigma/styles and ovaries at five different stages of petunia flower development as described previously. Figure 8.2b, shows that *Pet cyc1* transcript is absent during sepal development but constantly present during petal and stigma/style development up to stage 4 (Figure 8.3a–8.3b), thereafter the mRNA completely disappeared. As sepals are fully expanded during these stages of flower development, the lack of the cyclin can be explained as absence of cell division.

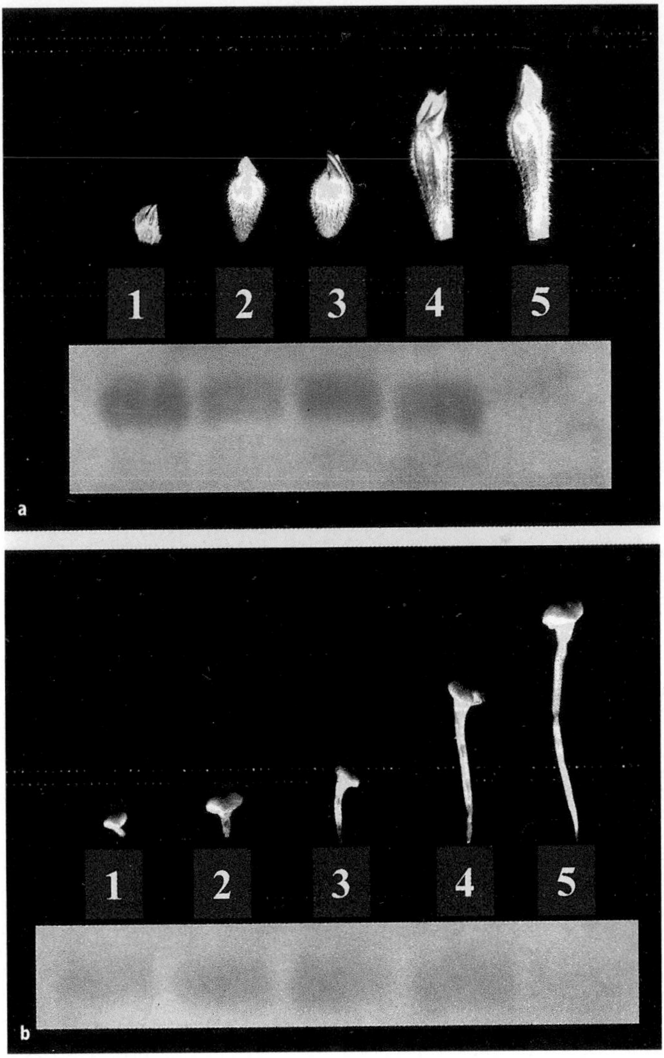

Fig. 8.3. (a) Accumulation of Pet cycl mRNA during petal development. **(b)** Accumulation of Pet cycl mRNA during style and stigma development

Petals and stigma/styles are actively growing during the first four stages, when they increase their size five times (from 4 to 20mm and 2 to 10 mm respectively). This correlates with the presence of cyclin transcript; cyclin expression and thus cell division ceases in these developing flower organs which corresponds at ~5–6 days before anthesis. At this point, the flower is only 40% of its final length. Further growth and unfolding of the flower bud, which shapes the final flower, may be, therefore, primarily a result of differential cell elongation. As will be seen in Fig. 8.4a the *Pet cyc1* transcript is present only in the first stage of anther development but is constantly expressed during ovary development throughout all the stages (Figure 8.4b).

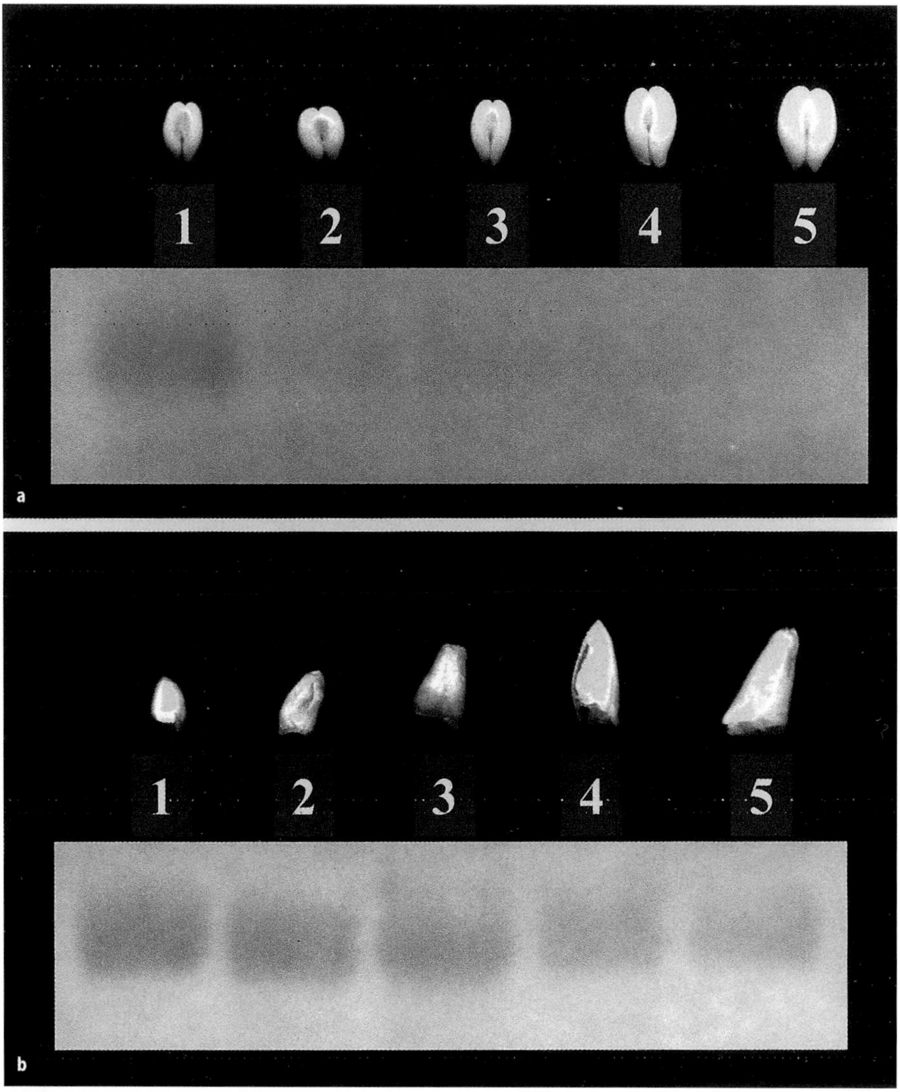

Fig. 8.4. a Accumulation of Pet cyc1 mRNA during anther development. **b** Accumulation of Pet cyc1 mRNA during ovary development

These results indicate that there is an early block in anther mitotic activity which does not occur in the ovaries. The block may be related to early maturation of male organs, seeing that specialised cell and tissue are already differentiated by stage 1 and from stage 2 to 5 grow further by enlargement while, female organs may probably need more cell divisions to complete differentiation and development of their reproductive structures.

In situ localisation of mRNA was performed to investigate the spatial expression pattern of *Pet cyc1* gene. The *Pet cyc1* S^{35} labelled antisense probe was hybridised to longitudinal sections of stage 4 and 5 ovaries. As shown in Figure 8.5, 2, the hybridisation signals were restricted to the ovules. The transcript was detected in only few cells within the ovule which were presumably the cells in the transition phase G2/M (Figure 8.5, 4). Because cell divisions are poorly synchronised and, in consequence, neighbouring cells are unlikely to be in the same phase of the cycle, isolated cells were labelled. There was no clear hybridisation signal on the other ovary structure, nor on the sections hybridised to the RNA probe used as a negative control (Figure 8.5, 1; 8.5, 3).

Fig. 8.5 see color plate at the end of the book.

In situ hybridisation showed that of *Pet cyc1* gene expression was exclusive to the ovule, the highly mitotic active structure of the ovary which will support macrogametogenesis, fertilisation and later embryogenesis.

Together, these results demonstrated that *Pet cyc1* gene is flower specific and temporally regulated during flower organ development .

Screening the library with a *Petunia cdc2* partial cDNA probe (Bergounioux et. al 1992) yielded 13 clones. The purified DNA from 7 of these clones enabled the cDNA encoded by two separate genes to be identified: *Pet cdc2a1* and *Pet cdc2a2*. DNA sequence revealed that the predicted protein products of *Pet cdc2a1* and *Pet cdc2a2* were 90% homologous. They were highly homologous with most plant cdc2 and carried a perfect PSTAIRE motif. We are currently studying the expression pattern of *Pet cdc2a1* and *cdc2a2* during flower development.

8.4 Seeking Petunia Meiotic Genes by Yeast Trans-complementation Analysis

There are many genetic and cytological descriptions of plant meiotic mutants which affect initiation of meiosis, pairing of homologous chromosomes, control of meiotic division and mitotic division within the gametophytes, but none has been molecularly characterised in plants, even though some genes which influence these processes have been cloned in other organisms (yeast, *Drosophila*).

Cells of the budding yeast *S. cerevisae* produce mitotic daughters whenever nutrients are plentiful. However, starvation causes cell growth and mitotic division to cease. One type of cell, the a/α diploid cell, then initiates a sporulation program that leads through meiosis to spore formation. Two nutritional conditions are required for sporulation. One is limitation for an essential nutrient. Nitrogen limitation causes efficient sporulation and is generally used in the laboratory to induce sporulation (Mitchell, 1994). The other condition is absence of a fermentable carbon source, such as glucose. Yeast mutants with specific meiotic defects for instance recombination, spore packaging, or reductional division have been identified. The yeast *spo13-1* mutant is of particular interest. The *spo13-1* mutation has a particularly intriguing meiotic phenotype with respect to its effect on the meiosis I chromosome distribution (Wang et al. 1987). The recessive *spo13-1* mutation causes cells to undergo an atypical meiosis consisting of one rather than two meiotic divisions. During this single division, *spo13-1* homozygous diploids execute some of the early landmark events of meiosis I, including premeiotic DNA synthesis as well as chromosome pairing and recombination. Proper segregation of homologs at meiosis I, which typically follows, and depends upon recombination, however, is eliminated. The cells do not terminate development at this stage but instead bypass completion of the first division and skip directly to a meiosis II-like division in which each chromosome divides equationally. Two diploid spores result from this unusual single division, which includes features of both meiosis I and meiosis II (Figure 8.6).

Fig. 8.6 see color plate at the end of the book.

In contrast to the genes which are responsible for both mitosis and meiosis, the *SPO13* gene is essential for normal meiosis and sporulation but is not essential for mitotic cell division. Such meiotic alterations that affect the female and lead to gametophytic apomixis (meiotic diplospory) have been described in plants.

We are using a trans-complementation approach to isolate meiotic genes in an attempt to understand the role meiotic genes play in regulating megasporogenesis and ovule development. We have constructed an at meiosis stages ovule cDNA library in yeast expression vector to transform different diploid yeast strain homozygous for the mutation *spo13*. Plant cDNA clones complementing yeast strains will be selected, purified and characterised.

Acknowledgements

We thank Judy Etherington for editorial assistance. This research was supported by the Ministero dell'Università e della Ricerca Scientifica e Tecnologica (funds 60% year 1997). Subject: Isolation and characterisation of *Petunia hybrida* cell-cycle genes (Project leader: Dott. Mario Pezzotti).

References

Angenent GC, Franken J, Bussscher M, Van Dijken A, Van Went JL, Dons H, Van Tunen AJ (1995) A novel class od MADS box genes is involved in ovule development in *Petunia*. Plant Cell 7:1569–1582

Bergounioux C, Perennes C, Hemerly A S, Qin LX, Sarda C, Inzè D, Gadal P (1992) A *cdc2* gene of Petunia hybrida is differentially expressed in leaves, protoplasts and during cell cycle phases. Plant Mol Biol 20:1121–1130

Bouman F (1984) The ovule. In: Johri BM (ed) Embryology of Angiosperm. Springer Verlag, New York, pp 123–158

Drews GN, Lee D, Christensen CA (1988) Genetic analysis of female gametophyte development and function. Plant Cell 10:5–18

Ferreira PCG, Hemerly AS, de Almeida Engler J, Van Montagu M, Engler G, Inzè D (1994) Developmental expression of the *Arabidopsis* cyclin gene *cyc1At*. Plant Cell 6:1763–1774

Hemerly AS, Ferreira PCG, de Almeida Engler J, Van Montagu M, Engler G, Inzè D (1993) Cdc2 expression in Arabidopsis is linked with competence for cell division. Plant Cell 5:1711–1723

Joppa LR, Williams ND, Mann SS (1987) The chromosomal location of a gene (*msg*) affecting megasporogenesis in durum wheat. Genome 29:578–581

Lindon RF (1990) Plant Development: The Cellular Basis (Winchester, MA: Unwin Hyman Inc.)

Mansfield SG, Briarty LG, Erni S (1990) Early embryogenesis in *Arabidopsis thailiana*

Minshull J, Blow JJ, Hunt T (1989) The role of cyclin synthesis and degradation in the control of maturation promoting factor activity. Nature 339:280–286

Mitchell AP (1994) Control of meiotic gene expression in *Saccharomyces cerevisae*. Microbiol. Review 58:56–70

Parrott WA, Hanneman RE Jr (1988) Megasporogenesis in normal and synaptic-mutant *(sy-2)* of *Solanum commersonni* Dun. Genome 30:536–539

Pines J, Hunter T (1989) Isolation of human cyclin cDNA; eveidence for mRNA and protein regulation in the cell cycle and for interaction with $p34^{cdc2}$. Cell 58:833–846

Redei G (1965) Non Mendelian megagametogenesis in *Arabidopsis*. Genetics 51:857–872

Reiser L, Fisher RL (1993) The ovule and the embryo sac. Plant Cell 5:5671–5675

Renaudin JP, Doonan JH, Freeman D, Hashimoto J, Hirt H, Inze D, Jacobs T, Kouchi H, Rouze P, Sauter M, Savoure A, Sorrell DA, Sundaresan V, Murray JA (1996) Plant cyclins: a unified nomenclature for A-, B- and D-type cyclins based on seuqence organization. Plant Mol Biol 32:1003–1018

Rodkiewicz B (1970) Callose in cell walls during megasporogenesis in angiosperm. Planta 93:39–47

Schneitz K, Hulskamp M, Pruitt RE (1995) Wild-type ovule development in *Arabidopsis thaliana*: A light microscope study of cleared whole-mount tissue. Plant J 7:731–749

Steeves TA, Sussex IM (1989) Patterns in Plant Development (Cambridge, UK: Cambridge University Press)

Wang HT, Franckman S, Kowalisyn J, Easton Esposito R, Elder R (1987) Developmental Regulation of SPO13, a gene required for separation of homologous chromosomes at meiosis I. Mol Cell Biol 7:1425–1435

Willemse MTM, Van Went JL (1984) The female gametophyte. In: Johri BM (ed) Embryology of Angiosperm. Springer Verlag, New York, pp 159–196

CHAPTER 9

Homomorphic Self-Incompatibility in Flowering Plants

D. de Nettancourt

Unité de Biochimie physiologique, Université Catholique de Louvain, 2, Place Croix du Sud, Bte 20, B-1348 Louvain-la-Neuve, Belgium
telephone: +32-2-3 58 22 27
fax: +32-2-3 58 22 27

Abstract. A description is provided of some of the most studied self-incompatibility (SI) systems which operate in homomorphic flowering plants. The control of recognition and rejection reactions is complex and little information is available on the nature of the gene products in the pollen which communicate with stylar RNases in gametophytic SI of the *Nicotiana* type or with stigmatic receptor kinases in certain sporophytic systems. The role of RNases and kinases against other intruders (pathogens or pollen tubes from related SC species) is discussed.

The very ancient origin of SI in certain families of plants has been demonstrated but the evolutionary processes which gave rise to the present diversity of rejection systems and the mechanisms involved in the generation of new SI alleles are not understood.

Some of the applications which may result from SI research are briefly discussed.

9.1 Early Work, Definition and Importance of SI

Self-incompatibility (SI) was first reported in *Verbascum phoeniceum (Scrophulariaceae)* by Kölreuter in 1764. Observations of the phenomenon in many different plant families by Darwin and scientists of the 19th and early 20th centuries were reviewed by Darwin (1892), Correns (1913), East and Park (1917), Stout (1920) and East (1924).

All early definitions implicitly or explicitly underlined the function of SI as an outbreeding mechanism but failed to establish a distinction from self-sterility or to exclude zygote lethality as a consequence of the process. Now that it is commonly established that SI is a prezygotic barrier between otherwise fertile pollen and pistil components of a same flower, the most appropriate definition appears to be that of Lundqvist (1964) who equated SI with "**the inability of a fertile hermaphrodite seed-plant to produce zygotes after self-pollination**".

SI occurs in half of the species in angiosperms (Darlington and Mather 1949). Brewbaker (1959), who confirmed this figure, found that SI operated in more than 70 families and in 250 genera of the 600 which he analysed. A list of SI wild species and cultivars which play an important role in agriculture or plant-breeding sciences has been prepared by de Nettancourt (1977).

9.2 Different SI Systems

The capacity of the fertile hermaphrodite seed-plant to reject its own pollen is inherited. One or several different di-allelic or poly-allelic genetic loci may be involved which control pollen-pistil recognition and rejection upon selfing and between plants with identical incompatibility phenotypes. Early models of the recognition phase were those of Lewis (1965), where pollen and pistil components are identical, and of Linskens (1965), who predicted complementary interactions.

The determination of the incompatibility phenotype in the pistil coincides with the opening of the flower and occurs in the stigma (**stigmatic incompatibility**), the style (**stylar incompatibility**) or, rarely, the ovary (**ovarian incompatibility**).

The incompatibility phenotype of the pollen grain (stigmatic incompatibility) or of the pollen tube (stylar incompatibility) can be determined by the diploid genotype of the mother plant (sporophytic SI, known as **SSI**) or by the haploid genotype of microspores, pollen grains or pollen tubes (gametophytic SI, known as **GSI**). It is generally considered that incompatibility substances are exine-held in SSI systems and intine-held in GSI species.

In certain SSI systems, the mechanism of pollen-pistil rejection is reinforced by a number of associated differences in floral morphology which accentuate the outbreeding potential of the self-incompatible plant and contribute to the prevention of self-pollination. Such systems are called **heteromorphic** (for information on features, distribution, mechanisms, evolution and key references, see Barrett and Cruzan 1994).

All other systems are **homomorphic** and constitute the subject of the present review. The gene (s) governing homomorphic SI in several sporophytic systems and in GSI are poly-allelic. While very large numbers (up to 200) of S-alleles have been found to segregate in certain colonies of clover (monofactorial GSI), the size of allelic series in populations usually fluctuates between 5 and 50 (Lawrence 1996). The number of different incompatibility pollen phenotypes in the population is always high in GSI systems governed by two or more complementary and unlinked polyallelic loci.

Through an analysis of about 1,000 SI and self-compatible (SC) species belonging to about 900 genera from 250 families, Heslop-Harrison and Shivanna (1977) were able to identify relationships between pollen type (**bi-nucleate** or **trinucleate**), stigma type (**wet** or **dry**), the site of rejection (**pollen** or **pollen tube**) and the SI system (gametophytic or sporophytic) involved. A summary of the relationships is provided in Table 9.1.

9.3 Recognition and Rejection Mechanisms

The function of SI in all plant families where it occurs is to ensure the prevention of consanguinity, the promotion of outbreeding and, perhaps also, as discussed elsewhere

Table 9.1. Usual features of self-incompatibility systems

	HOMOMORPHIC SI		HETEROMORPHIC SI
	Sporophytic	**Gametophytic**	**Sporophytic**
Reaction site	stigma	style	stigma, style or ovary
Stigma type	dry	wet	dry (or dry and wet)
Pollen cytology	tri-nucleate	bi-nucleate	tri-nucleate
Incompatibility loci	one or more	one or more	one or more
Incompatibility alleles at each locus	two, several or many	several or many	two
Determination of pollen SI phenotype	by SI gene(s) in tapetum or in PMCs	by SI gene(s) in pollen grains	probably not in tapetum

(de Nettancourt 1997), a protection against the pollen of self-compatible (SC) species. In contrast, differences appear between the molecular mechanisms which implement the SI function in the homomorphic systems most intensively studied: stigmatic SSI in the *Brassicaceae*, stigmatic GSI in the *Papaveraceae* and the *Graminaceae*, stylar GSI in the *Solanaceae*, *Rosaceae* and *Scrophulariaceae*.

9.3.1 Stigmatic SSI in *Brassica*

Two closely linked genes, SLG (S-locus glycoprotein) and SRK (S-receptor kinase), with multiple alleles organized in distinct haplotypes have been discovered in *Brassica*. The SLG gene was cloned by Nasrallah et al (1985), and sequenced by Nasrallah et al. (1985, 1987) and Takayama et al. (1987). The SRK gene was cloned and sequenced by Stein et al. (1991).

The data available on the SI mechanism (see Boyes and Nasrallah 1993; Nasrallah et al. 1994; Boyes et al. 1997) clearly indicate that the participation of SRK results, through one or more phosphorylation events, in the activation of a signalling pathway within the stigmatic papillae.

Several arguments, advanced by Nasrallah et al. 1994 and by Stein et al. 1996, confirm the hypothesis that the SLG product and SRK cooperate to function as molecules for the recognition of incompatible pollen on the stigmatic surface:

- the simultaneity of their transcription in open flowers,
- a close vicinity of sites (throughout the cell wall of the papillar cell for SLG; restricted to the area of the plasma membrane in the case of SRK)
- the apparent concerted evolution of the SLG/SRK gene pair within each haplotype.

Stein et al. (1996) consider that the presence of a thick cell wall in plants may represent an obstacle for plasma membrane-based signalling.The role of a soluble molecule like SLG could be, according to them, to shuttle a pollen ligand from the outer cell wall of the papilla and present it to the plasma membrane-localized SRK.

Very little information is available on the nature of this pollen-borne substance which is transferred for interaction with the kinase domain of SRK. The model currently considered (Boyes and Nasrallah 1995; Boyes et al. 1997) implies the delivery of a diffusible signal by an anther specific linked gene (SLA) from the pollen to the stigmatic surface, which activates the kinase receptor and initiates a signalling cascade leading to the inhibition of incompatible pollen. On the other hand, Dickinson's group (see Dickinson, this volume) observed in *Brassica* pollen a glycosylated polypeptide (7 kDA) which forms an intermediate product with stigmatic SLG proteins and may also play a role in the recognition phase.

Pollen recognition components are perhaps associated to members of the thioredoxin-h family which Bower and co-authors (1996) reported to interact specifically with the kinase domain of a *Brassica* S-locus receptor kinase. The standard role of thioredoxins is to modulate enzyme activity by reducing disulfide bridges (Holmgreen, cited by Li et al. 1994).

It will not be possible before the pollen component is identified to find out how pollen S-phenotypes are determined in SSI systems, from tapetum to pollen exine (Heslop-Harrison, 1968) or in PMCs (Pandey 1958).

9.3.2
Stigmatic GSI in the Grasses

The hypothesis that thioredoxins behave as effector molecules in the SI kinase cascade leading to the rejection of self-pollen in *Brassica* finds support in the discovery (Li et al. 1994) of a pollen gene in *Phalaris coelurescens* which co-segregates with the S-locus, one of the two loci governing stigmatic GSI in this species. A complete SC mutant of *Phalaris* produces an S-protein with reduced thioredoxin activity (Li et al. 1996). Sequence analyses of the probable S-gene in wild type and SC mutants suggest that it has two distinct sections, a variable N-terminus determining S-specificity and a conserved C-terminus with catalytic activity and strong similarities to thioredoxins (Li et al. 1994). It is proposed by Li et al. (1994, 1995) that the S and Z gene products in the pollen interact closely, possibly to form a dimer, within the germinating pollen, and that the stigmatic S and Z products are taken up by the pollen, also as dimers or as separate proteins. McCubbin and Kao (1996), in a review of this work suggested that S-thioredoxins modulate kinase activity in *Phalaris* SI and speculated, in the absence of information on mechanisms, that the Z gene-product may be the relevant kinase.

A somewhat similar mechanism may be operating in rye for which Wehling et al. (1994, 1995) have proposed the involvement of protein phosphorylation and Ca^{2+} as components of a signal transduction pathway. Their model predicts the presence of S and Z signal molecules acting as ligands in the stigmatic papillae. Recognition of the S and Z proteins by a membrane bound receptor would lead to the cytotoxic accumulation of Ca^{2+}, callose deposition or other events in cascade reaction, and the final inhibition of incompatible pollen. Wehling et al. (1994) found, in support of this model, that the phosphorylation of pollen proteins increased considerably in pollen exposed to eluates from incompatible stigmas. The sensitivity of the SI reaction of rye to protein kinase inhibitors and the Ca^{2+} dependence of the SI mechanism clearly confirmed, furthermore, the role of phosphorylation in the pollen-stigma incompatibility relationship of grasses.

9.3.3
Stigmatic GSI in *Papaver rhoeas*

A single polyallelic gene controls stigmatic GSI in the style-less *Papaver rhoeas* (Lawrence et al. 1978). The stigmatic component of the S-gene was first cloned by Foote et al. (1994). In vitro studies by Franklin-Tong et al. (1993, 1995) and by Rudd et al. (1996) have shown that the stigmatic S-protein specifically increases in incompatible pollen tubes the phosphorylation of a pollen phosphoprotein (p 26.1) which elicits, directly or indirectly, the arrest of pollen tube growth. The phosphorylation process appears to be Ca^{2+} and calmodulin dependent.

Rudd et al. (1996) consider that the SI mechanism operating in *Papaver* is conform to the model (see above) proposed for rye by Wehling et al. (1994). It would involve a Ca^{2+} dependent protein kinase requiring calmodulin-like domains, whose activation comprises an intracellular signal mediating the SI response in the pollen.

9.3.4
Stylar GSI

The first observations of GSI with the electron microscope were carried out in *Petunia hybrida* by van der Pluijm and Linskens (1966) and in *Lycopersicum peruvianum* by the

Siena group (de Nettancourt et al. 1973). They showed that SI pollen tubes undergo transformations, suggestive in certain cases, of a cessation of protein synthesis, which lead to the appearance of the endoplasmic reticulum as a whorl of concentric layers, the expansion of the tube outer wall and the degeneration of the cytoplasm often followed by the bursting of the tube tip.

It is now known, for at least three different families of angiosperms, *Solanaceae, Rosaceae* and *Scrophulariaceae,* that S-specific RNases participate directly in this rejection process and, probably also, in the initial recognition by the style of incompatible pollen tubes.

The cDNA of a stylar S-RNase allele was sequenced for the first time, in *Nicotiana alata,* by Clarke and co-workers (Anderson et al. 1986). McClure et al. (1989, 1990) provided evidence that the products of S-alleles were ribonucleases and that S-allele specific degradation of pollen RNA occurs in vivo after incompatible pollination. In transformation experiments, Lee et al. (1994), working with *Petunia inflata*, and Murfett et al. (1994), with the F1 and F2 populations of transgenic interspecific *Nicotiana* hybrids, established the final proof that S-proteins were necessary and sufficient for the rejection of self-pollen by the pistil. At the same time, Huang et al. (1994) demonstrated in *Petunia* the involvement of ribonucleases in the rejection process.

S-RNases comprise hyper-variable regions, shown in *Solanum chacoense* to include the recognition site (Matton et al. 1997), and conserved regions, which form the core of the protein. One of these regions (C4) was not found in the *Scrophulariaceae* (Xue et al. 1996) and in the *Rosaceae* (Broothaerts, 1995; Sassa et al. 1996).

The pollen component in stylar GSI is not known. Sassa et al. (1997) observed that the complete deletion of the S-RNase gene leads to the non-functionning of the S-allele in the style but does not affect the S-phenotype of the pollen tubes. Their conclusion that the S-locus region in the *Rosaceae* may contain several linked genes, including the S-RNase gene and an unidentified pollen S-gene, consolidates the old hypothesis of Linskens (1965) that the S products in pollen tubes and in style are encoded by different sub-parts of the S-locus.

As underlined by McCubbin and Kao (1966) much of the current work on stylar GSI centers upon the identification of the pollen component, its relations to the specificity determinants of stylar RNases and its function as the "gatekeeper" (receptor) of the tube entrance (allowing entry to "self-RNases" only) or the inhibitor, inside the tube, of S-RNases originating from different S-alleles.

There is also a need to test, as Lush and Clarke (1997) have recently done, the cytotoxic model of inhibition in the *Solanaceae* (McClure et al. 1990). The model assumes, from observations of *Tradescantia* pollen tubes in vivo (Mascarhenas 1993) and of *Nicotiana alata* pollen tubes in vitro (McClure et al. 1990), that there is no transcript of ribosomal genes in pollen tubes. A definitive, irreversible arrest of pollen tube protein synthesis and of tube growth can, therefore, be predicted to result from the degradation by ribonucleases of the limited amounts of rRNA initially present in the emerging tube. What Lush and Clarke (1997) did was to demonstrate, through grafting experiments involving incompatibly pollinated styles and compatible styles, that the incompatibility reaction, in at least a significant proportion of tubes, is reversible. The authors conclude that either the cytotoxic model, as it is now defined, is wrong or pollen tubes of *N. alata* are able to synthesise rRNA.

9.4 The Origin of SI Systems

9.4.1 References to Early Research

Whether SI originated once, and only once, as a primitive ancestral breeding system or arose de novo several times during the evolution of flowering plants has been extensively discussed in the fifties by Whitehouse (1950), Bateman (1952) and Stebbins (1957). At the time, many workers (Brewbaker 1957; Pandey 1958; Crowe 1964) shared the conviction of Whitehouse (1950) that polyallelic GSI was the ancestral form of SI and coincided with the expansion of angiosperms in the mid-Cretaceous. Self-compatibility, as it now occurs in wild plants, would in several instances represent a derived condition of SI in habitats initially hostile but succesfully colonized through a regime of cross-fertilization. A summary and discussion of arguments have been presented by de Nettancourt (1997).

9.4.2 The Old Age of S-Alleles

It is possible (for a review of approaches and of available methodologies, see Ioerger et al., 1991; Dwyer et al. 1991; Clark and Kao, 1994; Uyenoyama 1995; Richman and Kohn, 1996; Boyes and al. 1997) to quantify S-sequence divergences between plants, populations, species and families and to date S-alleles in reference to the branching of evolutionary trees. The results demonstrate that S-alleles originated many million years ago and outdated speciation in several of the plant families where SSI and GSI operate. Bell (1995), on the basis of other arguments, considers that GSI originated in *Gnetum*, a probable ancestor of the angiosperms, where interactions appear to take place between the pollen tube and the nucellus.

9.4.3 The Molecular Diversity of SI Systems

The ancient origin of GSI and SSI, such as it was studied in the *Solanaceae* and in the *Brassicaceae,* does not imply of course that one of the two systems gave rise, presumably through shifts in the timing of gene action, to the second one. In fact, the wide differences between the S-proteins discovered so far (for instance kinases and glycoproteins in the *Brassicaceae,* RNases in the *Solanaceae*, thioredoxins in the *Graminaceae*) are indications (Read et al. 1995) that SI arose several times independently in the past. At the same time it must not be forgotten that the molecular pictures in these families are still incomplete; as underlined by McCubbin and Kao (1996) in no single system have both the male and the female components been identified. It does seem, furthermore, in several cases of stigmatic SI, and not only in *Brassica*, that SI results from the activation of a receptor-like kinase. The molecular differences are perhaps less significant than initially foreseen (for an exchange of views, see Bell 1995 and Read et al. 1995).

9.4.4 Mutations at the S-Locus

It is often through the study of mutations that knowledge accumulates on gene structures, gene functions and structure-function relationships.

9.4.4.1
Mutations Leading to Function-Loss

The study of function loss, as resulting from deletions or specifically engineered at particular genetic sites, has greatly contributed to our understanding of the molecular biology of SI-systems (for examples of such achievements see de Nettancourt 1997).

9.4.4.2
The Generation of New S-Alleles

Artificial mutagens appear to lack the capacity to induce specificity changes at the S-locus. It is possible, however, as predicted by Fisher (1961), that new alleles remain incompatible, until further differentiation occurs, to each parental allele and thus escape detection in test-crosses and progeny tests. And indeed, working with the S11 and closely related S13 alleles of Solanum chacoense, Matton and the group of Cappadocia in Montreal (Matton, unpublished data) produced transgenic plants with S1l constructs in which key residues in the hypervariable region ranged from a complete conversion of S11 into S13 to dual specificities rejecting both S11 and S13 pollen.

Whatever the explanation may be for the absence of detectable new S-alleles after mutagenic treatment it is a fact that such new alleles appear sporadically in the progenies of inbred plants (see Table 9.2 for a list of cases). It is unlikely, since the new specificities usually originate from S-homogygous inbreds, that either equal crossing over or gene conversion may account for their occurrence (Lewis et al. 1988).

9.5
Relationships of S-Proteins to Other Plant Functions

9.5.1
Disease Resistance

Specific interactions such as they take place in SI between pollen and pistil also occur in the resistance-avirulence relationships of host-plants and pathogens (see Niks et al. 1993; Matton et al. 1994). The similarities between the two types of processes have led Dickinson (1994) to consider SI as a “social disease”.

Stein et al (1996) are of the opinion that the SRK gene originated from a group of genes which include the maize protein kinase gene (Walker and Zhang 1990), the *Arabidopsis* receptor kinase genes (Tobias et al. 1992; Dwyer et al. 1994) and *Arabidopsis* re-

Table 9.2. SI species where new alleles have been detected in inbred populations

	System	Source
Trifolium pratense	stylar GSI	Denward 1963 Anderson et al. 1974
Lycopersicum peruvianum	stylar GSI	de Nettancourt and Ecochard 1969 de Nettancourt et al. 1971 Hogenboom 1972 a, 1972 b
Nicotiana alata	stylar GSI	Pandey 1970
Solanum chacoense	stylar GSI	Saba-El-Leil 1974
Raphanus sativus	stigmatic SSI	Lewis et al. 1988

ceptor-like kinase genes (Walker 1993). These relatives of SRK, expressed in families as different as the *Brassicaceae* and the *Graminaceae*, are not involved in SI, and perhaps not even in the biology of the gametophyte, but presumably carry out cell-cell communication functions in vegetative tissues. Pastuglia et al. (1997), who reported, in the *Brassicaceae*, the rapid induction by wounding and bacterial infection of an S-gene family receptor-like kinase gene, reviewed, at the same time, other examples of the participation of protein kinases in the disease resistance of plants.

RNases have also been found to participate in plant defense (Green 1994). They appear, in particular, to express a multiple function in the protection of the pistil against a wide range of invaders, including pathogens and unwanted pollen tubes (from the same plant or, as outlined below, from related SC species).

A general increase in RNase activity often occurs in diseased plants which does not seem to represent a direct response to attack by a pathogen but a secondary effect of wounding or senescence (for a review, see Green 1994).

9.5.2 Unilateral Incompatibility

The involvement of SI in unilateral incompatibility (UI) between SI species, as pistillate parents, and SC relatives, as pollen sources, has been the subject of numerous studies and discussions (for a review, see de Nettancourt, 1977, 1997).

Murfett et al. (1996), through manipulations and transformation of different SI and SC *Nicotiana* species and their hybrids, produced plants, in SI or SC backgrounds, which expressed different concentrations of S-RNases. They were able with such material to test, through a series of different cross-pollinations, the involvement of S-RNases, and the consequences of their absence, upon the manifestation of UI. The results showed that S-RNases (and therefore the S-locus) were responsible in *Nicotiana* for the rejection by SI styles of the pollen tubes of SC species.

9.6 Contributions of SI Research to Plant Breeding and Biotechnology

9.6.1 Plant Breeding

SI is either an inconvenience or an advantage to the plant breeder. Any research directed towards agricultural objectives must of course be carefully adapted to the specificities of the material, the requirements of the market and the particular needs of the plant-breeder.

Several different types of information and modifications arising from SI research (for a review, see de Nettancourt 1977) may contribute to plant-breeding. Of these, one of the simplest and, to date, most useful accomplishments is the induction, through mutagenic treatment or genetic engineering, of the self-compatibility character in self-incompatible cultivars. Such a change may render possible a re-orientation of selection and breeding strategies or an adaptation to the absence, in a polluted environment, of cross-pollinating insects.

Far more difficult to achieve is the reverse operation, namely the incorporation of SI in SC cultivars.The transfer and expression of the *Brassica* SRK gene in tobacco stigmas and styles (see Moore and Nasrallah 1990; Kandasamy et al. 1990; Susuki et al. 1996;

Stein et al. 1996) and of S-ribonucleases from *N.alata* into SC *Nicotiana* species (see Murfett et al. 1996) are possibly the first steps towards the transformation, for hybrid seed production or other purposes, of autogamous crop-plants into cross-fertilizers. The objective could be difficult to reach, however, in certain material (for examples, see de Nettancourt 1997), because the full expression of the S-gene complex may possibly also require the introduction of unlinked S-related genes necessary for a proper pollen-pistil incompatibility relationship.

9.6.2
Biotechnology

No breakthrough has occured. The pollen-pistil reaction, particularly if it can be induced and observed, as in *P. rhoeas* for instance (Franklin-Tong et al. 1988; Foote et al 1994), in a reliable in vitro system, constitutes however a choice material for the analysis of a polymorphic master recognition genetic complex and the study of cell-cell communication, discrimination between self and non-self and activation of transduction cascades. It is perhaps through the systematic understanding of the structures and processes involved that progress will be made towards the control and exploitation, as model or tool, of complex molecular signalling in higher eukaryotes.

9.7
Conclusions

Considerable knowledge has been acquired on the morphology, genetics and diversity of SI in flowering plants. Homomorphic SI appears to form an integrated part of a general system of cell recognition, of very ancient origin, with a wide range of different signalling functions.

Current data on the SI mechanism itself are however incomplete and fragmentary. For instance, no information is available on the identity of the pollen component (perhaps a thioredoxin) which acts in certain stigmatic systems as a ligand to switch on, through the activation of a kinase-like domain, a signalling pathway leading to the rejection of incompatible pollen. Similarly, one does not know how stylar S-RNases in GSI systems are identified at their entry into incompatible tubes and why the pollen of related SC species display the same sensitivity to their presence as selfed SI pollen.

Many other questions are unanswered with regard, for example, to the possibility that SI arose de novo several times during evolution or to the processes leading, to the apparition of new alleles in S-homozygous inbreds.

As far as agricultural applications are concened, it remains to find out, among many other problems, if the introduction of SI into SC cultivars will require the transfer of the cohort of related genes which appear to be needed in some species for the functional expression of the SI character.

References

Anderson MA, Cornish EC, Mau S-L, Williams EG, Hoggart R, Atkinson A, Bönig I, Greg B, Simpson R, Roche PJ, Haley JD, Penschow JD, Niall HD, Tregear GW, Coghlan JP, Crawford RJ, Clarke AE (1986) Cloning of cDNA for a stylar glycoprotein associated with expression of self-incompatibility in *Nicotiana alata.* Nature 321:38–44

Anderson MK, Taylor NL, Duncan JF (1974) Self-incompatibility, genotype identification and stability as influenced by inbreeding in red clover (*Trifolium pratense L.).* Euphytica 23:140–148

Barrett SCH, Cruzan MB (1994) Incompatibility in heterostylous plants. In: Williams EG, Clarke AE, Knox RB (eds) Genetic control of self-incompatibility and reproductive development in flowering plants. Kluwer Academic Publishers, Dordrecht Boston London, pp 189–219

Bateman AJ (1952) Self-incompatibility systems in angiosperms. I. Theory. Heredity 6:285–310

Bell PR (1995) Incompatibility in flowering plants: adaptation of an ancient response. Plant Cell 7:5–16

Bower MS, Matias D, Fernandes-Carvalho, Mazzurco M, Gu T, Rothstein SJ, Goring DR (1996) Two members of the thioredoxin-h family interact with the kinase domain of a *Brassica* S-locus receptor kinase. Plant Cell 8:1641–1650

Boyes DC, Nasrallah JB (1993) Physical linkage of the SLG and SRK genes at the self-incompatibility locus of *Brassica oleracea.* Mol Gen Genet 236:369–373

Boyes DC, Nasrallah JB (1995) An anther-specific gene encoded by an S locus haplotype of *Brassica* produces complementary and differentially regulated transcripts. Plant Cell 7:1283–1294

Boyes DC, Nasrallah ME, Vrebalov J, Nasrallah JB (1997) The self-incompatibility (S) haplotypes of *Brassica* contain highly divergent and rearranged sequences of ancient origin. Plant Cell 9:237–247

Brewbaker JL (1957) Pollen cytology and incompatibility systems in plants. J Hered 48:217–277

Brewbaker JL (1959) Biology of the angiosperm pollen grain. Ind J Genet Plant Breed19:121–133

Broothaerts WJ, Janssens GA, Proost P, Broekaert WF (1995) cDNA cloning and molecular analyses of two self-incompatibility alleles from apple. Plant Mol Biol 27:499–511

Clark AG, Kao TH (1994) Self-incompatibility: theoretical concepts and evolution. In: Williams EG, Clarke AE, Knox RB (eds) Genetic control of self-incompatibility and reproductive development in Correns C (1913) Selbsterilität und Individualstoffe. Biol Centr 33:389–423

Crowe (1964) The evolution of outbreeding in plants. I. The angiosperms. Heredity 19:435–457

Darlington CD, Mather K (1949) The elements of genetics. Allen and Unwin Ltd

Darwin C (1982) De la variation des animaux et des plantes à l'état domestique. Vols I and II transl 2nd ed London J Murray 1980

Denward T (1963) The function of the incompatibility alleles in red clover (*Trifolium pratense* L.). Hereditas 49:189–334

Dickinson H (1994) Simply a social disease? Nature 367:517–518

Dwyer KG, Balent MA, Nasrallah JB, Nasrallah ME (1991) DNA sequences of self-incompatibility genes from B*rassica campestris* and *B.oleracea*: polymorphism predating speciation. Plant Mol Biol 16: 481–486

East EM (1924) Self-sterility. Bibliographica Genetica 5:331–370

East EM, Park JB (1917) Studies on self-sterility. I. The behavior of self-sterile plants. Genetics 2: 505–609

Fisher RA (1961) A model for the generation of self-sterility alleles. J Theoret Biol 1:411–414

Foote HCC, Ride JP, Franklin-Tong VE, Walker EA, Lawrence MJ, Franklin FCH (1994) Cloning and expression of a distinctive class of self-incompatibility (S-) gene from *Papaver rhoeas* L. Proc Natl Acad Sci USA 91:2265–2269

Franklin-Tong VE, Lawrence MJ, Franklin FCH (1988) An *in vitro* bioassay for the stigmatic product of the self-incompatibility gene in *Papaver rhoeas* L. New Phytol 110:109–118

Franklin-Tong VE, Ride JP, Read ND, Trewavas AJ, Franklin FCH (1993) The self-incompatibility response of *Papaver rhoeas* is mediated by cytosolic free calcium. Plant J 4:163–177

Franklin-Tong VE, Ride JP, Franklin FCH (1995) Recombinant stigmatic self-incompatibility protein elicits a Ca^{2+} transient in pollen of *Papaver rhoeas.* Plant J 8:299–307

Goring DR, Rothstein SJ (1992) The S-locus receptor kinase gene in a self-incompatible *Brassica napus* line encodes a functional serine/threonine kinase. Plant Cell 4:1273–1281

Green PJ (1994) The ribonucleases of higher plants. Annu Rev Plant Physiol Plant Mol Biol 45:421–445

Heslop-Harrison J (1968) Pollen wall development. Science 161:230–237

Heslop-Harrison Y, Shivanna KR (1977) The receptive surface of the angiosperms stigma. Ann Bot 41: 1233–1258

Hogenboom NG (1972a) Breaking breeding barriers in *Lycopersicon*; 2 Breakdown of self-incompatibility in *L. peruvianum* (L.) Mill. Euphytica 21:228–243

Hogenboom NG (1972b) Breaking breeding barriers in *Lycopersicon*.3. Inheritance of self-compatibility in *L. peruvianum (*L.) Mill. Euphytica 21:244–256

Huang S, Lee HS, Karunanandaa B, Kao Th (1994) Ribonuclease activity of *Petunia inflata* S-proteins is essential for rejection of self-pollen. Plant Cell 6:1021–1028

Ioerger TR, Clark AG, Kao TH (1990) Polymorphism at the self-incompatibility locus in *Solanaceae* predates speciation. Proc Nat Acad Sci USA 87:9732–9735

Ioerger TR, Gohlke JR, Xu B, Kao TH (1991) Primary structural features of the self-incompatibility protein in *Solanaceae*. Sex Plant Reprod 4:81–87
Kandasamy MK, Paolillo CD, Faraday JB, Nasrallah JB, Nasrallah ME (1989) The S-locus specific glycoproteins of *Brassica* accumulate in the cell wall of developing stigma papillae. Devel Biol 34:462–472
Kölreuter JG (1764) Vorläufige Nachricht von einigen das Geschlecht der Pflanzen betreffenden Versuchen und Beobachtungen, nebst Fortsetzungen, 1, 2 u. 3, pp. 266. Ostwald's Klassiker, Nr. 41. Leipzig: Engelmann, 1761–1766
Lawrence MJ (1996) Number of incompatibility alleles in clover and other species.Heredity 76:610–615
Lawrence MJ, Afzal M, Kenrick J (1978) The genetical control of self-incompatibility in *Papaver rhoeas* L. Heredity 40:239–285
Lee HS, Huang S, Kao Th (1994) S proteins control rejection of incompatible pollen in *Petunia inflata.* Nature 367:560–563
Lewis D (1965) A protein dimer hypothesis on incompatibility. Proc 11th Intern Cong Genetics, The Hague 1963. In: Genetics Today, Geerts SJ (ed) 3:656–653
Lewis D, Verma SC, Zuberi MI (1988) Gametophytic-sporophytic in the *Cruciferae* – *Raphanus sativus.* Heredity 61:355–366
Li X, Nield J, Hayman D, Langridge P (1994) Cloning a putative self-incompatibility gene from the pollen of the grass *Phalaris coerulescens.* Plant Cell 6:1923–1932
Li X, Niedl J, Hayman D, Langridge P (1995) Thioredoxin activity in the C terminus of *Phalaris* S-proteins. Plant J 8:133–138
Li X, Nield J, Hayman D, Langridge (1996) A self-fertile mutant of *Phalaris* produces an S protein with reduced thioredoxin activity. Plant J 10 (3):505–513
Linskens FH (1965) Biochemistry of incompatibility. Proc 11th Intern Cong Genet, The Hague 1963. In: Genetics Today, Geerts SJ (ed) 3:621–636
Lundqvist A (1964) The nature of the two-loci incompatibility system in grasses. I. The hypothesis of a duplicative origin. Hereditas 48:153–168
Lush WM, Clarke AE (1997) Observations of pollen tube growth in *Nicotiana alata* and their implications for the mechanism of self-incompatibility. Sex Plant Reprod 10:27–35
Mascarhenas JP (1993) Molecular mechanisms of pollen tube growth and differentiation. Plant Cell 5: 1303–1314
Matton DP, Nass N, Clarke A, Newbigin Ed (1994) Self-incompatibility: how plants avoid illegitimate offspring. Proc Natl Acad Sci USA 91:1992–1997
Matton DP, Maes O, Laublin G, Xike K, Bertrand C, Morse D, Cappadocia M (1997) Hypervariable domains of self-incompatibility RNases mediate allele-specific pollen recognition. Plant Cell 9: 1757–1766
McClure BA, Haring V, Ebert PR, Anderson MA, Simpson RJ, Sakiyama F, Clarke AE (1989) Style self-incompatibility products of *Nicotiana alata* are ribonucleases. Nature 342:955–957
McClure BA, Gray JE, Anderson MA, Clarke AE (1990) Self-incompatibility in *Nicotiana alata* involves degradation of pollen rRNA. Nature 347:757–760
McCubbin AG, Kao TH (1996) Molecular mechanisms of self-incompatibility. Current Opinion Biotech 7:155–160
Moore HM, Nasrallah JB (1990) A *Brassica* self-incompatibility gene is expressed in the stylar transmitting tissue of transgenic tobacco. Plant Cell 2:29–38
Murfett J, Atherton T, Mou B, Gasser C, McClure BA (1994) S-RNase expressed in transgenic *Nicotiana* causes S-allele-specific pollen rejection. Nature 367:563–566
Murfett J, Strabala TJ, Zurek DM, Mou B, Beecher B, McClure B (1996) S-RNase and interspecific pollen rejection in the genus *Nicotiana*: multiple pollen-rejection pathways contribute to unilateral incompatibility between self-incompatible and self-compatible species. Plant Cell 8:943–958
Nasrallah JB, Kao T-h, Goldberg ML, Nasrallah ME (1985) A cDNA clone encoding an S-specific glycoprotein from *Brassica oleracea.* Nature 318:263–267
Nasrallah JB, Kao TH, Chen CH, Goldberg ML, Nasrallah ME (1987) Amino acid sequences of glycoproteins encoded by three alleles of the S-locus of *Brassica oleracea.* Nature 213:617–619
Nasrallah JB, Rundle SJ, Nasrallah ME (1994) Genetic evidence for the requirement of the *Brassica* locus receptor kinase gene in the self-incompatibility response. Plant J 5:73–384
Nettancourt D de (1977) Incompatibility in angiosperms. Springer, Berlin Heidelberg New york
Nettancourt D de (1997) Incompatibility in angiosperms. Sex Plant Reprod 10:185–199
Nettancourt D de, Ecochard R (1969) New incompatibility specificities in the M3 progeny of a clonal population of *L. peruvianum*. Tomato Genet Coop Rep 19:16–17
Nettancourt D de, Ecochard R, Perquin MDG, Drift T van der, Westerhof M (1971) The generation of new S-alleles at the incompatibility locus of *L. peruvianum* L Mill. Theor Appl Genet 41:120–129
Nettancourt D de, Devreux M, Bozzini A, Cresti M, Pacini E, Sarfatti G (1973) Ultrastructural aspects of the self-incompatibility mechanism in *Lycopersicum peruvianum L* MillL. J Cell Sci 12:403–419
Niks RE, Ellis PR, Parlevliet JE (1993) Resistance to parasites. In: Hyward MD, Bosemark NO, Romagosa I (eds) Plant Breeding: principles and prospects. Chapman & Hall, London, pp 422–442
Pandey KK (1958) Time of S-allele action. Nature 181:1220–1221

Pandey KK (1970) Self-incompatibility alleles produced through inbreeding. Nature 227:689–690
Pastuglia M, Roby D, Dumas Ch, Cock JM (1997) Rapid induction by wounding and bacterial infection of an S gene family receptor-like kinase gene in *Brassica oleracea.* Plant Cell 9:49–60
Pluijm J van der, Linskens F (1966) Feinstruktur der Pollen Schläuche im Griffel von *Petunia.* Züchter 36:220–224
Read SM, Newbigin E, Clarke AE, McClure BA, Kao TH (1995) Disputed ancestry: comments on a model for the origin of incompatibility in flowering plants. Plant Cell 7:661–665
Richman AD, Kohn JR (1996) Learning from rejection: the evolution biology of single locus incompatibility. Trends Ecol Evol 11:497–502
Rudd JJ, Franklin FCH, Lord JM, Franklin-Tong VE (1996) Increased phosphorylation of a 26-kD pollen protein is induced by the self-incompatibility response in *Papaver rhoeas.* Plant Cell 8:713–724
Saba-El-Leil RK, Rivard S, Morse D, Cappadocia M (1994) The S11 and S13 self-incompatibility alleles in *Solanum chacoense* Bitt. are remarkably similar. Plant Mol Biol 24:571–583
Sassa H, Nishio T, Kowyama Y, Hirano H, Koba T, Ikehashi H (1996) Self-incompatibility (S) alleles of the *Rosaceae* encode members of a distinct class of the T2/S ribonuclease superfamily. Mol Gen Genet 250:547–557
Sassa H, Hirano H, Nishio, T, Koba T (1997) Style-specific self-compatible mutation caused by deletion of the S-RNase gene in Japanese pear (*Pyrus serotina*). Plant J 12(1):223–227
Stebbins GL (1957) Self-fertilization and population variability in the higher plants.Am Naturalist 91: 337–354
Stein JC, Howlett B, Boyes DC, Nasrallah ME, Nasrallah JB (1991) Molecular cloning of a putative receptor protein kinase of *Brassica oleracea.* Natl Acad Sci USA 88:8816–8820
Stein JC, Dixit R, Nasrallah ME, Nasrallah JB (1996) SRK, the stigma-specific S locus receptor kinase of B*rassica* is targeted to the plasma membrane in transgenic tobacco. Plant Cell 8:429–445
Stout AB (1920) Further experimental studies on self-incompatibility in hermaphroditic plants. J Genetics 9:85–129
Susuki G, Watanabe M, Toriyama K, Isogai A, Hinata K (1996) Expression of SLG9 and SRK9 in transgenic tobacco. Plant Cell Physiol 37:866–869
Takayama S, Isogai A, Tsukamoto C, Ueda Y, Hinata K, Susuki A (1987) Sequences of S-glycoproteins, products of the *Brassica campestris* self-incompatibility locus. Nature 326:102–105
Tobias CM, Howlett B, Nasrallah JB (1992) An *Arabidopsis thaliana* gene with sequence similarity to the S-locus receptor kinase of *Brassica oleracea.* Plant Physiol 99:284–290
Uyenoyama MK (1995) A generalized least-squares estimate for the origin of sporophytic self-incompatibility. Genetics 139:975–992
Walker JC (1993) Receptor-like protein kinase genes of *Arabidopsis thaliana.* Plant J 3:451–456
Walker JC, Zhang R (1990) Relationship of a putative receptor protein kinase from maize to the S-locus glycoproteins of *Brassica.* Nature 345:743–746
Wehling P, Hackauf B, Wricke G (1994) Phosphorylation of pollen proteins in relation to self-incompatibility in rye (*Secale cereale* L). Sex Plant Reprod 7:67–75
Wehling P, Hackauf B, Wricke G (1995) Characterization of the two factors self-incompatibility system in S*ecale cereale* L. In: Kuck G, Wricke G (eds) Genetic mechanisms for hybrid breeding. Adv Plant Breed 18:149–161
Whitehouse HLK (1950) Multiple-allelomorph incompatibility of pollen and style in the evolution of the angiosperms. Ann Bot New Series 14:198–216
Xue Y, Carpenter R, Dickinson HG, Coen ES (1996) Origin of allelic diversity in *Antirrhinum* S-locus Rnases. Plant Cell 8:805–814

CHAPTER 10

Cell Death of Self-Incompatible Pollen Tubes: Necrosis or Apoptosis?

A. Geitmann

PCM, Wageningen Agricultural University, Arboretumlaan 4, 6703 BD Wageningen, The Netherlands
e-mail: anja.geitmann@guest.pcm.wau.nl
telephone: +31-317-484864
fax: +31-317-485005

Abstract. During self-incompatibility reaction in flowering plants a cell-cell interaction takes place, which results in the rejection and eventual death of the incompatible pollen. An overview of different plant families demonstrates that along with the existence of various mechanisms of recognition in self-incompatibility there are manifold ways of pollen tubes to die. Depending on the plant species different kinds of cell degeneration, necrosis or programmed cell death, can be observed to determine the fate of incompatible pollen tubes.

10.1 Introduction

The phenomenon self-incompatibility (SI) was described as early as in the latter half of the nineteenth century by Charles Darwin who mentions in his volume "The Effects of Cross and Self Fertilization in the Vegetable Kingdom" that plants of some species are "completely sterile to their own pollen but fertile with that of any other individual of the same species", which he considered "one of the most surprising facts which I have ever observed" (Darwin 1862; 1877). Stout (1917) was the first to use the term SI for the description of this phenomenon. De Nettancourt (1977) defines SI as "inability of a fertile hermaphrodite seed-plant to produce zygotes after self-pollination". In plants with homomorphic self-incompatibility – plants that have no physical barrier to self-fertilization – a "self" pollen grain is not able to germinate or to form a normal pollen tube that reaches the ovary. Depending on the species the inhibition process takes place somewhere between the attachment of the pollen grain to the stigma, and the penetration of the pollen tube through the style towards the ovary.

Pollen tube growth *in situ* is in fact rather unusual behavior for plant cells, that is, as the pollen tip grows it actually has to migrate intercellularly through the tissues of the stigma and style of the receptive flower in order to arrive at its ultimate destination, the ovary. Morphologically this behavior resembles the penetration and invasion of plant tissues by fungal hyphae. Similar to hyphal growth the penetration of the pollen tube in the case of dry stigmas is facilitated by the secretion of a cutinase at the growing tip (Heslop-Harrison 1975; Heslop-Harrison et al. 1977; Hiscock et al. 1994). In plants with solid style the pollen tube has to force its way not only through the stigmatic but also through the stylar transmitting tissues during this process. In many cases this process results in the death of these tissue cells presumably by programmed cell death, the degeneration products of which provide nutrients necessary for the elongation of the pollen tube (Wilms 1980). This process of siphonogamy is rather particular in the plant kingdom and inevitably involves the direct contact between two organisms with individual genomic characteristics. As proposed by Linskens (1975) during this cell-cell interaction a recognition process takes place. Subsequently either a compatible reaction occurs, i.e. the pollen tube is allowed to pursue growth towards the ovary and is provided with nutrients on its way, or an incompatible reaction causes the pollen tube to arrest growth in general resulting in degeneration and eventually death of the tube. This recognition event has been compared to other mechanisms in plants where discrimination of compatible and incompatible takes place, for example during plant-pathogen interactions (Hodgkin et al. 1988). In this system an incompatible reaction causes the plant cells around the infection site to undergo cell death thus preventing further colonization of the plant by the pathogen (hypersensitive response). The recognition mechanism between pollen and pistil has also been compared to the immune system in animals. However, an important difference consists in the fact that the SI response represents a rejection of "self", whereas the principle of the animal immune response is based on the recognition and rejection of "non-self" (Burnet 1971).

Intensive research over the last decades has clarified many aspects of SI, but it has revealed even more questions which are still unanswered. One point of interest has been the morphological description of the inhibition process (Linskens and Esser 1956; van der Pluijm and Linskens 1966; Tupý 1959; de Nettancourt et al. 1973a; 1973b; 1974; Cresti et al. 1979; Herrero and Dickinson 1981; Pacini 1982; Heslop-Harrison 1983 to name only a few) and many authors discussed the morphological parallels between pollen tube penetration and recognition in the stigma and host-pathogen interaction in the case of pathogen infections of plants (Teasdale et al. 1974; Hogenboom 1983; Dumas and Knox 1983; Hodgkin et al. 1988). However, one of the major differences between these two systems consists in the fact that SI in pollen tube growth is a reaction strictly confined to the individual pollen tube, thus allowing compatible tubes to grow uninhibited in mixed pollinations, whereas growth of incompatible tubes is arrested (Sarker et al. 1988; Dickinson 1995). In contrast, during plant-pathogen interactions the local hypersensitive reaction to an incompatible pathogen prevents subsequent infection by a compatible pathogen, thus demonstrating a more global character of the rejection mechanism.

During the last two decades concentration was shifted to the understanding of the genetical control of the recognition process. The efforts made in this domain demonstrate the existence of a variety of controlling mechanisms serving the purpose of inhibition of inbreeding. The discovery of this variety of SI systems, only a small portion of which has been investigated so far, has initiated a lively discussion concerning the evolution of angiosperms (Pandey 1960; 1980; 1981; Bell 1995a; 1995b; Read et al. 1995).

Some of the recent contributions to SI research focused on the correlation between genetical control and biochemical mechanism of the inhibition process proper. The question arose as to whether the process of pollen tube inhibition and degradation shows typical features of apoptotic or necrotic cell death, and if so, whether there exists a common principle for all types of SI mechanisms. Especially with programmed cell death, apoptosis and necrosis being issues of current interest in cell biology it seems interesting to investigate whether incompatible pollen tubes might provide a representative example for one or several of these categories of cell death.

In order to discuss SI from the point of view of the degenerating pollen tube, the term cell death and its phenomena have to be clarified. As far as the definition and distinction of the different categories of cell death is concerned there is considerable confusion in the animal literature and the situation has not been improved by the recent contribution of observations made in plant systems. In the following an attempt is made to sum up the most important observations and clarify as far as possible those definitions of various ways of cell death which are relevant in this case.

10.2 Different Ways of Cells to Die

Following the first detailed description by Kerr et al. (1972) cell death has become an issue of enormous interest in the past two decades. Due to the variety of systems in which cell death occurs, the precise nomenclature of the terms "programmed cell death", "apoptosis" and "necrosis" is somewhat confusing. The term apoptosis, often inappropriately used as synonymous of programmed cell death, defines a type of cell death distinct from necrosis, which is characterized as unscheduled or accidental lysis. The following morphological criteria are characteristic for apoptosis in animal cells: condensation of the chromatin on the inner face of the nuclear membrane, cell shrinkage, membrane blebbing, nuclear and cytoplasmic condensation, fragmentation of the cell with formation of membrane-bound acidophilic globules (apoptotic bodies) containing nuclear material and condensed cytoplasm. Apoptosis is associated with *de novo* synthesis of Ca^{2+}- or Mg^{2+}-dependent endonucleases, which in turn cause cleavage of DNA at the linker regions between nucleosomes, thus resulting in the formation of DNA fragments multiple of 180 bp. This process can be prevented by inhibitors of the endonucleolytic DNA cleavage (Escargueil-Blanc et al. 1994). Necrosis on the other hand is a form of cell death during which DNA breakdown occurs randomly and chromatin disappears progressively (Wyllie et al. 1980). In some cases clumping of chromatin occurs as well as karyolysis. During necrosis the cells and organelles swell and autolysis takes place due to activation and/or release of lysosomal and proteolytic enzymes, a process which can be prevented by selective proteinase inhibitors (Escargueil-Blanc et al. 1994). The plasma membrane as well as organellar membranes are ruptured, in some cases electrondense calcium phosphate deposits are observed. (Buja et al. 1993). However, the principle difference between the two ways to die, necrosis and apoptosis, consists in the way the process is initiated. Whereas necrosis is basically the result of a cytotoxic activity of an external factor, thus being a rather passive way to die, apoptosis is a form of programmed cell death. In the latter case a genetically programmed process of cell degeneration is triggered upon an external or internal signal, requiring *de novo* gene expression and thus being an energy consuming process. In fact, the presence of ATP is a necessary prerequisite for apoptosis. Apoptosis has therefore been termed a "suicidal" way for the cell to die, whereas necrosis has been equated with "murder" (Dangl et al. 1996).

Apoptosis, however, is not the only way for a cell to commit suicide. At least two distinctive ways of programmed cell death have been distinguished in animal cells: Whereas the characteristic features of apoptosis mentioned above occur for example in T-cell hybridoma cells and in various other cell types affected by pathogens, a different way of programmed cell death has been observed in intersegmental muscle cells. In these cells the nuclei round up, the nuclear membrane swells, chromatin becomes pyknotic, but no deposition of electron-dense chromatin along the inner surface of the nuclear membrane was observed (Schwartz et al. 1993). However, as during apoptosis, *de novo* gene expression is a necessary requirement for these cells to die, thus indicating an active suicidal process. Yet another way of programmed cell death in animals was described by the term oncosis (Trump et al. 1997). Cells undergoing oncosis form cytoplasmic organelle-free blebs, they show chromatin clumping, dilatation of the ER and Golgi, and mitochondrial condensation followed by swelling. These different observations indicate that the term programmed cell death comprises various manners of cells actively killing themselves, only one of them being defined under the term apoptosis.

There exist two fundamental differences in programmed cell death between plants and animals: in plants the dead cells can in some cases become part of the very architecture of the plant performing crucial functions after their death. This is the case in xylogenesis, when xylem cells undergo autolysis as they differentiate and mature, the cell wall being the only remaining part of the cell, performing essential functions for the plant after cell death. Another difference between plants and animals consists in the fact that plant cells do not engulf their degrading neighbor cells which have undergone programmed cell death, as is the case in some animal systems (Greenberg 1996).

Both in plants and in animals further distinctions can be made in order to understand the various conditions under which a cell initiates the process of programmed cell death and thus eliminates itself. Three major categories have been characterized by Farber (1994):

1. Cell death can be developmentally programmed and thus occur at predictable times and places during normal development. In plants this kind of death occurs during xylogenesis, senescence and ripening. It is also involved in sexual reproduction, for example during sex determination (e.g. selective killing of female reproductive primordia), during the breakdown of the tapetum, during anther dehiscence, during suspensor degeneration and upon megaspore degeneration in the ovule (Greenberg 1996; Jones and Dangl 1996).

2. Alternatively programmed cell death can be elicited upon a physiological trigger. This occurs mainly in adult organisms during regression of hyperplastic organs. The death program does not start compulsorily, only if an excess of cells requires the reduction in numbers.

3. Another category of programmed cell death is biochemically induced. The application of pathological stimuli of chemical, physical or biological nature can trigger the initiation of a genetically programmed suicide process. Cell death is in this case the response to the toxicity of an external causative agent. The degeneration of the cells of the stylar transmitting tissue upon penetration of compatible pollen tubes has been claimed to be of apoptotic nature thus representing an example of an external trigger of cell death in plants (Greenberg 1996).

A phenomenon, the nature of which has been investigated and discussed intensively is the hypersensitive reaction occurring upon microbe-plant interactions as an expression of resistance, first described by Stakman (1915). On the one hand the hypersensitive reaction has been claimed to be necrotic, e.g. in the case of lettuce infected by *Pseudomonas syringae*. Bestwick et al. (1995) describe the degenerating cells affected by this pathogen to show swelling of ER and mitochondria, organelle disruption, and vacuolation. The authors put emphasis on the observed cell membrane damage which is attributed to the necrotic death of the cells (Bestwick et al. 1995; Fett and Jones 1995). In other systems the hypersensitive reaction has been claimed to be of apoptotic nature due to its apparent mechanistic similarities to animal apoptosis: DNA breaking at 3′OH ends, blebbing of the plasma membrane, nuclear and cytoplasmic condensation (Dangl et al. 1996). Also the occurrence of an oxidative burst supposedly indicates that these systems represent examples of programmed cell death. These results seem contradictory, but it has in fact been concluded that both kinds of death can occur in the same cell type, apoptosis preceding the final necrotic breakdown of the cells (Columbano 1995). Other studies showed that a certain cell type can undergo either apoptosis or necrosis, the latter supervening in the case of high intensity of the insult or lack of ATP. Thus according to a recent hypothesis, apoptosis and necrosis seem to represent different shapes of cell demise resulting from a more or less complete execution of the internal death program (Leist and Nicotera 1997).

Due to the fact that during programmed cell death the course of events is genetically predetermined, whereas the process of necrotic death is characterized by the properties of the active cytotoxic agent, it was claimed that proof for the programmed nature of cell death can be provided by the existence of mutants showing the same morphological symptoms as the organism under influence of a pathogenic factor (Greenberg 1996). Also the induction of the characteristic symptoms by factors other than the pathogen has been proposed to provide indication for a classification as programmed cell death (Dangl et al. 1996). In summary, the controversial issue of hypersensitive reaction in plant cells is representative for the entire field of cell death. The fact that the various categories of cell death are not easily distinguishable, since not all the symptoms are present in every case and some symptoms are not strictly confined to one of the categories, leaves the question unanswered, as to whether programmed cell death is a single core program with multiple manifestations or multiple programs initiated by different signals (Jones and Dangl 1996).

Another feature of the hypersensitive reaction makes it an important system to be discussed in parallel with cell death during pollen tube incompatibility reaction. Both systems are based on a cell-cell interaction during which death of one of the interaction partners is caused either by triggered suicide or by murder. In the hypersensitive reaction the "victim" is the infected plant tissue which does (upon incompatible reaction) or does not degenerate (upon compatible reaction). In contrast, in pollen-pistil interaction both interaction partners may potentially undergo degeneration depending on the nature of the reaction. Upon compatible pollination the stylar tissue decomposes, whereas it stays alive upon incompatible reaction during which however the other reaction partner, the pollen tube, dies (Herrero and Dickinson 1979; Hodgkin et al. 1988; Greenberg 1996).

For both phenomena, hypersensitive reaction and pollen-stigma interaction, it is of interest to identify the cause of death, the elucidation of both systems is apt to contribute important pieces of information to the understanding of cell death in plants.

10.3 Different Ways to the Rejection of Pollen Tubes

In order to understand how pollen tube growth is inhibited after self-incompatible pollinations, whether their death is of apoptotic, of necrotic or of yet a different nature, at least two aspects have to be considered: the morphology of the inhibited pollen tube and the pieces of knowledge available on the mechanism of recognition, signal transduction and rejection. Since self-incompatibility mechanisms vary among the families of angiosperms, the different systems have to be analyzed individually.

Two basic principles of SI are distinguished in homomorphic plants. In sporophytic SI the recognition mechanism is controlled by the genome of the diploid pollen donor, whereas in gametophytic SI the genome of the haploid pollen determines the outcome of the interaction. In the following, both sporophytic and gametophytic self-incompatibility mechanisms will be discussed, the focus will however be on the system present in the Solanaceae, this family having been studied the most intensively.

The genetic basis for the recognition mechanisms in these SI systems is discussed by de Nettancourt (1997; and this issue). One of the main questions in SI research is the identity of the factors responsible for the interaction between male gametophyte and pistil. In Brassicaceae and Solanaceae the responsible pistil compound, the S-proteins, have been identified. However, neither the precise mode of operation of these proteins nor the identity of the pollen S-factors is known. Regarding the mode of interference with pollen tube elongation following the recognition event, various mechanisms are known to be operating in the different plant families, some of which are outlined in the following.

10.3.1 Sporophytic Self-Incompatibility

10.3.1.1 *Brassicaceae*

The Brassicaceae exhibit sporophytic SI, which has been reviewed in numerous articles (Dickinson et al. 1992; Dzelkalns et al. 1992; Trick and Heizmann 1992; Sims et al. 1993; Nasrallah and Nasrallah 1993). In *Brassica* pollen tube growth of incompatible grains is arrested on the stigma, in most cases pollen germination as such is inhibited due to weak adherence of the self grain to the papilla (Stead et al. 1980) and to failure to achieve hydration (Zuberi and Dickinson 1985); if the grain does germinate, the resulting pollen tube fails to invade the papillar cell wall (Sears 1937; Ockendon 1972). The precise mechanism of inhibition remains elusive, various hypotheses range from the simple blockage of water flow (Roberts et al. 1980) to the production of inhibitors by the pollen itself (Ferrari and Wallace 1977), whereas recent results favor the coordinated activity of pollen factors and stigmatic kinases (Dzelkalns et al. 1992; Nasrallah and Nasrallah 1993).

In the Brassicaceae SI is governed by a single multiallelic S locus, rejection being controlled by the interaction of the SI genotype of the pistil with the genotype of the pollen parent. At least two multiallelic genes are found within the *Brassica* S locus, one of them encoding a glycoprotein (SLG), and the other a transmembrane receptor serine/threonine kinase (SRK), both of which are expressed in the stigmas (Nasrallah et al. 1985; 1994; Kandasamy et al. 1989; Stein et al. 1991; Goring and Rothstein 1992). Experiments based on the inhibition of serine/threonine protein phosphatases with okadaic acid suggested that protein-phosphorylation events in the pistil are involved in the inhibition of pollen germination (Nasrallah and Nasrallah 1993; Rundle et al. 1993; Scutt et al. 1993; Nasrallah et al. 1994). It is presumed that allelic specificity results from sequence differences in the SLGs and SRKs produced by the pistil. The signaling pathway is supposed to be triggered by binding of a pollen ligand, whose nature and way of interaction is hitherto unknown, to the SLG-SRK complex, thus activating SRK (Hiscock and Dickinson 1993; Nasrallah and Nasrallah 1993; Dickinson 1995). The search for the male determinant of SI has resulted in promising candidates, the characterization of which will hopefully soon contribute to the completion of the picture of the recognition mechanism in *Brassica* (Doughty et al. 1993; Boyes and Nasrallah 1995; Stanchev et al. 1996; Yu et al. 1996). In self-pollinations in *B. oleaceae* it has been observed that callose is synthesized in the stigmatic papillae, which in the past has been interpreted as structural barrier towards incompatible tubes induced by the recognition reaction (Kerhoas et al. 1983). However, more recent data showed that the operation of SI functions independently of stigmatic callose synthesis (Singh and Paolillo 1990), and its formation is suggested to be elicited secondarily by molecules released from the degenerating pollen protoplast rather than being a direct consequence of the recognition reaction (Dickinson 1995; Elleman and Dickinson 1996). It was however proposed that the callose deposition in the stigmatic cells might provide an osmotic sink capable of pulling water back into the cell and away from the pollen, thus producing a localized region of stigmatic wall that is partially dehydrated to prevent hydration of the incompatible pollen grain (Pruitt 1997). Furthermore, recent studies revealed that in *Brassica campestris* an aquaporin like molecule in the stigmatic papillar might be involved in the control of pollen hydration (Ikeda et al. 1997; Pruitt 1997).

After compatible pollinations an expansion of the zone of the outer papillar cell wall layer which is in contact with the pollen grain was observed, a process which has been interpreted as formation of a localized pollen adhesion site initiated by factors derived from the pollen coating (Elleman and Dickinson 1996) and which does not take place in incompatible pollinations as the result of a kinase activation (Elleman and Dickinson 1990; Hiscock et al. 1994). This lack of papillae preparation supposedly interrupts hydration of the pollen grain and thus prevents its further development. Self-grains seem therefore not to be inhibited metabolically, but they are physiologically isolated from the subjacent stigmatic papilla (Dickinson and Elleman 1994). This model is supported by the fact that incompatible pollen grains remain viable for quite a while, even though they do not germinate as long as they are in contact with the incompatible stigma, thus indicating a biostatic rather than a biocidal effect (Hodgkin et al. 1988). Removing pollen from an incompatible stigma and placing it on a permissive one allows in fact resumption of growth (Kroh 1966). In the context of cell death these findings imply, that in the case of sporophytic SI in Brassicaceae the pollen eventually dies neither by triggered suicide (programmed cell death) nor by murder (necrosis). Death is rather due to "old age" or "starvation" after being isolated and prevented from hydration. However, it cannot be ignored, that some grains succeed nevertheless to produce pollen tubes after incompatible pollination of *Brassica* flowers and that these are inhibited only after germination. This indicates that denial or removal of water might not be an entirely satisfactory explanation for inhibition of pollen tube growth in this species (Hiscock, personal communication).

10.3.2 Gametophytic Self-Incompatibility

Homomorphic gametophytic self-incompatibility represents the most widespread form of SI (reviewed by de Nettancourt 1977; 1997; Ebert et al. 1989; Haring et al. 1990; Mau et al. 1991; Thompson and Kirch 1992; Sims 1993; Clarke and Newbigin 1993; Dodds et al. 1996; Kao and McCubbin 1996 to name only a few). In contrast to sporophytic SI the recognition is determined by the haploid genotype of the male gametophyte. Unlike in the sporophytic system, in the gametophytic SI system the initial phase of pollen germination and elongation in most cases proceeds successfully after incompatible pollination, in some cases the pollen tube even reaches the ovary (East 1934; Ebert et al. 1989). The rejection mechanism takes place late in the process, somewhere on the way of the pollen tube through the stigma and the style or even as late as at the ovary. An exception to this has been observed in *Papaver* where the inhibition takes place on the stigma since poppy flowers do not possess a style.

It was recognized early, that some kind of stigmatic and/or stylar factor produced by the pistil of plants exhibiting gametophytic SI must interfere with growth of incompatible pollen tubes. De Nettancourt et al. (1973a) proposed the arrest of incompatible pollen tubes to result partly from cessation of protein synthesis and from binding of an "incompatibility protein" to wall-precursor containing bodies in the cytoplasm, thus suggesting a cytotoxic mechanism. The cytostatic alternative, a failure of transition from autotrophic to heterotrophic nutrition as proposed for incompatible lily pollen tubes (Ascher 1975) is unlikely to be responsible in Solanaceous species since cytochemical tests of pollen tubes of *Petunia* sp. revealed that the levels of protein and carbohydrate in incompatible tubes is equal, if not greater, compared to compatible tubes (Herrero and Dickinson 1981). In addition to the apparent abundance of nutrition in incompatible tubes other observations have been made which support the hypothesis that growth in incompatible pollen tubes is inhibited actively. *In vitro* growing pollen tubes of *Nicotiana alata* were found to reduce growth rate rather rapidly after addition of stylar extract of incompatible flowers (Sharma and Shivanna 1986) indicating that the stylar exudate actively interacts with the pollen tube metabolism.

Experiments were undertaken in order to elucidate the biochemical mechanism of recognition between style and pollen tube. Several models were established for the recognition and subsequent rejection of "self-pollen". Heslop-Harrison (1983) proposed a model of lectin-carbohydrate interaction. The model suggests that S-proteins in the stylar secretion have lectin-like properties with sites complementary to sugar sequences or arrays displayed by pectic wall components of incompatible tubes. The presence of soluble and membrane-bound lectins was indeed found in pollen and pistils of *Petunia* (Kovaleva 1992). The model by Heslop-Harrison (1983) suggested that by crosslinking of microfibrils apical growth would be disrupted, thus implicating a mechanism of direct interference of the stigmatic incompatibility factor with pollen tube growth. According to this model the effect on incompatible pollen tubes would be cytostatic.

However, observations made in the past decade were able to elucidate the character of the factors, the S-proteins, which are produced by the style, and which are responsible for the interference with pollen tube growth. It was found that there exist several different mechanisms of gametophytic SI. Whereas in Solanaceae as well as Scrophulariaceae and Rosaceae the SI system seems to be associated with an RNase activity of the stylar S-proteins, in the Papaveraceae no detectable ribonuclease activity correlates with the presence of the functional stigmatic S-gene product (Franklin-Tong et al. 1991; Foote et al. 1994). Liliaceae have a genetical control typical for the gametophytic SI system, but the molecular mechanism differs from the families mentioned above, even though experimental observations only provide a very preliminary picture so far.

In the families mentioned hitherto SI behavior seems to be governed by a single multiallelic S-locus. This is not always the case, as has been shown for some grasses (Hayman 1956; Lundqvist 1956), *Beta vulgaris* (Lundqvist et al. 1973; Larsen 1977), and *Ranunculus acris* (Osterbye 1975) which have up to four loci. In the following some of the plant families expressing gametophytic SI are discussed individually.

10.3.2.1
Solanaceae

The Solanaceae represent the plant family, the SI mechanism of which has been studied most intensively, it comprises the species *Petunia*, *Nicotiana*, *Lycopersicon*, *Brugmansia* and *Solanum* (reviewed for example in Ebert et al. 1989; Haring et al. 1990; Singh and Kao 1992; Sassa et al. 1992; 1994; Clarke and Newbigin 1993; Sims 1994; Xue et al. 1996). The progress made in recent years on the molecular basis of SI brought to light a picture of the rejection mechanism of incompatible pollinations in Solanaceous plants. Kovaleva and Musatova (1975) showed that the acquisition of SI during the development of *Petunia* flowers correlates with an increase in stylar ribonuclease activity and the appearance of a new stylar protein. In the following, glycoproteins produced in the pistil and apparently involved in gametophytic SI of Solanaceae were identified and characterized. These glycoproteins contain a signal peptide sequence at the 5′-end indicating their function as secretory proteins (Haring et al. 1990), their expression coincides with the pathway of the pollen tube (Jahnen et al. 1989; Mau et al. 1991; McFadden et al. 1992). Sequence analysis and assays testing enzymatic activity have revealed the ribonuclease activity of these proteins. Lee et al. (1994) and Murfett et al. (1994) showed that these S-proteins are necessary and sufficient for the rejection of self-pollen by the pistil.

The questions which arose from these results were manifold. It was of interest to find out whether the ribonuclease activity of the S-proteins is really functional in SI, whether the enzymes enter the cytoplasm of pollen tubes grown *in vivo* and if so, what their substrate is and how specificity is encoded. Details concerning these aspects can be found in de Nettancourt (this issue and references therein). In the present chapter, emphasis is put on the last step in the gametophytic SI of Solanaceous species: the interference with pollen tube growth and the eventual death of incompatible tubes. For this purpose attention is drawn to the morphological symptoms accompanying pollen tube inhibition in SI.

Descriptions of the morphological changes occurring during pollen tube inhibition have been given for several plants exhibiting gametophytic SI.

In most cases the first visible signs of inhibition did not occur until several hours after pollination. Cresti et al. (1980) showed that in *Lycopersicon peruvianum* no differences are observed between compatible and incompatible pollen tubes during the first 4 hours 30 minutes after pollination. In *Brugmansia suaveolens* the first visible signs of inhibition appeared between 6 and 8 hours after pollination (Geitmann 1993). Following this lag phase several ultrastructural changes mark the onset of the inhibition process, which eventually leads to the degradation of the incompatible pollen tubes.

The first visible sign of alteration is the appearance of circular endoplasmic reticulum (ER) in the cytoplasm (de Nettancourt et al. 1974; Cresti et al. 1979; Geitmann 1993). Big stacks of rough ER form cylindrical configurations, starting 6 to 8 hours after pollination in *Brugmansia suaveolens* (Fig. 10.1). This phenomenon has also been observed in young leaf buds of *Betula verrucosa* (Dereuddre 1971) and the buds of resting potato (Shih and Rappaport 1971). Because of its presence in resting tissues circular ER was interpreted as a symptom of arrested protein synthesis. This assumption was adopted by Cresti et al. (1979) who observed circular ER in the pollen tubes of *Petunia hybrida* and *Lycopersicum peruvianum* pollen tubes which were exposed to gamma-irradiation. De Nettancourt et al. (1973a; 1974) described this symptom in incompatible *Lycopersicon peruvianum* pollen tubes. The interpretation of circular ER as a sign for arrested protein synthesis is attractive since it would provide an explanation for the inhibition mechanism in gametophytic SI. If the interpretation is correct the appearance of circular ER would also indicate that the overall synthesis of proteins is inhibited and not the production of a specific protein. However, even though the appearance of circular ER is the first visible step in the inhibition process, it remains obscure, if the arrest of protein synthesis is the crucial factor for the inhibition or its consequence.

During the course of the inhibition events the deposition of callose is visibly influenced. Callosic plugs normally are formed in regular intervals in order to separate the vital part of the pollen tube cytoplasm from the distal part of the tube which is bound to degenerate. In incompatible pollen tubes these plugs are elongated and the distances

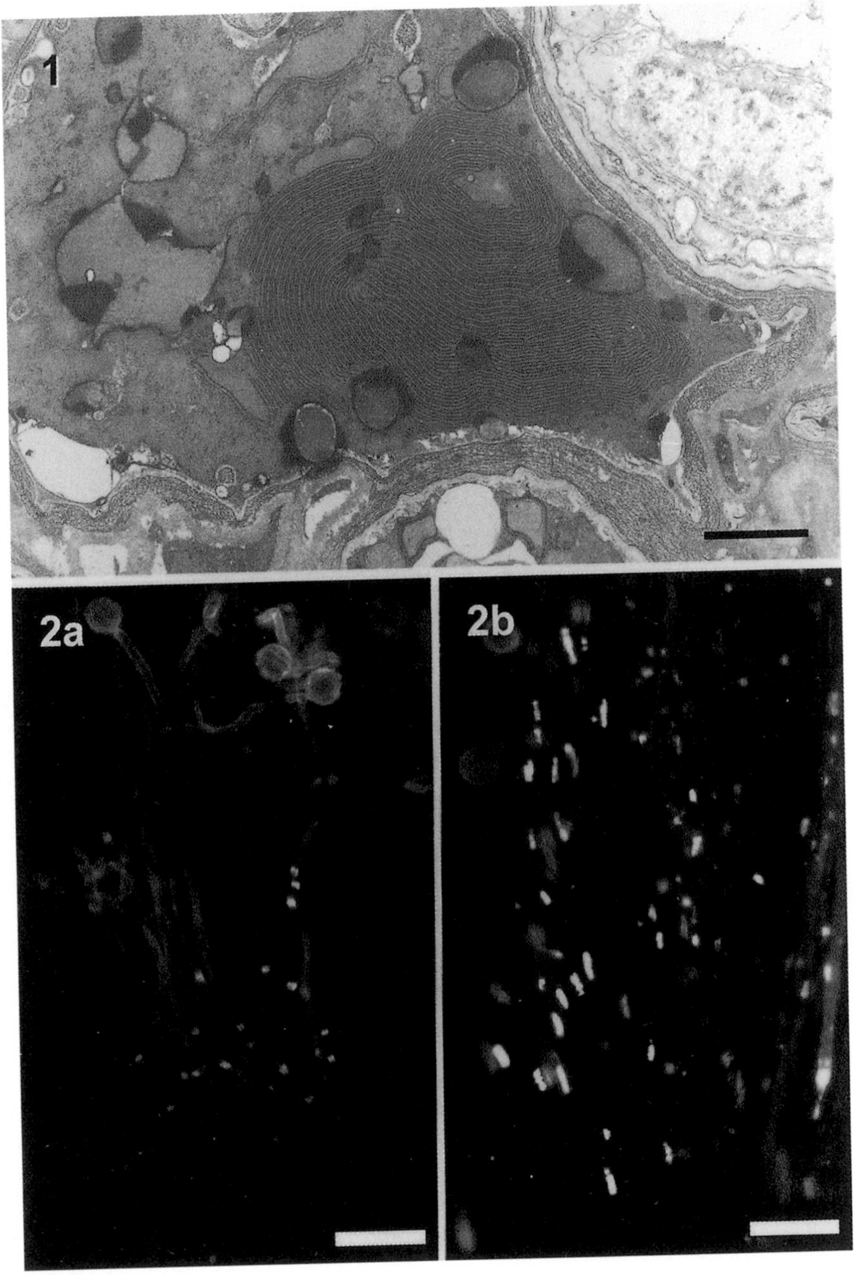

Fig. 10.1. Transverse section of incompatible *Brugmansia suaveolens* pollen tube showing circular ER. Bar = 2 μm

Fig. 10.2. Fluorescence micrographs of *Brugmansia suaveolens* stigma and style after compatible **(a)** and self-incompatible **(b)** pollination. Pistils were cut longitudinally and treated with decolorized aniline blue for labeling of callose. **(a)** Callose plugs are regularly distributed and have a length of ca 10 μm. **(b)** Callose plugs are irregularly distributed and their lengths vary between 20 and 40 μm. Bar = 100 μm between them are irregular as observed in *Pyrus malus, Nicotiana alata* (Tupý 1959; Lush and Clarke 1997) and *Brugmansia suaveolens* (Geitmann 1993; Geitmann et al. 1995) (Fig. 10.2). In the latter case the plugs in incompatible tubes were at least twice as long (20–40 μm) as in compatible tubes and had a somewhat irregular shape as well as less homogeneous ultrastructure (Geitmann 1993). An increased number of plugs was observed in incompatible pollen tubes of *Petunia hybrida*, whereas the length of the plugs appeared unchanged in this species (Linskens and Esser 1956; Cresti et al. 1979).

Callose is also deposited in the form of cell wall material. In incompatible pollen tubes of *Lycopersicum* the inner cell wall was reported to disappear (de Nettancourt et al. 1973a; 1974) and in exchange, fibrillar vesicles of bipartite structure appeared which were also reported for *Petunia* (Cresti et al. 1979). In other species the thickness of the callosic inner wall increases immensely. This begins 10 hours after self-incompatible pollination in *Brugmansia suaveoloens* (Fig. 10.3). In this species the diameter of the callosic wall layer can reach up to 600 nm 24 hours after pollination, whereas tubes in an advanced state of degeneration showed a thickness of the inner cell wall of up to 1.8 μm (Geitmann 1993; Geitmann et al. 1995). In specimens observed in the FM after treatment with decolorized aniline blue callose deposition at the apical part of the tube is conspicuous (Fig. 10.4a; Schlösser 1961). The tips in these tubes showed swelling achieving a diameter of 30% above the one of the tube shank (Fig. 10.4b). In many cases the tips also lost their regular round shape and seemed to be deflated. From these findings it is difficult to conclude whether the increase of rigidity of the tubular cell wall through the deposition of callose is the reason for the growth inhibition or whether the relationship between cause and consequence is the opposite. However, the fact that a marked thickening of the cell wall occurs after the appearance of circular ER, indicates that the alterations in the structure of the inner cell wall are caused secondarily (Shivanna et al. 1982; Harris et al. 1984). Tupý remarked already in his 1959 paper that the appearance of callose deposits might reflect a decreased rate of tube extension and therefore a secondary effect instead of an increased pace of callose synthesis.

Not only the inner but also the outer, fibrillar layer of the cell wall is subject to thickening in incompatible Solanaceous pollen tubes (van der Plujim and Linskens 1966; de Nettancourt et al. 1973a). The thickening can be considerable, however it starts later than the thickening of the callose layer. In *Brugmansia suaveolens* up to 12 hours after pollination the thickness of the fibrillar layer is around 120 nm. In samples taken after 24 and 48 hours outer cell wall layers of 200 to 300 nm were observed (Figs. 10.5) In most cases the thickened fibrillar layer did not show homogeneous structure, but seemed to consist of crescent shaped, pectin containing inclusions in the cell wall, probably secretory vesicles the contents of which were not correctly inserted in the present cell wall (Fig. 10.6) (Geitmann 1993; Geitmann et al. 1995). This observation might indicate that the final step of transport of the outer wall material is either inhibited or that the processing of these precursors is overstrained due to reduction of tube elongation rate. As for the callose depositions it is not obvious whether this process is a primary result of the incompatibility reaction as suggested by Heslop-Harrison (1983). However, given the fact that thickening of the outer cell wall layer occurs rather late in the sequence of events following incompatible pollination, it seems reasonable to conclude that the thickening is a secondary effect occurring in tubes which have already reduced their growth rate or arrested growth completely. The phenomenon could be explained by an ongoing liberation of secretory vesicles, which therefore contributes to wall thickness instead of pollen tube length, since tube elongation is reduced or stopped (Herrero and Dickinson 1981; Shivanna et al. 1982).

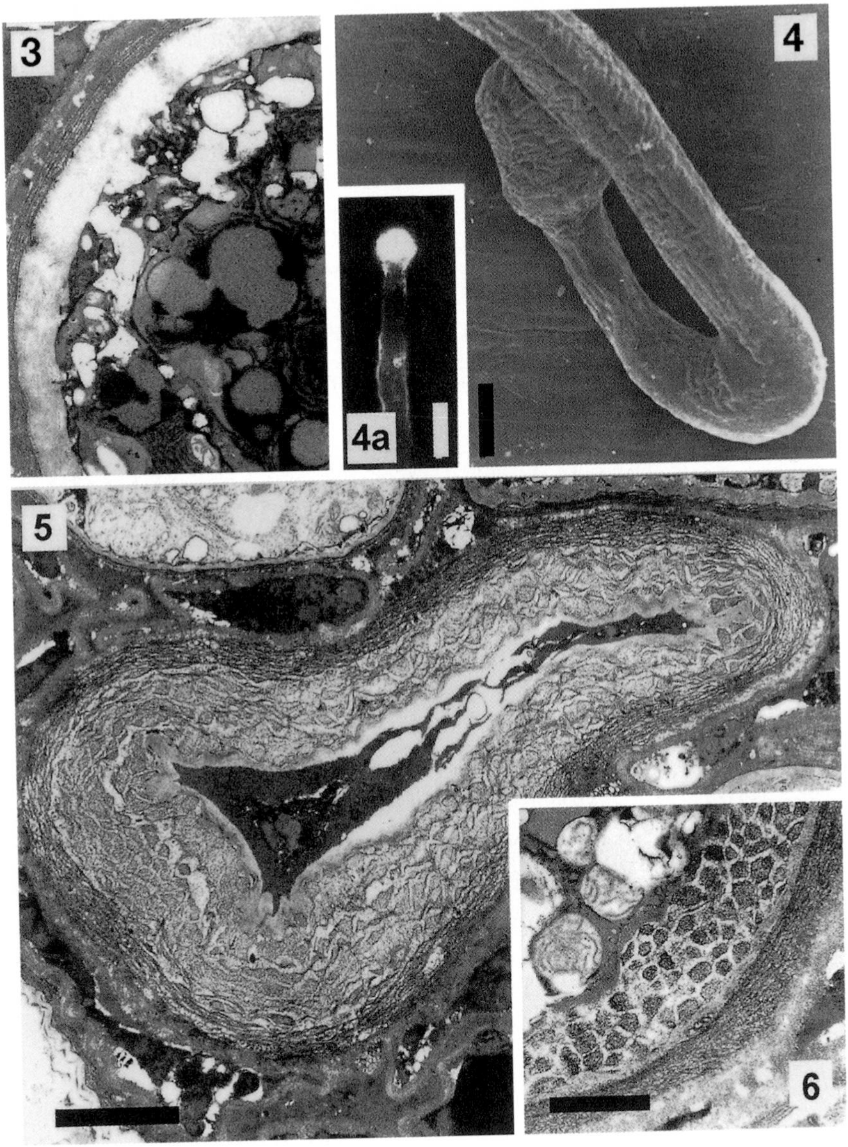

Fig. 10.3. Transverse section of incompatible *Brugmansia suaveolens* pollen tube showing thickened callose layer in the cell wall and electron dense deposits in the cytoplasm. Bar = 1 µm

Fig. 10.4. Apical tip of an incompatible *Brugmansia suaveolens* pollen tube in the scanning electron microscope **(a)** and in the fluorescence microscope after staining with decolorized aniline blue **(b)**. The tip appears swollen and contains deposits of callose. Bars = 20 µm **(a)**, 10 µm **(b)**

Fig. 10.5. Transverse section of incompatible *Brugmansia suaveolens* pollen tube showing thickening of the outer pectinaceous cell wall layer. Bar = 2 µm

Fig. 10.6. Cell wall of an incompatible *Brugmansia suaveolens* pollen tube demonstrating an accumulation of secretory vesicles at the inside of the wall. Bar = 1 µm

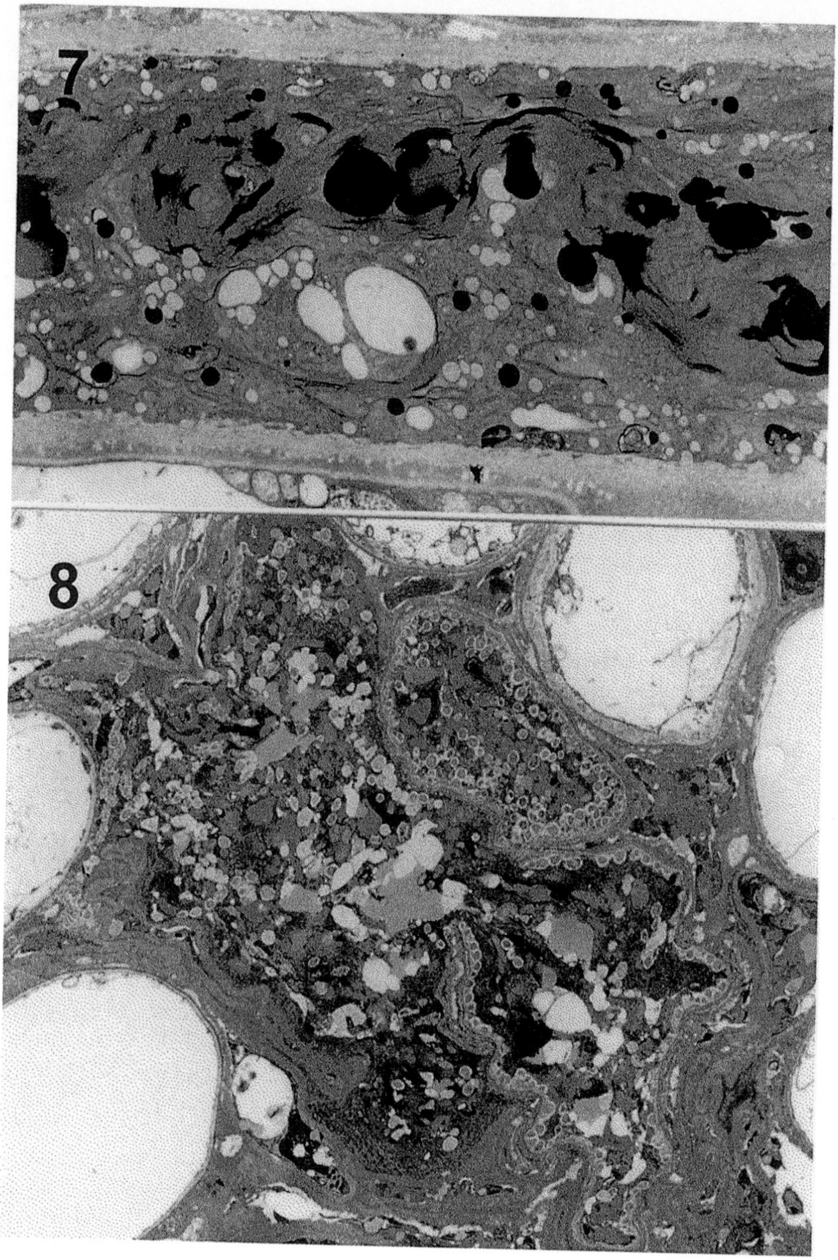

Fig. 10.7. Longitudinal section of an incompatible *Brugmansia suaveolens* pollen tube showing fusion and degeneration of ER. Bar = 2 µm

Fig. 10.8. Transverse section through the transmitting tissue of *Brugmansia suaveolens* after incompatible pollination. One of the pollen tubes has burst (arrow) and released its contents into the stylar matrix. Bar = 3 µm

During a later stage of pollen tube inhibition the cytoplasm of incompatible pollen tubes degenerates. Membrane integrity is deranged and the ER stacks seem to fuse (Fig. 10.7). These changes possibly imply alterations in membrane permeability (Dzelzkans et al. 1992). Furthermore electron dense depositions appear (Figs. 10.1, 10.3) and plasmolysis is observed in some cases. Cellular contents and remnants of pollen tube cell wall in the intercellular space between the cells of the transmitting tissue indicate that incompatible pollen tubes tend to burst (Fig. 10.8; de Nettancourt 1977; Lush and Clarke 1997).

There does not exist quite enough evidence yet for a clear-cut identification of the mode of death which incompatible Solanaceous pollen tubes are subjected to. On the one hand side indications have been found which favor the hypothesis of apoptotic cell death in these pollen tubes. Sharma and Shivanna (1986) state that the application of transcription inhibitors can overcome SI in *Nicotiana alata*, concluding that the reaction is dependent on new gene expression in the pollen. According to the characteristics of the various types of cell death, these results would support the model of a genetically programmed suicide program activated in pollen tubes upon contact with the incompatible stigmatic and/or stylar extract.

On the other hand, from the description of the morphological changes occurring in incompatible pollen tubes, in particular the degeneration of endoplasmic membrane system indicating necrotic death, it seems reasonable to speculate that incompatible tubes in Solanaceous plants die due to the cytotoxic influence of stigmatic or stylar factors. According to the current hypothesis the specificity of stylar S-proteins consists in their selective uptake by incompatible tubes. An alternative hypothesis is based on non-specific uptake followed by specific inactivation or other modification (Thompson and Kirch 1992). Whatever the way of uptake and the mode of specific recognition is, incompatible pollen tubes seem to be "murdered" directly through the action of these enzymes whereas triggering of a programmed suicide program does not seem to be involved.

Lush and Clarke (1997) postulate to contribute further evidence against the programmed nature of pollen tube death. The authors argue that the finding that growth inhibition of incompatible pollen tube can be reversed by grafting an incompatible stigma on a compatible style, indicates that the incompatible reaction is not a form of programmed cell death. However, even though the latter statement is likely to be correct, a conclusion cannot be drawn from these experiments since the process of programmed cell death is reported to be reversible up to a certain point of time. It has been shown in animal cells that if salvaged from the programmed death inducing conditions prior to this "point of no return" cells can indeed recover from initial morphological changes (Servomaa and Rytomaa 1988; Desoize and Sen 1992; Minguell and Hardy 1993; Abe and Watanabe 1995). The grafting might therefore have salvaged those pollen tubes which had not yet crossed the threshold to certain death. Nevertheless, the hypothesis favoring a necrotic kind of cell death in incompatible Solanaceous pollen tubes is still rather plausible. However, further study of the morphology of inhibited pollen tubes as well as the monitoring of DNA degeneration has confirmed that neither apoptosis nor another way of programmed cell death occurs in incompatible pollen tubes of these species.

Various attempts have been made to prove the hypothesis which claims that the RNase activity of the S-proteins in Solanaceous species is pivotal for their functioning.

A first indication for the essential role of the RNase activity for the SI mechanism is provided by the fact that the enzymatic activity is well conserved in the S-proteins of various Solanaceous species (Broothaerts et al. 1991; Kaufmann et al. 1991; Singh et al. 1991; Ai et al. 1992) and also in Rosaceae, a family only distantly related to the Solanaceae (Sassa et al. 1992; 1993; 1994; Broothaerts et al. 1995). With the help of pollen containing ^{32}P labeled RNA McClure et al. (1990) showed that the incompatible reaction resulted in a high degradation of rRNA. Furthermore it was indicated by site directed mutation that the ribonuclease activity of the S-proteins seems to be essential for the SI reaction (Huang et al. 1994). Similarly it was demonstrated that the change at an active-site histidine residue in a S-RNase of *Lycopersicon peruvianum* causes the plant to be self-compatible albeit producing the (non-functional) S-gene product (Kowyama et a. 1994; Rivers and Bernatzky 1994; Royo et al. 1994). These results strongly indicate a cytotoxic effect of the S-proteins via RNA degradation, even though the conclusion is not accepted by all researchers (Bell 1995).

S-RNases have been shown to degrade rRNA in a rather unspecific manner (Singh et al. 1991) and it was suggested that the expression of SI is mediated by a degradation of overall pollen rRNA, which should be highly effective since ribosomal genes are supposedly not translated in germinated pollen (McClure et al. 1989; Mascarenhas 1990; 1993). The pollen tube would therefore depend on the rRNA reserves stored during pollen development and the degradation of rRNA consequently causes the arrest of any protein synthesis in the incompatible pollen grain and tube. However, the absence of rRNA synthesis in pollen tubes has been disputed recently by Lush and Clarke (1997) who postulate that the maintenance of slow growth in incompatible pollen tubes of *Nicotiana alata* as well as the reversibility of tube growth inhibition after grafting of pollinated incompatible stigmata to compatible styles indicates that rRNA can be replaced and is therefore synthesized in pollen tubes. This finding alone does not however contradict the hypothesis that the function of the S-proteins is based on the RNase activity. If rRNA synthesis occurred in pollen tubes, the rate of RNA degeneration by the S-RNases would simply have to exceed the rate of RNA production in order to be effective.

However, skepticism towards the rRNA degradation theory has also been nourished by morphological observations. If the function of Solanaceous S-proteins was indeed the degradation of RNA, one should expect to find a reduced number of ribosomes and especially polysomes in the cytoplasm of incompatible pollen tubes. However, in incompatible pollen tubes of *Brugmansia suaveolens* as late as 24 or 48 hours after pollination ribosomes were highly abundant and polysomes were present in the cytoplasm indicating active protein synthesis. Similar observations were made by Bell (1995a) in the micrographs published by Herrero and Dickinson (1981). Contradiction is also provided by the results of RNA analyses in pollen tubes of *Petunia hybrida*. Van der Donk (1974) states that in incompatible pollen tubes the amount of synthesized RNA with a molecular weight similar to rRNA is higher than in compatible tubes. Bell (1995a) concludes from the apparent lack of ribosome degradation that the RNase activity of Solanaceous S-proteins might be irrelevant for the SI mechanism. In this context the remark by Dzelzkalns et al. (1992) should not be ignored in as much as "it is interesting to note that mammalian extracellular RNases have been implicated in such diverse activities as angiogenesis, the bundling of actin filaments, suppression of tumor growth, and immunosuppression, roles that may or may not depend upon the known catabolic activity of the associate ribonuclease (Benner and Alleman 1989; D'Alessio et al. 1991)". Nevertheless, one cannot ignore that "it seems unlikely that the RNase function would have been so tightly conserved if it were not functional in SI" (Carke and Newbigin 1993). Further research is likely to elucidate this essential issue in the coming years.

10.3.2.2
Papaveraceae

The genetic control of SI in the Papaveraceae is gametophytic, however, in *Papaver* flowers which lack a style, pollen tube growth is stopped during or just after germination on the stigma. Furthermore *Papaver* flowers have a dry stigma, which is why morphologically the self-incompatible reaction in *Papaver* resembles the one in the Brassicaceae rather than the one in the families displaying gametophytic SI.

In *Papaver rhoeas* no detectable ribonuclease activity correlates with the presence of the functional stigmatic S-gene product (Franklin-Tong et al. 1991; Foote et al. 1994). Investigation of the metabolic events occurring in the pollen have shown in *in vitro* experiments that on the pollen side RNA transcription and *de novo* glycosylation are necessary for full inhibition of pollen tube growth during SI reaction in *Papaver rhoeas* (Franklin-Tong and Franklin 1992; Franklin et al. 1990; 1992). Furthermore it was observed that purified S-proteins elicit a transient rise in cytosolic Ca^{2+} in incompatible pollen after which tube growth is arrested (Franklin-Tong et al. 1993; 1995). This finding together with the observation that during the SI reaction certain proteins are rapidly and transiently phosphorylated (Franklin-Tong and Franklin 1992; Rudd et al. 1996; 1997) indicates that in *Papaver* a signal transduction pathway is involved in the SI reaction. According to the latest model the observed Ca^{2+} increases are mediated, at least in part, by inositoltrisphosphate induced Ca^{2+} release (Franklin-Tong et al. 1996).

Considering the cell death point of view the question arises as to which function the transient increases in cytosolic Ca^{2+} has. Two scenarios can be envisioned. The wave of Ca^{2+} migrating through the pollen tube upon contact with the respective incompatible S-protein, in itself could be sufficient to dissipate the tip-focused Ca^{2+} gradient, which has been found to be pivotal for pollen tubes growth (Pierson et al. 1994; Malhó et al. 1994), and thus cause arrest of growth and inhibition of fertilization. In this case the effect could be termed cytostatic or cytotoxic. In fact, dissipation of the Ca^{2+} gradient in pollen tubes by a Ca^{2+}-chelating buffer causes the arrest of growth but also the deposition of callose at the apical tip, thus resulting in a similar phenomenon compared to the incompatible reaction (Chasan 1994). However, callose depositions have been observed as a consequence of various external influences interfering with pollen tube growth (Cresti et al. 1985; 1986; Read et al. 1993a; 1993b), which is why this alteration in cell wall morphology is more likely to be a secondary effect of growth inhibition than the primary cause as has been claimed by Bell (1995a; 1995b).

On the other hand, however, the Ca^{2+} transient could be part of the signal transduction pathway which might include the activation of calmodulin-dependent kinase(s), thus resulting in phosphorylation of specific proteins, which in turn would activate gene expression (Franklin-Tong et al. 1995; Rudd et al. 1996; 1997). This signaling pathway could be responsible for the final reaction of the pollen tube, including its death. *De novo* gene expression specific to the SI response was indeed demonstrated to occur in *Papaver* (Franklin-Tong et al. 1990). Therefore the process of pollen tube death in *Papaver* could be classified as triggered suicide and thus be termed programmed cell death. However to date the precise nature of the process, apoptotic or another kind of programmed cell death, remains elusive and can only be determined upon characterization of the morphological and metabolic changes as well as the identification of the activated genes in the inhibited pollen tubes. Possibly the entire rejection process in poppy might be a combination of both, the immediate cytotoxic effect on the Ca^{2+} gradient, followed by a programmed auto-destruction triggered by a signal transduction based on the observed Ca^{2+} wave. Further research is warranted to clarify these questions.

10.3.2.3
Liliaceae

It is known that the SI system in Liliaceae is gametophytic and details of its mechanism are under investigation. Based on the observation that lily pollen tube growth speeds up upon entry in the style, Ascher (1975) claimed that a high-velocity operon is activated exclusively in compatible tubes, which then assume heterotrophic growth. According to this model incompatible tubes would not be able to undergo this transition from autotrophic to heterotrophic nutrition and therefore stop growth as soon as the stored nutrients are used up by the elongating tube, thus suggesting a cytostatic mechanism. The fact that pollen tube inhibition takes place after entry into the style was confirmed for *Lilium longiflorum* cv. Hinomoto where pollen tubes cease to elongate in the stylar canal about 20 h after self-pollination (Hiratsuka et al. 1983; Tezuka 1990; Tezuka et al. 1993). Various observations of self-incompatible behavior in lilies include the finding that a functional cAMP-regulated system involving adenylate cyclase and phosphodiesterase might play a role in the control of incompatible pollen tube growth in this species (Tezuka et al. 1993). A study by Zhang et al. (1991) has shown that estrogens might be in some way involved in the incompatibility process, whereas testosterone despite being associated with development and ripening of the male gametophyte, does not seem to participate in the process of SI in lily (Yang et al. 1994).

Unlike in plant species with a solid style, where pollen tubes have to find their way by penetrating through the intercellular matrix of the transmitting tissue cells, pollen tubes of *Lilium longiflorum* grow through a stylar canal containing an exudate providing nutrients for the growing tube. A series of studies by Ichimura and Yamamoto (1991; 1992a; 1992b) revealed that the amount of exudate present in the stylar canal seems to influence pollen tube growth. The authors attribute SI to the lack of stimulation of stylar exudate production by a self pollen suggesting the availability of arabinogalactan to be one of the key factors. These latter findings indicate a cytostatic mechanism of SI interference with pollen tube growth, since the cell is not actively destroyed, but only prevented from growing further due to deprivation of nutrients. Contrary to this it was found that the stylar exudate of *Lilium longiflorum* secreted after self-pollination has an inhibitory effect on incompatible pollen tubes thus suggesting an active interference with cell elongation (Amaki and Higuchi 1991). In accordance with this latter finding it was observed that during SI reaction in *Lilium* a stress reaction is induced associated with the accumulation of free radicals and elevated levels of activity of superoxide-forming NADPH-dependent oxidase, xanthine oxidase, superoxide dismutase, catalase and ascorbate peroxidase in the pistils (Tezuka et al. 1997). The authors propose that the incompatible reaction corresponds to a stress response which eventually leads to the rejection reaction by influencing the metabolic system of the pollen tube. This model represents a fundamental difference to the SI systems operating in other plant families. Since any changes in the composition of the stylar exudate triggered by incompatible pollination should affect all pollen tubes growing through the style at this point of time, it seems that the SI system in lily operates in a global manner in contrast to the other systems investigated so far, in which the rejection mechanism is confined to the individual reaction between a single pollen and the pistil. One would therefore expect that in the case of mixed pollinations on a lily stigma all pollen tubes – compatible and incompatible – would be inhibited to the same degree, unlike the situation in mixed pollinations in other plant families, where compatible tubes grow uninhibited in close neighborhood to inhibited pollen grains or tubes. However, to my knowledge no indication for experiments testing this hypothesis was provided in the articles mentioned above.

It is evident from the limited results available on the SI in *Lilium*, that the picture of the system working in this plant species is very fragmented. A considerable amount of research will be necessary to produce a plausible model for the SI mechanism in this family.

10.3.2.4 *Poaceae*

The grasses show gametophytic SI, but no similarity was found between their S-proteins and the S-RNases in the Solanaceae or the S-proteins in the Papaveraceae. The carboxy-terminal end of an S-protein predicted from a cDNA sequence found in the pollen of *Phalaris coerulescens* shows similarity to thioredoxin H (Li et al. 1994; 1995; 1996). Thioredoxins are small ubiquitous proteins involved in numerous biological reactions, for instance as providers of hydrogen for ribonucleotide reductase or as regulatory factors for enzymes or receptors (Holmgreen 1995). An alteration in this thioredoxin domain in *Phalaris* causing the breakdown of SI (Li et al. 1996) and its high sequence conservation indicate the importance of its catalytic function in SI (Li et al. 1997). Although, the operation mechanism of these S-proteins in grasses is largely unknown, phosphorylation of pollen proteins might play a role as demonstrated in *Secale cereale* (Wehling et al. 1994). It is interesting to note that incompatible pollen tubes are rejected extremely quickly after emerging from the grain (Heslop-Harrison 1982; Shivanna et al. 1982). Since grass pollen grains usually germinate soon after contact with the receptive stigma, the entire SI response takes only a few minutes. This indicates that the SI reaction in grasses must be based on a mechanism different from the one depending on gene transcription, proposed for *Papaver*, and RNA breakdown, suggested for Solanaceae. Instead it has been proposed that the synthesis of the pollen tube cell wall is affected directly (Golz et al. 1995), but other mechanisms with immediate effects might be possible. Whatever the exact inhibition mechanism will turn out to be, it seems obvious from the observation of the rapidity of the response that rejection in the Poaceae is rather unlikely to represent a kind of programmed cell death, but must be cytotoxic or cytostatic.

10.4 Conclusions

The investigation of various plant families shows that different mechanisms of homomorphic SI exist. Concerning the genetical regulation one can distinguish between gametophytic and sporophytic control. The arrest of pollen tube growth can happen as early as during germination or later during pollen tube growth within the stigmatic or stylar tissue. The responsible S-proteins which directly or indirectly interfere with pollen tube growth were shown to be receptor kinases and glycoproteins in the Brassicaceae, whereas ribonucleases were found in Solanaceae and S-proteins without ribonuclease activity seem to be acting in Papaveraceae. The existence of these different mechanisms gave rise to the discussion about the origin of SI. On the one hand it was claimed that a primitive SI system arose early in angiosperm evolution before the taxa diverged and that this system contributed significantly to their radiation and evolutionary success over the gymnosperms (Whitehouse 1950). According to this theory the different rejection mechanisms developed along with the evolution of angiosperms from a common origin. In contrast to this, Bateman (1952) suggested that SI arose independently on more than one occasion. The controversy about the evolution of SI and about its monophyletic versus polyphyletic origin has been discussed by many authors, among others by Pandey (1960; 1980; 1981).

An interesting discussion was initiated by Bell (1995a) who proposed that the SI mechanism is the adaptation of an ancient response, the separation by a thickened cell wall barrier at the point of contact between the two generations of the sexual cycle, and thus existed before and represented a necessary prerequisite for the evolution of angiospermy. According to the author compatibility derived from incompatibility by mutation and thus would be a later phenomenon. This hypothesis was disputed by numerous researchers who emphasized that the existence of various SI mechanisms is due to the independent appearance during evolution of manifold devices preventing inbreeding in the different angiosperm families (Read et al. 1995). For a response to this paper see Bell (1995b).

The overview of various plant families shows that along with a variety of recognition and rejection mechanisms there exist different ways of pollen tubes to die as a consequence of the SI rejection. It might therefore be interesting to consider this final step in the SI reaction when discussing the evolutionary development of SI. For this purpose it is of course necessary to characterize the various SI mechanisms and the fate of the pollen tube in a much more detailed way. It will be important to better understand the nature of the pollen part of the S locus as well as the steps in the signaling mechanism. It will also be crucial to know how molecules are transported from the extracellular matrix across the wall of the pollen tube and how the required nutrient supply in the growing tube is maintained.

Finally, the variety of causes of death in the different SI mechanisms indicates that pollen tubes might be an important system in the study of cell death in plants. It was demonstrated here that both, necrosis and programmed cell death are likely to occur in incompatible pollen tubes. Further morphological investigations as well as studies using the instruments of cell death research such as the terminal deoxynucleotide transferase nick end labeling assay (TUNEL) or determination of DNA fragment length will certainly reveal the precise character of the fate of these cells.

Acknowledgements

The author would like to thank Simon Hiscock for helpful comments on the manuscript. The collaboration with the Botany Department of the Stockholm University, where the work on *Brugmansia suaveolens* was carried out, is gratefully acknowledged.

References

Abe K, Watanabe S (1995) Apoptosis of mouse pancreatic acinar cells after duct ligation. Arch Histol Cytol 58:221–229

Ai Y, Tsai D-S, Kao T-H (1992) Cloning and sequencing of cDNAs encoding two S-proteins of a self-compatible cultivar of *Petunia hybrida*. Plant Mol Biol 19:523–528

Amaki W, Higuchi H (1991) Effects of stylar exudates collected from pollinated pistils on pollen tube growth in *Lilium longiflorum*. Scientia Horticulturae 46:147–154

Ascher PD (1975) Special stylar property required for compatible pollen tube growth in *Lilium longiflorum* Thunb. Bot Gaz 136:317–321

Bateman AJ (1952) Self-incompatibility systems in angiosperms I. Theory. Heredity 6:285–310

Bell PR (1995a) Incompatibility in flowering plants: adaptation of an ancient response. Plant Cell 7: 5–16

Bell PR (1995b) Disputed Ancestry: Comments on a model for the origin of incompatibility in flowering plants. Reply. Plant Cell 664–665

Benner SA, Alleman RK (1989) The return of pancreatic ribonucleases. Trends Biochem Sci 14:396–397

Bestwick CS, Bennett MH, Mansfield JW (1995) Hrp mutant of *Pseudomonas syringae* pv *phaseolicola* induces cell wall alterations but not membrane damage leading to the hypersensitive reaction in lettuce. Plant Phys 108:503–516

Boyes DC, Nasrallah JB (1995) An anther-specific gene encoded by an S locus haplotype of *Brassica* produces complementary and differentially regulated transcripts. Plant Cell 7:1283–1294

Broothaerts W, Janssens GA, Proost P, Broeckaert WF (1995) cDNA cloning and molecular analysis of two self-incompatibility alleles from apple. Plant Mol Biol 27:499–511

Broothaerts W, Vanvinckenroye P, Decock XB, van Damme J, Vendrig JC (1991) *Petunia hybrida* S-proteins: ribonuclease activity and the role of their glycan side chains in self-incompatibility. Sex Plant Reprod 4:258–266

Buja LM, Eigenbrodt ML, Eigenbrodt EH (1993) Apoptosis and Necrosis: basic types and mechanisms of cell death. Arch Pathol Lab Med 117:1208–1214

Burnet FM (1971) Self recognition in colonial marine forms and flowering plants in relation to the evolution of immunity. Nature 332:230–238

Chasan R (1994) Pollen: arresting developments. Plant Cell 6:1693–1696

Clarke AE, Newbigin E (1993) Molecular aspects of self-incompatibility in flowering plants. Ann Rev Genet 27:257–279

Columbano A (1995) Cell death: current difficulties in discriminating apoptosis from necrosis in the context of pathological processes in vivo. J Cell Biochem 58:181–190

Cresti M, Ciampolini F, Mulcahy DLM, Mulcahy G (1985) Ultrastructure of *Nicotiana alata* pollen, its germination and early tube formation. Amer J Bot 72:719–727

Cresti M, Ciampolini F, Pacini E, Sarfatti G, van Went JL, Willemse MTM (1979) Ultrastructural differences between compatible and incompatible pollen tubes in the stylar transmitting tissue of *Petunia hybrida*. J Submicr Cytol 11:209–219

Cresti M, Ciampolini F, Sarfatti G (1980) Ultrastructural investigations on *Lycopersicum peruvianum* pollen activation and pollen tube organization after self- and cross-pollination. Planta 150:211–217

Cresti M, Ciampolini F, Tiezzi A (1986) Ultrastructural studies on *Nicotiana tabacum* pollen tubes grown in different culture medium (preliminary results). Acta Bot Neerl 35:285–292

D'Alessio G, di Donato A, Parente A, Piccoli R (1991) Seminal RNase: a unique member of the ribonuclease superfamily. Trends Biochem Sci 16:104–106

Dangl JL, Dietrich RA, Richberg MH (1996) Death don't have no mercy: cell death programs in plant-microbe interactions. Plant Cell 8:1793–1807

Darwin C (1862) J Linnean Soc (London) Bot 6:77–96

Darwin C (1877) The Different Forms of Flowers on Plants of the Same Species. John Murray, London

de Nettancourt D (1977) Incompatibility in Angiosperms. Springer-Verlag, Berlin

de Nettancourt D (1997) Incompatibility in angiosperms. Sex Plant Reprod 10:185–199

de Nettancourt D, Devreux M, Bozzini A, Cresti M, Pacini E, Sarfatti G (1973a) Ultrastructural aspects of the self-incompatibility mechanism in *Lycopersicum peruvianum* Mill. J Cell Sci 12:403–419

de Nettancourt D, Devreux M, Laneri U, Cresti M, Pacini E, Sarfatti G (1974) Genetical and ultrastructural aspects of self and cross incompatibility in interspecific hybrids between self-compatible *Lycopersicum esculentum* and self-incompatible *L. peruvianum*. Theor Appl Genet 44:278–288

de Nettancourt D, Devreux M, Laneri U, Pacini E, Cresti M, Sarfatti G (1973b) Ultrastructural aspects of unilateral interspecific incompatibility between *Lycopersicum peruvianum* and *L. esculentum*. Caryologia 25:207–217

Dereuddre MJ (1971) Sur la présence de groupes de saccules appartenant au réticulum endoplasmique dans les cellules des ébauches foliaires en vie ralentie de *Betula verrucosa* Ehrh. C R Acad Sc Paris 273:2239–2242

Desoize B, Sen S (1992) Apoptosis or programmed cell death: concepts, mechanisms and contribution in oncology. Bull Cancer Paris 79:413–425

Dickinson H (1995) Dry stigmas, water and self-incompatibility in *Brassica*. Sex Plant Reprod 8:1–10

Dickinson HG, Crabbe MJC, Gaude T (1992) Sporophytic self-incompatibility systems: S-gene producs. In: Russell SD, Dumas C (eds) International Review of Cytology. Sexual Plant Reproduction in Flowering Plants". Academic, New York, 140:525–561

Dickinson HG, Elleman CJ (1994) Pollen hydrodynamics and self-incompatibility in *Brassica oleraceae*. In: Stephenson AG, Kao T (eds) Pollen-Pistil Interactions and Pollen Tube Growth. Current Topics in Plant Physilogy Vol. 12, 45–61

Dodds PN, Clarke AE, Newbigin E (1996) A molecular perspective on pollination in flowering plants. Cell 85:141–144

Doughty J, Hedderson F, McCubbin A, Dickinson H (1993) Interaction between a coating-borne peptide of the *Brassica* pollen grain and stigmatic S(self-incompatibility)-locus-specific glycoproteins. PNAS 90:467–471

Dumas C, Knox RB (1983) Callose and determination of pistil viability and incompatibility. Theoret Appl Genet 67:1–10

Dzelkalns VA, Nasrallah JB, Nasrallah ME (1992) Cell-cell communication in plants: self-incompatibility in flower development. Dev Biol 153:70–82

East EM (1934) Norms of pollen tube growth in incompatible matings of self-sterile plants. PNAS 20: 225–230

Ebert PR, Anderson MA, Bernatzky R, Altschuler M, Clarke AE (1989) Genetic polymorphism of self-incompatibility in flowering plants. Cell 56:255–262

Elleman CJ, Dickinson HG (1990) The role of the exine coating in pollen-stigma interactions in *Brassica oleraceae* L. New Phytol 114:511–518

Elleman CJ, Dickinson HG (1996) Identification of pollen components regulating pollination-specific responses in the stigmatic papillae of *Brassica oleraceae*. New Phytol 133:197–205

Escargueil-Blanc I, Salvayre R, Nègre-Salvayre (1994) Necrosis and apoptosis induced by oxidized low density lipoproteins occur through two calcium-dependent pathways in lymphoblastoid cells. FASEB J 8:1075–1080

Farber E (1994) Programmed cell death: Necrosis versus apoptosis. Mod Pathol 7:605–609

Ferrari TE, Wallace DH (1977) Incompatibility in *Brassica* stigmas is overcome by treating pollen with cycloheximide. Science 196:436–438

Fett WF, Jones SB (1995) Microscopy of the interaction of *hrp* mutants of *Pseudomonas syringae* pv. *phasseolicola* with a nonhost plant. Plant Sci 107:27–39

Foote HCC, Ride JP, Franklin-Tong VE, Walker EA, Lawrence MJ, Franklin FCH (1994) Cloning and expression of a distinctive class of self-incompatibility (S) gene from *Papaver rhoeas* L. PNAS 91: 2265–2269

Franklin FCH, Franklin-Tong VE, Thorlby GJ, Howell EC, Atwal K, Lawrence MJ (1992) Molecular basis of the incompatibility mechanism in *Papaver rhoeas* L. Plant Growth Reg 11:5–12

Franklin-Tong VE, Atwal KK, Howell EC, Lawrence ME, Franklin FCH (1991) Self-incompatibility in *Papaver rhoeas*: There is no evidence for the involvement of stigmatic RNase activity. Plant Cell Environ 14:423–429

Franklin-Tong VE, Droebak BK, Allan AC, Watkins PAC, Trewavas AJ (1996) Growth of pollen tubes of *Papaver rhoeas* is regulated by a slow-moving calcium wave propagated by inositol 1,4,5-trisphosphate. Plant Cell 8:1305–1321

Franklin-Tong VE, Franklin FCH (1992) Gametophytic self-incompatibility in *Papaver rhoeas* L. Sex Plant Reprod 5:1–7

Franklin-Tong VE, Lawrence MJ, Franklin FCH (1990) Self-incompatibility in *Papaver rhoeas* L.: inhibition of incompatible pollen tube growth is dependent on pollen gene expression. New Phytol 116: 319–324

Franklin-Tong VE, Ride JP, Franklin FCH (1995) Recombinant stigmatic self-incompatibility (S-) protein elicits a Ca^{2+} transient in pollen of *P. rhoeas*. Plant J 8:299–307

Franklin-Tong VE, Ride JP, Read ND, Trewavas AJ, Franklin FCH (1993) The self-incompatibility response in *Papaver rhoeas* is mediated by cytosolic free calcium. Plant J 4:163–177

Geitmann A (1993) Structural and Functional Studies of the Gametophytic Self-Incompatibility System in *Brugmansia suaveolens* (Solanaceae). Masters Thesis, University of Konstanz, Germany

Geitmann A, Hudák J, Vennigerholz F, Walles B (1995) Immunogold localization of pectin and callose in pollen grains and pollen tubes of *Brugmansia suaveolens* – implications for the self-incompatibility reaction. J Plant Physiol 147:225–235

Golz FJ F, Clarke AE, Newbigin E (1995) Self-incompatibility in flowering plants. Curr Op Genet Develop 5:640–645

Goring DR, Rothstein SJ (1992) The S-locus receptor kinase gene in a self-incompatible *Brassica napus* line encodes a functional serine/threonine kinase. Plant Cell 4:1273–1281

Greenberg JT (1996) Programmed cell death: a way of life for plants. PNAS 93:12094–12097

Haring V, Gray JE, McClure BA, Anderson MA, Clarke AE (1990) Self-incompatibility: a self-recognition system in plants. Science 250:937–941

Harris PJ, Anderson MA, Bacic A, Clarke AE (1984) Cell-cell recognition in plants with special reference to the pollen-stigma interaction. Ox Sur Plant Mol Cell Biol 1:161–203

Hayman DL (1956) The genetic control of self-incompatibility in *Phalaris coerulescens*. Austral J Bot Sci 9:321–331
Herrero M, Dickinson HG (1979) Pollen-pistil incompatibility in *Petunia hybrida*: changes in the pistil following compatible and incompatible intraspecific crosses. J Cell Sci 36:1–18
Herrero M, Dickinson HG (1981) Pollen tube development in *Petunia hybrida* following compatible and incompatibile intraspecific matings. J Cell Sci 47:365–383
Heslop-Harrison J (1982) Pollen-stigma interaction and cross-incompatibility in the grasses. Nature 215:1358–1364
Heslop-Harrison J (1983) Self-incompatibility: phenomenology and physiology. Proc R Soc Lond B 218:371–395
Heslop-Harrison J, Heslop-Harrison Y, Barber J (1975) The stigma surface in incompatibility responses. Proc R Soc Lond B 188:293–303
Heslop-Harrison Y (1977) The pollen-stigma interaction: pollen tube penetration in *Crocus*. Ann Bot 41:913–922
Hiratsuka S, Tezuka T, Yamamoto Y (1983) Use of longitudinally bisected pistils of *Lilium longiflorum* for studies on self-incompatibility. Plant Cell Physiol 24:765–768
Hiscock SJ, Dewey FM, Coleman JOD, Dickinson HG (1994) An active cutinase from the pollen of *Brassica napus* closely resembles fungal cutinases. Planta 193:377–384
Hiscock SJ, Dickinson HG (1993) Unilateral incompatibility within the Brassicaceae: further evidence for the involvement of the self-incompatibility (S)-locus. Theoret Appl Genet 86:744–753
Hodgkin T, Lyon GD, Dickinson HG (1988) Recognition in flowering plants: a comparison of the *Brassica* self-incompatibility system and plant pathogen interactions. New Phytol 110:557–569
Hogenboom NG (1983) Bridging a gap between related fields of research: pistil-pollen relationships and the distinction between incompatibility and incongruity in nonfunctioning host-parasite relationships. Phytopathol 73:381–383
Holmgren A (1995) Thioredoxin structure and mechanism: conformational changes on oxidation of the active-site sulfhydryls to a disulphide. Current Biol 3:239–243
Huang S, Lee H-S, Karunanandaa B, Kao T-H (1994) Ribonuclease activity of *Petunia inflata* S proteins is essential for rejection of self-pollen. Plant Cell 6:1021–1028
Ichimura K, Yamamoto Y (1991) Effects of the quantity and composition of stylar canal exudate on *in vitro* and *in vivo* growth of pollen tubes in self-incompatible *Lilium longiflorum*. J Jap Soc Hort Sci 60:659–668
Ichimura K, Yamamoto Y (1992a) Correlation between the accumulation of stylar canal exudate and the diminuition of self-incompatibility in aging flowers of *Lilium longiflorum*. J Jap Soc Hort Sci 61: 159–165
Ichimura K, Yamamoto Y (1992b) Changes in the amount and composition of stylar canal exudate after self- or cross-pollination in self-incompatible *Lilium longiflorum* Thunb. J Jap Soc Hort Sci 61: 609–617
Ikeda S, Nasrallah JB, Dixit R, Preiss S, Nasrallah ME (1997) An aquaporin-like gene required for the *Brassica* self-incompatibility response. Science 276:1564–1566
Jahnen W, Batterham MP, Clarke AE, Moritz RL, Simpson RJ (1989) Identification, isolation and N-terminal sequencing of style glycoproteins associated with self-incompatibility in *Nicotiana alata*. Plant Cell 1:493–499
Jones AM, Dangl JL (1996) Logjam at the styx: programmed cell death in plants. Trends Plant Sci 1: 114–119
Kandasamy MK, Paolillo CD, Faraday JB, Nasrallah JB, Nasrallah ME (1989) The S-locus specific glycoproteins of *Brassica* accumulate in the cell wall of developing stigma papillae. Dev Biol 134:462–472
Kao T-H, McCubbin AG (1996) How flowering plants discriminate between self and non-self pollen to prevent inbreeding. PNAS 93:12059–12065
Kaufmann H, Salamini F, Thompson RD (1991) Sequence variability and gene structure at the self-incompatibility locus of *Solanum tuberosum*. Mol Gen Genet 226:457–466
Kerhoas C, Knox RB, Dumas C (1983) Specificity of the callose response in stigmas of *Brassica*. Ann Bot 52:597–602
Kerr JFR, Wyllie AH, Currie AR (1972) Apoptosis: a basic biological phenomenon with wide ranging implications in tissue kinetics. Br J Cancer 26:239–257
Kovaleva L (1992) Gametophytic incompatibility in *Petunia hybrida*: molecular aspects of male gametophyte and female sporophyte interactions. In: E. Ottaviano DM, M. Sari Gorla GBM (eds) Angiosperm pollen and ovules. Springer Verlag, pp 101–103
Kovaleva LV, Musatova NA (1975) Ribonuclease activity in the generative organs of *Petunia* in connection with the formation of the mechanism of auto incompatibility. Dokl Nauk Akad SSSR 223:762–765
Kowyama Y, Kunz C, Lewis I, Newbigin E, Clarke AE, Anderson M (1994) Self-compatibility in a *Lycopersicon peruvianum* variant (LA2157) is associated with a lack of style S-RNase activity. Theoret Appl Genet 88:859–864
Kroh M (1966) Reaction of pollen after transfer from one stigma to another. Züchter 36:185–189
Larsen K (1977) Self-incompatibility in *Beta vulgaris* L. I. Four gametophytic, complementary S-loci in sugar beet. Hereditas 85:227–248

Lee H-S, Huang S, Kao T-H (1994) S proteins control rejection of incompatible pollen in *Petunia inflata*. Nature 367:560–563
Leist M, Nicotera P (1997) The shape of cell death. Biochem Biophys Res Comm 236:1–9
Li X, Nield J, Hayman D, Langridge P (1994) Cloning of a putative self-incompatible gene from the pollen of the grass *Phalaris coerulescens*. Plant Cell 6:1923–1932
Li X, Nield J, Hayman D, Langridge P (1995) Thioredoxin activity in the C terminus of *Phalaris* S protein. Plant J 8:133–138
Li X, Nield J, Hayman D, Langridge P (1996) A self-fertile mutant of *Phalaris* produces an S protein with reduced thioredoxin activity. Plant J 10:505–513
Li XM, Paech N, Nield J, Hayman D, Langridge P (1997) Self-incompatibility in the grasses – evolutionary relationship of the S-gene from *Phalaris coerulescens* to homologous sequences in other grasses. Plant Mol Biol 34:223–232
Linskens HF (1975) Incompatibility in *Petunia*. Proc Royal Soc B 188:299–311
Linskens HF, Esser K (1956) Über eine spezifische Anfärbung der Pollenschläuche im Griffel und die Zahl der Kallosepfropfen nach Selbstung und Fremdung. Naturwiss 16
Lundqvist A (1956) Self-incompatibility in rye I. Genetic control in the diploid. Hereditas 42:293–348
Lundqvist A, Osterbye U, Larsen K, Linde-Laursen IB (1973) Complex self-incopmatibility systems in *Ranunculus acris* L. and *Beta vulgaris* L. Hereditas 74:161–168
Lush WM, Clarke AE (1997) Observations of pollen tube growth in *Nicotiana alata* and their implications for the mechanism of self-incompatibility. Sex Plant Reprod 10:27–35
Malhó R, Read ND, Salomé Pais M, Trewavas AJ (1994) Role of cytosolic free calcium in the reorientation of pollen tube growth. Plant J 5:331–341
Mascarenhas JP (1990) Gene activity during pollen development. Ann Rev Plant Phys Plant Mol Biol 41:317–338
Mascarenhas JP (1993) Molecular mechanisms of pollen tube growth and differentiation. Plant Cell 5: 1303–1314
Mau SL, Anderson MA, Heisler M, Haring V, McClure BA, Clarke AE (1991) Molecular and evolutionary aspects of self-incompatibility in flowering plants. In: Jenkins GJ, Schuch W (eds) Molecular biology of plant development
McClure BA, Gray JE, Anderson MA, Clarke AE (1990) Self-incompatibility in *Nicotiana alata* involves degradation of pollen rRNA. Nature 347:757–760
McClure BA, Haring V, Ebert PR, Anderson MA, Simpson RJ (1989) Style self-incompatibility gene products of *Nicotiana alata* are ribonucleases. Nature 342:955–957
McFadden GI, Anderson MA, Bönig I, Gray JE, Clarke AE (1992) Self-incompatibility: insights through microscopy. J Microsc 166:137–148
Minguell JJ, Hardy CL (1993) Restorative effect of IL-3 on adherence of cloned hemopoietic progenitor cell to stromal cell. Exp Hematol 21:55–60
Murfett J, Atherton TL, Mou B, Gasser CS, McClure BA (1994) S-RNase expressed in transgenic *Nicotiana* causes S-allele-specific pollen rejection. Nature 367:563–566
Nasrallah JB, Kao T-H, Chen CH, Goldberg ML, Nasrallah ME (1985) A cDNA clone encoding an S-specific glycoprotein from *Brassica oleraceae*. Nature 318:263–267
Nasrallah JB, Nasrallah ME (1993) Pollen-stigma signaling in the sporophytic self-incompatibility response. Plant Cell 5:1325–1335
Nasrallah JB, Rundle SJ, Nasrallah ME (1994) Genetic evidence for the requirement of the *Brassica* S-locus receptor kinase gene in self-incompabitibility response. Plant J 5:373–384
Ockendon DJ (1972) Pollen tube growth and the site of the incompatibility reaction in *Brassica oleraceae*. New Phytol 71:519–522
Osterbye U (1975) Self-incompatibility in *Ranunculus acris* L. Genetic interpretation and evolutionary aspects. Hereditas 80:91–112
Pacini E (1982) Pollen-stigma interactions in plants with gametophytically controlled self-incompatibility. Phytomorph 81:175–180
Pandey KK (1960) Evolution of gametophytic and sporophytic systems of self-incompatibility in angiosperms. Evolution 14:98–115
Pandey KK (1980) Evolution of incompatibility systems in plants: origin of 'independent' and 'complementary' control of incompatibility in angiosperms. New Phytol 84:381–400
Pandey KK (1981) Evolution of unilateral incompatibility in flowering plants: further evidence in favour of twin specificities controlling intra- and interspecific incomaptibility. New Phytol 89:705–728
Pierson ES, Miller DD, Callaham DA, Shipley AM, Rivers BA, Cresti M, Hepler PK (1994) Pollen tube growth is coupled to the extracellular calcium ion flux and the intracellular calcium gradient: effect of BAPTA-type buffers and hypertonic media. Plant Cell 6:1815–1828
Pruitt RE (1997) Molecular mechanics of smart stigmas. Trends in Plant Sci 2:328–329
Read SM, Clarke AE, Bacic A (1993a) Requirements for division of the generative nucleus in cultured pollen tubes of *Nicotiana*. Protoplasma 174:101–115
Read SM, Clarke AE, Bacic A (1993b) Stimulation of growth of cultured *Nicotiana tabacum* W38 pollen tubes by poly(ethylene glycol) and Cu(II) salts. Protoplasma 177:1–14

Read SM, Newbigin E, Clarke AE, McClure BA, Kao T-H (1995) Disputed ancestry: comments on a model for the origin of incompatibility in flowering plants. Plant Cell 661–664
Rivers BA, Bernatsky R (1994) Protein expression of a self-compatible allele from *Lycopersicon peruvianum*: introgression and behaviour in a self-incompatible background. Theoret Appl Genet 7: 357–362
Roberts IN, Stead AD, Ockendon DJ, Dickinson HG (1980) Pollen-stigma interactions in *Brassica oleraceae*. Theoret Appl Genet 58:241–246
Royo J, Kunz C, Kowyama Y, Anderson M, Clarke AE, Newbigin E (1994) Loss of a histidine residue at the active site of the S locus ribonuclease is associated with self-compatibility in *Lycopersicon peruvianum*. PNAS 91:6511–6514
Rudd JJ, Franklin FCH, Franklin-Tong VE (1997) Ca^{2+}-independent phosphorylation of a 68 kDa pollen protein is stimulated by the self-incompatibility response in *Papaver rhoeas*. Plant J 12: 507–514
Rudd JJ, Franklin FCH, Lord JM, Franklin-Tong VE (1996) Increased phosphorylation of a 26-kD pollen protein is induced by the self-incompatibility response in *Papaver rhoeas*. Plant Cell 8:713–724
Rundle SJ, Nasrallah ME, Nasrallah JB (1993) Effects of inhibitors of protein serine/threonine phosphatases on pollination in *Brassica*. Plant Phys 103:1165–1171
Sarker RH, Elleman C, Dickinson HG (1988) The control of pollen hydration in *Brassica* requires continued protein synthesis whilst glycosylation is necessary for intra-specific incompatibility. PNAS 85:4340–4344
Sassa H, Hirano H, Ikehashi H (1992) Self-incompatibility-related RNases in styles of Japanese pear (*Pyrus serotina* Rehd.). Plant Cell Physiol 33:811–814
Sassa H, Hirano H, Ikehaski H (1993) Identification and characterisation of stylar glycoproteins associated with self-incompatibility genes of Japanese pear, *Pyrus serotina* Rehd. Mol Gen Genet 2241: 17–25
Sassa H, Mase N, Hirano H, Idehashi H (1994) Identification of self-incompatibility-related glycoproteins in styles of apple (*Malus domestica*). Theroet Appl Genet 89:201–205
Schlösser K (1961) Cytologische und cytochemische Untersuchungen über das Pollenschlauch-Wachstum selbststeriler *Petunia*. Z Bot 49:266–288
Schwartz LM, Smith SW, Jones MEE, Osborne BA (1993) Do all programmed cell deaths occur via apoptosis? PNAS 90:980–984
Scutt C, Fordham-Skelton AP, Croy RRD (1993) Okaidic acid causes breakdown of self-incompatibility in *Brassica oleracea*: evidence for the involvement of protein phosphatases in signal transduction. Sex Plant Reprod 6:282–285
Sears ER (1937) Cytological phenomena connected with self-sterility in the flowering plants. Genetics 22:130–181
Servomaa K, Rytomaa T (1988) Suicidal death of rat chloroleukaemia cells by activation of the long interspersed repetitive DNA element (L1Rn). Cell Tissue Kintics 21:33–43
Sharma N, Shivanna KR (1986) Self-incompatibility recognition and inhibition in *Nicotiana alata*. In: Mulcay DL, BergaminiMulcahy G, Ottaviano E (eds) Biotechnology and ecology of pollen. Springer-Verlag, New York Berlin Heidelberg Tokyo, pp 179–184
Shih CY, Rappaport L (1971) Regulation of bud rest in tubers of potato, *Solanum tuberosoum* L. Plant Physiol 48:31–35
Shivanna KR, Heslop-Harrison Y, Heslop-Harrison J (1982) The pollen-stigma interaction in the grasses. 3. Features of the self-incompatibility response. Acta Bot Neerl 31:307–319
Sims TL (1993) Genetic regulation of self-incompatibility. Crit Rev Plant Sci 12:129–167
Sims TL (1994) Molecular genetics of gametophytic self-incompatibility in *Petunia hybrida*. In: Williams EG, Clarke AE, Knox RB (eds) Genetic control of self-incompatibility and reproductive development in flowering plants. Kluwer, Dordrecht, pp 19–41
Singh A, Ai Y, Kao T-H (1991) Characterization of ribonuclease activity of three S-allel associated proteins of *Petunia inflata*. Plant Phys 96:61–68
Singh A, Kao T-H (1992) Gametophytic self-incompatibility: biochemical, molecular genetic, and evolutionary aspects. In: Russell SD, Dumas C (eds) Sexual reproduction in flowering plants. Academic Press Inc, pp 449–483
Singh A, Paolillo DJ (1990) Role of calcium in the callose response of self-pollinated *Brassica* stigmas. Am J Bot 77:128–133
Stakman EC (1915) Relation between *Puccinia graminis* f.sp. *tritici* and plants highly resistant to its attack. J Agr Res 4:195–199
Stanchev BS, Doughty J, Scutt CP, Dickinson H, Croy RRD (1996) Cloning of PCP1, a member of a family of pollen coat protein (PCP) genes from *Brassica oleraceae* encoding novel cysteine-rich proteins involved in pollen-stigma interactions. Plant J 10:303–313
Stead AD, Roberts IN, Dickinson HG (1980) Pollen-stigma interaction in *Brassica oleraceae*: the role of stigmatic proteins in pollen grain adhesion. J Cell Sci 42:417–423
Stein JC, Howlett B, Boyes DC, Nasrallah ME, Nasrallah JB (1991) Molecular cloning of a putative receptor protein kinase gene encoded at the self-incompatibility locus of *Brassica oleracea*. PNAS 88: 8816–8820

Stout AB (1917) Am J Bot 4:375–395
Teasdale J, Daniels D, Davis WC, Eddy R, Hadwiger LA (1974) Physiological and cytological similarities between disease resistance and cellular incompatibility responses. Plant Phys 54:690–695
Tezuka T (1990) Self-incompatibility in lily. Plant Cell Technol 2:621–630
Tezuka T, Tsuruhara A, Suzuki H, Takahashi SY (1997) A connection between the self-incompatibility mechanism and the stress response in lily. Plant Cell Physiol 38:107–112
Thompson RD, Kirch HH (1992) Trends Genet 8:383–387
Trick M, Heizmann P (1992) Sporophytic self-incompatibility systems: *Brassica* S gene family. In: Russell SD, Dumas C (eds) International Review of Cytology. Sexual Reproduction in Flowering Plants. Academic, New York, 140:485–524
Trump BF, Berezesky IK, Chang SH, Phelps PC (1997) The pathways of cell death: oncosis, apoptosis, and necrosis. Toxicol Pathol 25:82–88
Tupý J (1959) Callose formation in pollen tubes and incompatibility. Biol Plantarum (Praha) 1:192–198
van der Donk JAWM (1974) Differential synthesis of RNA in self- and cross-polinated styles of *Petunia hybrida* L. Molec gen Genet 131:1–8
van der Pluijm J, Linskens HF (1966) Feinstruktur der Pollenschläuche im Griffel von *Petunia*. Züchter 36:220–224
Wehling P, Hackauf B, Wricke G (1994) Phosphorylation of pollen proteins in relation to self-incompatibility in rye (*Secale cereale*). Sex Plant Reprod 7:67–75
Whitehouse HLK (1950) Multiple-allelomorph incompatibility of pollen and style in the evolution of angiosperms. Ann Bot 14:199–216
Wilms HJ (1980) Ultrastructure of the stigma and style of spinach in relation to pollen germination and pollen tube growth. Bot Acta Neerl 29:33–47
Wyllie AH, Kerr JFR, Currie AR (1980) Cell death: the significance of apoptosis. Int Rev Cytol 68: 251–306
Xue Y, Carpenter R, Dickinson HG, Coen ES (1996) Origin of allelic diversity in *Antirrhinum* S-locus RNases. Plant Cell 8:805–814
Yang AH, Tang Y, Cao ZX, Tsao TH (1994) The changes of steriodal sex hormone-testosterone contents in reproductive organs of *Lilium davidii* Duch. Acta Bot Sinica 36:215–220
Yu K, Schafer U, Glavin TL, Goring DR, Rothstein SJ (1996) Molecular characterization of the S locus in two self-incompatible *Brassica napus* lines. Plant Cell 8:2369–2380
Zhang JS, Yang ZH, Tsao TH (1991) The occurrence of estrogens in relation to reproductive processes in flowering plants. Sex Plant Reprod 4:193–196
Zuberi MI, Dickinson HG (1985) Pollen-stigma interaction in *Brassica* III. Hydration of the pollen grains. J Cell Sci 76:321–336

CHAPTER 11

Aspects of the Cell Biology of Pollination and Wide Hybridization

J. S. Heslop-Harrison

John Innes Centre, Norwich NR4 7UH, UK
e-mail: Pat.Heslop-Harrison@bbsrc.ac.uk
telephone: +44 1603 452571
FAX: +44 1603 456844

11.1 Introduction

The description of the cell biology of pollination encompasses most aspects of the life of the haploid (n) organism represented by the pollen. While much of the active part of the life of the pollen is spent in close contact with diploid (2n) plants, control of its development and growth are ultimately autonomous, using many genes not expressed in the diploid (Mascarenhas 1992; Twell 1994; Ferrant et al. 1994). Because of the close interactions between the gametophytic, haploid, organism and the sporophyte, some aspects of the diploid organisms – particularlarly the stigma and style – will be described here. The processes involved in angiosperm meiosis and pollen development are reviewed elsewhere in this volume (Schwarzacher, Twell). Progress in understanding of the enveloped female gametophyte in angiosperms development has lagged behind that in pollen, but advances in microscopy, culture and molecular biology technologies have ensured that the gap has diminished in recent years (e.g. see chapters by Colombo, Kranz; Dumas et al. 1998; Huang et al. 1992; Wagner *et al.* 1990). While all plants show the alternation of haploid and diploid generations, the organisms resulting from the gametophytic stage in angiosperms are much reduced compared to some other taxonomic groups of plants. Fully developed gametophytes consist of three cells in the pollen (the vegetative cell, and the two sperm cells, the gametes) and, typically, seven cells in the embryo sac (the two synergids, egg, central cell with two nuclei, and several antipodals, but varying between species depending on the embryo sac developmental pathway; see Russell 1993). It is worth noting the much greater extent of the haploid organisms in the ferns, where the male gametophyte is a photosynthetic heart-shaped plant, typically 5 mm wide, producing free-living and swimming sperm cells, or the mosses where the gametophyte is the dominant vegetative organism.

As will also be clear from other chapters in this volume, the study of the cell and molecular biology of pollination is important. Firstly, it is critical to crops and the development of new crops through plant breeding. For the cereals, the seed, the product of pollination, is the crop; for many other crops, propagation is via the seed. Plant breeding uses meiosis and pollen transmission of the new combinations of genes to generate the variability required by plant breeders. While biotechnology offers the opportunity for direct gene transfer through transformation, further cycles of breeding are normally required to gain stability and multiply seed. Wide hybridization is used to introduce chromosomes, chromosome segments and genes into early stages of breeding programmes, increasing the range of variability available to breeders.

A clear understanding of the constraints on pollen viability, its dispersal, fertility and capacity to fertilize different genotypes, and perhaps long-term storage potential, is important to both *in situ* and *ex situ* germplasm or biodiversity conservation, and related topics of population ecology and gene-flow. The study of pollen has allowed and will continue to facilitate fundamental understanding of developmental biology and genetics. As a haploid, genetics may be more straightforward, and its relative simplicity combined with defined developmental pathways and easy accessibility means developmental and molecular studies can be carried out without some of the problems encountered in working with sporophytic tissue. Wide hybrids also provide a powerful tool to examine genomic interactions and define critical, often genetically-controlled, aspects of development.

In this review, I plan to highlight a few aspects of pollen development and pollen interactions with the sporophyte which show some of the range of results and differing techniques which can be used for the study of the cell biology of pollination.

11.2 Pollen Desiccation at the Time of Dehiscence

Dehisced pollen grains from different species have widely varying hydration status, ranging from 6% in *Populus* to 60% in grasses such as maize. Because most biological fixatives change hydration status rapidly and drastically, it is difficult to study mature dehisced pollen using cytochemistry or electron microscopy; unfixed free grains are essentially uninformative because of the high lipid content and refractile nature of the sculptured exine. The rehydration of pollen on the stigma may occur in a few minutes, again making study difficult. However, an understanding of the events of dehydration before or at the time of dehiscence, the hydration status during dispersal, and the rehydration following capture by the stigma are important for elucidating the constraints on pollen viability and fertility.

Grass pollens, which remain hydrated at the time of dehiscence, in general lack a dormancy period, remaining in a partially hydrated state in the anther after maturation with the contents of the vegetative cell in active motion. This intracellular movement persists after dispersal and throughout the period of viability, with both the large inclusions such as the vegetative nucleus and sperms and the smaller bodies such as the P-particles moving within the cell (Heslop-Harrison et al. 1997). Such grass pollen is short lived, and viability reduces greatly within an hour of anthesis. However, pollen of most other families is at least partially dehydrated at the time of dehiscence and is almost inactive metabolically (Hoekstra and Bruinsma 1980); like fungal spores, it can remain viable for extended periods, with a few reports of more than a year under good conditions (see Stanley and Linskens 1974).

Early work on the effects of dehydration on membrane structure (Heslop-Harrison 1979) suggested that the plasma membrane is porous and ineffective as an osmotic barrier in dehydrated pollen at the time of dispersal. The model was based on both theoretical considerations and observations, and a range of non-hydrating methods have been used to show that this is indeed the case: freeze substitution, vapour-phase fixation, Fourier transform infrared spectroscopy and nuclear magnetic resonance (see below and Crowe et al. 1989; Tiwari and Polito 1988; Tiwari et al. 1990). The hydration of the pollen grain on a stigma, in a moist atmosphere, or with an artificial medium is effected by relatively slow passage of water into the grain. If rehydration is too rapid, imbibitional leakage of cytoplasm occurs before the membranes rehydrate, leading to death of the pollen. However, under controlled conditions, Elleman et al. (1987) fixed *Brassica* pollen grains in osmium tetroxide vapour at various points during the hydration process. Striking changes in cytoplasmic organization of the grains were then observed by electron microscopy, showing the membranes changing from a micellar form to a hydrated, continuous lipid bilayers within ten minute periods.

11.3 Pollen Viability

Viable pollen can be defined as pollen that is competent to deliver two male gametes to the embryo sac. Testing this quality is not as straightforward as might be expected because of the wide range of interactions between the pollen and its environment, each of which may fail although the pollen itself is viable. Furthermore, it is usual for a proportion of pollen grains to be inviable, although enough grains may be viable that overall fertility of the plant is normal. Typically, between 10% and 90% of all grains are inviable because of developmental errors which may arise from expression of mutated genes in the haploid set of chromosomes carried, poor maturation in the anther, or errors in chromosome segregation at meiosis, which may be very frequent in ornamental cultivars.

Testing of pollen viability by scoring seed set after placing the pollen on the stigma is an obvious but often unsuitable or less than optimum test. Stigma and female embryo sac receptivity and viability, single-gene responses of self-incompatibility (see other chapters) and pollen rehydration are all tested, but do not relate to pollen viability. In a few species, including some of economic importance such as rubber, reliable hand-pollination has not been achieved, making the test inappropriate. The test has some advantages because the genotype of the pollen is preserved in the progeny, and very low levels of pollen viability can be found following heavy pollination of a stigma, although no measurement of percentage pollen viability can be made.

An important *in vitro* test of pollen viability is the fluorochromatic reaction (FCR; Heslop-Harrison and Heslop-Harrison 1972; Heslop-Harrison et al. 1984). It tests primarily the integrity of the vegetative cell plasma membrane, relying on the presence of non-specific esterases in the pollen cytoplasm. Membranes are permeable to the non-fluorescent, polar, molecule fluorescein diacetate, which enters the pollen grain. The active esterases within the pollen cleave the acetate residues, leaving the fluorescent molecule, fluorescein, which accumulates in an intact grain. Grains with no intact membrane do not accumulate the fluorescein. The method has proved sensitive, reliable and fast, and results correlate well with other methods. However, pre-conditioning of the pollen in a humid atmosphere or other suitable conditions may be required to ensure disrupted membranes are correctly rehydrated. A range of non-invasive methods has indicated that the test does indeed check plasma-membrane integrity (Kerhoas et al. 1987).

11.4 The Stigma and Style, and Interactions With the Pollen Grain and Pollen Tube

11.4.1 The Stigma and its Morphology

The angiosperm stigma, being the receptive surface of the style, is an exposed structure with efficient adaptations that enable pollen capture, hydration and germination. The stigma plays a critical role in the initial selection of pollen, largely rejecting pollen from

Table 11.1. A classification of Angiosperm stigma types based on the amount of secretion present during the receptive period and the morphology of the receptive surface. Abbreviations for each major type are given in the right hand column. (After Heslop-Harrison 1981)

Surface Dry				
Receptive cells (trichomes) dispersed on a plumose stigma	D	Pl		
Receptive cells concentrated in ridges, zones or heads				
Surface non-papillate	D	N		
Surface distinctly papillate				
Papillae unicellular	D	P	U	
Papillae multicellular				
Papillae uniseriate	D	P	M	Us
Papillae multiseriate	D	P	M	Ms
Surface Wet				
Receptive cells papillate; secretion moderate to slight, flooding interstices	W	P		
Receptive cells non-papillate; secretion usually copious	W	N		

alien species as well as fungal spores, and in some self-incompatibility responses at the stigma surface (see other chapters). In its morphology, the stigma surface shows as much variation as any other plant characteristic. Pollen morphology has also been extensively studied, and several guides to identification of pollen are available. Both pollen and stigma morphology are useful for taxonomy, while study of the co-adaptation of pollen and stigma reveals particular aspects of their interactions. However, the resistant and abundant pollen, with sculptured exine, is frequently preserved for thousands of years, while the stigma may become a flaccid and indistinct mass of cells within hours of reaching maturity.

Table 11.1 provides a framework for a classification of stigma morphology based on examination of nearly 1000 genera from more than 250 families, based on surveys of (Heslop-Harrison 1981; Heslop-Harrison and Shivanna 1977). The different groupings vary widely in their frequency, and dry stigmas are more common than those with surface secretions.

The dry plumose stigmas are characteristic of the grasses, and are illustrated in light and electron micrographs of pearl millet stigmas in the plate. Further examples of all the major classes are illustrated by Heslop-Harrison (1992).

Figs. 11.1 to **11.5** see color plate at the end of the book.

11.4.2 Mechanical Aspects of Pollen Capture

There are both mechanical and osmotic co-adaptations between stigma and pollen grain. The sizes of the receptive cells on the surface of dry papillate or plumose stigmas match that of the pollen in a species, and larger or smaller grains often fail to adhere to the surface. The stigma and pollen grain sizes are also co-adapted with respect to pollen rehydration, and alien pollen will not normally rehydrate and germinate on a foreign stigma. The importance of the regulated uptake of water to successful pollen germination was discussed above.

The plate shows the stigma of pearl millet, *Pennisetum glaucum* R.Br. (syn. *P. americanum* (L.) Leeke) after pollination with its own pollen (Fig. 11.1) and that of the related species, maize (*Zea mays*, Figs. 11.4, 11.5). The latter grains are some 60% larger in diameter, corresponding to a four-fold greater volume. They are not captured efficiently on the millet stigma, and readily fall off; even when the crops are grown in adjacent plots, few maize pollen grains are observed on millet stigmas. After artificial cross pollination, the pollen hydrates successfully (Fig. 11.4) and germinates, but the small relative size of the cells in the millet stigma trichomes relative to the large maize pollen tube limits correct penetration of the tube under the stigma pellicle and into the intracellular spaces.

11.4.3 Mechanical Aspects of Pollen Tube Directionality

The adaptation in size between pollen and stigma extends further to the stylar cells of the pollen tube transmitting pathway and the pollen tube itself (see Heslop-Harrison 1987). Taking the grasses as an example, the pollen tube grows between the cells into the intracellular spaces of the transmitting tract. This consists of cells with specializations to allow growth of the pollen and provide nutrition as required. The orientation of the cells is also critical, and it becomes clear that directionality is lost when a much larger pollen tube, from maize rather than millet, enters the stigma (Fig. 11.5 and legend).

11.4.4 Stigma Abscission Following Pollination

Normally, many times more pollen grains arrive at a stigma than are required for fertilization of all the available ovules – in the case of the grasses, only one per stigma. Furthermore, the stigma provides both the water and nutrients for pollen tube growth, and this function is liable to parasitism by fungal spores, as discussed by Willingale and Mantle (1985). Figs 11.1 and 11.2 illustrate one of the mechanisms which limits both the number of pollen tubes entering the ovule and prevents later entry of fungal hyphae. After pollination, an abscission zone comprising well-defined cells with thin walls near the base of the style, is activated by the growth of pollen tubes (Heslop-Harrison and Heslop-Harrison 1997). As the first tubes pass through the zone, the cells of the zone start to collapse and functionally occlude the transmitting tract within an hour. Soon after, the zone constricts and the stigma will abscise when the flowering head is disturbed.

11.5 Conclusions

The last ten years have seen many exciting developments in the cell biology of pollination. There has been increased interest in the critical events of sexual plant reproduction, and the importance of understanding the process from both fundamental and applied viewpoints has been increasingly recognized. New techniques in imaging and microscopy including the widespread use of fluorescence microscopy, image analysis, and other techniques combined with the ability to dissect most of the processes at the genetic and molecular level, have driven these discoveries forward. The next ten years promise to make more of the ability to understand and control aspects of plant reproduction.

References

Crowe JH, Hoekstra FA, Crowe FJ (1989) Membrane phase transitions are responsible for imbibitional damage in dry pollen. Proc Natl Acad Sci USA 86:520–523

Dumas C, Berger F, Faure J-E, Mattys-Rochon E (1998) Gametes, fertilization and early embryogenesis in flowering plants. Advances in Botanical Research 28:231–261

Elleman CJ, Wilson CE, Dickinson HG (1987) Fixation of *Brassica oleracea* pollen during hydration: a comparative study. Pollen Spores 29:273–289

Ferrant V, van Went J, Kreis M (1994) Ovule cDNA clones of *Petunia hybrida* encoding proteins homologous to MAP and shaggy/zeste-white 3 protein kinases. Society for Experimental Biology Seminar Series 55: Molecular and Cellular Aspects of Plant Reproduction. Eds RJ Scott, AD Stead. Pp 159–172. Cambridge University Press: Cambridge

Heslop-Harrison J, Heslop-Harrison Y, Heslop-Harrison JS (1997) Motility in ungerminated grass pollen:association of myosin with polysaccharide-containing wall-precursor bodies (P-particles). Sexual Plant Reprod 10:65–66

Heslop-Harrison J, Heslop-Harrison Y, Shivanna KR (1984) The evaluation of pollen quality, and a further appraisal of the fluorochromatic (FCR) test procedure. Theor Appl Genet 67:367–375

Heslop-Harrison J, Heslop-Harrison Y (1972) Evaluation of pollen viability by enzymatically induced fluorescence: intracellular hydrolysis of fluorescein diacetate. Stain Techn 45:115–120

Heslop-Harrison J (1979) An interpretation of the hydrodynamics of pollen. Am J Bot 66:737–743

Heslop-Harrison J (1987) Pollen germination and pollen-tube growth. International Review of Cytology 107:1–78

Heslop-Harrison JS (1992) The angiosperm stigma. In: Cresti M, Tiezzi A (eds) Sexual Plant Reproduction. Springer, Heidelberg, pp 59–68

Heslop-Harrison Y, Heslop-Harrison JS (1997) The pollen-tube activated abscission zone in the stigma of pearl millet: structural and physiological aspects. Can J Bot 75:1200–1207

Heslop-Harrison Y, Reger BJ, Heslop-Harrison J (1984) The pollen-stigma interaction in the grasses. 6. The stigma ('silk') of *Zea mays* L. as host to the pollen of *Pennisetum americanum* (L.) Leeke. Acta Bot Neerl 33:205–227

Heslop-Harrison Y, Shivanna KR (1977) The receptive surface of the angiosperm stigma. Ann Bot 41:1233–1258

Heslop-Harrison Y (1981) Stigma characteristics and angiosperm taxonomy. Nordic Journal of Botany 1:401–420

Hoekstra FA, Bruinsma J (1980) Control of respiration of binucleate and trinucleate pollen under humid conditions. Physiol Plant 48:71–77

Huang BQ, Pierson ES, Russell SD, Tiezzi A, Cresti M (1992) Video microscopic observations of living, isolated embryo sacs of Nicotiana and their component cells. Sex Plant Reprod 5:156–162

Kerhoas C, Gay G, Dumas C (1987) A multidisciplinary approach to the study of the plasma membrane in *Zea mays* pollen during a controlled dehydration. Planta 171:1–10

Mascarenhas JP (1992) Pollen expressed genes and their regulation. In: Ottaviano E, Mulcahy DL, Sari Gorla M, Bergamini Mulcahy G (eds) Angiosperm Pollen and Ovules. Springer, New York, pp 3–6

Russell SD (1993) The egg cell: development and role in fertilization and early embryogenesis. Plant Cell 5:1349–1359

Stanley RG, Linskens HF (1974) Pollen: Biology, Biochemistry and Management. Springer, Berlin

Tiwari SC, Polito VS, Webster BD (1990) In dry pear (*Pyrus communis* L.) pollen, membranes assume a tightly packed multilamellate aspect that disappears rapidly upon hydration. Protoplasma 153:157–168

Tiwari SC, Polito VS (1988) Spatial and temporal organization of actin during hydration, activation and germination of the pollen in *Pyrus communis* L: a population study. Protoplasma 147:5–15

Twell D (1994) The diversity and regulation of gene expression in the pathway of male gametophyte development. Society for Experimental Biology Seminar Series 55: Molecular and Cellular Aspects of Plant Reproduction. Eds RJ Scott & AD Stead. Pp 83–135. Cambridge University Press: Cambridge

Wagner VT, Dumas C, Mogensen HL (1990) Quantitative three-dimensional study on the position of the female gametophyte and its constituent cells as a prerequisite for corn (*Zea mays*) transformation. Theor Appl Genet 79:72–76

Willingale J, Mantle PG (1985) Stigma constriction in pearl millet, a factor influencing reproduction and disease. Ann Bot 56:109–115

CHAPTER 12

Pollen Coat Signals With Respect to Pistil Activation and Ovule Penetration in *Gasteria Verrucosa* (Mill.) H. Duval

M. T. M. Willemse

Laboratory of Plant Cytology and Morphology, Agricultural University, Arboretumlaan 4, 6703 BD Wageningen NL
e-mail: Michiel.Willemse@algem.pcm.wau.nl
telephone: ++3 17 48 24 55
fax: ++3 17 48 50 05

Abstract. In the pistil of *Gasteria* there is an activation by pollen coat proteins leading to water uptake of the pistil. This enhances the growth of pollen tubes of cross pollen. It also activates the ovule to attract the pollen tube. This may be a result of the exposition of the micropylar exudate with the attractant. The micropylar exudate attracts one or more pollen tubes. In vitro tubes may stop before they enter the micropyle which is marked by a swelling of the pollen tube tip. Thereafter they penetrate the funnel shaped micropyle.

12.1 Introduction

The succulent plant *Gasteria verrucosa* stores water in its leaves but needs water supply to realise sexual reproduction. When flowering, this South African plant makes a long inflorescence with about 40–60 flowers. After flower opening, at first the anthers with pollen appear. The pollen stick together in the presence of a pollen coat. After anther opening and the release of the pollen, the style elongates and the tip passes the anthers. Pollen can be placed on the stigma and self pollination takes place. During the receptive period a droplet of exudate appears on the stigma. Pollen germinate in this stigmatic exudate and the pollen tubes penetrate the solid part present in the tip of the hollow style. When the pollen tubes reach the open stylar canal, the pollen tube pathway consists of the fluid in the stylar canal and the placental fluid in the ovary locule. The final goal of the pollen tube is the ovular micropyle in which the micropylar exudate is present(Franssen-Verheijen and Willemse 1992).

To realise the arrival of the pollen tubes to the micropyle the pollen tube pathway needs water supply as a solvent for nutrients. Also the septal nectar glands produce 2–4 droplets of nectar during flowering. The flower needs water supply for the pollen tube pathway and the attraction of pollinators.

12.2 Ovular Incompatibility

Gasteria needs cross pollination for seed set. Although self pollination is common, the ovular self-incompatibility system hampers the seed development. In *Gasteria* this ovular incompatibility system was discovered by Sears in 1937. It concludes that self as well as cross pollen germinate on the stigma, pass the fluid stylar canal and placental fluid and penetrate into the micropyle. Sperm cells are transferred to a synergid and the egg cell and central cell are fertilised. After cross and self pollination the first mitosis in the endosperm takes place but in the case of self pollination further seed development will stop. The inner integument starts to degenerate followed by the degeneration of the outer integument, the embryo sac and the arillus. Compared with the normal pollen tube growth of cross pollen, the pollen tube has a retarded tube growth rate after self pollination. It contains a higher number of callose plugs and it has less callose in its wall (Willemse and Franssen-Verheijen 1988). These phenomena indicate similar reactions as described for a gametophytic self-incompatibility system. Because of the incompatibility reaction in the ovule and late response of the incompatibility reaction, this type of incompatibility is considered as ovular gametophytic incompatibility or as late acting incompatibility (Seavey and Bawa 1986). The ovular incompatibility system of *Gasteria* is genetically based on two or more loci (Brandham and Owens 1978; Naaborg and Willemse 1992). The incompatibility reactions are considered to be a result of signals between the pollen and the pistil. In gametophytic incompatibility, the germinating pollen form substances which react with products present in the transmitting tissue. Signal molecules present in the pollen coat mix with substances on the stigmatic surface, as occurs in sporophytic incompatibility (Brownlee 1994).

12.3 Pollen Coat Substances

The pistil receptivity of *Gasteria* is marked by a stigmatic droplet on expanded papilla cells, figure 12.1a,b.

The tip of the style is slightly bowed and the stigma is outside the flower or just inside the flower opening. The perianth starts to wilt and during this shrinkage the nectar in the basal part of the flower is pressed out of the flower. In nature pollen are transferred by small honey birds which climb over the inflorescence, sucking the nectar. Pollen stick on the bird's head and are transferred to other flowers of the same or other plants to promote cross pollination.

Clumps of pollen are transferred to the stigma and mix with the stigmatic exudate. The lipophilic components of the pollen coat are spread over the outside of the stigmatic droplet. The hydrophilic pollen coat components, such as proteins, are mixed in the hydrophilic part of the stigmatic exudate. Such separation of pollen coat substances occurs in the same way in vitro when a germination medium is used. The lipophilic part of the pollen coat covers the surface of the germination medium and its hydrophilic part mixes with the medium.

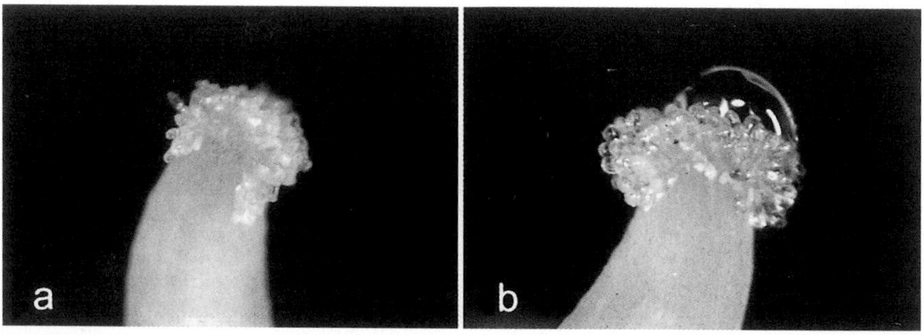

Fig. 12.1. Stigma before **(a)** and after receptive period **(b)** with stigmatic exudate on the widened papillar cells. ×40

Pollen coat substances are of sporophytic origin. They contain remnants of the locular fluid but they mainly contain products synthesised by the tapetal cells and remnants of degenerated tapetal cells. Pollen coat substances are considered to contain the products of early recognition between self or cross pollen and play a role in the recognition reactions on the stigma in case of sporophytic incompatibility.

The functional significance of the pollen coat can be established by pollination experiments. Using a high number of flowers the result of seed set after pollination with normal coated or washed cross pollen can be expressed in the yield of fruits. From such experiments with *Gasteria*, pollination with normal coated pollen has about 84% fruit set while the cross pollen washed in germination medium lead to a lower fruit set of about 14%. Cross pollen were able to set fruit but without the pollen coat the yield was strongly decreased (Willemse 1996). This means that the pollen coat contains signals to enhance fruit set after cross pollination. The presence of signal substances in the pollen coat is mentioned as recognition signal in case of sporophytic incompatibility but it seems that the signal found in *Gasteria* also is related to a successful fruit set after cross pollination.

12.4 Pollen Coat Signals

In previous experiments it became clear that during the collection of placental fluid for protein analysis, the contents of the placental fluid increased after cross pollination in the receptive period of the pistil and this results in a lower osmolarity of the placental fluid. The conclusion was that cross pollination enhances the increase of water in the placental fluid (Willemse and Franssen-Verheijen 1992a).

Based on these findings, self and cross pollinations were carried out in *Gasteria* using a high number of plants with unpollinated stigmas. Besides cross and self pollen, a small droplet of hydrophilic components, solved from the pollen coat, was also put on the stigma. As a control an unpollinated stigma was used. After 3 hours of pollination the weight of a) the whole style, b) the stylar canal fluid, c) the ovary and d) the placental fluid was measured. After weighting the style, the stylar fluid was pressed out and the placental fluid was collected by centrifugation of the ovary (Willemse 1996). The graphs a and b represent the results of the weights in 10^{-5} g of the pistil, the stylar fluid (=a), the ovary and the placental fluid (=b).

The weight of the style is about ten times lower than that of the ovary. The data of the unpollinated and self pollinated stigmas are lower than those of the cross pollinated

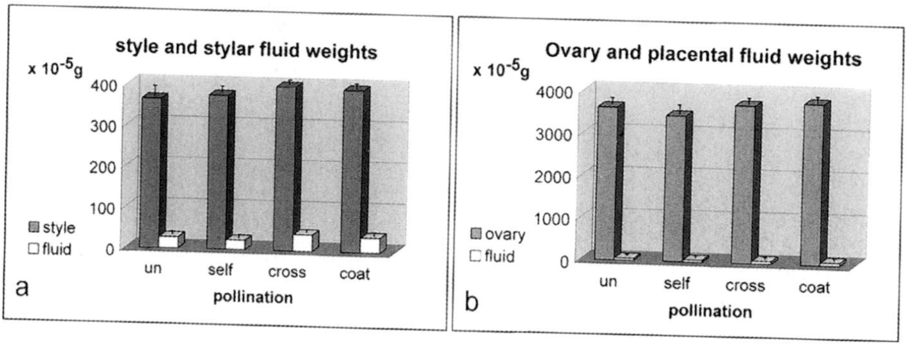

and pollen coat pollinated stigmas. The weights of the style and ovary of the self and unpollinated pistils are both about 7% lower than the weights of the cross and pollen coat pollinated pistils. The weights of the stylar and placental fluid of the self and unpollinated pistil are both about 30% lower then the cross and coat pollinated pistil. The style and ovary and their fluid parts increase in weight due to water uptake related to the cross and coat pollination.

Cross pollination evokes water uptake of the pistil. After cross pollination the fluid pollen tube pathway becomes more diluted which promotes the cross pollen tube growth. The retarded pollen tube growth of the self pollen can be explained as a consequence of the unchanged condition of the stylar and the placental fluid with still a higher osmolarity. The signal of water uptake by the pistil is present in the hydrophilic fraction of the pollen coat containing the proteins. This signal promotes the pollen tube growth of cross pollen.

Both pollen tubes from self and cross pollen will reach the micropyle and penetrate into the synergid. But the pollen tubes of self pollen are strongly retarded in their pollen tube growth. Therefore, the pollen coat signal can be characterised as an activation signal preparing the fluid pollen tube pathway for cross pollen but also as a recognition signal. The signal substance is present in the pollen coat and, in case of recognition, the self-incompatibility reaction is determined by the signal substances in the pollen coat which are of sporophytic origin. On the stigma, molecules should be present to accept the signal molecules but the nature of these products is still unknown.

12.5
Pollen Coat Proteins

In the hydrophilic part of the pollen coat, collected from the germination medium, several proteins and glycoproteins are present. Compared with the protein pattern of germinated pollen, the medium contains proteins which are also present in the germinating pollen. Only five groups of glycoproteins are specifically present in the medium. These glycoproteins are considered to come from the pollen coat (Willemse and Vletter 1995).

The pollen coat can be extracted with cyclohexane, a lipid solvent. Pollen were collected from the anthers of 20 flowers and shacked for 10 minutes in 1 ml cyclohexane. After filtration the cyclohexane was centrifuged for 1 minute at 500 g to remove the pollen. To the cyclohexane, 0.05 ml distilled water was added and the cyclohexane was evaporated by a strong air flow at 0° C. From the remaining water a sample of 2 ml was used for two dimensional electrophoresis with the Phast system (Pharmacia, Lund, Sweden).

For cyclohexane extraction two types of pollen were used, one sample of pollen from different plants, called "cross pollen", and another pollen sample of plants with a pollen mutation, called "self pollen". The self pollen are able to set seed, due to the overcoming of the self incompatibility system. Figure 12.2a, b represent the data.

This way of pollen coat extraction by cyclohexane cannot be considered to be representative for the pollen coat only. The sample contains proteins also present in germinated pollen. It is possible that the used pollen sample contained damaged pollen or that the pollen will burst during the procedure. Microscopical inspection shows that the pollen can swell in the cyclohexane.

Both protein patterns have a group of glycoproteins, marked as E in the figure, which is also present in the protein pattern of the hydrophilic part of the pollen coat. The short period of pollen coat extraction indicate the presence of these glycoproteins in the pollen coat only.

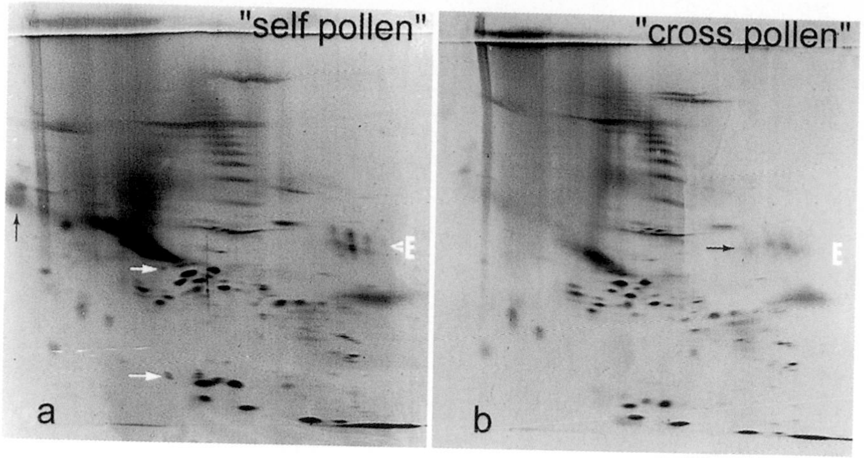

Fig. 12.2. Pollen coat cyclohexane extracted protein pattern of "self pollen" **(a)** and "cross pollen" **(b)**. groups of glycoproteins (*E*), the black arrow on **(a)** indicates the highly concentrated protein, the white arrows point to the other concentrated protein spots. The black arrow on **(b)** indicates the higher concentrated glycoprotein

The differences between the pollen samples are small. In comparison with the "cross pollen", the "self pollen" have another three proteins. In the pattern of the "self pollen" there is one higher concentrated spot and two more concentrated spots . These differences may characterise the "self pollen" mutation. In the group of glycoproteins of the "cross pollen" there is a more concentrated spot which may indicate a glycoprotein as candidate involved in the incompatibility recognition reaction.

12.6 Ovule Activation

Pollen activates the pistil after landing on the stigma. Some types of activation are known, such as enhanced protein synthesis in the ovules (Deurenberg 1977), or ethylene production of the pistil (Hoekstra and Weges 1986). The in-vitro micropylar penetration technique enables pollen tube growth into the micropyle. Ovular penetration takes place when isolated ovules and pollen are put together on a pollen tube growth adapted agar medium (Willemse et al 1995). With the in-vitro pollination system in *Gasteria,* ovule activation, expressed as a high percentage of pollen tube penetration in the micropyle, could be demonstrated as a result of cross pollination.

Unpollinated stigmas were pollinated with self pollen and cross pollen. As a control, unpollinated pistils were used. After three hours of pollination, when pollen tubes are halfway through the stylar canal, the ovules from self-, cross- and unpollinated pistils were isolated and put in a Petri dish with a mixture of pollen. The penetration percentage of the ovules from the different pollination type was counted. The unpollinated pistils have a penetration percentage of 15 and 16%, the self-pollinated have 21% and the cross-pollinated 39% penetration percentage(Willemse 1996). The higher micropylar penetration after the cross pollination is a sign that the ovules are better prepared to attract the pollen tubes (Willemse 1996).

In *Gasteria* the micropylar exudate is considered to function in the attraction of pollen tubes. Probably, proteins present in the filiform apparatus of the synergids are responsible for the pollen tube attraction (Willemse et al. 1995).

Before pollen tube arrival in the micropyle, a change in the shape of the embryo sac can be observed. The embryo sac is narrowing especially in the area of the egg apparatus. Figure 12.3a,b indicates the observed morphological change in the shape of the embryo sac.

The morphological change of the embryo sac coincides with a change in the filiform apparatus and some other changes in the embryo sac (Franssen-Verheijen and Willem-

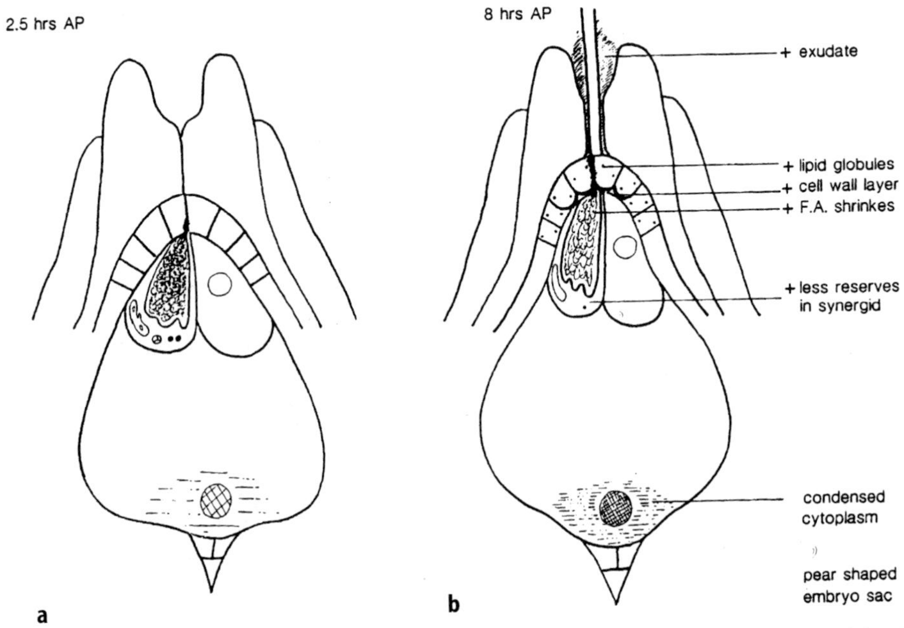

Fig. 12.3. Drawing of the ovule expressing the shape of the embryo sac 2.5 hrs after pollination **(a)** and just before pollen tube penetration **(b)**. In Figure 12.3b other changes in the pear shaped ovule are also noted

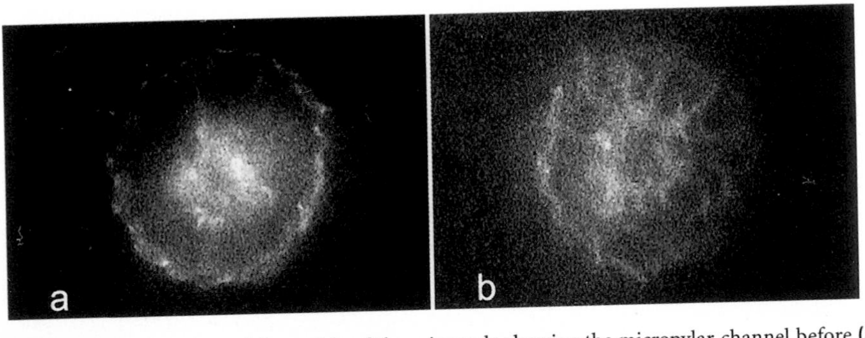

Fig. 12.4. Autofluorescence of the cuticle of the micropyle showing the micropylar channel before **(a)** and during the receptive period **(b)**, note the larger opening in **(b)**. × 150

se 1992b) The change can be a result of water uptake in the ovule after pollination. The integuments surrounding the embryo sac may swell and this may result in a pressure on the embryo sac which becomes pear shaped. This pressure can cause the export in the micropylar channel of a part of the filiform apparatus. As a consequence the micropyle fills itself with the micropylar exudate which is coming from the synergid filiform apparatus and the attractant is better presented to attract the pollen tubes.

The swelling of the inner integument can be observed in the morphology of the micropyle. The micropyle of *Gasteria* is funnel shaped. A narrow channel of inner integument is positionned against the nucellar tissue and this channel widens towards the outside of the micropyle bordered by the inner integument. The morphology of the micropyle can be marked by the autofluorescence of the cuticle and the border of the upper part of the micropylar channel can be observed. Figure 12.4a,b shows two micropyles, one of a normal micropyle (=a) before and one during the receptive period (=b).

In comparison with a normal micropylar channel an enlarged micropylar opening can be observed during the receptive period. The observed morphological change of the micropyle can be explained by the swelling of the inner integument which borders the micropylar channel, as a consequence of the water uptake by the pistil after pollination due to pollen coat signals. However the micropylar widening is not always clear and may be absent. Besides, the micropylar opening depends on the moment of ripening of the ovules which starts in the basal part of the ovary and gradually runs in the direction of the style.

12.7 Micropylar Penetration by the Pollen Tube

In vivo pollen tube penetration in the micropyles can be easily demonstrated by aniline blue stained pollen tubes and visualised by UV microscope. Figure 12.5a, b shows the penetration with one (=a) or more pollen tubes in a micropyle (=b).

In case of the penetration of one pollen tube in the micropyle, some torsion of the tube or a local swelling in the tube can be observed. The penetration of more tubes in a micropyle, called "multiple pollen tube penetration", happens frequently. In such a case some of the pollen tubes look more straight and some of them make a kind of loop. The penetration process of pollen tubes in the micropyle in vivo is seldom reported, but it seems reasonable that pollen tubes growth is directed towards the micropyle.

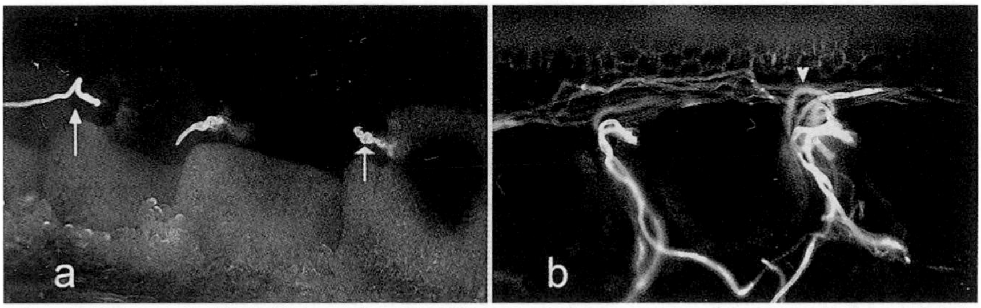

Fig. 12.5. Pollen tube penetration in vivo of aniline stained fluorescent pollen tubes. Single pollen **(a)** and multiple pollen **(b)** penetrations. In **(a)** the small arrow indicates the torsion in the pollen tube and the wide arrow indicates a swelling. In **(b)** the arrowhead points to a loop in the pollen tube. ×150

The in-vitro penetration system makes it possible to show the process of pollen tube penetration. The penetration of one tube or more tubes can be observed by using video microscopy. Some events of pollen tube behaviour in vitro can be registered: the penetration of a pollen tube in the micropyle, the multiple penetration of pollen tubes and the micropylar refusal of pollen tube.

12.7.1 Pollen Tube Penetration in the Micropyle

In Figure 12.6, a registration is given of a pollen tube near but lateral from a micropyle. This penetration was registered and within about 4 minutes the tube penetrates the micropyle.

Because of the lateral approach of the pollen tube from behind the micropyle, the tube has to turn for micropylar penetration. This means a change in the direction of the pollen growth. These events are arguments to postulate that an attractant is present in the micropyle.

During the approach of the pollen tube to the micropyle there is a delay in growth and a small swelling of the pollen tube tip appears, just above the opening of the micropyle. Such a delay or stop in the pollen tube elongation is observed frequently. Most of the pollen tubes wait for a moment before they penetrate the opening of the micropyle. During this delay, some pollen tube tips will swell and from this swelling there is again a further growth of the pollen tube tip into the micropyle. The stop or slower elongation of the pollen tube can be explained by two reasons. The micropylar exudate of *Gasteria* coming out of the micropyle consists of a more compact substance. This means that

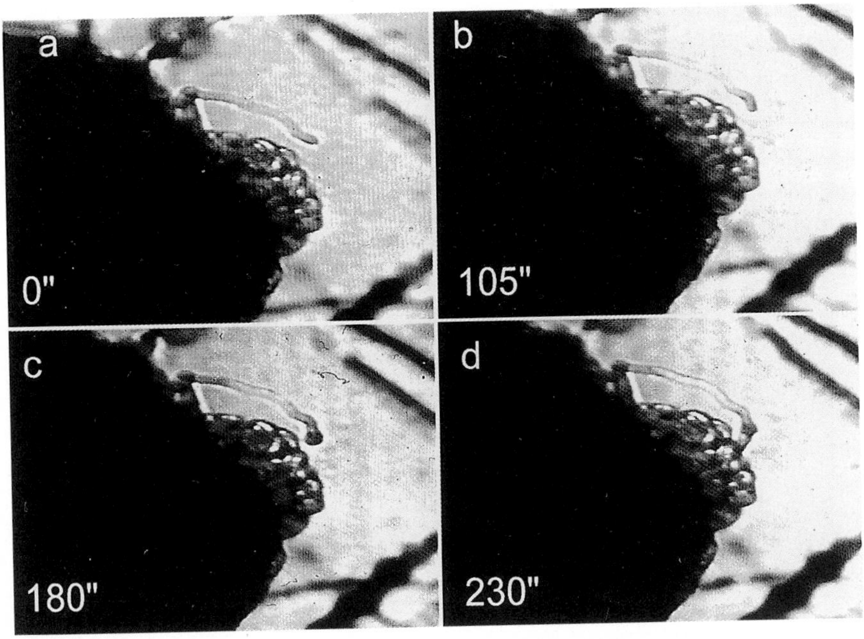

Fig. 12.6a–d. Pollen tube penetration in vivo visualised by a series of pictures from video microscopy. The seconds are noted on the picture. ×100

there may be a kind of mechanical or resistant barrier. Another possibility is the loss of the osmotic pressure which causes the elongation of the pollen tube. The last effect seems to be more common since after the stop of the pollen tube the exocytosis proceeds, forming a new membrane and the pollen tube wall and the tip swells. After the stop and swelling, the osmotic pressure is repaired and the pollen tube growth continues by forming a tip directed to the micropylar opening.

12.7.2
Multiple Pollen Tube Penetration

In case of multiple pollen tube penetration other phenomena in and around the micropyles can be observed. Figure 12.7a–i represents different steps of the multiple pollen tube penetration within one hour.

Figure 12.7a gives a view of a micropyle with a penetreted pollen tube. In front of the micropyle a local swelling of the pollen tube is visible, a sign of the short arrest of the pollen tube before penetration. After penetration as presented in Figures 12.7b–h, this pollen tube (A) continues its growth and, as a consequence, the tube elongates at the tip which stays in the micropyle. As a result, the pollen tube starts to make a small loop outside the ovule, see Figure 12.7c. Continuing growth more and more loops are formed. The pollen tube cytoplasm remains in the tip area. Such extensive curling of the pollen tube means that the pollen tube tip growth in the micropyle is not retarded and probably a further penetration through the narrow canal is disturbed.

In the meantime a second pollen tube (B) arrives from the lateral side of the micropyle, as is shown in Figure 12.7d . This tube stops and swells, see Figure 12.7e, thereafter, visible in Figures 12.7f, g the tip begins to penetrate. But during the penetration of this tube, a third tube (C) arrives as can be seen in Figure 12.7g. This third tube arrests and waits for a moment. But the still growing second pollen tube (A) presses this new coming tube (C) away from the micropyle, see Figure 12.7h and i. A fourth fast growing pollen tube (D) passes the micropyle but retards its growth, see Figure 12.7j. In the meantime, the pressure of pollen tube (A) is no longer hampering the third pollen tube (C). Now this third pollen tube (C) succeeds in entering the micropyle and the pollen tube tip makes a small swelling as can be seen in Figure 12.7k. The Figure 12.7l shows that the fourth pollen tube (D) is stopped after its pass of the micropyle and forms a swelling at the tip and stays arrested next to the micropyle. Finally four of the pollen tubes are positioned in front of the micropyle, three of them penetrated and one arrested.

The events of multiple penetration reveal influence of the attractant present in the micropyle. Considering the fast growing pollen tube (D) which passes the micropyles the attractant is not a very fast acting signal, nevertheless it seems that the pollen tube becomes influenced and stops. Therefore the signal given by the micropyle is relatively strong. This can also be noted of the third pollen tube (C) which was pushed away but still able to penetrate. A micropyle is able to attract more than one pollen tube, a sign that the attractant remains active, also after the first penetration. The attractant can be strong, can be present for a longer period and influences or stops the pollen tube elongation. After the stop of the pollen tube the attractant directs the pollen tube to enter the micropyle.

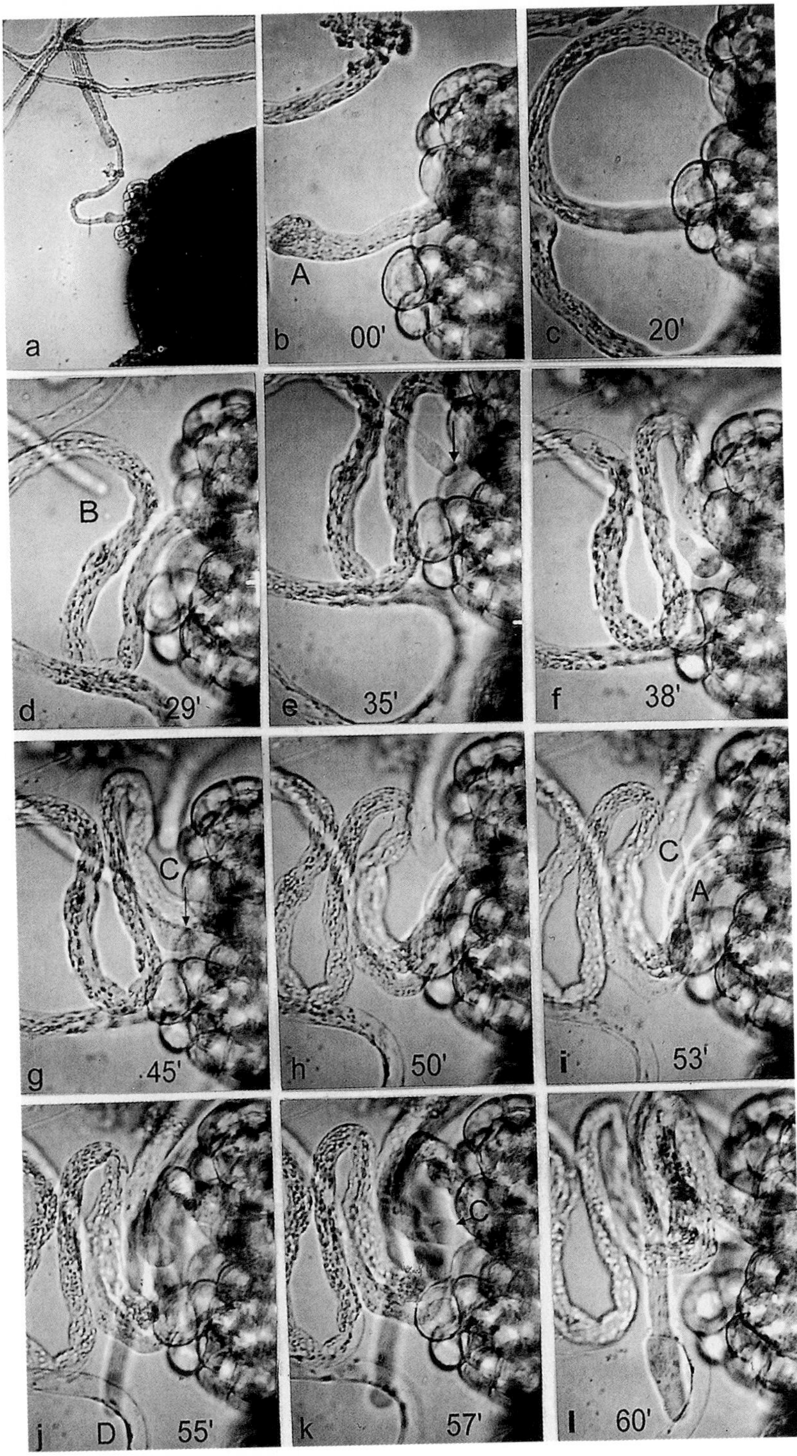

Fig. 12.7a–l. Multiple penetration of four pollen tubes represented by a series of pictures from video microscopy. The minutes are given on the pictures. *A* = the first, *B* = the second *C* = the third and *D* = the fourth pollen tube. The arrow on **(e)** indicates the swelling of the pollen tube tip of B and the arrow on **(g)** indicates the penetration of this tip. ×400

12.7.3 Refusal of Pollen Tubes

In vitro other events can be observed. Figure 12.8a, b collects some examples of pollen tubes which do not penetrate the micropyle.

In some cases, see Figure 12.8a, the pollen tubes stop growing further in front of the micropyle and swell but they remain in such a condition and do not penetrate the micropyle. In this Figure 12.8a the shape of the micropyle is somewhat contracted and the inner integument bordering the micropyle has an incomplete swelling.

After refusal another phenomena can occur. Pollen tubes pass the micropyle or are all positioned in front of the micropyle and none of them have penetrated the micropyle but all are turned away from the micropyle, see Figure 12.8b.

This behaviour in vitro indicates that the penetration of the pollen tube can be divided into two parts. First the pollen tubes arrive and stop, which can be influenced by the presence of the consistent micropylar exudate and the change of pollen tube osmotic pressure. Thereafter the pollen tube tip swells, such swelling can also be observed in vivo. The pollen tube growth restarts from the swollen tip and the tube orientates itself in the direction of the micropyle. This last effect is due to the presence of the attractant which influences and orientates the pollen tube growth.

12.8 Final Remark

Signals during the progamic phase from pollination to penetration in the micropyle create interactions between pollen and pistil in the interface. Signal substances present in the pollen coat, in the pollen tube path way and in the micropyle lead to recognition, activation and attraction of pollen and pollen tubes. Between pollen and pistil complex interactions occur.

The pollen tube pathway in *Gasteria* is heterogeneous, this means that the pollen tube growth occurs through different part of the pathway with different conditions.

In *Gasteria* the final incompatibility reaction takes place in the ovule but from the beginning of the pollen tube germination signs of gametophytic incompatibility are present.

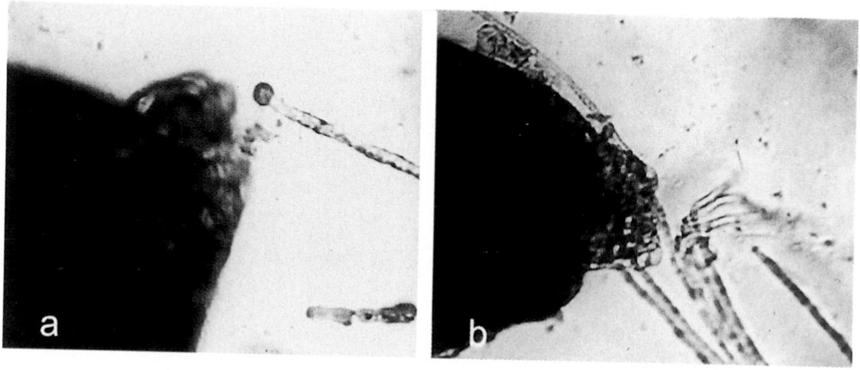

Fig. 12.8. Refused pollen tube penetration. The pollen tube tip swells but stops and does not penetrate the micropyle **(a)**. Pollen tubes near but turning away from the micropyle **(b)**. × 100

The pollen coat signal involves the pistil activation as well as pollen recognition; this signal is of sporophytic origin and related to sporophytic incompatibility.

The pollen tube attractant in the micropyle influences the growth and orientation of the pollen tube.

Acknowledgements
The author thanks Drs. B. Willemse for corrections and Dr. J. H. N. Schel for the critical reading of the manuscript. Mr. S. Massalt for digitalizing the figures.

References

Brandham PE, Owens SJ (1977) The genetic control of self-incompatibility in the genus *Gasteria* (Liliaceae). Heredity 40:155–169
Brownlee C (1994) Signal transduction during fertilization in algae and vascular plants. New Phytologist 127:399–418
Deurenberg JJM(1977)Differentiated protein synthesis with polysomes from *Petunia hybrida* ovaries before fertilisation. Planta 133:201–206
Franssen-Verheijen MAW, Willemse MTM (1993) Micropylar exudate in *Gasteria* (Aloaceae) and its possible function in pollen tube growth. Am J Bot 80:253–262
Hoekstra FA, Weges R (1976) Lack of control by early pistillate ethylene of the accelerated wilting in *Petunia hybrida* flowers. Plant Physiol 80:403–408
Naaborg AT, Willemse WTM (1992) The ovular incompatibitlity system in *Gasteria verrucosa* (Mill.) H. Duval. Euphytica 57:231–240
Sears ER (1937) Cytological phenomena connected with self-incompatibility in flowering plants. Genetics 22:130–171
Seavey SR, Bawa KS (1976) Late-acting self-incompatibility in Angiosperms. Botanical Rev. 52:195–219
Willemse MTM (1996) Progamic phase and fertilisation in *Gasteria verrucosa* (Mill.) H. Duval: pollination signals. Sex Plant Reprod 9:347–352
Willemse MTM, Franssen-Verheijen MAW (1988) Pollen tube growth and its pathway in *Gasteria verrucosa* (Mill.) H. Duval. Phytomorphology 38:127–132
Willemse MTM, Franssen-Verheijen MAW (1992a) Pollen tube growth in the ovary of *Gasteria*. In: Batygina TB (ed) Embryology and seed production. Academy of sciences of Russia, St. Petersburg. Proceedings of the XI international Symposium, Nauka Leningrad, pp 613–614
Willemse MTM, Franssen-Verheijen MAW (1992b) Pollen tube growth and and ovule penetration in Gasteria verrucosa (Mill.) H. Duval. In: Ottaviani E, Mulcahy DL, Sari Gorla M, Bergaminni Mulcahy G (eds) Angiosperm pollen and ovules. Springer, New York Berlin Heidelberg, pp 168–173
Willemse MTM, Plyush TA, Reinders MC (1995) In vitro micropylar penetration of the pollen tube in the ovule of *Gasteria verrucosa* (Mill.) H. Duval and *Lilium longiflorum* Thunb: conditions, attraction and application. Plant Sci 108:201–208
Willemse MTM and Vletter A (1995) Appearance and interaction of pollen and pistil pathway proteins in *Gasteria verrucosa* (Mill.) H. Duval. Sex Plant Reprod 8:161–167
Willemse MTM (1996) Progamic phase and fertilisation in *Gasteria verrucosa* (Mill.) H. Duval: pollination signals. Sex Plant reprod 9:348–352

Tapetum and Orbicules (Ubisch Bodies): Development, Morphology and Role of Pollen Grains and Tapetal Orbicules in Allergenicity

G. El-Ghazaly

Swedish Museum of Natural History, Stockholm, Sweden
e-mail: gamal.el-ghazaly@nrm.se
telephone: +46 8 666 41 93
fax: + 46 8 16 77 51

13.1 Introduction

13.1.1 Tapetum and Orbicules (Ubisch Bodies)

13.1.1.1 *Tapetum*

The tapetum is the innermost layer of anther wall and surrounds the sporogenous tissue. It plays an important role in pollen development, namely nourishment of microspores and formation of exine. At the final stages of pollen maturation, the tapetal products take part in the deposition of tryphine and pollenkitt.

The cells of the tapetum are generally multinucleate and polyploid. In angiosperms the tapetum is of two main types: a – glandular or secretory type in which the tapetal cells may remain intact and persist in situ until their degeneration, b – the periplasmodial or amoeboid type, where the inner tangential and radial walls of tapetal cells break down, and the protoplasts protrude into the locule and fuse to form a coenocytic plasmodium.

Generally, the tapetum comprises a single layer; exceptionally it may divide and become biseriate throughout as in *Pyrostegia*, *Tecoma* and in *Magnolia* (Kapil and Bhandari 1964). A multiseriate condition is known in *Combretum grandiflorum* and *Oxystelma esculentum* (Maheswari Devi and Kakshminarayana 1976).

13.1.1.2 *Orbicules*

Most angiosperms with a secretory tapetum produce orbicules. Orbicules are resistant to acetolysis and react similar to the exine with different histochemical stains providing ample evidence for sporopollenin composition of the orbicule wall (El-Ghazaly and Jensen 1987). Skvarla and Larson (1966) interpreted orbicules to represent sporopollenin deposition upon membranes aggregated at the tapetal surface.

Pro-orbicules were believed to appear between the tapetal plasmalemma and the degenerating walls of the tapetal cells (Horner and Lersten 1971; Christensen et al. 1972; Steer 1977). The timing of pro-orbicules formation varies slightly between the different species. They form at early stages of meiosis II in *Avena* (Steer 1977), late tetrad stage in *Sorghum* (Christensen et al. 1972) to microspores release from the tetrads in *Citrus*

(Horner and Lersten 1971). Generally, pro-orbicules formation is considered to be intra-tapetal (e.g. Echlin and Godwin 1968; El-Ghazaly and Jensen 1986a). Suarez-Cervera et al. (1995) described the so-called 'grey bodies' in the tapetal cytoplasm to be the progenitors of the Ubisch bodies .

The mechanism of formation of Ubisch bodies wall is mediated by the plasmalemma coating "glycocalyx" (Rowley and Skvarla 1974; Abadie and Hideux 1979; El-Ghazaly and Jensen 1985; Clément and Audran 1993a, b).

In monocots the orbicular wall seems to develop on a specific reticulate pattern on the plasmalemma. For example, Rowley and Skvarla (1974) reported a plasmalemma glycocalyx in pro-orbicules of *Phleum pratense*, in *Triticum* (El-Ghazaly and Jensen 1985) and in *Lilium* (Clément and Audran 1993b). The coat of the pro-orbicules, which appears to guide the deposition of sporopollenin, is positively stained for polysaccharides and proteins, similar to the pollen wall (Rowley 1976; Rowley and Dahl 1982; El-Ghazaly and Jensen 1987).

Orbicules are variable in shape and size (El-Ghazaly 1989). Generally they are less than 5 microns in diameter. Frequently, similarities between morphology of orbicules and the respective pollen sexine have been reported (Christensen et al. 1972, for *Sorghum;* El-Ghazaly and Jensen 1986a, for *Triticum*; Clément and Audran 1993a, b, for *Lilium*; Suarez-Cervera et al. 1995, for *Platanus acerifolia*). Although this is not always true (El-Ghazaly and Nilsson 1991; Huysmans et al. 1997).

The possible taxonomic value of orbicules has been mentioned by (Heslop-Harrison 1962; Rowley 1963; Banerjee 1967). Banerjee argued the possibility of using orbicules as criterion in taxonomy of grasses since their size, shape and abundance vary for different species. Despite this awareness only few studies were published on the systematic importance of orbicules characteristics.The work of Ueno (1959, 1960) in gymnosperms was pioneering in this respect. The taxonomic applicability of orbicules in angiosperms has been demonstrated in Chloanthaceae (Raj and El-Ghazaly 1987) and in the genus *Euphorbia* (El-Ghazaly 1989; El-Ghazaly and Chaudhary 1993). In many cases pollen morphological similarity between species is also reflected in their orbicules. However, conflicting results may occur. In a recent paper (Huysmans et al. 1997) character states of orbicules have been delimited in the Cinchonoideae-Rubiaceae on the basis of SEM-observations. Evidence is given that, at least in this group, orbicule characteristics are systematically useful on generic or tribal level. Striking analogies between orbicules and the respective sexine seem to be common in higher plants (Hesse 1986). The similarities are not only macro-morphological but also in the finest micromorphological characters of the structural elements in orbicules and the respective sexine (e.g. El-Ghazaly and Jensen 1986a). These parallelisms are explained by Hesse (1985) as rooted in the homology of tapetum and sporogenous tissue.

Ontogenetic disturbance in the structure and/or function of the tapetum leads to degeneration of pollen, so characteristic of the cytoplasmic male sterile angiosperms (Gifford and Foster 1989). In a comparative study on the tapetal behaviour in male fertile and male sterile *Iris pallida*, Lippi et al. (1994) found no orbicules at all in the sterile clone. In the male fertile line the inner anther surface was completely covered by orbicules. The same pattern was observed for *Zea mays* (Colhoun and Steer 1981) and *Beta vulgaris* (Hoefert 1971; Nakashima 1975). Young et al. (1979) studied anther and pollen development in male sterile intermediate wheat grass plants. They found different expressions of male sterility including orbicular wall malformations. De Vries and Ie (1970) have shown that male sterility does not necessarily affect orbicule formation. They investigated anther tissue and pollen grains of cytoplasmic male sterile and fertile wheat (*Triticum aestivum*) for differences in cytoplasmic structures and concluded that orbicules occur in the anthers of both male sterile and fertile plants, with no visible differences.

13.1.2 Allergenicity of Pollen Grains and Tapetal Orbicules

Pollen grains, like other plant cells, contain different types of proteins, which are located in the cytoplasm and the outer wall. Only a small number of the proteins are allergenic. Airborne pollen has been measured for many years in several European countries to get information on daily quantities of allergenic pollen and their relations to symptoms of pollen allergy, and to make forecasts of coming airborne pollen concentration in the air. Mostly the aeroallergen information service is based on the volumetric sampling of airborne pollen grain on adhesive tape of the Burkard pollen and spore trap. So far, very little is known about the structure and characteristics of submicronic allergens in the air. Pollen grains of grasses, ragweed, olive trees and white birch are among the most allergenic types. For example pollen grains of the white birch (*Betula pendula*) is one of the main causes of allergic reactions in Middle and Northern Europe, North America and Russia. The birch allergies are a major threat to public health in these countries, since 10-15 % of the population suffers from these diseases. Earlier studies have demonstrated the presence of allergens in association with cytoplasm of pollen grains of birch trees (Grote 1991; Grote et al. 1993). The major allergen, Bet v 1, as well as some minor allergens, Bet v 2 and Bet v 3, have recently been identified using the technique of gene cloning (Seiberler et al. 1994). However, the information on the precise distribution of allergens in the cytoplasm has been rarely investigated. Knox et al. (1970) assumed that the pollen allergens were stored in the exine or intine of the pollen grains. Belin and Rowley (1972) concluded that the allergens in *Betula* are probably stored inside the pollen protoplasm and they diffuse through the aperture during germination. Staff et al. (1990), Taylor et al. (1994), and Knox and Suphioglu (1996) localized allergens in the starch granules and cytosol of rye-grass pollen. El-Ghazaly et al. (1996) localized the main allergen Bet v1 in the starch granules of *Betula pendula* pollen grains

Non-pollen sources also might contribute to airborne allergens. For example the anther sacs are lined by tapetal cells, which provide the developed pollen grains with essential nutrients. In many species the tapetal cells are covered with numerous bodies called Ubisch bodies or Orbicules. These structures are generally less than ca 5 µm in diameter and formed during microspores development. The question as to whether orbicules also contain allergens was first raised by Davis (1966). Orbicules are often much more numerous than pollen grains and extremely small so they can pass easily through the pores of most protective masks (Miki-Hirosige et al. 1994) and airfilters. Intracellular localization of the two major allergens in *Lolium perenne* using specific monoclonal antibodies showed no labelling on the orbicules or in the tapetal cells (Taylor et al. 1994). On the other hand, Miki-Hirosige et al. (1994), using both monoclonal and polyclonal antibodies, localized the major allergen responsible for Japanese cedar pollinosis in the orbicules and on the tapetal materials remaining in the young anther. Moreover, El-Ghazaly et al. (1995) using polyclonal antibodies presented the possible allergenicity of orbicules in *Betula*. It is assumed that some of the allergenic protein is produced in the tapetum and transferred to the orbicules and pollen wall during maturation.

13.2 Localization of Allergens in Pollen Grains and Orbicules

Pollen grains from *Betula pendula* were collected at the beginning of May1995, 1996 in the Stockholm area from catkins that had just begun to release their pollen. They were directly fixed with 1.5% paraformaldehyde and 1.5% glutaraldehyde containing 0.2% picric acid and buffered by 0.05 M sodium cacodylate at pH 7.5 (Maldonado et al. 1986), dehydrated in an alcohol series and embedded in Spurr's resin, and sectioned with a diamond knife. In addition mature anthers of *B. pendula* were rapid freeze fixed and substituted in an alcohol series and embedded in LRW. Immunogold labelling was performed by the method of Probert et al. (1981). The ultrathin sections were treated with a drop of the mouse monoclonal antibody (anti BV-10) against Bet v 1 (IgG1, ALK Laboratories, Hörsholm, Denmark), prepared at dilution of 1:500 in 0.05M Tris-buffered saline (pH 7.2), and incubated for 24 h at 4°C, sections were then incubated in 10 nm colloidal gold-labelled goat anti-mouse IgG (GAM, 1:20, Amersham International PLC, Amersham, UK) for 60 min. at room temperature. For control staining, to demonstrate the specificity of immunogold-labelling, the antiserum was replaced with normal goat serum. After immunogold staining, ultrathin sections were stained with uranyl acetate and then lead citrate and observed with a JEM 100B electron microscope (JEOL Co, Tokyo, Japan).

13.3 Tapetum and Orbicules (Ubisch Bodies)

13.3.1 Tapetum

The tapetal cells form a well distinguished layer which surrounds the microsporogenous tissue in the anther (Fig. 13.1).

13.3.1.1 *Characteristics of the Tapetal Cells*

1. They are distinctly enlarged and always ephemeral.
2. The cytoplasm is rich in ribosomes and active organelles.
3. Tapetal cells are comparatively rich in DNA and they may be multinucleate or polyploid.

Fig. 13.1. *Betula pendula.* Freeze fracture of anthers showing anther loculi full of young microspores surrounded by the tapetal cells (arrows). SEM. × 100. Scale = 100 µm ▶

Fig. 13.2. *Triticum aestivum.* Part of orbicular wall (arrowheads) and two orbicules with central core and thick wall interrupted by microchannels. TEM. × 36000. Scale = 1 µm

Fig. 13.3. *Betula pendula.* Tapetal cells covered with orbicules (arrowheads) and expanding between the microspores. SEM. × 5000. Scale = 2 µm

Fig. 13.4. *Ephedra foliata.* Tapetal markers developed on the abaxial walls between tapetal cells (arrow). Orbicules have a central core zone and a white line. TEM. × 52000. Scale = 1 µm

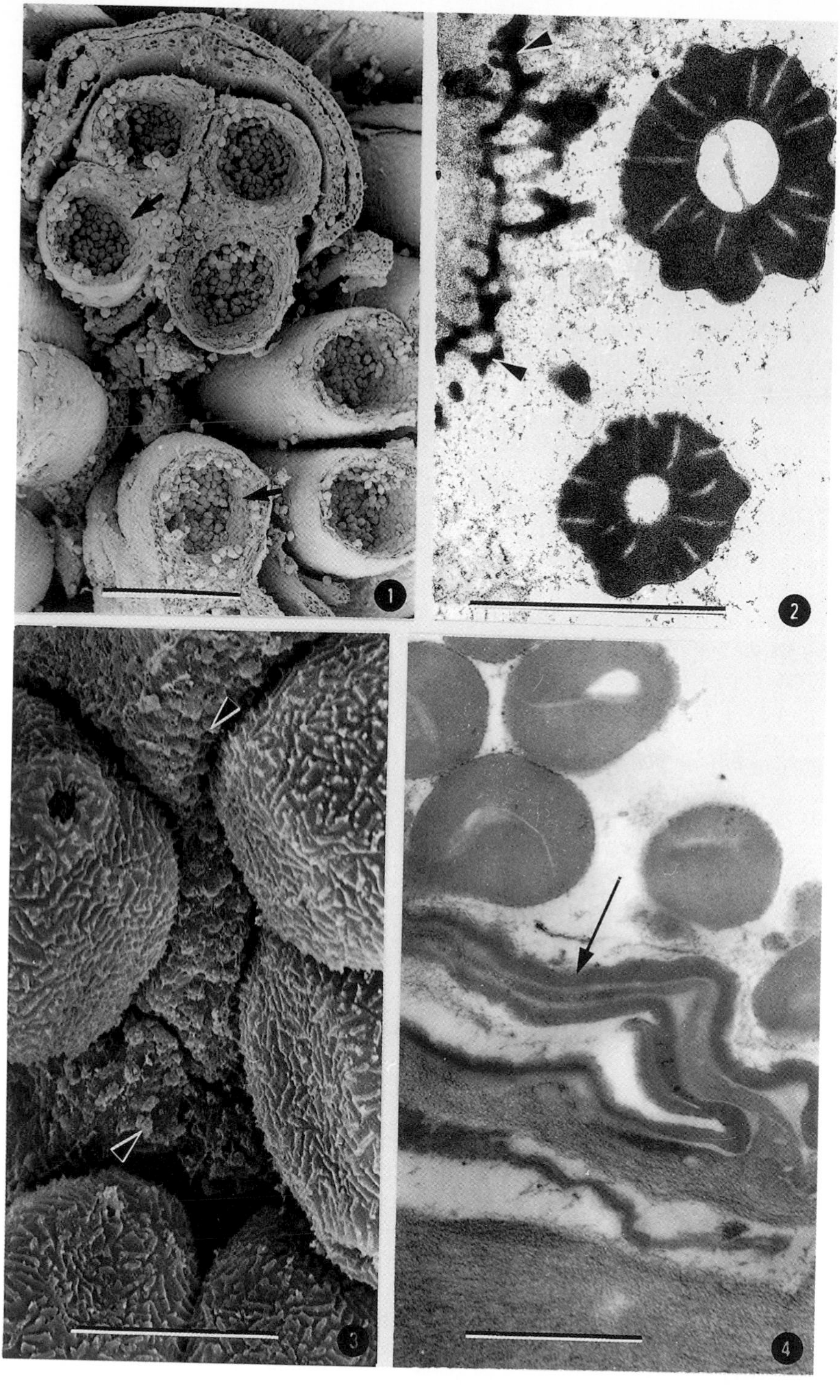
1
2
3
4

4. Various irregular mitotic divisions and nuclear fusion take place in the tapetal cells.
5. They are characterised by rapid and intense activity ending with degeneration of their cytoplasm.

13.3.1.2
Types of Tapetum

The tapetum tissue is mainly divided into two main types

Glandular type (secretory), in which the cells remain in their original position, and later disintegrate.
Amoeboid type (invasive or periplasmodial), in which the tapetal cells fuse among themselves to form a tapetal periplasmodium. The protoplasts of the fused tapetal cells penetrate between the pollen mother cells or the developing pollen grains.

In several cases the tapetal cells repeatedly enter into periods of hypersecretory activity until near maturity of pollen grains. For e.g. in *Nemphaea* (Rowley et al. 1992; Gabarayeva and El-Ghazaly 1997), there are two distinct intervals during which tapetal cells protrude into anther loculi. Following each interval of locular invasion tapetal cells retracted and arranged as a palisade around the loculus.

As information has been built up, the border between different types of tapetum tissue has become less clear. Variations in the morphology of the tapetum may be an adaptation to pollen grain size, shape, type of dispersal, pollenkitt, and probably other factors (Pacini 1990).

13.3.1.3
Functions of Tapetum

The main function of tapetum tissue is supplying nutrients to the developing microspores. In addition the tapetal cells may take part in the following activities:

1. Secretion of enzyme callose to dissolve the callosic wall of the tetrad (Stieglitz 1977) and set them free.
2. Secretion of polysaccharides into the loculus during free microspore stage, which are absorbed by the microspores (Pacini and Franchi 1983).
3. Formation of exine precursors, which according to the recent results of degradation experiments (Schulze Osthoff and Wiermann 1987) could be phenols or p-coumaric acid (Wehling et al. 1989).
4. Tapetal cells might have other activities that result in formation of different structures, which characterise some families. For example in Onagraceae tapetal cells play a role in formation of fine flexible threads, known as viscin threads, continuous with the outer layer of the exine (Hesse 1984). In family Compositae (Asteraceae) the tapetum forms an acetolysis resistant membrane outside the sporogenous tissue (Heslop-Harrison 1969). This membrane is known as the culture sac or peritapetal membrane (Shivanna and Johri 1985).
5. Formation of orbicules (Ubisch bodies).
6. Formation of pollenkitt, which is a hydrophobic layer composed mainly of lipids and carotenoids (Pacini and Casadoro 1981; Rowley and El-Ghazaly 1992). Pollenkitt is generally deposited on the exine surface and helps to bind pollen grains together, probably for efficient insect pollination.

7. Formation of tryphine, which is a mixture of hydrophobic and hydrophilic substances. Tryphine is usually deposited on the pollen surface and possibly help in protection of pollen grains and in insect pollination (Dickinson and Lewis 1973a, b). Tryphine is mainly distinguished from pollenkitt by ist hydrophilic nature. Besides, in the case of tryphine the tapetal cells membranes are ruptured and their contents released to cover the pollen surface, as in the case of *Brassica*.

13.3.2 Orbicules (Ubisch Bodies)

Orbicules were first observed by Rosanoff (1865) in close association with the tapetum. Schnarf (1923) and Ubisch (1927) have published useful observations on orbicules with LM. Schnarf (1923) apparently first described such exine-like bodies in *Lilium martagon*. Kosmath (1927) suggested that these bodies should be called "Ubisch granules", because of the contribution of Ubisch toward their character. Heslop-Harrison (1962) used the term plaques to describe such bodies; and Erdtman et al. (1961) proposed the term orbicules. The orbicules are very small structures which could once be investigated solely by light microscopy, however with the advent of the electron microscope it is now easier to identify them and study their morphology and ontogeny.

13.3.2.1 *Morphology of Orbicules*

Orbicules are generally spheroidal structures found in the anthers of many genera of angiosperms, both monocotyledons, e.g. *Triticum aestivum* (Fig. 13.2) and dicotyledons, e.g. *Betula pendula* (Fig. 13.3) and many gymnosperms, e.g. *Ephedra foliata* (Fig. 13.4). The shape of orbicules may vary in different species. They might be spheroidal with spinulose surface, rod-like, doughnut shape, oval or rounded triangular or plate-like with undulating or jagged margin, etc. (Figs. 13.5–13.10). Orbicules frequently fuse into large compound aggregates as, for example, in *Euphorbia caputmedusae* (Fig. 13.10). Orbicules generally have a central core and thick wall with microchannels (Fig. 13.11). When anthers of *Betula pendula* were freezed in liquid nitrogen, fractured and examined with SEM, orbicules were connected together by filamentous structures (Fig. 13.12). In addition orbicules were also connected to the wall of the microspores (Fig. 13.13). The wall of orbicules resists acetolysis and apparently consists of sporopollenin. Both orbicules and pollen wall have similar stainability for protein (Fig. 13.14) and acidic polysaccharides (Fig. 13.15). The surface pattern of orbicules and exine can be remarkably similar (El-Ghazaly and Jensen 1986a). The morphological variation of orbicules could be of taxonomic importance (Raj and El-Ghazaly 1987; El-Ghazaly 1989; Huysmans et al. 1997).

13.3.2.2 *Functions of Orbicules*

Several investigators have suggested many different functions of orbicules. Echlin (1971), Bhandari (1984) and Pacini et al. (1985) provided a review of the early literature on the different hypotheses that have been proposed to attribute a function to orbicules. So far none of the suggested functions of orbicules has been definitely proved.

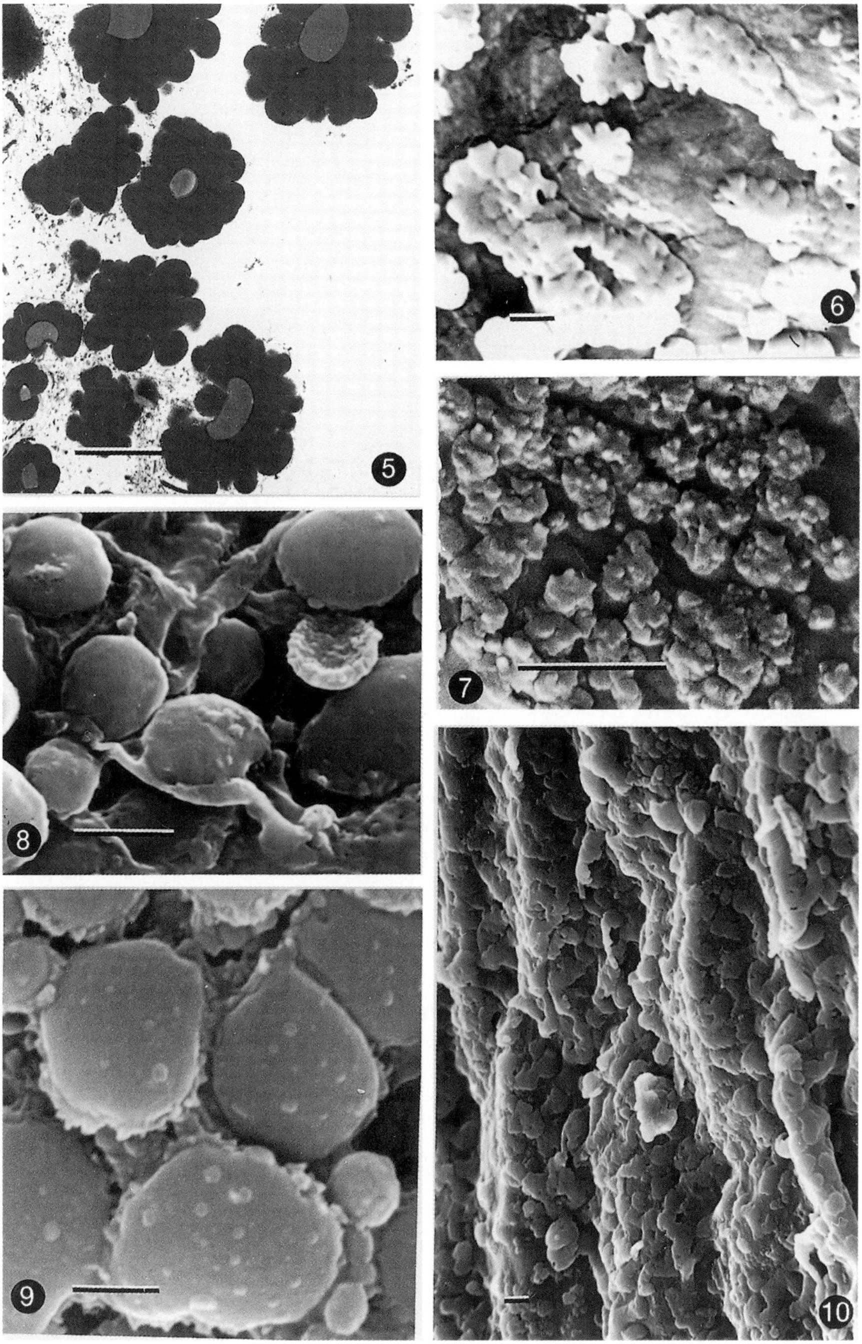

Figs. 13.5–10. Variations in the morphology of orbicules. **13.5.** *Lilium pyrenaicum*. TEM. ×6000. **13.6.** *Avicenia marina*. SEM. ×3000. **13.7.** *Betula pendula*. SEM. ×10,000. **13.8.** *Euphorbia prostrata*. SEM. ×7000. **13.9.** *Euphorbia monteiri*. SEM. ×6000. **13.10** *Euphorbia caputmedusae*. SEM. ×1500. (Scale = 2 μm)

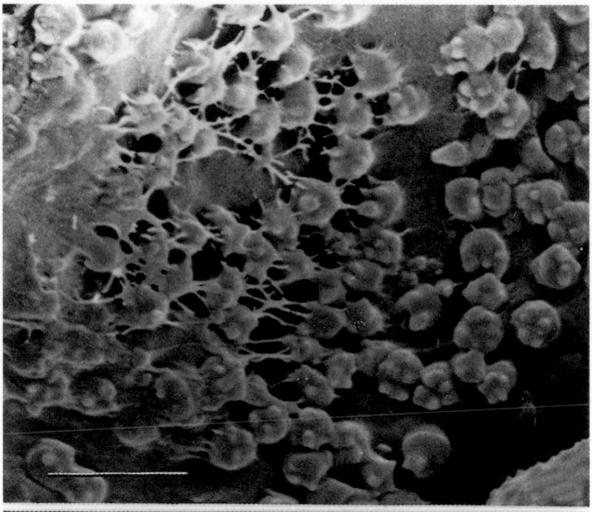

Fig. 13.11. *Betula pendula.* A part of tapetal cell surface covered with orbicules. Note, orbicules are spinulose and connected together by filamentous unites. SEM of a fresh anther, frozen in liquid nitrogen and fractured. × 14.000. Scale = 2 μm

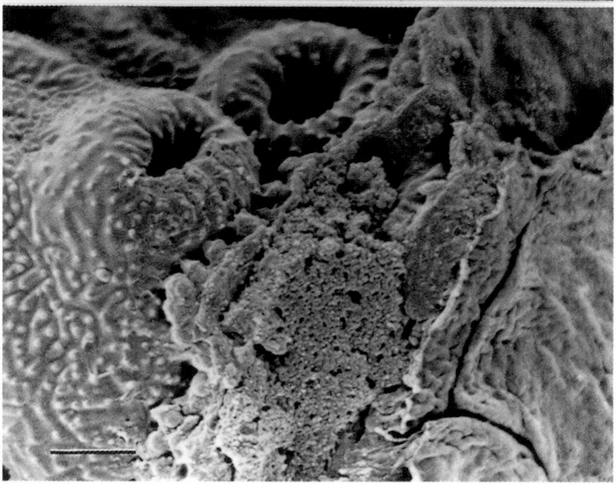

Fig. 13.12. *Betula pendula.* Orbicules are connected to the microspores by filamentous unites. SEM of a freeze fractured anther. × 8000. Scale = 2 μm

Examples of suggested functions of orbicules

1. A transport mechanism for the conveyance of sporopollenin between the tapetum and the developing microspores thus taking part in sporoderm formation (Maheshwari 1950; Rowley 1962; Banerjee and Barghoorn 1971; El-Ghazaly and Jensen 1986a).
2. Provide a non-wettable layer lining the anther sac from which pollen grains are readily detached or they may be associated with pollen dispersal.
3. Temporary packaging of sensitive material for transport through the locular sap. Rowley and Walles (1987) suggested this hypothesis in their work on *Pinus sylvestris.* The locular fluid probably contains exocellular enzymes (see Herdt et al. 1978) against which the sporopollenin wall of the orbicules forms an effective barrier.
4. Orbicules have no specific function; they are a by-product of tapetal cells metabolism (Heslop-Harrison 1968a; Echlin and Godwin 1968; Dickinson and Bell 1972).

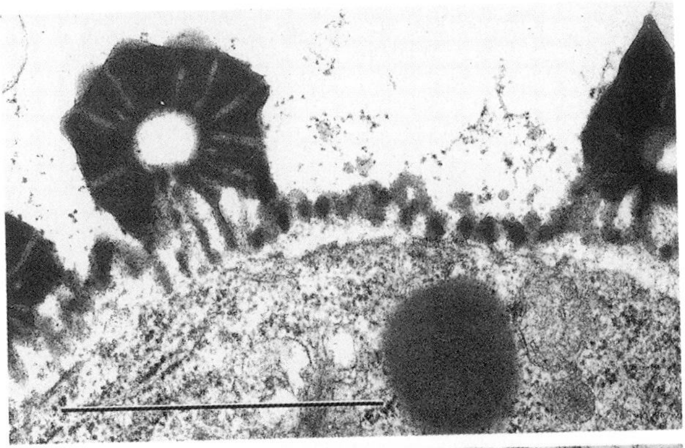

Fig. 13.13. *Triticum aestivum*. Part of a tapetal cell covered with orbicular wall and orbicules. TEM. × 55 000. Scale = 1 μm

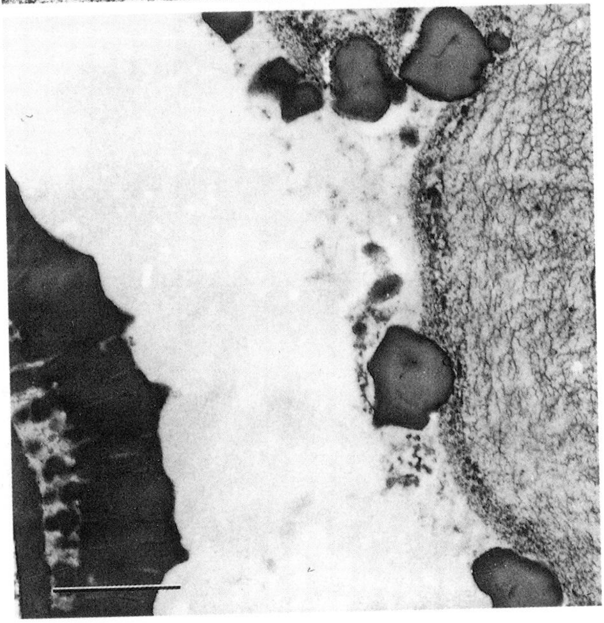

Fig. 13.14. *Betula pendula*. Parts of two tapetal cells covered by thick fibrillar layer and orbicules sunken in the fibrillar layer. Fibrillar material and surface of orbicules are stained positively for protein. Stain: PTA-acetone. TEM. × 25 000. Scale = 1 μm

5. Rowley and Erdtman (1967) suggested that orbicules might actively participate in lysis and degradation of tapetal cells. Herich and Lux (1985) observed that in *Lilium henryi* the orbicules get connected with the tapetal plasmalemma and then become lytically active.
6. Prevent osmosis and collapse of the developing microspores.

Part of sporopollenin precursors, passing from the secretory tapetum to the microspores, is deposited and polymerised on the pro-orbicules. This process may decrease the risk of high concentrations of sporopollenin precursors coming into contact with the developing microspores (El-Ghazaly and Nilsson 1991)

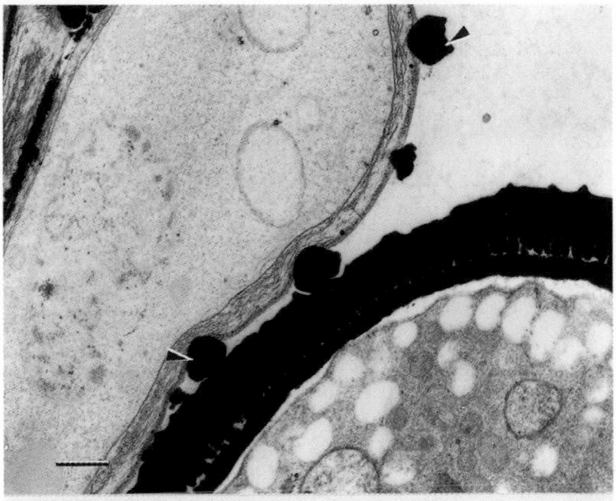

Fig. 13.15. *Betula pendula.* Part of microspore wall and tapetal cell. The exine, and orbicules (arrowheads) are densely stained for acidic polysaccharide. Stain: PTA-chromic. TEM. × 10 000. Scale = 1 µm

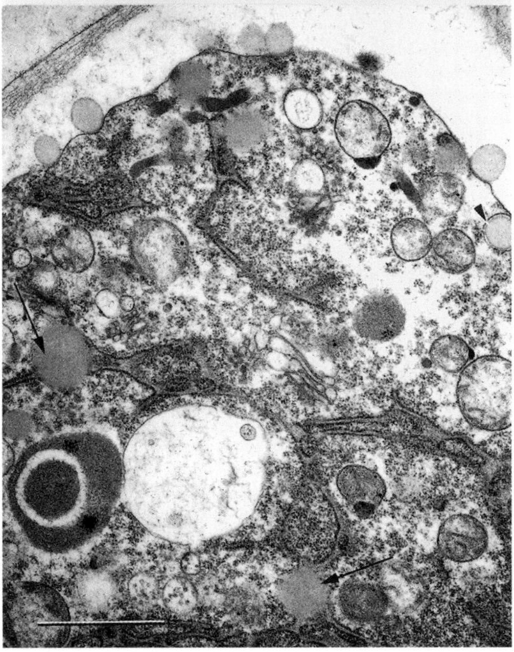

Fig. 13.16. *Triticum aestivum.* Tapetal cell at late tetrad. There are small globular bodies (pro-orbicules) at depressions on the plasma membrane. Globular bodies without membrane appear released from the endoplasmic reticulum (arrows) and others with membrane (arrowheads) are present in the cytoplasm of the tapetal cell. TEM. × 29 000. Scale = 1 µm

13.3.2.3
The Development of Orbicules

The development of orbicules begins, in the cytoplasm of tapetal cells on small globular bodies (pro-orbicules) with a diameter of ca. 0.1 µm.

The origin of the pro-orbicules seems to be associated with the endoplasmic reticulum of the tapetal cells (Fig. 13.16). At the sporogenous stage numerous membrane-bound pro-orbicules appear in the cytoplasm of the tapetal cells. The membrane coat of each pro-orbicule fuses with the plasmalemma of the tapetal cell and the pro-orbicule passes through the plasmalemma and attaches to its surface. Sporopollenin starts to accumulate on the surface of the pro-orbicules until their wall is developed (Figs.

13.17–20). The orbicules apparently remain attached to the plasmalemma of the tapetal cells. They are located in the outer and inner tangential layers and in radial sides of the tapetal cells (El-Ghazaly and Jensen 1986a, b; Hesse 1986; Rowley and Walles 1987). Tapetal endoplasmic reticulum as site of origin of pro-orbicules has also been proposed by Reznickova and Willemse (1980), Herich and Lux (1985) for *Lilium*, Chen et al. (1988) for *Anemarrhena asphodeloides*, and El-Ghazaly and Nilsson (1991) for *Catharanthus roseus*.

In monocots the orbicular wall seems to develop on a specific reticulate pattern (glycocalyx). Rowley and Skvarla (1974), for example, reported a plasmalemma glycocalyx in pro-orbicules of *Phleum pratense*, in *Triticum aestivum* (El-Ghazaly and Jensen 1985) and in *Lilium* (Clément and Audran 1993a, b). They stated that "the specific form of the orbicular wall which is identical with the pollen exine surface, is determined by a plas-

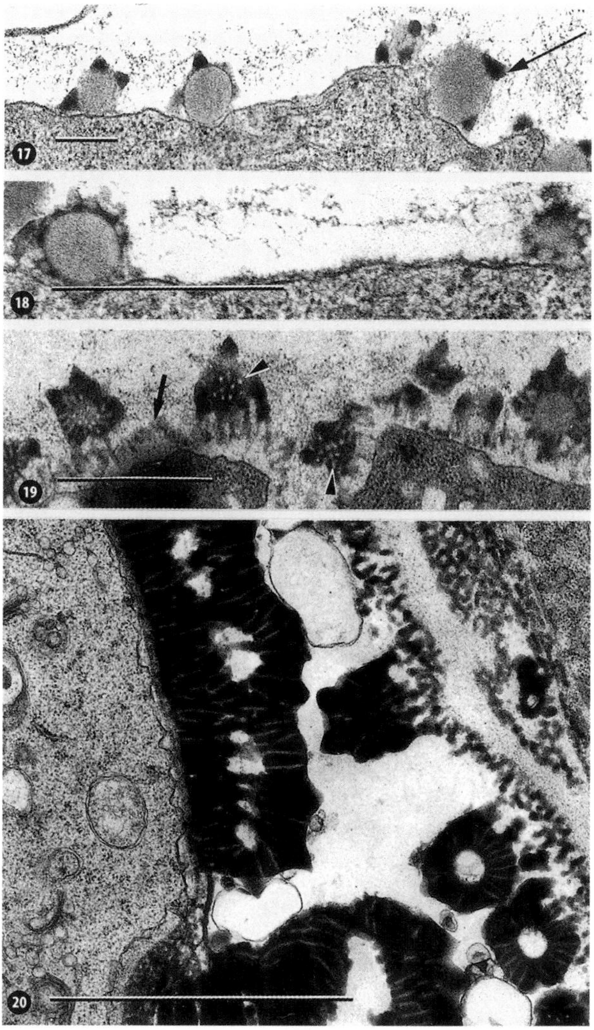

Fig. 13.17–13.20. *Triticum aestivum*. **17** Part of a tapetal cell and pro-orbicules coated with a thin layer and in some sites with spurs of electron dense material (arrow). Young free microspore stage. TEM. × 120 000. Scale = 0.1 µm. **18** Part of a tapetal cell and pro-orbicules in tangential section showing a honeycomb pattern between the spurs. TEM. x 45 000. Scale = 1 µm. **19** Oblique tangential section of two tapetal cells covered with inter-orbicular wall (arrow) and orbicules. Orbicules increase in thickness and microchannels develop in the lumina of the honeycomb pattern (arrowheads). TEM. × 30 000. Scale = 1 µm. **20** A part of the exine, orbicules and inter-orbicular wall that are thicker than before. The central cores of orbicules are lucent and microchannels in their wall are distinct. TEM. × 59 000. Scale = 1 µm

malemma-glycocalyx". The coat of the pro-orbicules which appears to guide the deposition of sporopollenin is positively stained for polysaccharides and proteins, similar to the pollen wall (Rowley 1976; Rowley and Dahl 1982; El-Ghazaly and Jensen 1987).

13.4 Allergenicity of Pollen Grains

The presence of protein in the intine was reported in pollen of birch (*Betula pendula*, *B. nana* and *B. pubescens*) by many authors, e.g. Knox and Heslop-Harrison (1970); Grote and Fromme (1984). The protein in birch pollen is considered to be one of the main causes of type I allergic reaction in Sweden (Wihl et al.1988) and in large parts of the Northern hemisphere (Swoboda et al. 1994). Bet v 1, a polypeptide of 17 kDa, has been identified as the major birch pollen allergen responsible for IgE binding in more than 95% of birch pollen allergic patients (Swoboda et al. 1994). El-Ghazaly and Grafström (1995) demonstrated that the stainability for protein is positive in Zwischenkörper of late microspores and in pollen wall and intine of mature pollen grain before anthesis. Swoboda et al. (1994) in their study of major birch pollen allergen during anther development found that " Bet v 1 proteins are encoded by late pollen-specific genes, whose transcription is activated only during the final stages of pollen maturation. We conclude that the protein detected in the intine of mature pollen grain of *Betula* may be the reason for allergic reactions. The earlier formed protein in Zwischenkörper of *B. pendula* is apparently not allergenic. This is based also on the study of Seiberler et al. (1994) where Bet v lll is expressed preferentially in mature birch pollen. This observation may add a clear histochemical difference between Zwischenkörper and the intine, regarding the type and timing of protein formation in these structures. It also points towards different functions of the two layers. On the basis of quantitative immunoassays and immunocytochemistry, Southworth et al. (1988) determined that purified exines are antigenic and that the polyclonal and monoclonal antibodies bind to exines of some selected angiosperms and gymnosperms. From this it seems that allergenic protein may exist in both material of the intine and the exine.

The stainability, for protein, of orbicules, pollen wall and starch grains in pollen cytoplasm was examined by LM and TEM. The protein is mainly localized at the apertures and starch grains in the cytoplasm of pollen and in the core and on the surface of orbicules. The nitrocellulose membrane test (El-Ghazaly et al. 1995) indicates possible allergenicity of orbicules as well as of the pollen. Unlike pollen grains, these small orbicules (ca 2 µm) have a high probability of entering the lower human airways constricting the bronchi and causing an attack of asthma.

For electron microscopic immunocytochemistry, we applied rapid freeze fixation and substitution followed by LRW embedding to minimise diffusion artefacts. Monoclonal antibodies have been used to locate the subcellular sites of *Betula pendula* (syn. *B. verrucosa*) pollen and tapetal allergens. The allergen is predominantly located in the starch grains and to a lesser extent in the exine and orbicules. The application of rapid freezing prevented relocation of allergens from their native sites and the use of immunogold labelling with monoclonal antibodies improved molecular specificity.

Although intact pollen grains are assumed to be the primary carrier of pollen allergens, specific immuno-reactive components have been found in other aerosol fractions. Observation of allergens associated with particles sized below 2 µm has been confirmed by using molecular membrane filters that were cleared and examined microscopically or extracted with aqueous solvents for immunochemical assay.

The question of relative contribution of small aerosol-associated fractions to total pollen allergen remains to be answered. Also the mechanism that might promote formation of small allergen bearing particles from living plant cells remain to be defined. We tested interaction between mature birch pollen and human saliva, human nostrils fluid and fluid from eyes. Our aim was to mimic more closely the in vivo situation. The preparations were examined by LM after 10 minutes. In all cases we observed several pollen grains that were burst, releasing their cytoplasmic contents.

The study of El-Ghazaly et al. (1996) using the monoclonal antibodies Bet v 1 reveals a significant pattern of immune stain associated with starch granules in the cytoplasm of *Betula* pollen grains (Fig. 13.21). We detected in some cases a minor labelling in the pollen wall and intine. The finding of the allergens on starch grains adds new precise data on the occurrence of the allergens in *Betula* pollen. In previous studies, allergens in birch were reported in the cytoplasm without specification of precise localization (Belin and Rowley 1972; Grote et al.1993). So far we do not know, precisely, how the allergens are liberated from the pollen grains of *Betula*. We believe there must be a combination of environmental factors at the time of pollen release, which lead to the rupture of pollen grains and dispersal of cytoplasmic organelles such as starch grains that carry the allergens of *Betula*. In grass pollen starch granules with allergens on their surface are liberated in to the air when pollen grains are ruptured osmotically during rainfall or a period of high relative humidity (Knox and Suphioglu 1996). Allergenic starch grains have been associated with epidemics of asthma during thunderstorms and the grass pollen season (Packe and Ayres 1985; Bellomo et al. 1992). The above mentioned observations indicate the necessity of a complement to the traditional forecasting of allergenic pollen grains using Burkard traps. A more relevant parameter may be the actual concentration of airborne allergen particles rather than pollen grains. For this purpose a

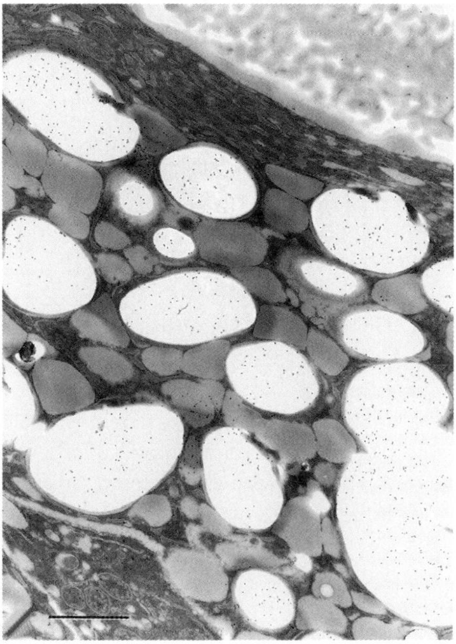

Fig. 13.21. *Betula pendula.* Immunogold localisation with specific monoclonal antibodies to the major allergen Bet v1. Bet v1, detected by monoclonal antibody anti BV-10 (shown by 10nm diameter gold particles) is isolated primarily within the starch granules. TEM. × 20 000. Scale = 1 µm

device for monitoring small allergenic particles such as starch grains and orbicules (ca 0.5 µm in diameter) should be used.

References

Abadie M, Hideux M (1979) l'anthère de *Saxifraga cymbalaria* L. ssp. *Huetiana* (Boiss.) Engl. And Irmsch. En microscopie électronique (MEB & MET). 1. Généralités. Ontogénèse des orbicules. Ann Sci Nat Bot Biol Vég 1:199–233
Banerjee U C (1967) Ultrastructure of the tapetal membranes in grasses. Grana palynol 7:365–377
Banerjee UC, Barghoorn ES (1971) The tapetal membranes in grasses and Ubisch body control of mature exine pattern. In: Heslop-Harrison J (ed) Pollen: development and physiology. Butterworths, London, pp 126–127
Belin L, Rowley JR (1972) Demonstration of birch pollen allergen from isolated pollen grains using immunofluorescence and single radial immunodiffusion technique. Int Arch Allergy 40:754–769
Bellomo R, Gigliotti P, Treloar A, Holmes P, Suphioglu C, Singh MB, Knox RB (1992) Two consecutive thunderstorm associated epidemics of asthma in the city of Melbourne. Med J Aust 156:834–837
Bhandari NN (1984) The Microsporangium. In: Johri BM (ed) Embryology of Angiosperms. Springer, Berlin Heidelberg, pp 53–121
Chen ZK, Wang FH, Zhou F (1988) On the origin, development and ultrastructure of the orbicules and pollenkitt in the tapetum of *Anemarrhena asphodeloides* (Liliaceae). Grana 27:273–282
Christensen JE, Horner HT and Lersten N R (1972) Pollen wall and tapetal orbicular wall development in *Sorghum bicolor* (Gramineae). Amer J Bot 59:43–58
Clément and Audran (1993a) Orbicule wall surface characteristics in *Lilium* (Liliaceae). An ultrastructural and cytochemical approach. Grana 32:348–353
Clément and Audran (1993b) Electron microscope evidence for a membrane around the core of the Ubisch body in *Lilium* (Liliaceae). Grana 32:311–314
Colhoun CW and Steer MW (1981) Microsporogenesis and the mechanism of cytoplasmic male sterility in maize. Ann Bot 48:417–424
Davis G L (1966) Systematic embryology of the angiosperms. John Wiley & Sons, Inc. New York
De Vries APA, Ie TS (1970) Electron-microscopy on anther tissue and pollen of male sterile and fertile wheat (*Triticum aestivum* L.). Euphytica 19:103–120
Dickinson HG (1973) The role of plastids in the formation of pollen grain coatings. Cytobios 8:24–40
Dickinson HG. Bell PR (1972) the identification of sporopollenin in sections of resin-embedded tissues by controlled acetolysis. Stain technol 48:17–22
Dickinson HG, Lewis D (1973a) Cytochemical and ultrastructural differences between intraspecific compatible and incompatible pollinations in *Raphanus*. Proc R Soc London B 183:21–38
Dickinson HG, Lewis D (1973b) The formation of the tryphine coating the pollen grains of *Raphanus* and its properties relating to the self-incompatibility system. Proc R Soc London B 184:149–165
Echlin P (1971) The role of the tapetum during microsporogenesis of angiosperms. In: Heslop-Harrison J (ed) Pollen: development and physiology. Butterworths, London, pp 41–61
Echlin P, Godwin H (1968) The ultrastructure and ontogeny of pollen in *Helleborus foetidus*. L. II. Pollen grain development through the callose special wall stage. J Cell Sci 3:175–186
El-Ghazaly G (1989) Pollen and orbicule morphology of some *Euphorbia* species. Grana 28:243–259
El-Ghazaly G, Chaudhary R (1993) Morphology and taxonomic application of orbicules (Ubisch bodies) in the genus *Euphorbia*. Grana Suppl 2:26–32
El-Ghazaly G, Grafström E (1995) Morphological and histochemical differentiation of the pollen wall of *Betula pendula* Roth. During dormancy up to anthesis. Protoplasma 187:88–102
El-Ghazaly G, Jensen W (1985) Studies of the development of wheat (*Triticum aestivum*) pollen: III. Formation of microchannels in the exine. – Pollen Spores 27:5–14
El-Ghazaly G, Jensen W (1986a) Studies of the development of wheat (*Triticum aestivum*) pollen: I. Formation of the pollen wall and Ubisch bodies. – Grana 25:1–29
El-Ghazaly G, Jensen W (1986b) Studies of the development of wheat (*Triticum aestivum*) pollen: formation of the pollen aperture. Can J Bot 64:3141–3154
El-Ghazaly G, Jensen W (1987) Development of wheat (*Triticum aestivum*) pollen. II. Histochemical differentiation of wall and Ubisch bodies during development. – Amer J Bot 74(9):1396–1418
El-Ghazaly G, Nilsson S (1991) Development of tapetum and orbicules of *Catharanthus roseus* (Apocynaceae). In: Blackmore S, Barnes SH (eds) Pollen and Spores: Patterns of Diversification. Syst Assoc Sp Vol 44, Clarendon Press, Oxford, pp 317–329
El-Ghazaly G, Takahashi Y, Nilsson S, Grafström E, Berggren B (1995) Orbicules in *Betula pendula* and their possible role in allergy. Grana 34:300–304
El-Ghazaly G, Nakamura S, Takahashi Y, Cresti M, Walles B, Milanesi C (1996) Localization of the major allergen Bet VI in *Betula* pollen using monoclonal antibody labelling. Grana 35:48–53

Erdtman G, Berglund B, Praglowski J (1961) An introduction to a Scandinavian pollen flora. Almqvist & Wiksell, Uppsala
Gabarayeva N, El-Ghazaly G (1997) Sporoderm development in *Nymphaea mexicana* (Nymphaeaceae). Pl Syst Evol 204:1–19
Gifford EM, Foster AS (1989) Morphology and evolution of vascular plants.3rd ed. Freeman, New York
Grote M (1991) Immunogold electron microscopy of soluble proteins: localization of Bet v I major allergen in ultra-thin sections of birch pollen after anhydrous fixation techniques. J Histochem Cytochem 39:1395–1401
Grote M, Fromme HG (1984) Immunoelectronmicroscopic localization of diffusible birch pollen antigens in ultrathin sections using the protein-A/cold techniques. Histochemistry 81:489–492
Grote M, Vrtala S, Valenta R (1993) Monitoring of two allergens, *Bet vi* and profilin, in dry and rehydrated birch pollen by immunogold electron microscopy and immunoblotting. J Histochem Cytochem 41:745–750
Herdt ER, Stütfeld , Wiermann R (1978) The occurrence of enzymes involved in phenylpropanoid metabolism in the tapetum fraction of anthers. Cytobiologie 17:433–441
Herich R, Lux A (1985) Lytic activity of Ubisch bodies (orbicules). Cytologia 50:563–569
Heslop-Harrison J (1962) Origin of exine. Nature 195:1069–1071
Heslop-Harrison J (1968a). Tapetal origin of pollen-coat substances in *Lilium*. New Phytol 67:779–786
Heslop-Harrison J (1969) An acetolysis-resistant membrane investing tapetum and sporogenous tissue in the anthers of certain Compositae. Can J Bot 47:541–542
Hesse M (1984) An exine architecture model for viscin threads. Grana 23:69–75
Hesse M (1985) Hemispheric surface processes of exine and orbicules in *Calluna* (Ericaceae). Grana 24:93–98
Hesse M (1986) Orbicules and the ektexine are homologous sporopollenin concretions in Spermatophyta. Plant Syst Evol 153:37–48
Hoefert LL (1971) Ultrastructure of tapetal cell ontogeny in *Beta*. Protoplasma 73:397–406
Horner HTjr, Lersten NR (1971) Microsporogenesis in *Citrus limon* (Rutaceae). Amer J Bot 58:72–79
Huysmans S, El-Ghazaly G, Nilsson S, Smets E (1997) Systematic value of tapetal orbicules: a preliminary survey of the Cinchonoideae (Rubiaceae). Can J Bot 75:815–826
Kapil RN, Bhandari NN (1964) Morphology and embroyology of *Magnolia* Dill.ex Linn. Proc Natl Inst Sci India B Biol Sci 30:245–262
Knox RB, Heslop-Harrison J (1970) Pollen wall proteins: Localization and enzymatic activity. J Cell Sci 6:1–27
Knox RB, Heslop-Harrison J, Reed C (1970) Localization of antigens associated with the pollen wall by immunofluorescence. Nature 225:1066
Knox RB, Suphioglu C (1996). Environmental and molecular biology of pollen allergens. Elsevier Trends Js 1(5):156–164
Kosmath L von (1927) Studien über das antherentapetum. Österreichische Botanische Zeitschrift 76: 235–241
Lippi M, Cimoli F, Maugini E, Tani G (1994) A comparative study of the tapetal behaviour in male fertile and male sterile *Iris pallida* Lam. during microsporogenesis. Caryologia 47:109–120
Maheshvari Devi H, Lakshminarayana K (1976) Embryology of *Oxystelma esculentum* R.B. In: Proc Indo – Sov Symp Embryology of crop plants. Ind Natl Sci Acad, Univ., Delhi, pp 20–21
Maheshwari P (1950) Introduction to the embryology of angiosperms. Mc Graw-Hill, New York
Maldonado CA, Saggau W, Forssmann WG (1986) Cardiodilatin munoreactivity in specific atrial granules of human heart revealed by the immunogold stain. Anat Embryol 173:295–298
Miki-Hirosige H, Nakamura S, Yasueda H, Shida T , Takahashi Y (1994) Immunocytochemical localization of the allergenic proteins in the pollen of *Cryptomeria japonica*. Sex Plant Reprod 7:95–100
Nakashima H (1975) Histochemical studies on the cytoplasmic male sterility in some crops. IV Electron microscopic observations in sugar-beet anther. Mem Fac Agric Hokkaido Univ Ser B 9:247–252
Pacini E (1990) Tapetum and microspore function. In: Blackmore S, Knox RB (eds) Microspores: Evolution and Ontogeny. Academic Press, London, pp 213–237
Pacini E, Casadoro G (1981) Tapetum plastids of *Olea europaea* L. Protoplasma 106:289–296
Pacini E, Franchi GG (1983) Pollen grain development in *Smilax aspera* and possible functions of the loculus. In: Pollen biology and implications for plant breeding, Elsevier, pp 183–190
Pacini E, Franchi GG, Hesse M (1985) The tapetum: its form, function and possible phylogeny in Embryophyta. Plant Syst Evol 149:155–185
Packe GE, Ayres JG (1985) Asthma outbreak during a thunderstorm. The Lancet 1:199–203
Probert L, De Mey J, Polak J.M (1981) Distinct subpopulations of enteric p type neurones contain substance P and vasoative intestinal polypeptide. Nature 294:470–471
Raj B, El-Ghazaly G (1987) Morphology and taxonomic application of orbicules (Ubisch bodies) in Chloanthaceae. Pollen Spores 29:151–166
Reznickova SA, Willemse MTM (1980) Formation of pollen in the anther of *Lilium*. II. The function of the surrounding tissues in the formation of pollen and pollen wall. Acta Bot. Neerl. 29:141–156

Rosanoff S (1865) Zur kenntniss des baues und der entwickelungsgeschichte des pollens der Mimosaceae. Jahrb. Wiss. Bot. 4:441–450
Rowley J, El-Ghazaly G (1992) Lipid in wall and cytoplasm of *Solidago* pollen. - Grana 31:273–283
Rowley JR (1962) Stranded arrangement of sporopollenin in the exine of microspores of *Poa annua*. Science 137:526–528
Rowley JR (1963) Ubisch body development in *Poa annua*. Grana Palynol. 4:25–36
Rowley JR (1976) Dynamic changes in pollen wall morphology. In: Ferguson IK, Muller J (eds) The evolutionary significance of the exine.Linn Soc Symp Ser 1, Acad Press, London, pp 39–66
Rowley JR, Dahl AO (1982) A similar substructure for tapetal surface and exine "tuft"-units
Pollen Spores 24:5–8, pl 1
Rowley JR, Erdtman G (1967) Sporoderm in *Populus* and *Salix*. Grana Palynol 7:518–567
Rowley JR, Gabarayeva NI, Walles B (1992) Cyclic invasion of tapetal cells into loculi during microspore development in *Nymphaea colorata* (Nymphaeaceae). Amer J Bot 79:801–808
Rowley JR, Skvarla J (1974) Plasma membrane-glycocalyx origin of Ubisch body wall. Pollen Spores 16:441–448
Rowley JR, Walles B (1987) Origin and structure of Ubisch bodies in *Pinus sylvestris*. Acta Soc Bot Poloniae 56:215–227
Schnarf K (1923) Kleine beiträge zur entwicklungsgeschichte der angiospermen. IV. Über das verhalten des antherentapetums einiger pflanzen. Österreichische Botanische Zeitschrift 72:242–245
Schulze Osthoff K, Wiermann (1987) Phenols as integrated compounds of sporopollenin from *Pinus* pollen. J Plant Physiol 131:5–15
Seiberler S, Scheiner O, Kraft D, Lonsdale D, Valenta R (1994) Characterization of a birch pollen allergen, Bet v II, representing a novl class of calcium-binding proteins, Specific expression in mature pollen and dependence of patients ige-binding on protein bound calcium. In: Heberle-Bors.., Hesse M, Vicente O (eds) Frontiers in sexual plant reproduction research. University Press, Vienna, p 100
Shivanna KR, Johri BM (1985) The Angiosperm pollen: Structure and function. Wiley Eastern, New Delhi
Skvarla JJ, Larson DA (1966) Fine structural studies of *Zea mays* pollen. I. Cell membranes in exine ontogeny. Amer J Bot 52:1112–1125
Southworth D, Singh MB, Hough T, Smart IJ, Taylor P, Knox RB (1988) Antibodies to pollen exines. Planta 176:482–487
Staff IA, Taylor PE, Smith PM, Singh MB, Knox RB (1990) Cellular localization of water-soluble, allergenic proteins in rye-grass (*Lolium perenne*) pollen using monoclonal and specific IgE antibodies with immuno-gold probes. Histochem J 22:276–290
Steer MW (1977) Differentiation of the tapetum in *Avena*. I. The cell surface. J Cell Sci 25:125–138
Stieglitz H (1977) Role of b-1,3-glucanase in postmeiotic microspore release. Develop Biol 57:87–97
Suarez-Cervera M, Marquez J, Seone-Camba J (1995) Pollen grain and Ubisch body development in *Platanus acerifolia*. Rev Palaeobot Palynol 85:63–84
Swoboda I, Dang TCH, Heberle-Bors E, Vicente O (1994) Expression of Bet v 1, the major birch pollen allergen, during anther development. An in situ hybridization study. In: Heberle-Bors E, Hesse M, Vicente O (eds) Frontiers in sexual plant reproduction research, abstract book, Univ. Vienna, p 111
Taylor PE, Staff IA, Singh MB, Knox B (1994) Localization of the major allergens in rye-grass pollen using specific monoclonal antibodies and quantitative analysis of immunogold labelling. Histochem J 26:392–401
Ubisch G von (1927) Kurze mitteilungen zur entwicklungsgeschichte der antheren. Planta 3:490–495
Ueno J (1959) Some palynological observations of Taxaceae, Cupressaceae and Araucariaceae. J Inst Polytech Osaka City Univ, D, 10:75–87
Ueno J (1960) On the fine structure of the cell wall of some gymnosperm pollen. Biol J Nara Women's Univ 10:19–25
Wehling K, Niester Ch, Boon JJ, Willemse MTM, Wiermann R (1989) *p*-Coumaric acid – a monomer in the sporopollenin skeleton. Planta 179:376–380
Wihl JÅ, Ipsen B, Nüchel Petersen B, Munch EP, Janniche H, Lövenstin H (1988) Immunotherapy with partially purified and standarized tree pollen extracts. Allergy 43:363–369
Young BA, Schulz-Schaeffer J, Carroll TW (1979) Anther and pollen development in male-sterile intermediate wheatgrass plants derived from wheat × wheatgrass hybrids. Can J Bot 57:602–618

CHAPTER 14

Development and Substructures of Pollen Grains Wall

G. El-Ghazaly

Swedish Museum of Natural History, Stockholm, Sweden
e-mail: gamal.el-ghazaly@nrm.se
telephone: +46 8 666 41 93
fax: +46 8 16 77 51

14.1 Introduction

Pollen grains and spores are the male reproductive units of many types of plants. They vary in shape and ornamentation and have a very resistant outer wall called the exine and an inner polysaccharide layer called the intine. There are several terms that have been applied to the layers of the exine. The most common terms are those of Erdtman (1952) and Fægri and Iversen (1975).

The exine is known to be composed mainly of sporopollenin, a highly resistant polymer of carotenoids and carotenoid esters (Brooks and Shaw 1968). Recently by using the high resolution solid state ^{13}C Nuclear Magnetic Resonance, Guilford et al. (1988) demonstrated that sporopollenin is not a unique substance, but rather a series of related biopolymers derived from largely saturated precursors such as fatty acids. Schulze Osthoff and Wiermann (1987) have shown that phenols represent an integral component of exine. Wehling et al. (1989) assumed that p-coumaric acid is a genuine structural unit in the exine skeleton (sporopollenin).

The pollen wall or the exine is much more than a protective shield. In angiosperms, for example, the pollen grains walls of many species carry specific self-incompatibility proteins that promote out-breeding. Pollen wall may also contain compounds that protect the gametophyte against microbial attack (Paul et al. 1992).

The development of pollen wall, origin and control of the wall pattern and the source and nature of sporopollenin, are among the subjects that attracted attention in the past as well as currently. Such developmental studies of the angiosperm pollen wall are of interest because they represent a complex of surface patterning event that is mainly controlled by the sporophyte rather than by the gametophyte (Heslop-Harrison 1971). On the other hand, Frova and Pè (1992) stated that pollen development and pollen function are under the control of a very complex genetic system, ruled by both the sporophyte and the gametophyte. The diversity in the exine ontogeny indicates the potential of information on development for research on phylogeny and evolution.

Although many informative papers have somewhat clarified the situation of pollen wall stratification, the terminology used in pollen and spore wall studies remains confusing. More than one valid classification of wall layers can be made and the classification depends upon the source of criteria used, which may be structural, microchemical, or ontogenetic.

In this chapter the main details of pollen wall development and exine subunits are analyzed and several original illustrations are included. The information is based on historical and recent studies in pollen wall development and personal communication and is to be regarded as an attempt to present different ideas and interpretation of pollen wall development and substructure in different species.

14.2 Material and Methods

14.2.1 Fixations

Both conventional chemical fixation and rapid freeze fixation and substitutions were applied. In chemical fixation the anthers were treated with phosphate buffer in 2% GA in 0.05 M Na-cacodylate buffer, pH 7.4, room temp., 24 hours, some anthers were post-fixed with OsO_4 while others were not. All anthers were then dehydrated in an acetone series, embedded in Spurr's (Spurr 1969) medium and sectioned. Ultrathin sections were stained to elucidate the histochemistry of the pollen wall.

Using the rapid freeze fixation and substitutions for transmission electron microscopy (TEM) has considerably improved ultrastructure preservation of microspores, pollen grains and the delicate tapetal tissue.The development of microspores, pollen grains wall and cytoplasmic organells and tapetal tissue in different species is investigated at different stages of development. Regions of germinal aperture and surface coat of the wall are thoroughly examined. Results obtained by conventional chemical fixation are compared with those of freeze fixation.

14.2.2 LM and TEM Stains

General pollen wall stain: Sections treated with 1% aqueous uranyl acetate for 5 min, washed with distilled water, and floated on lead citrate stain for 5 min.

Stain for protein: Sections on gold grids were stained with 5% phosphotungstic acid (PTA) in 10% acetone for 15 min (Benedetti and Bertolini 1963; Marinozzi 1968).

Stain for acidic polysaccharides: Sections on gold grids were stained with 0.1% PTA in 10% chromic acid for 5 min (Pease 1968; Marinozzi 1967).

Stain for neutral polysaccharides and other carbohydrates: Sections on gold grids were treated for 30 min in 0.2% periodic acid (PA), 20 min in 0.2% thiocarbohydrazide (TCH), and 20 min in 1% Silver proteinate (Thiéry 1967; Rowley and Dahl 1977).

As a control, sections were stained with TCH-Silver proteinate without prior oxidation with PA and other sections were stained with only Silver proteinate. TCH-Silver proteinate stain following fixation with Osmium was used to estimate the location of unsaturated lipids (Rowley and Dahl 1977).

14.2.3 Atomic Force Microscopy (AFM)

For AFM studies pollen and spores were acetolyzed (Erdtman 1960) to remove the cytoplasm and surface material. The isolated exines were washed and dehydrated with acetone. The dry spores were dusted onto a Scanning Tunneling Microscopy (STM) stub coated with a thin layer of silver paint. When the silver coating was well hardened the spore exines were fractured by applying adhesive tape onto the exines then lifting the tape off. To ensure conductivity for STM imaging a 50 Å Au layer was deposited on the sample surfaces. The non-invasive resolution of 5 to 10 nm gives scanning probe microscope (SPM) systems great potential for studies of pollen grain and spore wall substructure (Rowley et al. 1995; Wittborn et al. 1996). Efforts to obtain such information using transmission and scanning electron microscopes (TEM & SEM) have involved destructive preparations with unknown physical and chemical side effects (e.g., Rowley et al. 1980, 1995; Southworth 1985, 1986; Kedves 1990; Kedves et al.1993).

The AFM can be equipped to operate in liquid as well as in air or vacuum and Bustamante and Keller (1995) illustrate its capacity to follow processes of macromolecular assembly in biological material. The principle by which all scanning probe microscopes function is that by letting a fine tip closely follow the surface of a sample, a topographical measurement of a certain area is recorded. The measurement is turned into an image by letting the height value measured at each point be represented by a certain degree of whiteness on a grey-scale or by a colour in a pre-selected colour spectrum that can be chosen freely to enhance the structures depicted. Resolution is proportional to the accuracy with which the tip follows the sample surface. This means that an exact positioning of a sharp tip is necessary, or, at least, ideal. For AFM the tip is mounted at the end of a flexible cantilever having a spring constant of about 1 N/m. As the sample is scanned beneath the tip, small forces of interaction with the sample cause the cantilever to deflect. This deflection is measured and the piezotube positions the sample so that the cantilever is not deflected. The voltage over the piezotube giving this position then gives the height of that point.

14.3 Development of Pollen Grains Wall

14.3.1 Callose Special Cell Envelope

The pollen grains are developed from the sporogenous tissue in the anther through meiotic division where microspore mother cells (microsporocytes) are formed. After that, meiosis takes place and, tetrads of haploid microspores are developed. The tetrads are then enveloped by a callosic wall and are arranged peripherally in the locule close to the tapetum (Figs. 14.1, 2). Callosic wall (a β-1,3-polyglucan) can be detected around the microspore mother cells during initiation of meiosis. It forms a layer between the cytoplasm and pollen mother cell wall. Additional callose is formed after the second meiotic division and isolates the young microspores from each other (Heslop-Harrison 1968a).

The callose special cell envelope persists until it is enzymatically digested at the end of the tetrad stage.

14.3.1.1 *Function of the Callose Special Cell Envelope*

The possible functions of the callose envelope are:

1. To control major features, such as the arrangement of apertures, which are probably related to the tetrad geometry imposed by the callose wall, Wodehouse (1935).

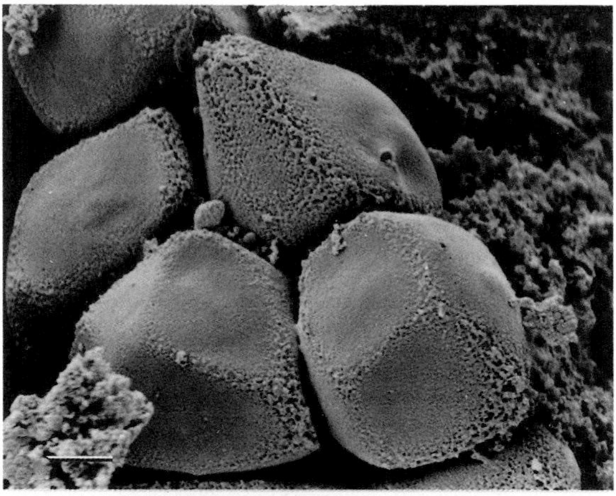

Fig. 14.1. *Rondeletia odorata.* Freeze fracture of a young anther showing tetrads covered with callosic envelope. SEM. ×6000. Scale = 2 μm

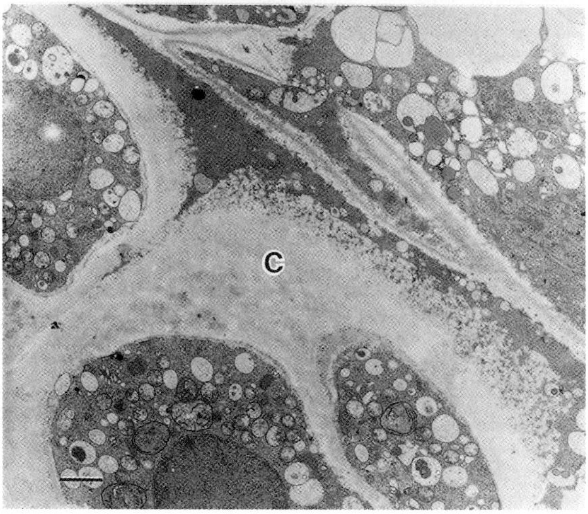

Fig. 14.2. *Rondeletia odorata.* Early tetrad stage of microspores enveloped in thick callosic envelope (C). TEM. ×8000. Scale = 1 μm

2. To isolate the young microspores from influences from the tapetal cells during early stages of development (Heslop-Harrison 1968a, b; Hideux and Abadie 1981). This isolation enables the young microspores to deposit a primexine without interaction between them and the tapetum. Knox (1984) emphasized that the callose wall thus serves to separate the gametophyte from the sporophyte. Evidence of tapetal activity during the tetrad period, which might disrupt primexine organization, is seen in the deposition of a layer of sporopollenin on the outside of tetrads (Horner and Pearson 1978). Rodriguez-Garcia (1978) and Dahl (1986) noted that, at late telophase each tetrad is usually surrounded by a very tenuous residual wall bounded by scattered electron dense, osmiophilic globules.
3. Participation in the development of wall ornamentation. Godwin et al. (1967) and Dickinson (1970) described the action of the callose wall as a template defining the position of apertures and the deposition of the primexine in *Ipomoea*.
4. In plants whose pollen is dispersed in permanent tetrads the callose wall is either absent or greatly reduced. This indicates that the presence of a callose wall is not an absolute prerequisite for successful pollen development.

14.3.2 Primexine and Glycocalyx

Primexine (Heslop-Harrison 1962, 1963) is a blue-print for the exine. It is formed while the microspores are still enclosed in the callosic special envelope. It has a matrix, presumably made up of cellulose, and radially oriented rods, the probaculae.

Glycocalyx (Dahl and Rowley 1974; Rowley and Dahl 1977) is a glycoprotein surface coating of microspores plasmalemma. Glycocalyx extends between the plasmalemma and the outer surface of the exine . The glycocalyx is initiated as radially oriented cylindrical units of uniform size that form the probaculae. The early deposition of the exine material takes place arround these cylindrical units.

The growth of the pollen wall involves synthesis of wall elements at positions specified by the latent pattern information. The first readily observable event is the synthesis of a plasmalemma surface coating (primexine, Heslop-Harrison 1963, 1968a and glycocalyx, Rowley and Dahl 1977) following the completion of meiosis II (Fig. 14.3). This is followed by the formation of tubular protrusions about, 70 nm in diameter, originat-

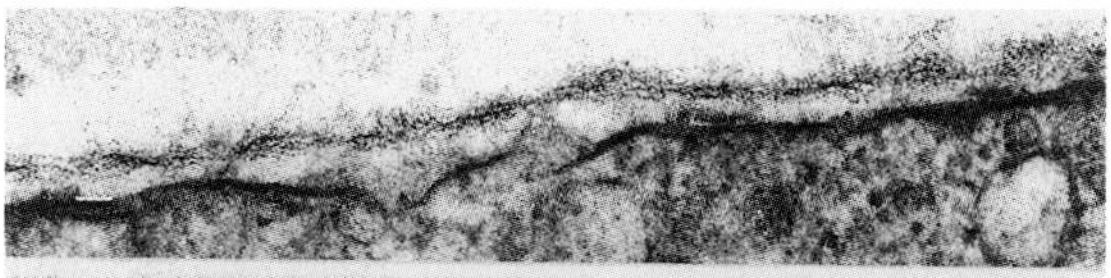

Fig. 14.3. *Triticum aestivum.* Early tetrad stage. The plasmalemma is covered with a thin fibrillar electron dense layer. ×80000. Scale = 0.5 µm

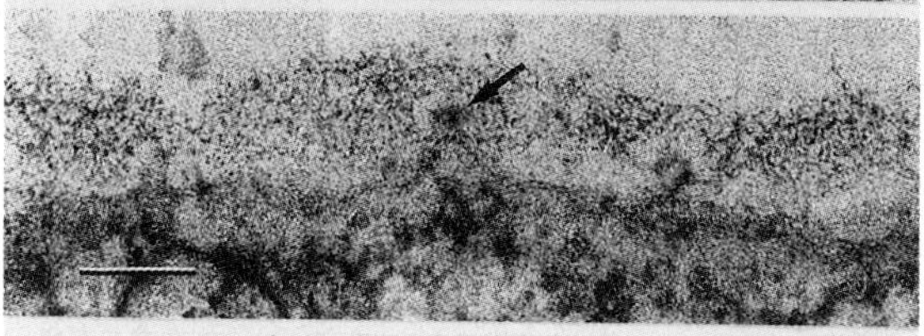

Fig. 14. 4. *Triticum aestivum.* The plasmalemma is convoluted and small patches of electron dense material develop. These patches (arrows) represent the probaculae. TEM. ×145000. Scale = 0.1 µm

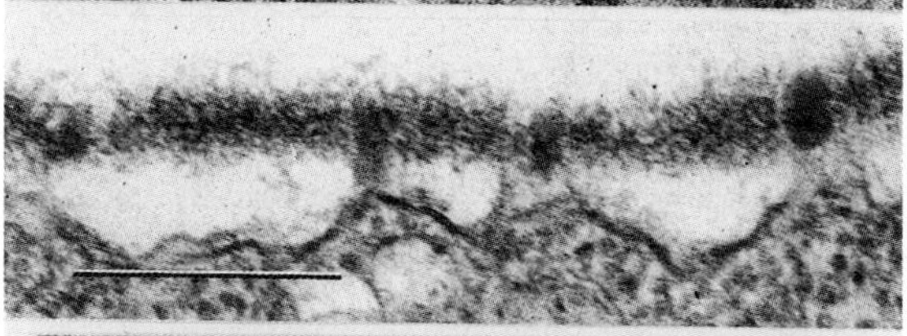

Fig. 14.5. *Triticum aestivum.* Late tetrad microspore stage. The probaculae increase in density, thickness and length. The protectal fibrillar reticulate pattern is evident. TEM. ×70000. Scale = 0.5 µm

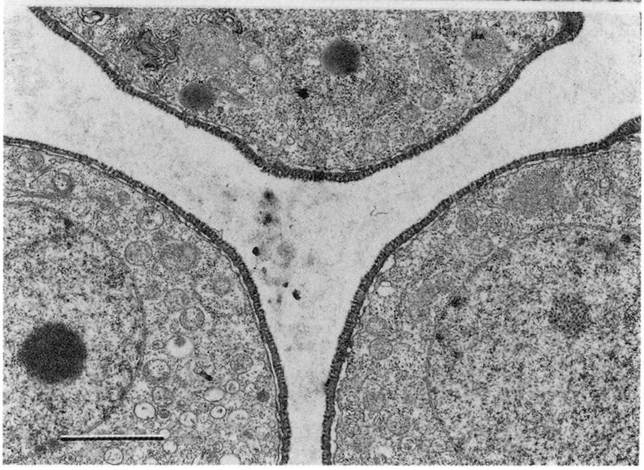

Fig. 14.6. *Betula pendula.* Late tetrad microspore stage. Plasmalemma is coated with a thick layer containing radially oriented, electron dense units that represent the probaculae . ×20000. Scale = 1 µm

ing from the plasmalemma (Figs. 14.4–14.7). The plasmalemma retreats except where it is apparently anchored within the surface coating by the tubular probaculae. According to Dickinson (1970) the tubular probaculae increase in thickness and length and consequently force the retreat of the plasmalemma.

The progenitor (primexine or glycocalyx) of the electron -dense exine of the mature sporoderm was visible in late tetrad stages as a complement of delicate protuberances radiating from the plasmalemma of the microspore protoplasts into the adjacent callose. Echlin and Godwin (1968), reported that primexine seems to condense in a space generated between the inner face of the callose and the plasmalemma of the new microspore at a stage of meiosis.

In early tetrad stages of *Echinodorus* microspores, when the tectum is first evident the exine units show a honeycomb pattern resulting from close packing and interdigitation of the units (Fig. 14.8). El-Ghazaly and Jensen (1985 and 1986a) show the occurrence of such a honeycomb arrangement in *Triticum* (Poaceae) in early stages of microspore development (Fig. 14.9). Huysmans et al. (in prep.) observed a similar pattern (Fig. 14.10) in *Rondeletia odorata* (Rubiaceae). Wodehouse (1935) pointed out that a

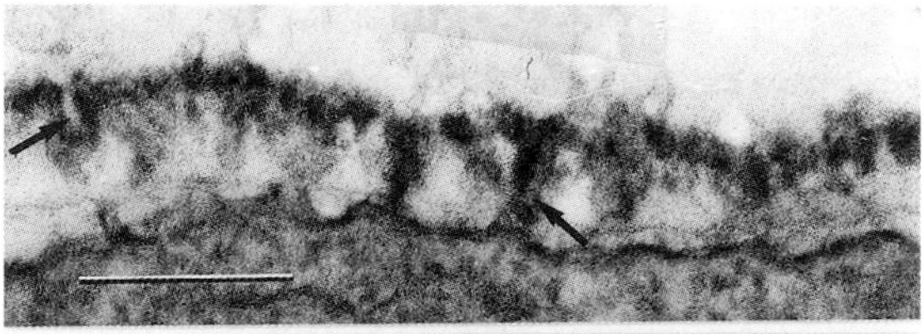

Fig. 14.7. *Betula pendula*. Part of microspores at late tetrad stage. The sides of the cylindrical probaculae are visible in many sites in this tangential section (arrows). TEM. ×55 000. Scale = 0.5 μm

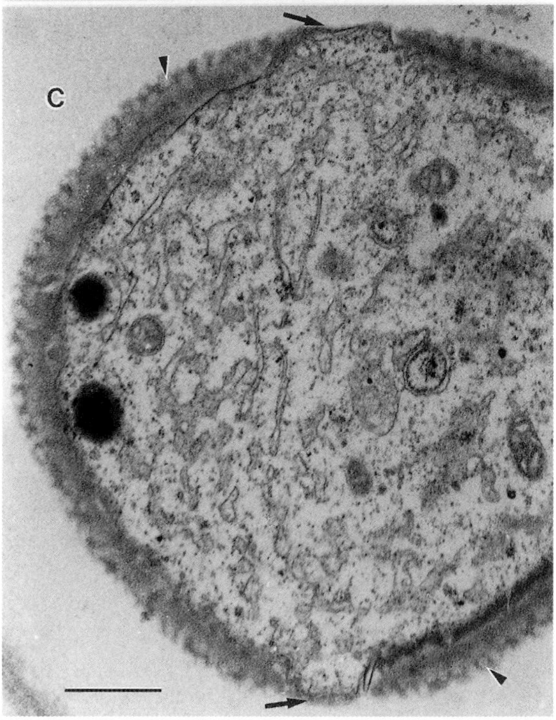

Fig. 14.8. *Echinodorus cordifolius* Oblique section of microspore tetrad stage. The two pores (arrows) in the section show cytoplasm and plasmalemma protruding to the surrounding callose (C).The plasmalemma coated with thin primexine and the exine units appear circular in this oblique section(arrowheads) TEM. ×20 000. Scale = 1 μm

reticulate pattern is a common theme in nature, and is formed where even shrinkage occurs within a uniform matrix.

Patterned arrays of tubular probaculae then condense within the primexine and these subsequently develop into mature wall elements by the accumulation of sporopollenin (Fig. 14.11). This sequence gives the idea that the primexine contains receptors that promote sporopollenin polymerization. It is not easy to define the term primexine, as its differentiation into exine is a continuous process (Blackmore and Barnes 1990). Primexine has been considered to function primarily as a matrix into which sporopollenin precursors are transported and polymerized (Heslop-Harrison 1963). In *Cosmos bipinnatus* (Dickinson and Potter 1976) the primexine is considered to function not only as a matrix for controlled sporopollenin deposition but also as a pathway for the diffusion of other substances such as enzymes, including those responsible for digestion of the callose wall.

There are taxa, including many marine angiosperms, which lack primexine and consequently do not develop normal exines (Pettitt et al. 1981). In the microspore tetrad of *Echinodorus* during the callose period the exine begins as rods originate from the plas-

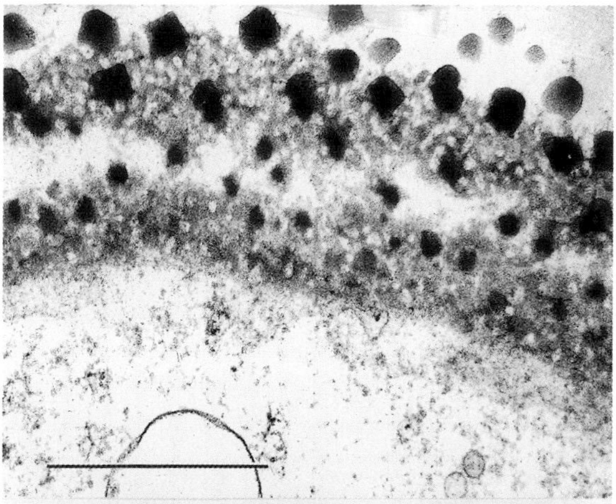

Fig. 14.9. *Triticum aestivum.* Early free microspore stage. The early exine units (baculae) are sectioned tangentially and show honeycomb pattern at their top and base. TEM. ×84 000. Scale = 0.5 μm

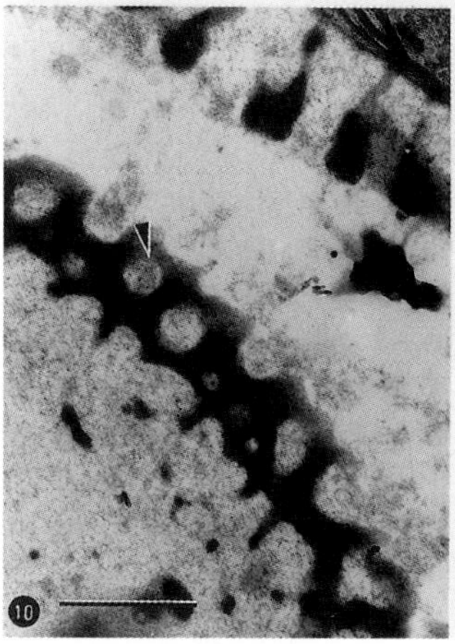

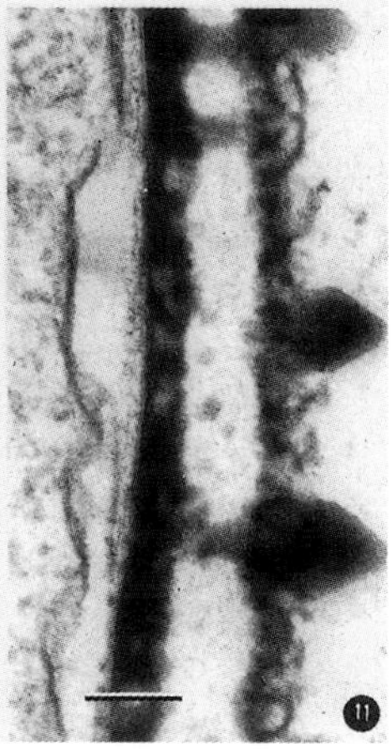

Fig. 14.10. *Rondeletia odorata.* Early free microspore stage. One microspore sectioned tangentially and the other radially. The baculae are radial with thick heads in radially sectioned microspore. In tangential section the honeycomb pattern is obvious (arrowhead). TEM. ×40 000. Scale = .05 μm

Fig. 14.11. *Triticum aestivum.* Free microspore stage. The exine units (tectum, baculae and foot layer) increased in height and thickness. TEM. ×120 000. Scale = 0.1 μm malemma. These rods are exine units that upon further development become baculae as well as part of the tectum and foot layer. Oblique sections of the early exine show that the tectum consists of the distal portions of close-packed exine units. The patterned sexine (tectum, baculae and foot layer) is formed within the primexine. Scott (1994) mentioned that "the primexine acts as a loose scaffold on to which sporopollenin monomers (fatty acids and phenols) are covalently attached by the localized action of superoxide radicals generated at the plasmalemma".

14.3.3 Sites of Apertures

The presumptive germinal apertures of pollen grain are already demarcated during the microspore tetrad stage. The sites of preapertures are distinguished by the absence of the primexine matrix (Heslop-Harrison 1963; Skvarla and Larson 1966).

In early stages of development of *Echinodorus* microspores, the apertures are evident (Fig. 14.12) as regions where the plasmalemma and its surface coating are without exine units and adjacent to the callose envelope (El-Ghazaly and Rowley in press). The exine

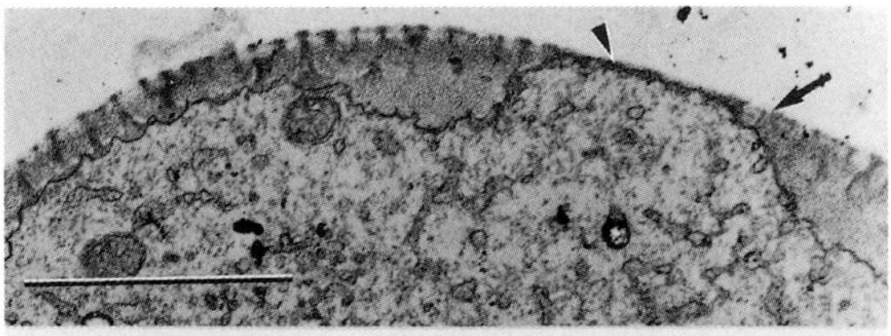

Fig. 14.12. *Ehinodorus cordifolius.* Microspore tetrad stage. Exine units are ca 0.2 µm in height between pores. The plasmalemma over the pore (arrowhead) has a thin coat but no exine units. The pores at this stage are surrounded by a thick fibrillar body. The exine units (arrows) decrease in height near the pore margin. × TEM. × 35 000. Scale = 1 µm.

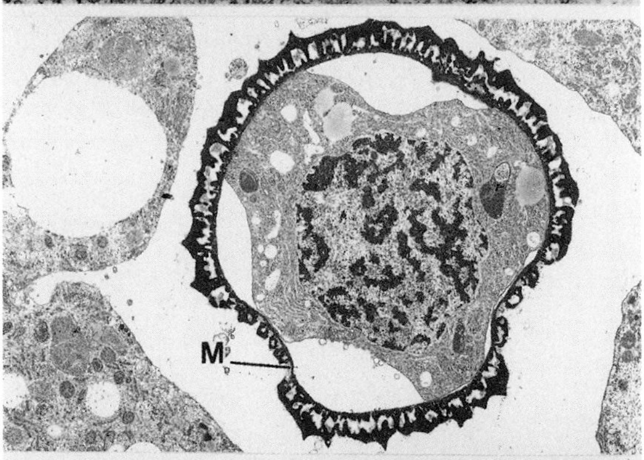

Fig. 14.13. *Ehinodorus cordifolius.* Young free microspore stage. Medium section of two pores. The central portion of the pores have few and small "exine flecks". The large alveolar-like exine components on apertures are mainly located on the central part of the aperture and connected with the interapertural part of the exine by the aperture margin (M) which consists only of a thin foot layer and endexine. The tapetum surrounding the microspore. TEM. × 14 000. Scale = 1 µm

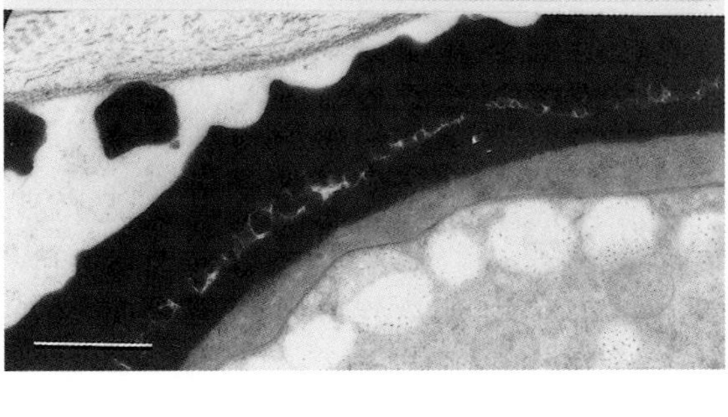

Fig. 14.14. *Betula pendula.* Free microspore stage. The exine and orbicules are stained dense for acidic polysaccharides. Stain: PTA-chromic. TEM. × 19 000. Scale = 1 µm

around the aperture margin is characterized by units of reduced height. According to Heslop-Harrison (1968b) the primexine is not deposited in areas destined to become an aperture. Occasionally the plasmalemma in the pre-apertures areas is associated with an underlying plate of endoplasmic reticulum (Heslop-Harrison 1963). Dover (1972), Sheldon and Dickinson (1986) concluded it is likely that the meiotic spindle plays a role in aperture positioning.

It seems, however, that apertures in fresh pollen grains have at least a thin covering exine (El-Ghazaly 1982; Rowley 1996). This exine covering is self-evident in the many taxa having pollen with apertures ornamented by granules or spinules in the apertures as for example in *Echinodorus* (Fig. 14.12).

14.3.4 Sporopollenin

Probably no structures in the plant cell are so little understood, either structurally or biochemically, as those that are made up of the sporopollenin. One of the principal barriers to evolving a satisfactory model for pollen wall ontogeny has been the failure to determine the biochemical composition, structure and mode of biosynthesis of sporopollenin.

The evolutionary history of sporopollenin, and that of spores of land plants, is indivisible (Chaloner 1976). Fossil green algae dating back to the Devonian period have been shown to contain sporopollenin (Wall 1962) and there are reports that sporopollenin also occurs in fungi (Shaw 1971). Zetzsche et al. (1937) determined that sporopollenin is an oxygenated hydrocarbon and contains hydroxyl and C-methyl groups and substantial levels of unsaturation. Shaw (1971) re-investigated the structure of sporopollenin and confirmed the result of Zetzsche et al. (1937). Shaw also obtained straight and branched chain monocarboxylic acids, which are characteristic breakdown products of fatty acids. Besides this he found a mixture of phenolic acids. Shaw and Yeadon (1966) proposed that sporopollenin is composed of a lipid fraction of 55–65%, consisting of molecules with a chain length of up to C_{16}, and a lignin fraction representing 10–15% of the total mass. The suggestion that sporopollenin was a mixed polymer of lipids and phenylpropanoid units was rapidly replaced by a chemical structure based on the polymerisation of carotenoids and carotenoid esters (Brooks and Shaw 1968; Shaw 1971). The apparent inconsistency with the contemporary analytical data was the origin of the phenolic degradation product of sporopollenin. To test the validity of their finding Brooks and Shaw (1968) polymerized various carotenoid mixtures extracted from lily anthers, under oxidative conditions in the presence of a metallic catalyst. They produced a series of synthetic sporopollenin, which shared many of the physicochemical properties of natural sporopollenin. Recently a further re-evaluation of the chemical structure of sporopollenin, has cast doubt on the role of carotenoids and favours the idea that sporopollenin, is a mixed polymer of phenylpropanoid and fatty acid derivatives. Prahl et al. (1985), demonstrate that a potent inhibitor (Norflurazon) of carotenoid biosynthesis, had little effect on the formation of sporopollenin in *Cucurbita pepo*. Schulze Osthoff and Wiermann (1987) showed that p-coumaric acid might constitute the main aromatic component of sporopollenin. In 1988, Guilford et al. applied ^{13}C NMR spectroscopy and concluded that sporopollenin is not a unique substance, but a series of related biopolymers derived from largely saturated precursors such as long chain fatty acids and oxygenated aromatic rings. The reassertion that sporopollenin is potentially a mixed polymer of phenolics and fatty-acid derivatives prompted Schulze

Osthoff and Wiermann (1987) to propose a structural homology with cutin and suberin. Cutin and suberin are lipid polyesters found mainly in the protective outer coverings of plants to provide mechanical strength to the cell wall and restrict water movement. These are quite similar to the functions of sporopollenin exine.

It is likely that, in most species, the majority of the pollen wall sporopollenin is contributed by the tapetum (Heslop-Harrison 1962) and therefore the pattern of sporopollenin accumulation largely reflects the activity of enzymes located in the tapetum. Long chain fatty acids, which may constitute the major monomer component of sporopollenin (Guilford et al. 1988), accumulate to high levels in both the tapetum and in pollen grains (Evans et al. 1992). Cytochemical studies indicate that substantial quantities of lipid accumulate within the tapetum by the mid-vacuolate stage of pollen development and continue until the tapetum ruptures at the mid-maturation stage (El-Ghazaly and Jensen 1987; Evans et al. 1992). This suggests that there is considerable lipid biosynthetic activity within the tapetum throughout pollen development, including the phase of sporopollenin biosynthesis. A great portion of the lipid biosynthesis activity in pollen grain is involved in the provision of storage lipids (Evans et al. 1992; Roberts et al. 1993). Similarly, in the tapetum, before anthesis, lipid accumulates in elioplasts and is transferred to coat the surface of the pollen grains (Dickinson 1973).

TEM studies (Wattendorf and Holloway 1980; Holloway 1982) indicate that the cuticle is composed of a regular arrangement of alternate electron – opaque and electron-translucent lamellae. Similar lamellated layers are observed in suberized cell wall (Kolattukudy 1980). For detailed descriptions and illustrations see Scott (1994). Lamellae do not occur in lignin, which may distinguish the composition of lignin from that of cutin and suberin. Since cutin and suberin contain high levels of fatty acids, which are absent from lignin, the formation of lamellae is apparently associated with the polymerization of aliphatic monomers (Scott 1994).

14.3.5 Patterning of The Exine

Dickinson and Sheldon (1986) discussed deposition of exine polymers and the determination of patterning in the pollen wall. Information and hypotheses about causes of form and ornamentation in exines of spores were presented by Pettitt (1979), and reviewed by Buchen and Sievers (1981), Raghavan (1989) and Tryon and Lugardon (1991).

A reticulum formed of hexagonal units is one of the most common patterns encountered in biological systems. This pattern also occurs in the pollen wall generation mechanism e.g., in *Lilium* (Dickinson 1970), in *Triticum* (El-Ghazaly and Jensen 1985), and in *Echinodorus* (El-Ghazaly and Rowley: in press). A hexagonal pattern generally distinguishes the surface coating of the plasmalemma (El-Ghazaly and Jensen 1985). Illustrations of hexagonal patterns are presented in Figs. 14.8 -14.10. According to Dickinson and Sheldon (1986) the hexagonal pattern occurs within the condensing element of the primarily fibrillar layer of the plasmalemma and, at a later stage, in the organization of the primexine. There is no firm information as to the source of the pattern including material. Dickinson and Sheldon (1986) suggested that a mechanism of 'self-assembly' results in the reorganization of the membrane mosaic so as to generate a reticulate pattern reflected in components of the bimolecular leaflet.

Sheldon and Dickinson (1983) proposed that the so-called coated vesicles are the only component of the prophase cytoplasm that constitutes a good candidate for determinant of pattern formation. These small protein coated vesicles are formed from prophase onwards from smooth endoplasmic reticulum, and migrate to the cell surface where they fuse with the plasmalemma. The implication is that the coated vesicles insert protein into the plasmalemma, which probably influences the position of the probaculae (Scott 1994). Heslop-Harrison (1972) and Sheldon and Dickinson (1983) proposed the role of physical force in the specification of exine pattern. Sheldon and Dickinson (1983) mentioned the difficulty of explaining the mechanism of protein insertion to the plasmalemma at positions subsequently occupied by probaculae. They proposed that the vesicles fuse randomly with the plasmalemma, and the deposited protein undergoes self-assembly into pattern-specifying units. They also suggest that pattern formation could be due to a physical force. This force initiates from the tendency of the material of the probaculae to aggregate into circular plates or cylindrical rods. When these units of the probaculae aggregate together they become hexagonal (Figs. 14.12, 14.13). This physical force guarantees a dual constraint of maximum space-filling and minimum free energy.

Rowley and Dahl (1977) presented a different interpretation of the early development of the microspores wall. They consider the exine as a part of the plasmalemma, a glycocalyx coated with sporopollenin. Pre-exine glycocalyx formation accumulates between microspores and callosic envelope. The distinctive pre-exine begins to be built up during a period of extreme cell surface activity, presumably involving uptake. As development progresses in the tetrad the cytoplasm of microspores becomes dense. Globules and whorls of lamellations between the unit membrane and glycocalyx together with the increased compactness of the cytoplasm offer indirect evidence for continued uptake. It seems that units of pre-exine are continuous with structures in the cytoplasm, which are surely cytoskeletal (Dickinson and Sheldon 1984; Hideux and Abadie1985; Gabarayeva and El-Ghazaly 1997). According to Hideux and Abadie (1985) an initial arranged tangentially on the surface of the plasmalemma is elaborated upon a microfilamentous matrix. Other initials are radially oriented to the plasmalemma surface and based upon a tubular arrangement deeply rooted in the cytoplasm.

The pre-exinous units consist, at late tetrad stage, of a stainable rod surrounded by a zone of low contrast (Dunbar and Rowley 1984, for *Betula*; Rowley 1983, in *Epilobium*; Rowley and Rowley 1986, for *Ulmus*). Rowley and Dahl (1977) concluded that plasmalemma glycocalyx could be with or without receptors for sporopollenin. Glycocalyx components lacking the capacity to accumulate sporopollenin, presumably because receptors are absent, seem to exist at very many sites on the microspore plasmalemma as well as on apertures, for example those destined to form the exine arcade and lumina of reticulate and other gaps in tectal systems. For *Ulmus*, Rowley and Rowley (1986) mentioned that "the specific form of the pre-exine is directed by the plasmalemma. The receptors for the cell surface code appear to be made up of proteins and the code letters themselves, the legends, are inscribed in sugar molecules (mucopolysaccharide)". Rowley and Dahl (1982) found that exine units have helical substructures; they referred to these exine units as tufts. Rowley (1987) also found that plasmodesmata between the tapetal cells, which are the nutritive cells in support of pollen grains, were very similar to tufts.

14.3.6
Baculae (Columellae)

The probaculae appear to condense around the tubular protrusion of the plasmalemma. As a result of accumulation of polymerizing agent (peroxidase, sporopollenin precursors) on the protrusion of the plasmalemma the probaculae become evident. Later during the late tetrad – early free microspore stages, the probaculae become discretly differentiated into electron – dense baculae. The baculae increase in height and electron density, but there is little or no change in the number of baculae per mesocolpium.

In *Echinodorus*, El-Ghazaly and Rowley (in press) observed probaculae with an unstained, hollow appearing, core zone. In an earlier stage the central part (core) of probaculae is positively contrasted while the surrounding zone is only weakly contrasted. There are other observations of stain reversal from early to less early stages of development in the primexine of microspores. The work of Rowley and Dahl (1977: Pls. VI, 2, 3; XIII, 4) on *Artemisia* shows early stages with the central part of probaculae solid in appearance. Rowley and Claugher (1996), and earlier Dunbar and Rowley (1984), found that in young stages of microspores development of *Epilobium* and *Betula,* the central part (their core zone) is contrasted positively by section stains while the outer ("binder") zone contrast so weakly that care is required for its detection. In a later stage there was a stain reversal where the core was more or less without contrast and appeared to be empty while binders were darkly contrasted.

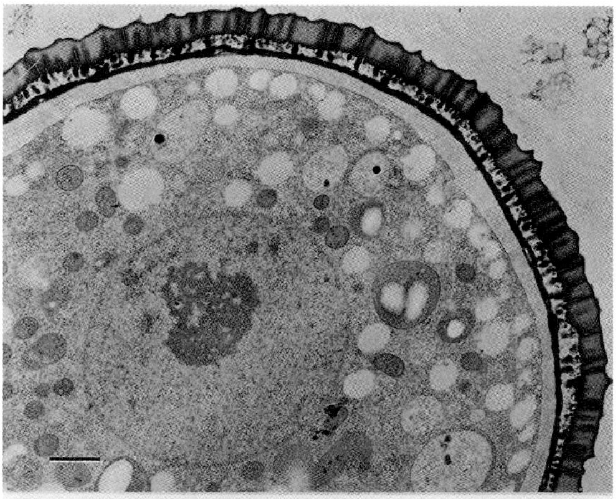

Fig. 14.15. *Betula pendula.* Free microspore stage. Due to plan of sectioning the baculae may appear granular or unattached to the foot layer or the tectum. Baculae appearing unattached to a foot layer or to a tectum in a part of the section are likely to be attached in another part of the same section. Fixation: rapid freezing. TEM. ×10 000. Scale = 1 μm

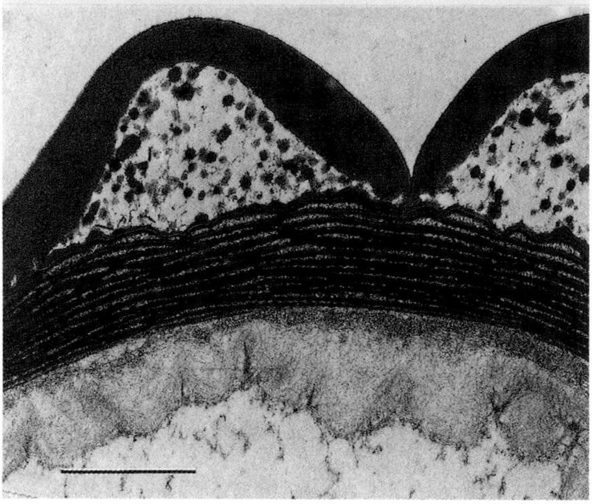

Fig. 14.16. *Ephedra foliata.* Pollen grain stage. The baculae appear granular due to plan of section. Foot layer thin, endexine lamellated and comparatively thick and intine thick. TEM. ×27 000. Scale = 1 μm

Blackmore (1990: Figs. 14.5, 14.6) showed that in freeze fractured material of *Echinops* the primexine during the early tetrad stage consisted of units having a hollow tubular construction.

In *Epilobium angustifolium* and *E. montanum*, Rowley (1973), in *Triticum aestivum*, El-Ghazaly and Jensen (1987) and in *Echinodorus*, the baculae developed round rods – like cylindrical membranes. In the exine of mature pollen grains, traces of these rods occur as microchannels. The baculae in free microspore stages are phosphotungstic acid (PTA-chromic) positive (Fig. 14.14), but their stainability decreases at maturity.

Baculae in several species have been described as granular. Granular infrastructure may be formed of anastomosing rods or irregular-shaped columns (Zavada and Dilcher 1986). The baculae in *Betula* (El-Ghazaly and Grafström 1995) and in *Ephedra* exines (El-Ghazaly and Rowley 1997) are examples of a bacular system misrepresented by being called granular (Figs. 14.15, 14.16). Due to plan of sectioning for TEM studies the baculae may appear granular or unattached to a foot layer or a tectum. Baculae appearing unattached to a foot layer or to a tectum in a sectioned exine are likely to be attached in another section of the same pollen grain (Fig. 14.15). Additional studies are needed to know more about the chemistry and origin of the probaculae and the material that thickened them within the tetrad.

14.3.7
Trilamellated Structures or White Line Center Lamellae

Following microspore release, additional sporopollenin is added to the exine units. The sporopollenin at this and the following stages is apparently of tapetal origin.

Sporopollenin deposition is generally associated with trilamellated structures that bear a striking homology to the lamellae of cutin and suberin. These lamellae are localized close to the plasmalemma. They have molecular dimensions, unit-membrane, 5–6 nm in thickness and electron lucent layer of 4–8 nm in thickness. The appearance of these lamellae is however, dependent on the developmental stage of the microspores and on the species (Figs. 14.17, 14.18). Sporopollenin-associated lamellae are found in spore walls of almost all-living plant taxa. This very broad phylogenetic distribution suggested an ancient origin, perhaps in association with the spores of early land plants or the algal progenitors of embryophytes (Atkinson et al. 1972).

In angiosperms lamellae are most readily observed in the endexine. The lamellae are particularly obvious in certain primitive plants, such as algae (Atkinson et al. 1972), bryophytes (Lugardon 1990) and gymnosperms (Kurmann 1990; El-Ghazaly and Rowley 1997). Figure 14.16, shows thick lamellated endexine in *Ephedra foliata*.

The sequence of events at the cell surface (Plasmalemma) that is responsible for the formation of the lamellated layer is not well understood. The lamellated structure may begin with the formation of the lucent layer on the outside of the plasmalemma which then synthesises a layer of sporopollenin on both its inner and outer faces (Scott 1994). As a result a trilamellated structure or white line centre lamella is developed. The trilamellated structure apparently moves away from the membrane and is included in the growing endexine; a new lamella forms at the plasmalemma, and the process is repeated. In a model of suberin structure proposed by Kolattukudy and Köller (1983), the lucent layer is interpreted as being composed of the hydrophobic regions of long-chain fatty acids, whilst the hydrophilic ends of these molecules comprise the opaque layers in a complex with the phenolic component.

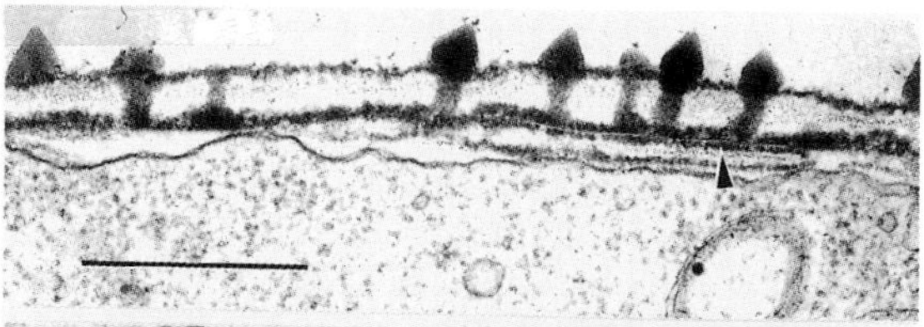

Fig. 14.17. *Triticum aestivum.* Young free microspores. Trilamellated structure or white line centred lamella separates the foot layer from the endexine (arrow). TEM. ×58 000. Scale = 0.5 μm

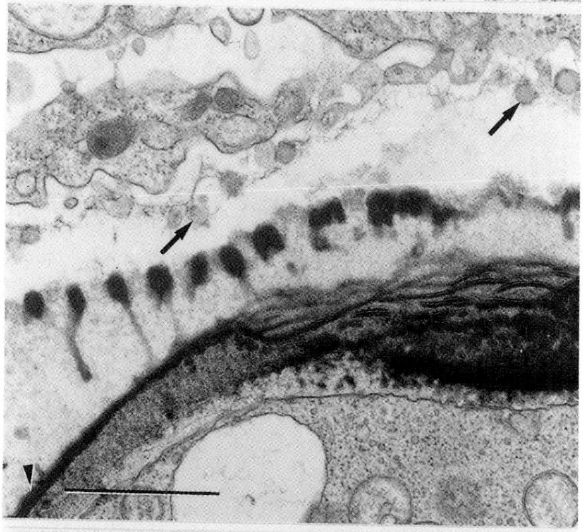

Fig. 14.18. *Rondeletia odorata.* Free microspore stage. Part of tapetum , few orbicules at its surface (arrows) and part of a microspore. White line centred lamella (arrowhead) separates thin foot layer from the endexine. The lamellae increase in number and thickness towards the aperture.TEM. ×32 000. Scale = 1 μm

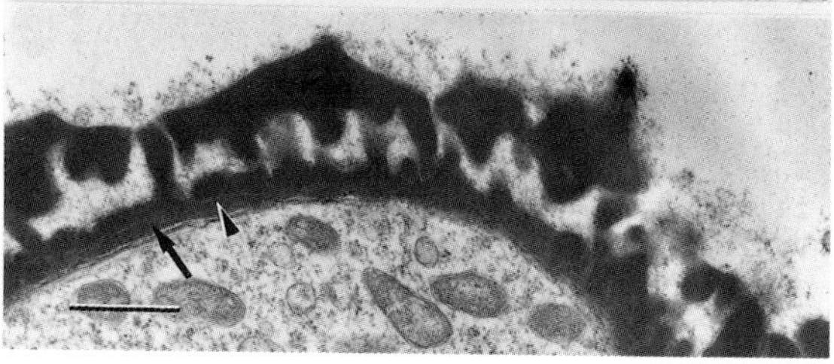

Fig. 14.19. *Echinodorus cordifolius.* Free microspore stage. The foot layer is discontinuous. There is a junction plane (arrowhead) that separates the foot layer from the endexine(arrow). TEM. ×15 000. Scale = 1 μm

14.3.8 The Foot Layer and Transitory Endexine

In *Echinodorus* the foot layer and all portions of the exine are made up initially of units of about equal size. When the foot layer forms in *Echinodorus*, after the microspores are released from callose envelopment, there are white-lines at its inner surface (Fig. 14.19). A white-line or plane persists through development of what Rowley and Dunbar (1996) called a "transitory endexine" in their study of *Centrolepis* pollen. In *Echinodorus* there is a transitory endexine often seen in thin sections to be separated from the foot layer by a white-line (Fig. 14.19).

Simpson (1983) in studies of pollen ultrastructure of the Haemodoraceae called the white-line between the inner and outer portions of the exine in most members of this family a "commissural line". He found that the commissural line (plane) was most prominent early in pollen development and in some genera its detection was difficult or not possible in mature pollen (Simpson 1983). Components below the commissural line (endexine) appear less solid than above and show a distinct radial orientation. Argue

(1972) in work with pollen grains of some Alismataceae, found that there were white-lines on short strands just outside the plasmalemma in microspores recently released from callose envelopment. As the early exine thickened white-lines were seen at the inner surface of the foot layer and still later in microspore development between the foot layer and a zone which he called "secondary exine" or a "basal layer". El-Ghazaly and Rowley (in press) consider a similar layer " in *Echinodorus* (Alismataceae) pollen to be a "transitory endexine". Other similarities to the transitory endexine were observed by Argue (1974 and 1976) in *Alisma, Baldellia, Caldesia, Echinodorus, Sagittaria* and *Lophotocarpus* (possibly the most primitive of the Alismataceae).

In addition to the above reports of transitory endexines or endexine-like zones in monocot pollen there are other examples such as: *Poa* (Poaceae) Rowley (1964); *Zea* (Poaceae) Skvarla and Larson (1966); *Anthurium* (Araceae) Rowley and Dunbar (1967); *Triticum* (Poaceae), El-Ghazaly and Jensen (1986a, b, 1987); *Callitriche* (Callitrichaceae) Martinsson (1993).

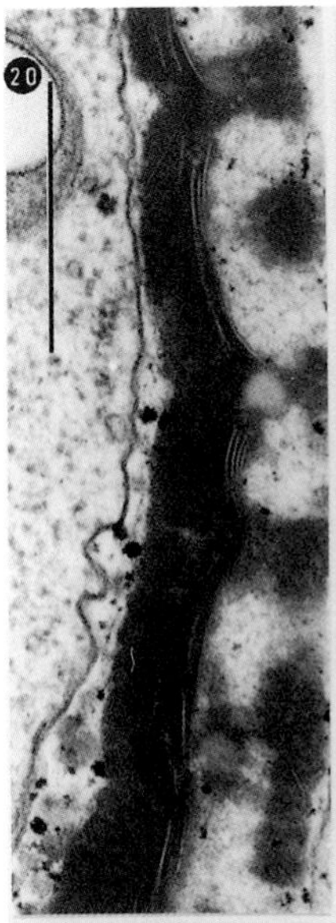

Fig. 14.20. *Rondeletia odorata.* Free microspore stage. The unite of the microspore wall increased in thickness (cf. Fig. 38). The foot layer is thin and separated from the endexine by the white line centred lamellae. TEM. ×68 000. Scale = 0.5 µm

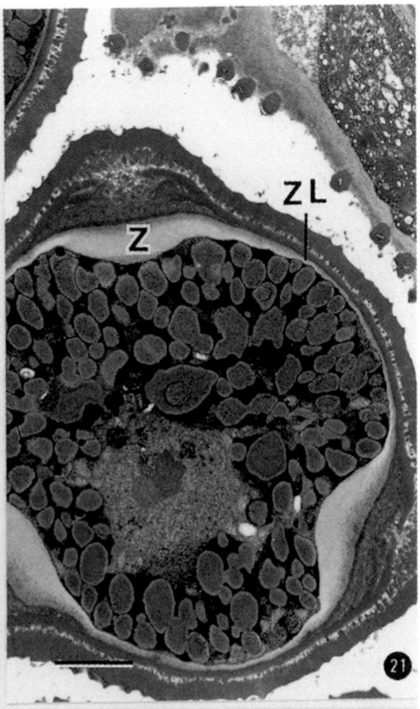

Fig. 14.21. *Betula pendula.* Young microspore stage. The cytoplasm contains numerous globules of lipids. The exine and the annulus are well developed. Note that the exine is positively stained, the Zwischenkörper (*Z*) and the Z-layer (*ZL*) are almost negatively stained. Stain: UA-Pb. TEM. ×6000. Scale = 2 µm

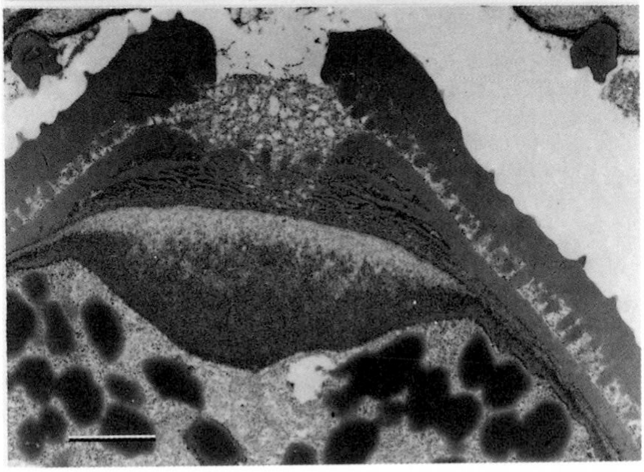

Fig. 14.22. *Betula pendula.* Free microspore stage. Magnified pore showing thick annulus with numerous endexinous lamellae. The spinules are with sharp end, occasionally appearing blunt because of plan of section. Baculae are densely spaced. Note positive stainability of the proximal portion of Zwischenkörper and Z-layer. Stain: PTA-acetone. TEM. ×14 000. Scale = 1 µm

In some cases as in grasses a rudimentary endexine becomes appressed against the proximal surface of the foot layer, and can no longer be seen (El-Ghazaly and Jensen 1986b). The development of the endexine and the intine may occur during the vacuolate period, e.g. in *Vicia* (Audran and Willemse 1982) and *Lilium* (Heslop-Harrison and Dickinson 1969). The formation of the foot layer and endexine generally takes place by the formation of lamellae at the plasmalemma. These lamellae form the sporopollenin-deposition surface (Fig. 14.20).

It is apparent that the common view that there is no endexine in monocot pollen needs to be re-examined. The endexine reported in monocot pollen have been best observed in young stages. Tomlinson's (1995: p. 590) suggestion that "Comparison among adult forms may be inappropriate; one has to appreciate how changes in ontogeny, beginning with seeding mutation, may affect final form" might just as well be applicable to pollen grains.

14.3.9
Endexine

The endexine has some morphological features and functional aspects that have not been commonly considered in the context of angiosperm and gymnosperm homology. In both groups the endexine is a transport system throughout the life of the pollen. Traces are taken up from external environment into the endexine and cytoplasm (Rowley 1996). Throughout intervals of rapid growth and presumably considerable uptake, e.g. during the vacuolated stages, endexine have many and extensive irregular channels (Rowley 1988, 1990). Rowley (1990, 1996) mentioned that irregular channels that exist during developmental stages of angiosperm pollen grain will return to an orderly state when the temperature is lowered in living pollen to 1–4° C. Thus when metabolic energy is turned off the endexine goes back together, implying that the endexine is elastic and the subunits remain bonded to the tubular or "lamellar" components to allow reassembly into an active or dormant condition. White lines are common within the lamellations (long tubular components) between these irregular channels. The lamellations are, or can be, rodlets or tubules. Rowley (1996) discussed the possibility that these tubules have extensive lateral cross-linking into sheets. A tubular organization with reversible lateral cross-linking offers versatility for growth and internal transport (Rowley1990, 1996). The often-irregular channels that exist between the lamellae of the endexine do not favour a system of fixed sheets.

The junction between the endexine and the foot layer appears as a white line in many of angiosperms and gymnosperms pollen wall (Rowley 1988; Xi and Wang 1989; El-Ghazaly and Rowley: in press). Rowley (1988) referred the term junction plane to this line, while Simpson (1983) used the term "commissural line". Dickinson and Heslop-Harrison (1971) and Dickinson and Sheldon (1986) observed double membrane inclusions in the cytoplasm of developing microspores of *Lilium* and suggested that they may play a very active part in the development of the nexine 2 layer (endexine).

14.3.10
Oncus (granular thick body)

Onci are bodies of granular material underlying apertures, which are present in some developing microspores, generally before the vacuolate stage. Onci may be highly modified or absent at maturity.

Onci may function during germination as a pre-formed tip to the emerging pollen tube (Rowley and Dahl 1977) or in the control of rehydration during germination since the oncus lie below the aperture which form the main pathway for rehydration (Heslop-Harrison 1979).

14.3.11 Zwischenkörper

In *Betula pendula* interstitial body (Zwischenkörper, sensu Fritzsche 1837, Beer 1906) is developed in young free microspores, between the cytoplasm and the granular-fibrillar layer at the sites of the pores (Fig. 14.21). Zwischenkörper is usually described as restricted below the pores (Christensen and Horner 1974; Dunbar and Rowley 1984; El-Ghazaly and Jensen 1986b, 1987; Charzynska et al. 1990). In *Betula* the Zwischenkörper is continuous in the interapertural wall as a thin layer (Z-layer). At the free microspore stage, the Zwischenkörper is stained positively for protein, particularly its proximal portion (Fig. 14.22). When anthers of *Betula* are fixed by rapid freezing and substitution, and PTA-acetone stain for protein is applied the Zwichenkörper stains densely and shows a characteristic network pattern (Fig. 14.23). The early development and histochemical tests support differentiation of Zwischenkörper in the apertures and Z-layer in interapertural wall from the intine (El-Ghazaly and Jensen 1986b; El-Ghazaly and Grafström 1995; El-Ghazaly et al. 1996). Zwischenkörper shows a negative stain with UA-Pb and faint stain for neutral and acidic polysaccharides, while the membranous channels and the fibrillar material of the intine are generally stained positively (Fig. 14.24). Because of the above mentioned morphological and histochemical variations

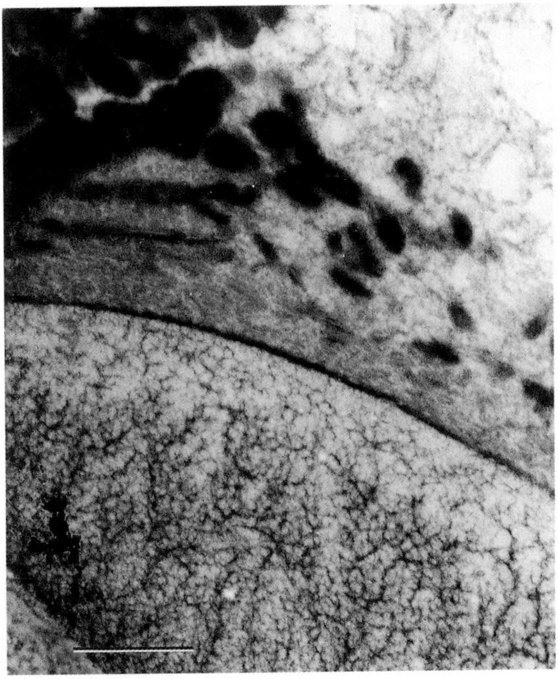

Fig. 14.23. *Betula pendula*. The exine and the proximal portion of the Zwischenkörper are darkly stained. Not the dense fibrillar material that are obvious in Zwischenkörper after fixation by rapid freezing method. Stain: PTA- acetone. TEM. ×56 000. Scale = 0.5 µm

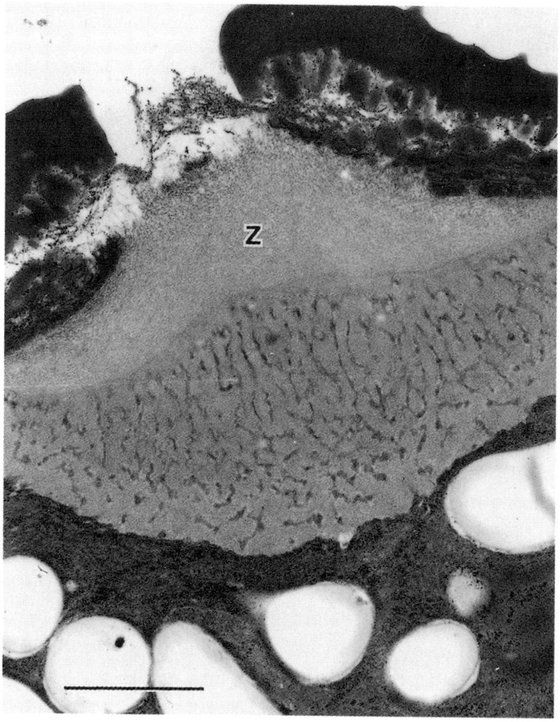

Fig. 14.24. *Betula pendula*. Pollen grain stage. The lamellae of the annulus become more compact than before. The Zwischenkörper (*Z*) is reduced in size and the intine is thick at the pores. The membranous units of the intine are stained densely, while the Zwischenkörper and Z-layer are moderately stained. Stain: UA-Pb. TEM. ×30000. Scale = 1 μm

between the intine and Zwischenkörper, the suggestion of Charzynska et al. (1990) in considering Zwischenkörper and Z-layer as outer layer of the intine is not applicable.

The Zwischenkörper in *Betula* reduces much in size just before anthesis and its upper surface stains positively for protein and carbohydrates. These components of Zwischenkörper may indicate a role of this layer in the male-female recognition reaction. This proposed function of Zwischenkörper was also mentioned by Grote and Fromme (1984) on birch pollen. In addition Charzynska et al. (1990) suggested that Zwischenkörper in *Secale* might facilitate the diffusion of solutes from the tapetum to the microspore. The early development of Zwischenkörper in close contact with the tapetum in *Betula* may be in favour of this suggestion. Additional experiments are needed to demonstrate possible routes for transfer of materials between sporophyte and gametophyte.

The complexities illustrated in the histochemistry of Zwischenkörper in *Betula* are similar to a great extent to that of *Triticum aestivum* (El-Ghazaly and Jensen 1986a) and to *Corylus avellana* described by Heslop-Harrison and Heslop-Harrison (1991). Zwischenkörper in *C. avellana* is distinguished from the intine by its staining properties. Heslop-Harrison and Heslop-Harrison (1991) referred that the pectin nature of Zwischenkörper reflects specialization for a critical early function in germination.

Additional histochemical tests and different techniques for e.g. SEM cryo-fracture at different stages of *Betula* wall development are needed to solve the chemistry and morphology of Zwischenkörper, and to prove its functions.

14.3.12 Development of The Intine

The intine usually starts to develop at the vacuolate stage, beneath the apertures. It increases in thickness under the pores and later on starts to develop under the interapertural parts as a thin layer.

The intine development could be associated with the plasmalemma (Fig. 14.25), as in *Triticum aestivum* (El-Ghazaly and Jensen (1986a, b) and in *Rondeletia odorata* (Fig. 14.26). In some pollen types, Golgi bodies are frequent during intine synthesis e.g. Ranunculaceae (Roland 1971), while in other, endoplasmic reticulum and polyribosomes are abundant e.g. *Cosmos* (Knox and Heslop-Harrison 1970).

In case of monoporate pollen grains e.g. in *Triticum* and in triporate pollen such as *Betula*, the morphology of Z-layer and the thin intine at interapertural part of the wall appears as irregular ingrowths. The pronounced irregular ingrowths of intine in *Triti-*

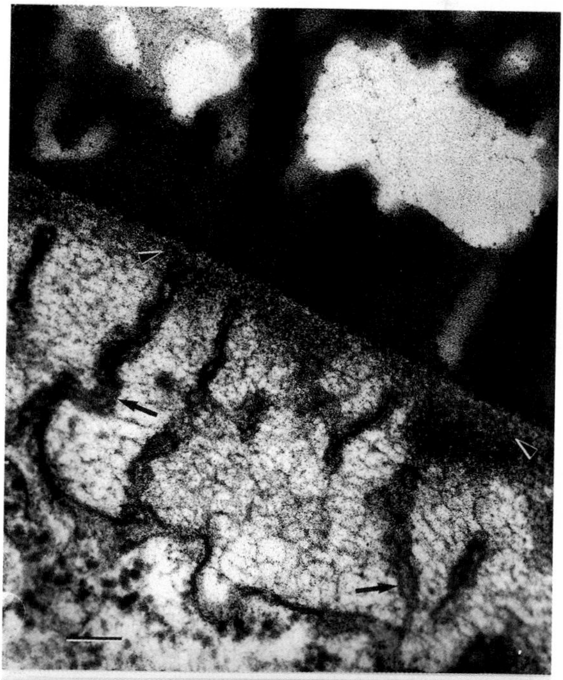

Fig. 14.25. *Triticum aestivum.* Young pollen grain stage. Magnified portion of the pollen wall showing the Z-layer (arrowheads), extensions of the plasmalemma (arrows) and reticulate fibrils of the intine between the plasmalemma membranes. TEM. ×120000. Scale = 0.1 µm

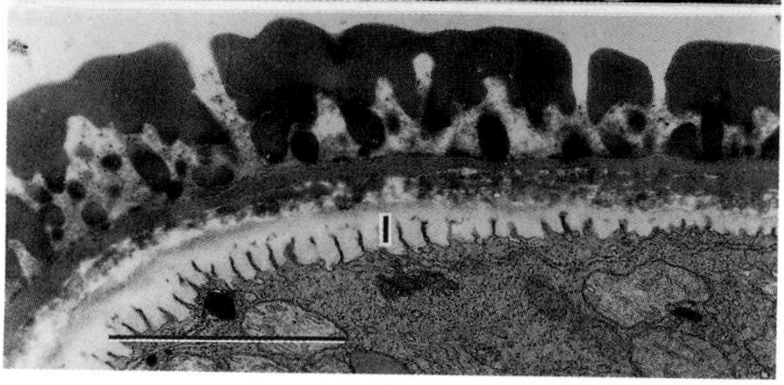

Fig. 14.26. *Rondeletia odorata.* Pollen grain stage. Thick intine (I) fully developed. Note numerous electron dense membranous units embedded in the intine. ×35000. Scale = 1 µm

cum aestivum (El-Ghazaly and Jensen 1986a), *Betula pendula* (El-Ghazaly and Grafström 1995) and in *Rondeletia odorata* (Huysmans et al. in prep) is in agreement with the observation of Charzynska et al. (1990) in *Secale cereale*, and may support their conclusion that these ingrowths form a transfer girdle in contact with the cytoplasm of the external pole of the microspore towards the tapetum.

In porate pollen grains, the fibrillar-granular layer generally observed above Zwischenkörper and between Z-layer and the exine is stained densely for protein. Similar layers were described by Ciampolini et al. (1993) in *Cucurbita pepo* as sporophytic proteins. This layer developed in *Betula* at early free microspore stage and is thicker at the pore sites. It may take part with the gametophytic proteins, localized in the intine, in pollen-stigma recognition and interaction. The intine beneath the pores becomes comparatively very thick and is provided with fibrillar material and radially arranged membranous units. The fibrillar and membranous unites of the intine are stained positively for acidic polysaccharides and protein. The deposition of protein in the exine and intine embedded with units of plasmalemma, prior to pollination, may be also a prerequisite for successful pollen-stigma recognition and interactions.

14.4 The Substructures (Exine Subunits) of Pollen Grains and Spores

The substructures (exine subunits) of pollen grains and spores are similar (Southworth 1986; Kedves 1990; Rowley 1996). Guédès (1982) and Kurmann (1989, 1990) indicated that the ectexine and endexine in gymnosperm pollen are structurally homologous to the ectexine and endexine in angiosperm pollen.

The bacular arcades do not change in height during development of the exines.

Baculae in exines that have thick, continuous tectum are folded down because of the outward pressure of the growing cytoplasm and the restraining force of the thick tectum (Rowley 1996). The microspore of *Strelitzia* is ca 10 µm in diameter and has 1 µm thick wall, and undergoes exceptional sequences of growth. At maturity the microspore is blown up to a diameter of 100 µm and the 1 µm thick exine is stretched, becoming ca 0.1 µm thick (Kronestedt-Robards and Rowley 1989). The ultrathin, acetolysis resistant exine of *Strelitzia* covers the pollen surface without interruption (Hesse and Waha 1983). Thus it may be assumed that a continuous covering of sporopollenin is permeable to solvent and nutrients for growth and ultrathin covering of sporopollenin is sufficient to function as an exine. In many examples of pollen wall development it is not clear how expanding exines both increase in thickness and retain specific form and ornamentation. For *Betula* (Dunbar and Rowley 1984; El-Ghazaly and Grafström 1995) and *Triticum* (El-Ghazaly and Jensen 1986a), new exine-units are apparently inserted between those initially formed.

J. Rowley and his collaborators have proposed ultrastructural aspects of pollen wall in several species. For example, Rowley et al. (1981) presented the substructure in exines of *Artemisia vulgaris* after partial oxidation methods. They proposed a model of glycocalyx unit-complex (tuft) consisting of a super-coiled helical subunit (binder) wound around several axially straight helical subunits. Tuft units are radial in the baculae and throughout the nexine (Fig. 14.29).

14.4.1 Atomic Force Microscopy and Scanning Tunneling Microscopy

The exine substructure of the *Alnus*, *Betula*, *Fagus*, and *Rhododendron* pollen grain and *Lycopodium* spores have been investigated by atomic force microscopy and scanning tunneling microscopy. Wittborn et al. (1996) find that these pollen and spores, despite their very distant relation and big difference in structure and morphology on a micrometer scale, have very similar substructure on a nanometer scale. The substructure appears to consist of a multi-helix, i.e. helices which in turn consists of helical structures with a self similarity at smaller and smaller length scales (Figs. 14.27, 14.28).

Our preparations graded from intact or fractured fresh pollen to pollen that was acetolyzed, chemically fixed and epoxy resin embedded. While our knowledge of the exact radial/lateral orientation of most of our scans is less than perfect, there were in all cases substructures or cross connections generally perpendicular to the prominent axis of exine units.

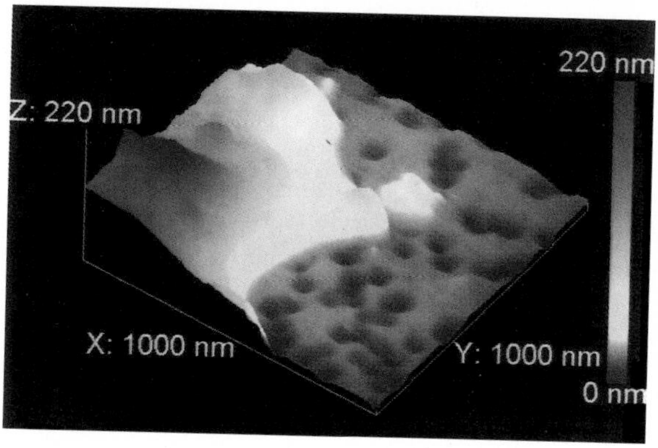

Fig. 14.27. *Betula pendula.* AFM scan of acetolyzed and freeze sectioned exine ($1000 \times 1000\ nm^2$). There is an exine area interpreted as the inside of the exine in the right hand part of the picture, the depressions are ca 100 nm in diameter. There is also a view of the endexine or underside of the foot layer to the left. (After Wittborn et al. 1996)

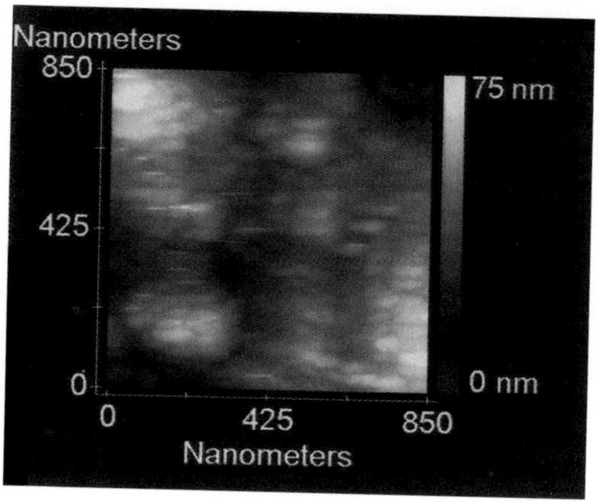

Fig. 14.28. STM image of the exine of *Betula pendula* ($850 \times 850\ nm^2$). The sample was acetolyzed, freeze sectioned and sputter coated with ca 100 Å of gold. The approximately 30–50 nm wide and 70–120 nm long subunits are shaped as elongated spheroids, which may be interpreted as part of helices seen sideways. (After Wittborn et al. 1996)

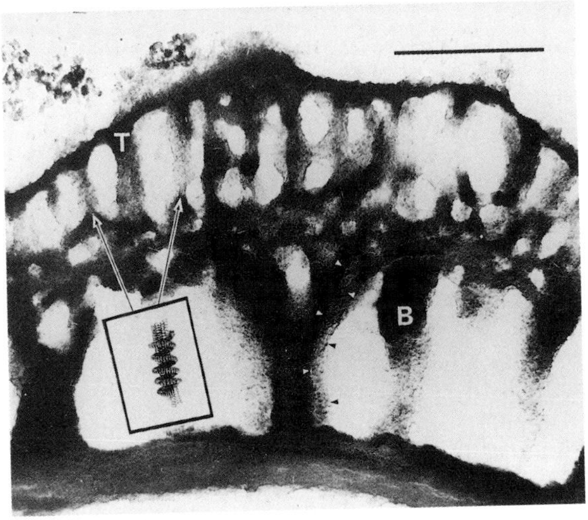

Fig. 14.29. TEM, showing thick section (ca. 150 nm) of partially oxidized exine shadowed with pd-Au after complete removal of epon-araldite. The nonradial binder subunits at the surface of bacules (*B*) and tectal processes (*T*) are more evident than the cores. The tectal process (*T*) consists of two tuft units and two or more units are exposed in bacule (*B*). The lines from the tuft unit model (insert) extend to tectal processes, which may consist of but one tuft unit and have cores and binders in evidence. Scale = 1 µm. (After Rowley et al. 1981)

References

Argue CL (1972) Pollen of the Alismataceae and Butomaceae. Development of the nexine in *Sagittaria lancifolia* L. – Pollen Spores 14:5–16

Argue CL (1974) Pollen studies in the Alismataceae. Bot Gaz 135:338–344

Argue CL (1976) Pollen studies in the Alismataceae, with special reference to taxonomy. Pollen Spores 18:161–201

Atkinson AW, Gunning BES, John PCL (1972) Sporopollenin in the cell wall of *Chorella* and other algae: Ultrastructure, chemistry, and incorporation of ^{14}C-acetate, studied in synchronous cultures. Planta 107:1–32

Audran JC, Willemse MTM (1982) Wall development and its autofluorescence of sterile and fertile *Vicia faba* L. Pollen. Protoplasma 110:106–111

Beer R (1906) On the development of the pollen grain and anther of some Onagraceae. Beih Bot Zbl (Erste Abt) 19:286–313

Benedetti EL, Bertolini B (1963) The use of phosphotungstic acid as a stain for the plasma membrane. J R Microsc Soc London 81:219–222

Blackmore S (1990). Sporoderm homologies and morphogenesis in land plants, with a discussion of *Echinops sphaerocephala* (Compositae). – Pl Syst Evol (Suppl 5):1–12

Blackmore S, Barnes SH (1990) Pollen wall development in angiosperms. In: Blackmore S, Knox RB (eds) Microspores: Evolution and Ontogeny. Academic Press, London, pp 173–192

Brooks J, Shaw G (1968) Chemical structure of the exine of pollen walls and a new function for carotenoids in nature. Nature 219:532–533

Buchen B, Sievers A(1981) Sporogenesis and pollen grain formation. In: Kiermayer O (ed) Cytomorphogenesis in Plants. Springer, Wien, pp 349–376

Bustamante C, Keller D (1995) Scanning Force Microscopy in Biology. – Physics Today 47:32–38

Chaloner WG (1976) The evolution of adaptive features in fossil exines. In: Ferguson IK, Muller J (eds) The Evolutionary Significance of the Exine. Academic Press, London, pp 11–14

Charzynska M, Murgia M, Cresti M (1990) Microspore of *Secale cereale* as a transfer cell type. Protoplasma 158:26–32

Christensen J E, Horner HT (1974) Pollen pore development and its special orientation during microsporogenesis in the grass *Sorghum bicolor*. Amer J Bot 61:604–623

Ciampolini F, Nepi M, Pacini E (1993) Tapetum development in *Cucurbita pepo* (Cucurbitaceae). In: Hesse M, Pacini E, Willemse M (eds) The tapetum cytology, function, biochemistry and evolution. Plant Syst Evol (Suppl) 7:13–22

Dahl AO (1986) Observations on pollen development in *Arabidopsis* under gravitionally controlled environments. In: Blackmore S, Ferguson IK (eds) Pollen and spores: Form and Function. Academic Press, London, pp 49–59

Dahl AO, Rowley JR (1974) A glycocalyx embedded within the pollen exine. J Cell Biol 63:75a

Dickinson HG (1970) Ultrastructural aspects of primexine formation in the microspore tetrad of *Lilium longiflorum*. Cytobiology 1:437–449

Dickinson HG (1973) The role of plastids in the formation of pollen grain coatings. Cytobios 8:24–40

Dickinson HG, Heslop-Harrison J (1971) The mode of growth in the inner layer of the pollen grain exine in *Lilium*. Cytobios 4:233–243

Dickinson HG, Potter U (1976) The development of patterning in the alveolar sexine of *Cosmos bipinnatus*. New Phytol 76:543–550

Dickinson HG, Sheldon JM (1984) A radial system of microtubules extending between the nuclear envelope and the plasma membrane during early male haplophase in flowering plants. Planta 161:86–90

Dickinson HG, Sheldon JM (1986) The generation of patterning at the plasma membrane of the young microspore of *Lilium*. In: Blackmore S, Ferguson IK (eds) Pollen and Spores: Form and Function. Academic Press, London, pp 1–17

Dover GA (1972) The organisation and polarity of pollen mother cells of *Triticum aestivum*. J Cell Sci 11:699–711

Dunbar A; Rowley JR (1984) *Betula* pollen development before and after dormancy: exine and intine. Pollen Spores 26:299–338, pl 1–14

Echlin P, Godwin H (1968) The ultrastructure and ontogeny of pollen in *Helleborus foetidus*. L. II. Pollen grain development through the callose special wall stage. J Cell Sci 3:175–186

El-Ghazaly G (1982) Ontogeny of pollen wall of *Leontodon autumnalis* (Hypochoeridinae, Compositae). Grana 21:103–113

El-Ghazaly G, Grafström E (1995) Morphological and histochemical differentiation of the pollen wall of *Betula pendula* Roth. During dormancy up to anthesis. Protoplasma 187:88–102

El-Ghazaly G, Jensen W (1985) Studies of the development of wheat (*Triticum aestivum*) pollen: III. Formation of microchannels in the exine. – Pollen Spores 27:5–14

El-Ghazaly G, Jensen W (1986a) Studies of the development of wheat (*Triticum aestivum*) pollen: I. Formation of the pollen wall and Ubisch bodies. – Grana 25:1–29

El-Ghazaly G, Jensen W (1986b) Studies of the development of wheat (*Triticum aestivum*) pollen: formation of the pollen aperture. Can J Bot 64:3141–3154

El-Ghazaly G, Jensen W (1987) Development of wheat (*Triticum aestivum*) pollen. II. Histochemical differentiation of wall and Ubisch bodies during development. – Amer J Bot 74(9):1396–1418

El-Ghazaly G, Rowley J (1997) Pollen wall of *Ephedra foliata*. Palynology 22

El-Ghazaly G, Rowley J (in press) Pollen wall and tapetum development in *Echinodorus cordifolius*. Nord J Bot

El-Ghazaly G, Cresti M, Walles B (1996) Ultrastructure of birch (*Betula pendula*) microspores and tapetum after rapid freeze fixation and substitution. 14th International Conference on Sexual Reproduction in plants, Melbourne

Erdtman G (1952) Pollen Morphology and Plant Taxonomy. – Almqvist & Wiksell, Stockholm

Erdtman G (1960) The acetolysis method. Sv Bot Tidskr 54:561–564

Evans DE, Taylor PE, Singh MB, Knox RB (1992) The interrelationship between the accumulation of lipids, protein and the level of acyl carrier protein during the development of *Brassica napus* L. Pollen. Planta 186:343–354

Fægri K & Iversen J (1975) Textbook of Pollen Analysis. Munksgaard, Copenhagen

Fritzsche CJ (1837) Über den pollen. – Hebdm Savant Etrang Acad St Petersbourg 3:649–672

Frova C, Pè ME (1992) Gene Expression during pollen development. In: Cresti M, Tiezzi A (eds) Sexual Plant Reproduction. Springer, Wien, pp 31–40

Gabarayeva N, El-Ghazaly G (1997) Sporoderm development in *Nymphaea mexicana* (Nymphaeaceae). Pl Syst Evol 204:1–19

Godwin H, Echlin P, Chapman B (1967) The development of the pollen grain wall in *Ipomoea purpurea* (L.) Roth. Rev Palaeobot Palynol 3:181–195

Grote M, Fromme HG (1984) Ultrastructural demonstration of a glycoproteinic surface coat in allergenic pollen grains by combined cetylpyridinium chloride precipitation and silver proteinate staining. Histochemistry 81:171

Guédès M (1982) Exine stratification ectexine structure and angiosperm evolution. Grana 21:16–170

Guilford WJ, Schneider DM, Labovitz J, Opella SJ (1988) High resolution solid state ^{13}C NMR spectroscopy of sporopollenins from different taxa. Plant Physiol 86:134–136

Heslop-Harrison J (1962) Origin of exine. Nature 195:1069–1071

Heslop-Harrison J (1963) An ultrastructural study of pollen wall ontogeny in *Silene pendula*. Grana Palynol 4:7–24

Heslop-Harrison J (1968a) Pollen wall development. Science 16:230–237

Heslop-Harrison J (1968b) Wall development within the microspore tetrad of *Lilium longiflorum*. Can J Bot 46:1185–1192

Heslop-Harrison J (1969) An acetolysis-resistant membrane investing tapetum and sporogenous tissue in the anthers of certain Compositae. Can J Bot 47:541–542

Heslop-Harrison J (1971) Wall pattern formation in angiosperm microsporogenesis. In: Control Mechanisms of Growth and Differentiation. Symp Society Exp Biol 25:277–300

Heslop-Harrison J (1972) Pattern in plant cell walls: morphogenesis in miniature. Proc R Inst G Britain 45:335–351

Heslop-Harrison J (1979) Aspects of the structure, cytochemistry and germination of the pollen of rye (*Secale cereale* L.). Ann Bot 44:1–47
Heslop-Harrison J, Dickinson HG (1969) Fine relationship of sporopollenin synthesis associated with tapetum and microspore in *Lilium*. Planta 84:199–214
Heslop-Harrison J, Heslop-Harrison Y (1991) Structural and functional variation in pollen intines. In: Blackmore S, Barnes SH (eds) Pollen and Spores: Patterns of Diversification. Syst Assoc Sp Vol 44, Oxford Sci Publ, Oxford, pp 331–343
Hesse M, Waha M (1983) The fine structure of the pollen wall in *Strelitzia reginae* (Musaceae). Plant Syst Evol 141:285–298
Hideux M, Abadie M (1981) The anther of *Saxifraga cymbalaria* L. ssp. *Huetiana* (Boiss.). A study by electron microscopy (SEM & TEM). 3. Dynamics of the relationships between tapetal and sporal cells. Ann Sci Nat Bot Biol Vég 2/3:27–37
Hideux M, Abadie M (1985) Cytologie ultrastructurale de l'anthère de *Saxifraga*. I. Période d'initiation de préceurceurs des sporopollénines au niveau des principaux types exiniques. Can J Bot 63(1): 97–112
Holloway PJ (1982) Structure and histochemistry of plant cuticular membranes: an overview. In: Cutler DF, Alvin KL , Price CE (eds) The Plant Cuticle. Academic Press, London, pp 1–32
Horner HT, Pearson CB (1978) Pollen wall and aperture development in *Helianthus annuus* (Compositae: Helianteae). Amer J Bot 65:293–309
Huysmans S, El-Ghazaly G, Smets E (In prep) Pollen wall development of *Rondeletia odorata* (Rubiaceae)
Kedves M (1990) Quasi-crystalloid basic molecular structure of the sporoderm. Rev Palaeobot Palynol 64:181–186, pl 1
Kedves M, Toth A, Farkas E (1993) An experimental investigation of the biopolymer organization of both recent and fossil sporoderms. – Grana Suppl 1:40–48
Knox RB (1984) The pollen grain. In: Johri BM (ed) Embryology of Angiosperms. Springer, Berlin, pp 197–271
Knox RB, Heslop-Harrison J (1970) Pollen wall proteins: Localization and enzymatic activity. J Cell Sci 6:1–27
Kolattukudy PE (1980) Biopolyester membranes of plants: cutin and suberin. Science 208:990–1000
Kolattukudy PE, Köller W (1983) Fungal penetration of the first line defensive barriers of plants. In: Callow JA (ed) Biochemical Plant Pathology, New York, J Wiley, pp 79–100
Kronestedt-Robards E C, Rowley J R (1989) Pollen grain development and tapetal changes in *Strelitzia reginae* (Strelitziaceae). Amer J Bot 76:856–870
Kurmann MH (1989) Pollen wall formation in *Abies concolor* and a discussion on wall layer homologies. Can J Bot 67:2489–2504
Kurmann MH (1990) Exine ontogeny in conifers. In: Blackmore S, Knox RB (eds) Microspores: Evolution and Ontogeny. Academic Press, London, pp 157–172
Lugardon B (1990) Pteridophyte sporogenesis: a survey of spore wall ontogeny and fine structure in a polyphyletic plant group. In: Blackmore S, Knox RB (eds) Microspores: Evolution and Ontogeny. Academic Press, London, pp 95–120
Marinozzi V (1967) Reaction de l'acide phosphotungstique avec la mucine et les glycoprotéines des plasmamembranes. J Microsc 6:682–692
Marinozzi V (1968) Phosphotungstic acid (PTA) as a stain for polysaccharides and glycoproteins in electron microscopy. Proc. 4th Eur Reg Conf, Rome
Martinsson K (1993) The pollen of Swedish *Callitriche* (Callitrichaceae) – trends towards submergence. Grana 32:198–209
Paul W, Hodge R, Smartt S, Draper J, Scott RJ (1992) Isolation and characterisation of the tapetum-specific *Arabidopsis thaliana* A9 gene. Plant Molec Biol 19:611–622
Pease, DC (1968) Phosphotungstic acid as an electron stain. 26th Annual EMSA Meeting. Calitor's, Baton Rouge, LA, pp 36–37
Pettitt JM (1979) Development mechanisms in heterospory: cytochemical demonstration of spore wall enzymes associated with β-lectins, polysaccharides and lipids in water ferns. J Cell Sci 38:61–82
Pettitt JM, McConchie CA, Ducker SC, Knox RB (1981) Submarine pollination. Sci Amer 244:134–143
Prahl AK, Springstubbe H, Grumbach K, Wiermann R (1985) Studies on sporopollenin biosynthesis: the effect of inhibitors of carotenoid biosynthesis on sporopollenin accumulation. Z Naturforsch 40c: 621–626
Raghavan V (1989) Developmental biology of fern gametophytes; Cambridge University Press, Cambridge
Roberts MR, Hodge R, Sorenson A-M, Ross J, Murphy DJ, Draper J, Scott R (1993) Characterisation of a new class of oleosins suggests a male gametophyte-specific lipid storage pathway. Plant J 3:629–636
Rodriguez-Garcia MI (1978) Elektronenmikroskopische Untersuchungen von Tapetum und Meiocyten während der Mikrosporogenesis bei *Scilla nonscripta*. Pollen Spores 20:467–484
Roland F (1971) Characterization and extraction of the polysaccharides of the intine and of the generative cell wall in the pollen grains of some Ranunculaceae. Grana 11:101–106
Rowley J, El-Ghazaly G (1992) Lipid in wall and cytoplasm of *Solidago* pollen. – Grana 31:273–283

Rowley JR (1964) Formation of the pore in pollen of *Poa annua*. In: Linskens HF (ed) Pollen physiology and fertilization. North-Holland Publ Co, Amsterdam, pp 59–69
Rowley JR (1973) Formation of pollen exine bacules and microchannels on a glycocalyx. Grana 13: 129–138
Rowley JR (1983) Plasma membrane surface processes a construction units of the exine of *Epilobium* (Onagraceae). In: Fertilization and Embryogenesis in Ovulated Plants. Vedy Slov Acad Sci, Bratislava
Rowley JR (1987) Plasmodesmata-like processes of tapetal cells. La Cellule 74:229–241
Rowley JR (1988) Substructure within the endexine, an interpretation. J Palynol 24:29–42, 3 pl
Rowley JR (1990) The fundamental structure of the pollen exine. Plant Syst Evol Suppl 5:13–29, 3 pl
Rowley JR (1996) Chapter 14D. In situ pollen and spores in plant evolution. Exine origin,development and structure in pteridophytes, gymnosperms and angiosperms. In: eds. Jansonius J, McGregor DC Palynology: Principles and Applications. Am AssocStratigrPalynolsFound Vol1, AASP, Houston, pp 443–462
Rowley JR, Claugher D (1996) Structure of the exine of *Epilobium angustifolium* (Onagraceae). Grana 35:79–86
Rowley JR, Dahl AO (1977) Pollen development in *Artemisia vulgaris* with special reference to glycocalyx material. Pollen Spores 19:169–284
Rowley JR, Dahl AO (1982) A similar substructure for tapetal surface and exine "tuft"-units. Pollen Spores 24:5–8, pl 1
Rowley JR, Dahl AO, Rowley JS (1980) Coiled construction of exinous units in pollen of *Artemisia*.
Rowley JR, Dahl AO, Sengupta S, Rowley JS (1981) A model of exine substructure based on dissection of pollen and spore exines. – Palynology 5:107–152
Rowley JR, Dunbar A (1967) Sources of membranes for exine formation. – Sv Bot Tidskr 61:49–64
Rowley JR, Dunbar A (1996) Pollen development in *Centrolepis asistata* (Centrolepidaceae). Grana 35: 1–15
Rowley JR, Flynn JJ, Takahashi M (1995) Atomic force microscope information on pollen exine substructure in *Nuphar*. Bot Acta 108:300–308
Rowley JR, Rowley JS (1986) Ontogenetic development of microspores of *Ulmus* (Ulmaceae). In: Blackmore S, Ferguson IK (eds) Pollen and Spores: form and function. Linn Soc Symp Ser 12, Academic Press, London, pp 19–33
Schulze Osthoff K, Wiermann (1987) Phenols as integrated compounds of sporopollenin from *Pinus* pollen. J Plant Physiol 131:5–15
Scott RG (1994) Pollen exine – the sporopollenin enigma and the physics of pattern. In: Scott RJ, Stead MA (eds) Molecular and cellular Aspects of plant Reproduction. Soc Exp Biol Semin Ser 55. Cambridge Univ Press, pp 49–81
Shaw G (1971) The chemistry of sporopollenin. In: Brooks J, Grant PR, Muir M, van Gijzel P, Shaw G (eds) Sporopollenin. Academic Press, London, pp 305–348
Shaw G, Yeadon A (1966) Chemical studies on the constitution of some pollen and spore membranes. J Chem Soc C:16–22
Sheldon JM, Dickinson HG (1983) Determination of patterning in the pollen wall of *Lilium henryi*. J Cell Sci 63:191–208
Sheldon JM, Dickinson HG (1986) Pollen wall formation in *Lilium*: the effect of chaotropic agents, and the organisation of the microtubular cytoskeleton during pattern development. Planta 168:11–23
Simpson MG (1983) Pollen ultrastructure of the Haemodoraceae and its taxonomic significance. Grana 22:79–103
Skvarla JJ, Larson DA (1966) Fine structural studies of *Zea mays* pollen. I. Cell membranes in exine ontogeny. Amer J Bot 52:1112–1125
Southworth D (1985) Pollen exine substructure I. *Lilium longiflorum*. – Amer J Bot 72:1274–1283
Southworth D (1986) Substructural organization of pollen exines. In: Blackmore S, Ferguson IK (eds) Pollen and Spores. Form and function. Linn Soc Symp Ser12, Academic Press, London, pp 61–69
Spurr AR(1969) A low-viscosity epoxy resin embedding medium for electron microscopy. J Ultrastruct Res 26:31
Thiéry JP (1967) Mise en évidence des polysaccharides sur coupes fines au microscope électronique. J Microscop 6:987–1018
Tomlinson PB (1995) Non-homology of vascular organisation in monocotyledons and dicotyledons. In: Rudall PJ, Cribb PJ, Cutler DF, Humphries CJ (eds) Monocotyledons: systematics and evolution. R Bot Gard, Kew, pp 589–622
Tryon AF, Lugardon B (1991) Spores of the Pteridophyta. Springer, N York
Wall D (1962) Evidence from recent plankton regarding the biological affinities of *Tasmanites*, Newton 1875, and *Leiosphaeridia*, Eisenack 1958. Geol Mag 99:353–362
Wattendorf J, Holloway PJ (1980) Studies on the ultrastructure and histochemistry of plant cuticles: the cuticular membranes of *Agave americana* L. *In situ*. Ann Bot 46:13–28
Wehling K, Niester Ch, Boon JJ, Willemse MTM, Wiermann R (1989) *p*-Coumaric acid – a monomer in the sporopollenin skeleton. Planta 179:376–380

Wittborn J, Rao KV, El-Ghazaly G, Rowley JR (1996) Substructure of spore and pollen grain exines in *Lycopodium, Alnus, Betula, Fagus* and *Rhododendron* investigated with Atomic Force and Scanning Tunnelling Microscopy. Grana 35(4):185–198
Wodehouse RP (1935) Pollen Grains. Their Structure, Identification and Significance in Science and Medicine. McGraw-Hill, New York
Xi YZ, Wang FH (1989) Pollen exine ultrastructure of extant Chinese gymnosperms. Cathaya 1:119–142
Young BA, Schulz-Schaeffer J, Carroll TW (1979) Anther and pollen development in male-sterile intermediate wheatgrass plants derived from wheat × wheatgrass hybrids. Can J Bot 57:602–618
Zavada MS, Dilcher DL (1986) Comparative pollen morphology and its relationship to phylogeny of pollen in the Hamamelidae. Ann Mo Bot Gard 73:348–381
Zetzsche F, Kalt P, Liechti J, Ziegler E (1937) Zur Konstitution des *Lycopodium*-sporonins des tasmanins und des lange-sporonins. J Prakt Chem 148:267–286

Mechanisms of Microspore Polarity and Differential Cell Fate Determination in Developing Pollen

D. Twell

Department of Biology, University of Leicester, Leicester LE1 7RH, UK
e-mail: twe@le.ac.uk
telephone: +44-116-2522281
fax: +44-116-2522791

Abstract. In prokaryotic and eukaryotic organisms asymmetry in the fate of daughter cells is commonly established by the unequal division of intrinsically polar mother cells. First microspore division represents a striking example of such an intrisically asymmetric division which has dramatic and determinative consequences for the differentiation or fate of the two unequal daughter cells. The aims of this chapter are to discuss the significance of asymmetric cell division for correct pollen differentiation, how asymmetric division leads to differential cell fate, how microspores develop the necessary polarity required for asymmetric division, and to summarize the results of new approaches that are being adopted to identify key genes involved in these processes.

15.1 Pollen Cell Lineage and Development

The completion of male meiosis marks the initiation of a unique process of differentiation which ultimately leads through a simple cell lineage to the formation of the mature male gametophytes or pollen grains. In all angiosperms examined this pathway involves a single asymmetric division of the microspore termed pollen mitosis I (PMI), which serves as a developmental marker for the termination of microspore development and the initiation of pollen development. This highly asymmetric division is a key determinative event in the differential fates of the vegetative and generative cells. The larger vegetative cell (VC), which remains undivided, constitutes the bulk of the pollen cytoplasm and accumulates stored metabolites (starch and/or lipid) required for rapid pollen tube extension. In contrast, the diminutive generative cell (GC) migrates into the VC cytoplasm, divides once to form two sperm cells and contains highly reduced amounts of cytoplasm and organelles and storage products. Details of major differentiation steps in the pathway of pollen development in *Arabidopsis* are illustrated in Fig. 15.1.

The extreme dimorphism in cytoplasmic volume, organelle, protein and RNA contents demonstrates that vegetative and generative cells have very different synthetic and genetic activities (reviewed in Mascarenhas, 1975; Sunderland and Huang, 1987). A number of pollen-specific genes have been isolated which are first activated during microspore (early genes) or pollen development (late genes). Their complex patterns of regulation involving transcriptional and translational (Bate et al., 1996) controls, including the identification of multiple pollen-specific *cis*-regulatory elements, have been comprehensively reviewed (Mascarenhas, 1992; Twell, 1994). Direct evidence for differential gene expression between the vegetative and GCs, was first obtained by producing transgenic tobacco plants harbouring the promoter of the tomato lat52 gene driving expression of a β-glucuronidase (*gus*) fusion gene containing the tobacco etch virus NIa nuclear-targeting signal (Twell, 1992). The *lat52* promoter was shown to be activated

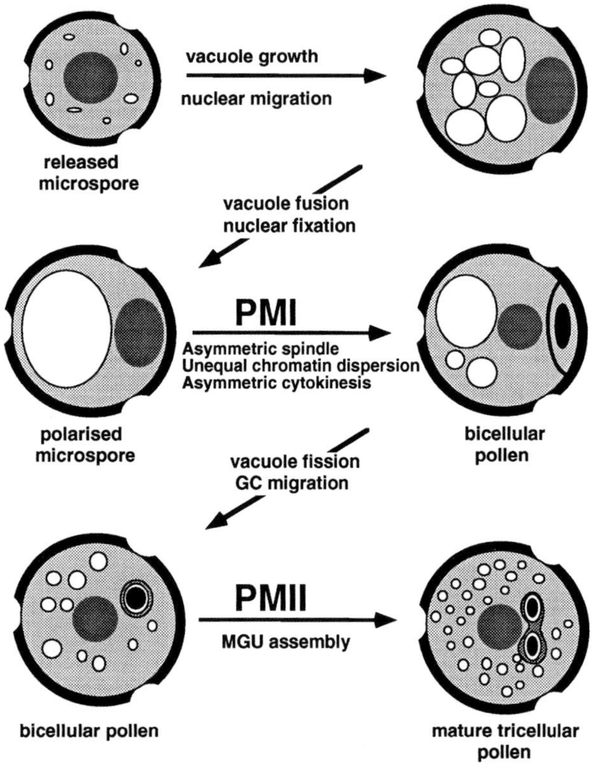

Fig. 15.1. Microspore and pollen differentiation in Arabidopsis. Major cytological events which occur during microspore and pollen development are illustrated based on electron microscopic studies (Owen and Makaroff, 1995). PMI = pollen mitosis I; PMII = pollen mitosis II; MGU = male germ unit

specifically in the vegetative nucleus, and not in the GC, neither during pollen development or tube growth. Evidence for specific gene expression in the GC is very limited. Isolated GCs express unique basic histone-like proteins (Ueda and Tanaka, 1995a, b) and posses their own pool of translatable mRNAs some of which encode proteins unique to the GC (Blomstedt et al., 1996). Definitive identification of a gene expressed specifically in the GC has recently been demonstrated by in situ hybridization of a cDNA designated clone #231 isolated from a cDNA library prepared from embryogenic microspores of Brassica napus (Havlicky et al., 1996).

15.2 The Importance of Asymmetric Division for Differential Cell Fate

Experimental evidence which demonstrates the importance of division asymmetry at PMI for differential cell fate has come from studies in which symmetrical divisions have been induced by microtubule inhibitors or by centrifugation (Tanaka and Ito, 1981; Zaki and Dickinson, 1991, Terasaka and Niitsu, 1987). The two apparently identical cells resulting from symmetrical division do not show the characteristic condensed nuclear chromatin of the GC, and more closely resemble VCs. Further insight into the differentiated state of the daughter cells resulting from unequal and equal division at PMI has been gained through the application of transgenic plants expressing gus activity under the control of the VC-specific *lat52* promoter to monitor cell fate (Twell, 1992; Eady et al., 1994, 1995).

When symmetrical divisions at PMI are induced *lat52-gus* is activated in both daughter cells, and both nuclei show a dispersed 'vegetative-like' chromatin appearance (Eady et al., 1995). Therefore symmetrical division produces apparently identical daughter cells which express vegetative cell fate. These data firmly established that the symmetry of division at PMI has a pivotal role in determining GC differentiation, including the repression of VC-specific genes and chromatin dispersion. However, when division at PMI was blocked in tobacco microspores matured in vitro, uninucleate 'pollen grains' were produced which still activated the *lat52-gus* marker and germinated in vitro (Eady et al., 1995). The production of mature uninucleate pollen grains demonstrates that VC-specific gene activation can be uncoupled from the division process. VC-specific gene activation is therefore likely controlled by a gametophytically-expressed transcription factor(s) (GF) which accumulates in the developing microspore and which normally reaches a threshold of activity at PMI by a pathway independent of cytokinesis (see Fig. 15.2).

15.3 Models of Differential Cell Fate Determination

Two general models have been proposed to account for the differential activation of VC-specific genes as a result of asymmetric cell division (Eady et al., 1995). Both models rely upon the underlying proposition that the activation of VC-specific genes is the default pathway resulting from the accumulation of a gametophytic transcription factor (GF) to threshold levels at PMI. These models provide alternative mechanisms, both involving the polarised distribution of gametophytic regulatory factor(s) to explain how VC-specific genes are normally repressed in the GC as a result of asymmetric division. In the passive repression model (PRM) gametophytic factor (GF) is excluded from a cytoplasmic domain at the GC-pole (Fig. 15.2). Repression of VC-specific gene activity then results from the absence of GF in the nascent GC. In the active repression model (ARM) a generative cell repressor (GCR) which is concentrated at the GC-pole acts to block VC-specific gene activation by GF in the GC (Fig. 15.2).

The ARM and PRM must also take into account the highly condensed nature of the GC chromatin. The compact GC chromatin structure is likely to limit the access of transcription factors to many genes and may therefore be directly responsible for the maintenance of its transcriptionally repressed condition. One clue as to the origin of this differential chromatin structure comes from the observation that differential chromatin dispersion is observed from late anaphase to late telophase of mitosis (Terasaka and Tanaka, 1974; Terasaka, 1982). Chromatin dispersion therefore appears to be differentially controlled by processes occurring within the same cytoplasm before the chromosomes are enclosed within the daughter nuclei. This suggests polarity in the microspore cytoplasm and that during microspore development a 'GC domain' is established at the future GC pole. Within this 'GC domain' specific factors must exist which act to limit chromatin dispersion. Unequal chromatin behaviour could therefore involve a cytoplasmic gradient of a GC pole localized 'condensation factor' which may be equivalent to the GCR proposed in the ARM. Histone or histone-associated factors are putative candidates for such a GCR, in particular because of their known ability to regulate chromatin packaging (Spiker, 1985). Recently two histone variants gH2B and gH3 purified from GC nuclei of lily have been shown to be present specifically in GC and sperm cell nuclei (Ueda and Tanaka, 1995a, b). Histone H1 on the other hand was shown to gradually decrease specifically in the vegetative nucleus after microspore mitosis (Tana-

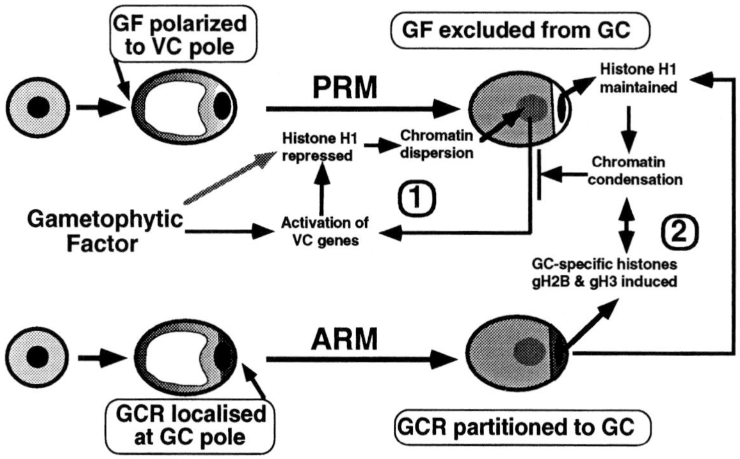

Fig. 15.2. Models of cell fate determination in developing pollen. In both models, VC-specific gene expression is activated through the accumulation of a gametophytic transcription factor (GF). The repression of VC-specific genes and chromatin condensation in the GC then results either from exclusion of GF from the GC pole (Passive Repression Model (PRM)), or from the asymmetric localisation of a generative cell repressor (GCR) protein at the GC pole (Active Repression Model (ARM)). Attenuation of histone H1 levels in the VC is proposed to contribute to the maintenance of VC-specific gene expression through chromatin dispersion remodelling via a feedback loop (*1*). According to the PRM maintenance of histone H1, through the absence of GF, would limit chromatin dispersion in the GC and contribute to repression of VC-specific genes and activation of GC-specific genes (2). According to the ARM GC-specific histones gH2B and gH3 induced by GCR would contribute to the reprogramming of the GC nucleus to induce GC-specific genes by maintaining chromatin condensation (*2*)

ka, 1997). This selective decrease in histone H1 in the vegetative nucleus could facilitate activation of VC-specific genes, through relieving higher order chromatin structure. Extending the cell fate models discussed above, downregulation or turnover of histone H1 could be a direct or indirect activity of GF which would maintain chromatin dispersion and through a feedback loop maintain GF expression and VC-specific gene expression (Fig. 15.2). Conversely, absence of GF in the GC (or repression of GF by GCR) would maintain histone H1 in the GC, which in turn would lead to chromatin condensation and repression of VC genes (Fig. 15.2). Detailed studies of the localization and expression of histone H1 and the GC-specific histones, together with manipulation of their expression are now required to establish the functional relationships proposed in this model.

15.4 The Determination and Expression of Microspore Polarity

15.4.1 Expression of Microspore Polarity

Microspore polarity is ultimately realized at the completion of asymmetric division. Two key processes must be closely coordinated for the expression of the polarity required for correct asymmetric division; (1) nuclear migration and (2) spindle axis determination.

15.4.1.1
Nuclear Migration

Given the role of the nucleus in dictating the site of spindle assembly, a key process in generating polarity is the control of nuclear migration to the future GC pole. There is compelling evidence that stage-specific MT arrays play an important role in controlling polar nuclear migration. In species such as *Tradescantia* (Terasaka and Niitsu, 1990) and tobacco (Twell et al., 1993; Eady et al., 1995) two nuclear movements occur (NM-1 and NM-2), with the second migration being towards the future GC pole. Whilst NM-1 was shown to be insensitive to MT inhibitors, and therefore may result from displacement by the developing vacuole, NM-2 was MT-dependent (Terasaka and Niitsu, 1990).

In *Tradescantia* and *Lilium* MT arrays have been shown to shift dynamically over the nucleus during NM-2 as a perinuclear cap at the future vegetative pole (Tanaka, 1997). MT localization studies in *Brassica napus* also showed that perinuclear MT arrays developed and cortical MT bundling occurred in association with nuclear migration (Hause et al., 1991). Compelling evidence for a specialized MT system that may be involved in nuclear migration comes from studies of orchid species. In *Phalaenopsis* a specialised MT array termed the generative pole microtubule system (GPMS) appears at the future GC pole prior to nuclear migration (Brown and Lemmon, 1991c, 1992). Here vacuolation does not occur so nuclear migration is most likely controlled by interactions of nuclear associated proteins with the GPMS.

Premitotic nuclear migration in cultured tobacco BY2 cells is dependent on MTs, but maintenance of nuclear and spindle position also involves actin microfilaments (MFs) (Katsuta et al., 1990). The application of improved immunolocalization procedures and the anti-MF drug cytochalasin D, has suggests that MFs are also necessary for maintenance of nuclear position in cultured *Brassica* microspores (Gervais et al., 1994). MFs initially radiate out from the centrally located microspore nucleus but then become asymmetrically localized to the vegetative face of the acentric nucleus. MTs have also been strongly implicated in the maintenance of acentric microspore nuclear position (Tanaka and Ito, 1981;Terasaka and Niitsu, 1990) and have been observed tethering the nucleus to the plasma membrane (Hause et al., 1991; Brown and Lemmon, 1991c, 1992).

In species which shed pollen in permanent tetrads it has been possible to determine the polarity of nuclear migration. Nuclear migration can either be towards the inner (radial) or to the outer (distal) walls of the tetrad, but is usually constant within a species (Geitler, 1935). So although the axis of polarity may differ between species nuclear migration appears to be predetermined towards a fixed site. This raises the question of whether the ability of the microspore to achieve asymmetric division is dependent upon the nucleus arriving at this site? Asymmetric division in *Pinus* microspores was shown to occur at any site along the radial or distal walls following nuclear displacement by centrifugation or cold treatments (Terasaka and Niitsu, 1987). Therefore, the formation of an asymmetric spindle and asymmetric division was not dependent on arrival of the nucleus at a particular location. The proposed role of the MT dependent system generating regular polarity then is first, to ensure nuclear migration to the wall and second, to ensure that the correct spindle axis is established to achieve asymmetric division.

15.4.1.2
Spindle Axis Determination

The orientation of the spindle axis has a key role in determining microspore division asymmetry. If the acentric spindle was rotated 90° from its normal axis perpendicular to the microspore wall symmetrical division would result. Terasaka and Niitsu (1987) showed that repositioning the nucleus in *Pinus* microspores by centrifugation, cold and caffeine treatments produced symmetric divisions resulting from a centrally located spindle, and symmetrical divisions in which both daughter nuclei were positioned near the microspore wall. The second type of symmetrical divisions are consistent with an asymmetric spindle position with its axis parallel to the microspore wall. Recent MT localization studies in *Tradescantia* have shown that the spindle begins to construct itself in the vegetative pole, and that all of the available tubulin in the form of MTs is localised to the cytoplasm at the vegetative face of the acentric nucleus immediately before spindle assembly (Terasaka and Niitsu, 1990, 1995). Spindle axis determination, as a critical component of microspore polarity, may therefore result from the polarised distribution of spindle precursors and/or assembly factors to the VC pole. Spindle growth in the 'GC domain' may be delayed by the limited availability of such factors and the close association of the microspore nucleus with the plasma membrane. This would result in the initiation of spindle growth at an internal 'vegetative pole' position defining the spindle axis, followed by completion of bipolarity in the GC half.

15.4.2
Determination of Microspore Polarity

15.4.2.1
The Nature of Polarity Determinants

This leaves the important question of the nature of the polarity determinants and signals which execute early events in generating polarity. First, it is highly unlikely that polarity is determined by directional external signals acting at the future GC pole, since nuclear migration is regular to a fixed pole, and microspores are usually randomly orientated with regard to the anther locule and tapetum. Therefore it is proposed that an intrinsic factor localized in the microspore wall, plasma membrane or in a cortical cytoplasmic domain at the future GC pole acts as a polarity determinant (Fig. 15.3). The activation of such a localised polarity determinant by an external signal in the anther locule could then initiate a signal transduction cascade leading to polarisation of the MTs and nuclear migration. Given the importance of MT systems in directing nuclear migration the polarity determinant could be a cytoplasmic gradient of a protein such as a microtubule motor or a regulator of MT assembly localised at the GC pole. Such a gradient could in turn be established by the GC pole localisation of its mRNA. There is conclusive evidence in animal systems for the determination of protein gradients by the MT directed transport or translation of mRNA species (Evans et al., 1994; Pokrywka and Stephenson, 1995; Kloc and Etkin, 1995). In plants, developing *Fucus* embryos show MF dependent-actin mRNA localization at the thallus pole of the polarized zygote (Bouget et al., 1996). Furthermore, cell fate determinants are localized in the cell wall in *Fucus* zygotes and act to maintain the differentiated state of thallus and rhizoid cells following asymmetric division (Berger et al., 1994). Cell polarization in animal cells and yeast commonly involves reorganization of both MFs and MTs which are regulated by a hierarchy of GTPases and protein kinases (Glotzer and Hyman, 1995). Similar signalling cascades can be envisioned to operate to signal polarity and establish MT reorganisation in the microspore.

15.4.2.2
The Origins of Microspore Polarity

The localisation of a microspore polarity determinant could be an early event generated during meiosis or tetrad formation or it could be generated during microspore development as a result of intrinsically determined events associated with early gametophytic gene expression. With regard to the first possibility, meiotic cytokinesis in most dicot species is simultaneous with all intersporal walls developing together. This results in the quadripartitioining of the postmeiotic cytoplasm into four equal 'spore domains' (Brown and Lemmon, 1991a, b and references therein). Intrinsic polarity within the spore domains, which could ultimately direct nuclear migration, could be generated early as part of the process of generating the spore domains (Fig. 15.3).

Studies of the cytoskeleton and phragmoplast development have revealed radial arrays of MTs emanating from the postmeiotic, but precleavage, nuclei which interact and define the boundaries of the spore domains (Brown and Lemmon, 1991a, b). The localisation of actin microfilaments in the multipolar spindle and MFs radiating from the four nuclei also indicates their involvement in phragmoplast development (Van Lammeren et al., 1989; Schopfer and Hepler, 1991). These cytosleletal elements could establish polarity by generating a cortical cytoplasmic domain or by delivering polarity determinants to membrane or wall locations as discussed above. Evidence for early cytoplasmic polarity is provided by observations which show that future aperatural sites are defined by the presence of endoplasmic reticulum (ER) closely associated with the plasma membrane in the tetrad. These polar ER configurations reflect the tetrad cleavage

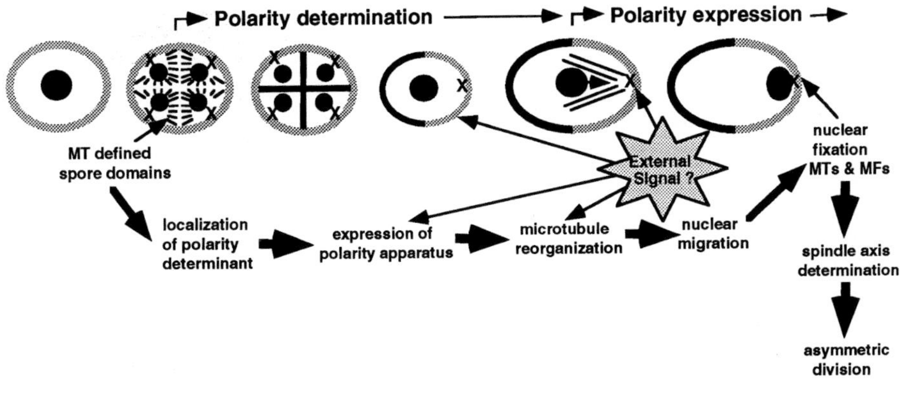

Fig. 15.3. Schematic model of the origins and mechanisms involved in microspore polarity determination and expression. Polarity is proposed to operate in two phases; the first determination phase involves the localization of a polarity determinant during the genesis of the spore domains and the expression of the polarity apparatus (MTs, MT-motors etc.); in the second phase polarity is expressed via the reorganisation and polarization of MTs, nuclear migration, fixation of nuclear position and spindle axis (involving MTs and MFs), followed by synthesis of the hemispherical cell plate and asymmetric division. External signals could play a role in both phases by activating the expression of genes encoding the polarity apparatus and/or by activating a prelocalised precursor of the polarity determinant

planes and appear to be dependent upon spindle orientation at meiosis I and II, since disruption of meiotic spindles with colchicine and/or centrifugation either prevents aperture formation or leads to apertures in incorrect positions (reviewed in Heslop-Harrison, 1971). A schematic model of the origin and localization of polarity determinants and their action in generating nuclear migration, spindle axis determination and asymmetric division is proposed in Fig. 15.3.

15.5 Genetic Screening for Polarity and Cell Fate Mutants

Given the success of the genetic approach in identifying molecular components of the polarity generating mechanisms and cell fate determinants in other systems (Horvitz and Herskowitz, 1992; Rhyu and Knoblich, 1995; Kraut et al., 1996) this should be a powerful approach for the dissection of microspore polarity and cell fate determination.

15.5.1 Direct Visible Screening

At maturity Arabidopsis pollen is tricellular containing a large vegetative nucleus and two smaller sperms cells with highly condensed intensely staining nuclei (Fig. 15.4A). We have screened for visible mutants that affect the stereotypical pollen cell divisions by examining mature pollen grains after DAPI staining using fluorescence microscopy. Approximately 10,000 M2 plants from an EMS mutagenised population (Nossen (No) background) were screened and 15–20 independently induced mutants were obtained that affected the stereotypical pollen cell divisions. These division mutants were classified into three groups based upon the proportion of pollen grains showing particular phenotypes. All three groups resulting from gametophytically acting mutations since; (1), phenotypes are observed in 50% or less of the pollen population, (2), heterozygous individuals show the phenotype and (3), transmission of the mutant allele is blocked or reduced through the pollen parent.

15.5.1.1 *Solo Pollen Mutants (Unicellular, Failing to Divide at PMI)*

In *solo pollen* mutants up to 50% of mature pollen grains contain a single nucleus, indicating failure of cytokinesis at PMI (Figs. 15.4 and 15.5). These mutations affect the ability of the microspore to divide and may affect a structural or regulatory component in the cell cycle or division process. Components of the spindle apparatus including factors regulating the assembly and/or function of the microtubule system are putative candidates. Therefore, *solo pollen* mutants may be thought of as cell cycle mutants; however if *solo* mutants also fail to express polarity *SOLO* genes may provide a link between the expression of cell polarity and cell cycle progression.

Despite the possession of a single nucleus solo microspores undergo maturation including cell expansion and the synthesis of a dense cytoplasm (Fig. 15.4C). The GUS positive staining of mutant solo pollen grains carrying the *lat52-gus/nia* gene shows that solo grains correctly activate the VC-specific promoter *lat52* (Fig. 15.4D), which confirms that VC fate is the default condition of the microspore nucleus in the absence of division at PMI (Eady et al., 1995).

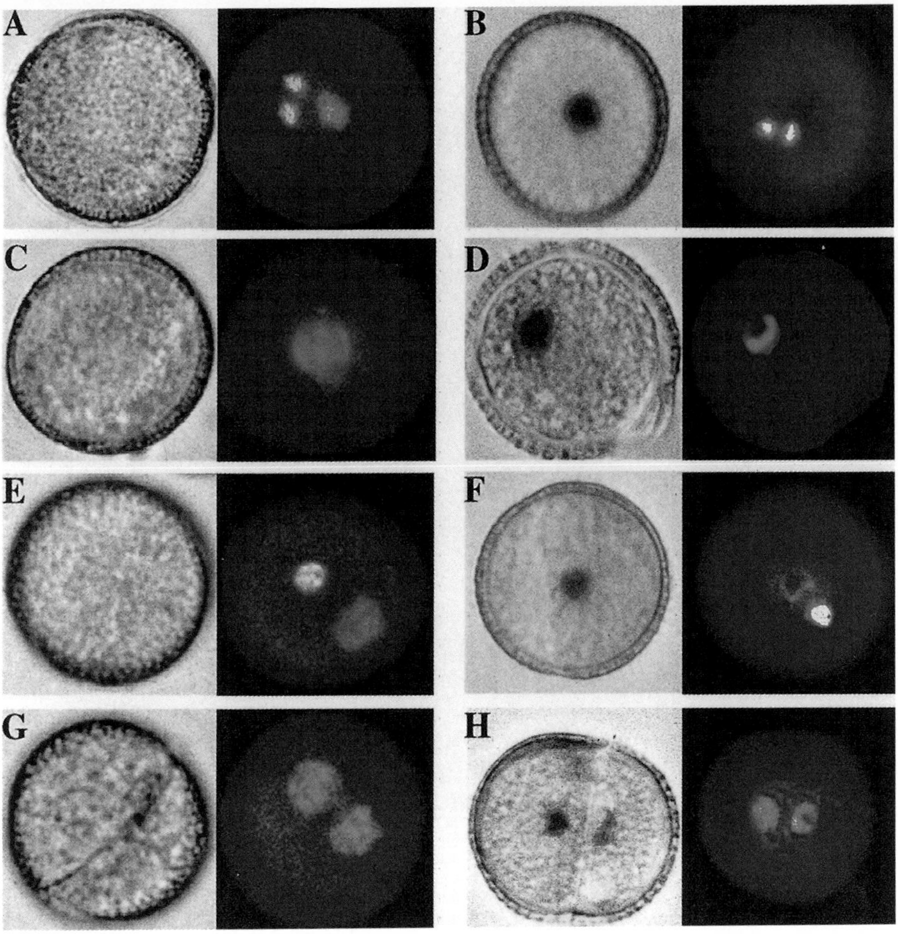

Fig. 15.4. Phenotypes and analysis of pollen cell fate in Arabidopsis male gametophytic mutants. Captured bright field and corresponding DAPI stained epifluorescence images of mature pollen are shown in the left and right panels of **A–H** respectively. **A, B** wild type pollen; **C, D** solo pollen mutant; **E, F** duo pollen mutant; **G, H** gemini pollen mutant. **B, D, F** and **G** show nuclear-localized GUS staining in wild type and mutant pollen, expressing the lat52-gus/nia VC-specific fate marker

15.5.1.2
Duo Pollen Mutants (Bicellular, Failing to Divide at PMII)

The *duo pollen* phenotype (Fig. 15.4) is indicative of pollen which has failed to enter or complete division at PMII. In duo116 approximately 50% of the pollen population shows the duo phenotype consistent with a gametophytically acting mutation (Fig. 15.5). These phenotypes suggest that the mutation may be a structural or regulatory component required for cell division, but that such components would specifically function in the symmetrical second division of the generative cell. However, the low frequency of symmetrically divided pollen grains observed in *duo116* (Fig. 15.5) further suggests that *DUO116* may also participate in regulating division asymmetry at PMI.

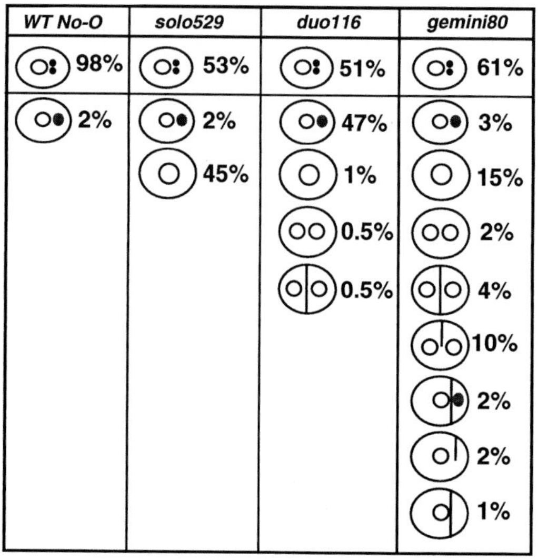

WT No-O
solo529
duo116
gemini80
98%
2%
53%
2%
45%
51%
47%
1%
0.5%
0.5%
61%
3%
15%
2%
4%
10%
2%
2%
1%

Fig. 15.5. Diagram showing the frequencies of aberrant pollen phenotypes observed in wild type and a representative of each of the phenotypic mutants classes solo, duo and gemini pollen

15.5.1.3
Gemini Pollen Mutants (Twin-Celled, More Symmetrical Division at PMI)

The gemini class of mutants display the widest range of pollen phenotypes which are illustrated in Fig. 15.5. In *gemini* mutants the symmetry of division at PMI is affected such that symmetrical and partial divisions frequently occur (Figs. 15.4 and 15.5). In symmetrically divided cells both daughter cells correctly activate the *lat52-gus/nia* marker and show dispersed nuclear chromatin (Fig. 15.4G and H) in accordance with previous observations which demonstrate that induced symmetrical divisions result in the default VC fate in both daughter cells (Eady et al., 1995). In partially or fully symmetric divisions in *gemini pollen*, the cell plate is frequently orientated parallel to long axis of the aperatural slits. Furthermore, the persistent internal cell walls are often seen to connect with the pollen grain wall even if these internal walls are incomplete. Such phenotypes may be expected if the spindle position or axis is altered. Normally the spindle axis is set up perpendicular to the wall such that the cell plate is initiated parallel to the outer wall and is subsequently fused to the wall by centrifugal growth at its margins (Hause et al., 1991; Owen and Makaroff, 1995; Tanaka, 1997). If the spindle axis is rotated in *gemini pollen*, as the division axis suggests, becoming parallel to the microspore wall, growth of the phragmoplast would first connect with the outer wall at the GC pole and the inward growth of the cell plate would most likely be inefficient and fail.

A further interesting phenotype in the gemini mutants is that nuclear division (karyokinesis) is often uncoupled from cellular division (cytokinesis). This commonly results in a smaller enucleate portion of the microspore cytoplasm being walled off from a larger cell containing one or two nuclei (Fig. 15.5). These 'uncoupling' phenotypes demonstrate that although cytokinesis and nuclear division are normally closely coupled these processes can operate independently. *GEMINI* genes may therefore function as 'checkpoint' genes ensuring that phragmoplast activity is suppressed until karyokinesis (telophase) is complete. These genes could also act in a dominant manner to precociously activate cell plate synthesis. Alternatively, due to the altered position of the internal cell walls, it is possible that *gemini* phenotypes result more directly from defects in components of the intracellular targeting machinery required for the asymmetric positioning of the cell plate.

Sporophytic mutants which affect cell plate synthesis but act in the early embryo and developing seedling have been identified. These include the mutants *gnom/emb30* (Mayer et al., 1993), and *knolle* (Lukowitz et al., 1996). Both the gnom and knolle mutants result from defects in putative components of the secretory apparatus and disrupt the normally regular planes and symmetry of division in the developing embryo. *knolle* in particular shows cells with partial walls and multi-nucleate cells which appear to result from incomplete cytokinesis. The *GNOM* protein shows sequence similarity to yeast *Sec7* which is involved in protein transport through the golgi (Shevell et al., 1994), while *KNOLLE* is similar to T-SNARE proteins which are involved in vesicle docking (Lukowitz et al., 1996). If similar proteins function in the positioning and synthesis of the cell plate at PMI it is possible that *GEMINI* genes encode gametophyte-specific versions of these proteins.

It is apparent that *GEMINI* genes could encode gametophytic proteins which function at several points in the pathway which signals and executes microspore polarity, including polarity determination, nuclear migration, spindle axis determination and cytokinesis (Fig. 15.6). Further detailed phenotypic analysis and the molecular cloning of *GEMINI* genes is now required to determine their identities and precise roles in this pathway.

15.5.1.4
Sidecar Pollen

In a similar direct screen a male-specific gametophytic mutant termed *sidecar pollen* has been isolated, which is distinct from the *solo*, *duo* and *gemini* mutants, but also affects microspore division and cell fate (scp; Chen and McCormick, 1996). Plants heterozygous for the *scp* mutation in the Nossen background shed 7% in a divided condition with an extra uninucleate vegetative-like cell. Other than being joined, the extra vegetative cell appears to mature normally in that the *lat52-gus/nia* (Twell, 1992) and *lat59-gus* (Twell et al., 1990) markers are activated and germination occurs in vitro. This mutant first undergoes a symmetrical division at the microspore stage followed by one of the daughter cells dividing asymmetrically eventually to form two sperm cells. Equally divided cells appear prior to the onset of asymmetric division in the population, apparently by premature division of an unpolarised microspore (Fig. 15.6). If detailed studies confirm these observations it is likely that *SCP* may encode a repressor of cell division normally active until the microspore achieves polarity. Alternatively, the *scp* mutation may act in a dominant (gain of function) manner such that a cell cycle activator is expressed precociously. Either way asymmetric division by only one of the two daughter cells in *scp* demonstrates that first micropore division in *scp* is in fact unequal with regard to their ability to express polarity. This supports the proposal that an asymmetrically localized polarity determinant is present in the microspore before polarity is expressed (Figs. 15.3 and 15.6).

15.5.1.5
Stud/Tetraspore

The phenotypes of a new class of sporophytic reduced fertility mutants of *Arabidopsis* which produce large multicellular pollen grains shed light on the origins and expression

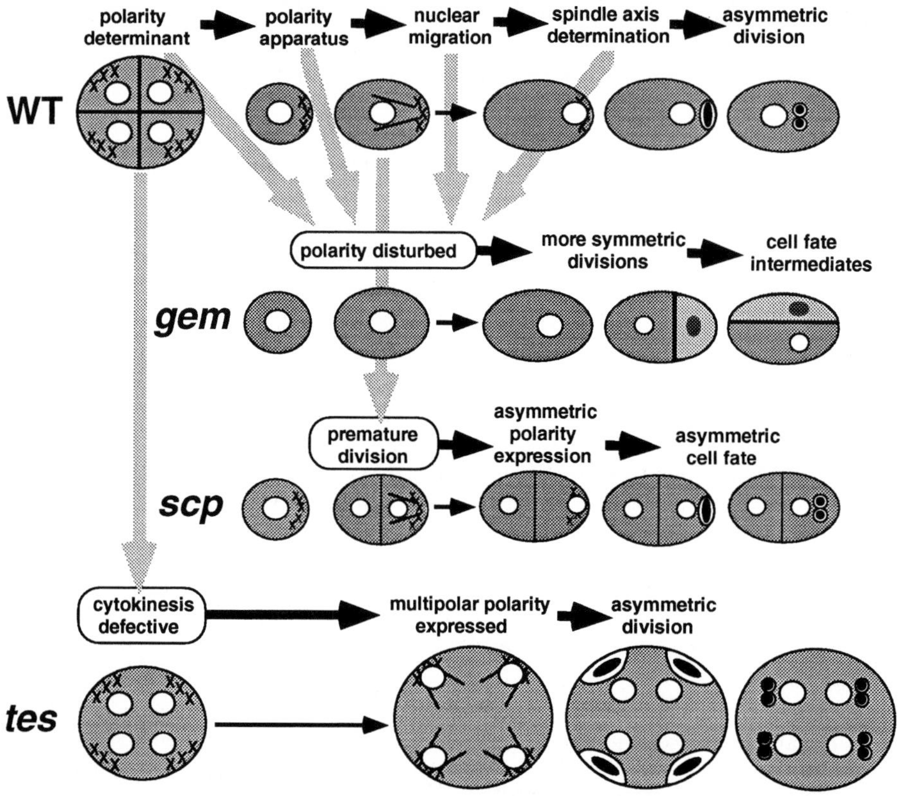

Fig. 15.6. Schematic figure summarizing the components of microspore polarity leading to asymmetric division and differential cell fate. How these components may be affected in gemini pollen (*gem*), sidecar pollen (*scp*) and tetraspore (*tes*) mutants is illustrated. The maintenance of polarity and correct cell fate in tes and scp mutants is consistent with the concept of a cortically localized polarity. Premature division and differential cell fate in scp suggests that microspore polarity is initially asymmetrically determined, but not expressed, and requires signals and/or components expressed later during microspore development. In contrast, partial division asymmetry at the correct time in gem mutants suggests that these mutants more directly affect the polarity determinant or polarity apparatus. The fate of the smaller daughter cell in gem mutants, which remains undivided with intermediate chromatin condensation and lat52 expression, either suggests that GEM genes also act after PMI, or that GC division and cell fate are blocked by abberrant segregation of GF or GCR (see Fig. 15.2) as a result of disturbed polarity and division asymmetry

of microspore polarity. These mutants have been independently classified as two allelic series in genes termed *STUD* (Hulskamp et al., 1997) and *TETRASPORE* (TES; Spielman et al., 1997) respectively. *stud/tes* mutations prevent the formation of separating callose walls after meiosis II. This initially creates coenocytic microspores (CM) composed of four nuclei within a common cytoplasm (Fig. 15.6). Up to four sperm cell pairs have been observed in mature stud/tes pollen grains suggesting that the four nuclei can each undergo nuclear migration and asymmetric division within the common cytoplasm. The *stud/tes* mutants first demonstrate that cytokinesis and microspore isolation after meiosis II are not prerequisites for microspores to achieve polarity, and further suggest that polarity determinants can operate locally and independently within each spore domain. These mutants eliminate the possibility that polarity determinants are localised to the new radial intersporal walls or plasmamembranes formed at meiosis II and support the localization of polarity determinants to the distal (outer) wall according to the model of micropore polarity described above (Figs. 15.3 and 15.6).

15.5.2
Screening for T-DNA and Transposon-Tagged Gametophytic Mutants

A systematic, although labour intensive, approach for the isolation of important genes involved in gametophytic development is provided by random gene-tagging by insertional inactivation. Since the inactivation of many important genes controlling gametophytic development is likely to lead to reduced transmission through the male or female gametes, it is possible to screen for the reduced transmission of a seedling expressed resistance marker carried by a T-DNA or a transposon. For example, if a single locus insertion inactivates a postmeiotically-expressed gene essential for male development, then the ratio of resistant sensitive seedlings in self progeny would be 1:1 rather than the expected 3:1. This approach initially does not allow one to distinguish between mutations affecting transmission through male or female gametes, but is attractive in that: (1) it will identify important gametophytically expressed genes required for the complete developmental pathway of male and female gametophyte development, including gamete recognition and fusion; (2) genes identified in this way are very likely to be tagged or very closely linked to a molecular maker; and (3) it is simple to carry out given the availability of large populations of transgenic lines containing a high proportion of single locus T-DNA or transposon insertions.

We have screened approximately 2,000 T-DNA lines and 1,500 transposon containing lines. In both populations approximately one percent of the lines showed segregation distortion (~1:1) for antibiotic resistance. This suggests that both T-DNA and transposon tagging of important gametophytic genes is similarly efficient depending on the availability or ease with which seed from large numbers of individual hemizygous for single locus insertions can be generated. T-DNA tagged mutants have been identified which affect the development and/or function of the male, female or both gametes. Among those affecting male development the *limpet pollen* (*lip*) mutant, while not affecting polarity and subsequent cell fate, does affect important events closely associated with PMI. Heterozygous *lip* plants produce 25% of mature pollen with the vegetative nucleus in a central position and the GC and/or sperm cells closely associated with the pollen wall. *Lip* prevents detachment of the GC and its migration into the VC cytoplasm after PMI and appears to act through maintenance of the separating hemispherical cell wall. Cloning of *lip* through T-DNA tagging will define help to define whether it is directly involved in regulating synthesis or degradation of the hemispherical cell wall or in the cytoplasmic reorganisation required for GC migration.

15.6
Perspective

Pollen mitosis I is a striking example of the importance of asymmetric division in determining differential cell fate. In this chapter, the aim has been to unify current cellular and molecular information into models of the mechanisms by which the microspore achieves asymmetric division (polarity) and how asymmetric division determines differential cell fate. We are now at the point where we need to identify the genes involved so that these models can be rigorously examined and a more detailed understanding developed. Several strategies designed to identify genetic components of the polarity and determination systems have been evaluated. This has led to the identification of novel classes of gametophytic mutants which affect cell cycle progression, division asymmetry at PMI and GC migration. The identification of such mutants, together with progress in positional cloning and gene-tagging strategies, means that it is now feasible to isolate molecular components involved in determining cell polarity, asymmetric division and the subsequent differentiation events which define pollen cell fate.

Acknowledgements
I gratefully acknowledge The Royal Society for their continuing support over the years and BBSRC for funding recent work on screening for gametophytic polarity mutants under the Cell Commitment and Determination Initiative. I would also like to thank Ross Howden and Soonki Park for their contributions to the unpublished work summarised in this review.

References

Bate N, Spurr C, Foster GD, Twell D (1996) Maturation-specific translational enhancement mediated by the 5′-UTR of a late pollen transcript. Plant J 10:101–111
Berger F, Taylor A, Brownlee C (1994) Cell fate determination by the cell wall in early Fucus development. Science 263:1421–1423
Blomstedt CK, Knox RB, Singh MB (1996) Generative cells of Lilium longiflorum possess translatable mRNA and functional protein synthesis machinery. Plant Mol Biol 31:1083–1086
Bouget F-Y, Gerttula S, Shaw S, Quatrano RS (1996) Localization of actin mRNA during the establishment of cell polarity and early cell divisions in Fucus embryos. Plant Cell 8:189–201
Brown RC, Lemmon BE (1991a) Pollen development in orchids. 1. Cytoskeleton and the control of division plane in irregular patterns of cytokinesis. Protoplasma 163:9–18
Brown RC, Lemmon, BE (1991b) Pollen development in orchids. 2. The cytokinetic apparatus in simultaneous cytokinesis. Protoplasma 165:155–166
Brown RC, Lemmon BE (1991c) Pollen development in orchids. 3. A novel generative pole microtubule system predicts unequal pollen mitosis. J Cell Sci 99:273–281
Brown RC, Lemmon BE (1992) Pollen development in orchids. 4. Cytoskeleton and ultrastructure of the unequal pollen mitosis in Phalaenopsis. Protoplasma 167:183–192
Chen Y-C, McCormick S (1996) sidecar pollen, an Arabidopsis thaliana male gametophytic mutant with aberrant cell divisions during pollen development. Development 122:3243–3253
Eady C, Lindsey K, Twell D (1994) Differential activation and conserved vegetative cell-specific activity of a late pollen promoter in species with bi- and tricellular pollen. Plant J 5:543–550
Eady C, Lindsey K, Twell D (1995) The significance of microspore division and division symmetry for vegetative cell-specific transcription and generative cell differentiation. Plant Cell 7:65–74
Evans TC, Crittenden SL, Kodoyianni V, Kimble J (1994) Translational control of maternal glp-1 mRNA establishes an asymmetry in the C. elegans embryo. Cell 77:183–194
Geitler L (1935) Beobachtungen Über die erste teilung im pollenkorn der angiospermen. Planta 24: 361–386
Gervais C, Simmonds DH, Newcomb W (1994) Actin microfilament organization during pollen development of Brassica napus cv. Topas. Protoplasma 183:67–76
Glotzer M, Hyman AA (1995) The importance of being polar. Curr Biol 5:1102–1105
Hause G, Hause B, van Lammeren AAM (1991) Microtubular and actin-filament configurations during microspore and pollen development in Brassica napus L. cv. Topas. Can J Bot 70:1369–1376
Havlicky T, Hause G, Hause B, Pechan P, van Lammeren AAM (1996) Subcellular localization of cell-specific gene expression in Brassica napus L, Arabidopsis thaliana L. Microspores and pollen by non-radioactive in situ hybridization. PhD thesis, G. Hause and B. Hause, Wageningen Argricultutal University, ISBN 90-5485-614-9
Heslop-Harrison J (1971) Wall pattern formation in angiosperm microsporogenesis. Symp Soc Exp Biol 25:277–300
Horvitz RH, Herskowitz I (1992) Mechanisms of asymmetric cell division: Two Bs or not two Bs, that is the question. Cell 68:237–255
Hulskamp M, Parekh NS, Grini P, Scheitz K, Zimmermann I, Lolle SJ, Pruitt RE (1997) The STUD gene is required for male-specific cytokinesis after telophase II of meiosis in Arabidopsis thaliana. Dev Biol 187:114–124

Katsuta J, Hashiguchi Y, Shibaoka H (1990) The role of the cytoskeleton in positioning of the nucleus in premitotic tobacco BY-2 cells. J Cell Sci 95:413–422
Kloc M, Etkin LD (1995) Two distinct pathways for the localization of RNAs at the vegetal cortex in Xenopus oocytes. Development 121:287–297
Kraut R, Chia W, Jan LY, Jan YN, Knoblich JA (1996) Role of inscuteable in orientating asymmetric cell divisions in Drosophila. Nature 383:50–55
Lukowitz W, Mayer U, Jurgens G (1996) Cytokinesis in the Arabidopsis embryo involves the syntaxin-related KNOLLE gene product. Cell 84:61–71
Mascarenhas JP (1975) The biochemistry of angiosperm pollen development. Bot Rev 41:259–314
Mascarenhas JP (1992) Pollen gene expression. In: Russell SD, Dumas C (eds) International Review of Cytology: Sexual Reproduction in Flowering Plants. Academic Press, San Diego, pp 3–18
Mayer U, Buttner G, Jurgens G (1993) Apical-basal pattern formation in the Arabidopsis embryo: studies on the role of the gnom gene. Development 117:149–162
Owen HA, Makaroff CA (1995) Ultrastructure of microsporogenesis and microgametogenesis in Arabidopsis thaliana (L.) Heynh. ecotype Wassilewskija (Brassicaceae). Protoplasma 185:7–21
Pokrywka NJ, Stephenson EC (1995) Microtubules are a general component of mRNA localization systems in Drosophila oocytes. Dev Biol 167:363–370
Rhyu MS, Knoblich JA (1995) Spindle orientation and asymmetric cell fate. Cell 82:523–526
Schopfer CR, Hepler PK (1991) Distribution of membranes and the cytoskeleton during cell plate formation in pollen mother cells of Tradescantia. J Cell Sci 100:717–728
Shevell DE, Leu W-M, Gilimor CS, Xia G, Feldman KA, Chua N-H (1994) EMB30 is essential for normal cell division, cell expansion, and cell adhesion in Arabidopsis and encodes a protein that has similarlity to Sec7. Cell 77:1051–1062
Spielman ML, Preuss D, Li F-L, Browne WE, Scott R, Dickinson HG (1997) TETRASPORE is required for male meiotic cytokinesis in Arabidopsis thaliana. Development 124:2645–2657
Spiker S (1985) Plant chromatin structure. Annu Rev Plant Physiol 36:235–253
Sunderland N, Huang B (1987) Ultrastructural aspects of pollen dimorphism. In International Review of Cytology. Pollen: Cytology and Development, K. L. Giles and J. Prakash, eds (London: Academic Press Inc.), pp 175–220
Tanaka I (1997) Differentiation of generative and vegetative cells in angiosperm pollen. Sex Plant Reprod 10:1–7
Tanaka I, Ito M (1980) Induction of typical cell division in isolated microspores of Lilium longiflorum and Tulipa gesneriana. Pl Sci Lett 17:279–285
Tanaka I, Ito M (1981) Control of division patterns in explanted microspores of Tulipa gesneriana. Protoplasma 108:329–340
Terasaka O (1982) Nuclear differentiation of male gametophytes in gymnosperms. Cytologia 47:27–46
Terasaka O, Niitsu T (1987) Unequal cell division and chromatin differentiation in pollen grain cells. I. centrifugal, cold and caffeine treatments. Bot Mag Tokyo 100:205–216
Terasaka O, Niitsu T (1990) Unequal cell division and chromatin differentiation in pollen grain cells. II. Microtubule dynamics associated with the unequal cell division. Bot Mag Tokyo 103:133–142
Terasaka O, Niitsu T (1995) The mitotic apparatus during unequal microspore division observed by a confocal laser scanning microscope. Protoplasma 189:187–193
Terasaka O, Tanaka R (1974) Cytological studies on the nuclear differentiation in microspore division of some angiosperms. Bot Mag Tokyo 87:209–217
Twell D (1992) Use of a nuclear-targeted §-glucuronidase fusion protein to demonstrate vegetative cell-specific gene expression in developing pollen. Plant J 2:887–892
Twell D (1994) The diversity and regulation of gene expression in the pathway of male gametophyte development. In: Scott RJ, Stead AD (eds) Molecular and Cellular Aspects of Plant Reproduction. Society for Experimental Biology, Cambridge, United Kingdom: Cambridge University Press, Seminar series 55, pp 83–135
Twell D, Patel S, Sorensen A, Roberts M, Scott R, Draper J, Foster G (1993) Activation and developmental regulation of an Arabidopsis anther-specific promoter in microspores and pollen of Nicotiana tabacum. Sex Plant Reprod 6:217–224
Twell D, Yamaguchi J, McCormick S (1990) Pollen-specific gene expression in transgenic plants: Coordinate regulation of two different tomato gene promoters during microsporogenesis. Development 109:705–713
Ueda K, Tanaka I (1995a) Male gametic nucleus-specific H2B and H3 histones. Designated gH2B and gH3, in Lilium longiflorum. Planta, 197:289–295
Ueda K, Tanaka I (1995b) The appearance of male gamete-specific histones gH2B and gH3 during pollen development in Lilium longiflorum. Dev Biol 169:210–217
Van Lammeren AAM, Bednara J, Willemse MTM (1989) Organization of the actin cytoskeleton in Gasteria verrucosa (Mill.) H. Duval visualized with rhodamine-phalloidin. Planta 178:531–539
Zaki MAM, Dickinson HG (1991) Microspore-derived embryos in Brassica: the significance of division asymmetry in pollen mitosis I to embryogenic development. Sex Plant Reprod 4:48–55

CHAPTER 16

Genetic Control of Pollen Development and Function

M. Sari-Gorla · M. E. Pè

Department of Genetics and Microbiology, University of Milano, Via Celoria 26, 20133, Milano, Italy
e-mail: SARI@IMIUCCA.CSI.UNIMI.IT
telephone: +39 2 26 60 52 04
fax: +39 2 26 60 52 04

Abstract. Although the pollen grain is a rather simple two- or three-celled organism, the construction of a functional male gametophyte involves a complex system of developmental processes, which requires the activity of a rather large pool of genes. In order to identify these genes, two main strategies can be adopted: i) starting from pollen characters, to identify the involved gene/s and then isolating them; ii) to isolate pollen-specific genes and then to determine their function. In this chapter, the main findings that have been obtained through these approaches are reviewed, with regard to genes specifically involved in the control of pollen development and to genes controlling pollen function: grain germination, tube growth and pollen-pistil interaction. From the large number of studies carried out in the last years, it emerges that the developmental pathway followed by the male gametophyte is a process involving precise control over cell division and gene expression by the haploid genome. About 20,000 gametophytic genes have been detected, at least 2000 of which are specifically expressed in the mature pollen grain; however, so far, just a small fraction of these genes has been characterized, and the function of most of them is still unknown. Strategies for elucidating these aspects are discussed.

16.1 Introduction

Although the pollen grain is a rather simple two- or three-celled organism, the construction of a functional male gametophyte requires the activity of a rather large pool of genes. Microspore maturation is the result of a precisely timed series of developmental steps, including an early stage, from meiosis to microspore mitosis, and a late stage, from first mitosis to mature grain. Once on a female receptive surface, mature pollen germinates and produces a pollen tube that grows at a very fast rate and responds to a sophisticated cell-cell communication system in a fundamentally different way as compared to the standard model of cellular growth in plants. Pollen function ends with the release of sperm cells into the embryo sac, achieving fertilization.

Several types of indirect evidence suggest that all these processes have to be controlled by a large number of genes.

Kindiger et al. (1991), utilising maize hypoploid stocks (TB-A: translocation between A and B chromosomes) that produced 50% aborted pollen, demonstrated that microspore development arrests at different stages, depending on which chromosome arm is deleted; this indicates that there are gametophytically-acting genetic elements located on various chromosome arms that affect discrete steps in microsporogenesis. Sari-Gorla et al. (1992, 1994, 1995) demonstrated by linkage analysis between RFLP markers and the expression of different pollen traits that sets of precisely timed genes control the different stages of pollen function and pollen-pistil interaction.

Molecular studies have demonstrated that a large number of genes are expressed in pollen (Mascarenhas 1990), and it has been estimated that close to 25% of the pollen cDNA represents genes expressed only in pollen (Bedinger et al. 1994). Even though the precise function of most of the pollen-expressed sequences isolated is still unknown, many of these genes are expected to control microspore development, pollen function and pollen-pistil interaction.

A number of independent mutations affecting nuclear genes and resulting in male sterility have been identified in many plant species; for example, there are about 60 genes that affect male fertility for maize, 55 for tomato, 48 for barley, 60 for rice (Horner and Palmer 1995). Although there is very little information on the molecular defect responsible for the mutation, this data clearly indicates that impairment of a large number of different genes affects the correct sequence of events leading to a functional pollen grain, and supports the idea that pollen development requires a coordination of gene expression in the sporophytic cells of the anther and in the developing pollen grains.

In the following discussion, an overview of the current knowledge on the genetic control of pollen development and function will be given, with emphasis on gametophytically acting genes. The biological relevance of this topic resides in its amenability for basic research and for pollen biotechnology; in fact, pollen offers an excellent system to study developmental processes such as cell fate determination and cellular differentiation, and its manipulation represents a powerful tool for the production and improvement of crops and related products.

In order to identify genes specifically involved in the control of pollen development, and function, two main strategies can be adopted: i) starting from pollen characters, to identify the involved gene/s and then isolating them; ii)isolating pollen-specific genes and then determining their function.

16.2
Pollen Development

Microsporogenesis involves different distinctive phases (see Chang and Neuffer 1989, for detailed illustration of maize microsporogenesis). Pollen mother cell undergoes meiosis, yielding four haploid microspores enclosed in a callose sheath (the quartet or tetrad stage); after callose dissolution, microspores are released from the quartet as free uninucleate cells. At a second stage, the microspore increases in size, a large vacuole and the germ pore are formed, and then the nucleus enters in a resting stage, ready for microspore mitosis. This is an asymmetric, determinative division, in that the resulting two nuclei have very different cell fates. The vegetative cell nucleus does not undergo another mitosis, its role being to drive on the development of the pollen grain and the growth of the pollen tube until delivery of the sperm cells into the ovary. On the contrary, the generative cell, which is enclosed within the cytoplasm of the vegetative cell, appears to be transcriptionally silent (Mascarenhas 1990) but undergoes the second mitosis, producing two sperm cells. In some species, such as maize and *Arabidopsis*, at anthesis mature pollen contains three cells: one vegetative and two sperm cells. In other species (such as tomato and lily) the second mitosis takes place only after shedding, in the pollen tube.

There are several reports that support the different functional role of vegetative and generative nuclei. Since 1953, Bishop and McGowan demonstrated that uninucleate pollen of *Tradescantia*, obtained after colchicine treatment of *in vitro* grown microspores, was still capable of germinating and growing pollen tubes. More recently, the same result on tobacco pollen was reported by Eady et al. (1995) that further demonstrated, on the basis of its morphology and of the capability of expressing the vegetative cell-specific *LAT52* gene, that the nucleus of the uninucleate pollen was the vegetative nucleus. This data was confirmed by observation on isolated tobacco microspores maturing *in vitro* (Touraev et al. 1995). Tobacco microspores can be induced to undertake a symmetrical division (for instance, by starvation), instead of the normal asymmetric first pollen mitosis, and the two daughter cells so formed have the characteristics of vegetative cells. On a germination medium, one of the two vegetative-like cells forms a pollen tube. Therefore, apparently normal gametophytic development can be maintained after symmetrical microspore division, provided at least one vegetative nucleus is present.

Vegetative cell-specific ablation was used to examine the role of the vegetative cell in controlling generative cell behaviour and differentiation during pollen development (Twell 1995). This technique involves the targeted expression of a cell autonomous cytotoxic protein under the control of cell-specific regulatory sequence. The results showed that the first sign of ablation of vegetative cell occurred immediately after the first pollen mitosis; in contrast, generative cell retained viability for several days, but progressively lost viability. These results directly demonstrate the dependence of the generative cell on vegetative cell functions.

16.2.1
From Characters to Genes: Male Gametophytic Mutants

A direct approach to detect genes controlling pollen development is based on the study of single gene mutants affecting pollen characteristics and gametophytically expressed. Few cases have been so far identified and none of them has been characterised with regard to its specific function. Mangelsdorf (1932) described a gametophytic mutant, *small pollen* (*sp*) that reduces the size of pollen grain in maize. The finding that wild type pollen from segregating +/*sp* heterozygous plants are bigger than wild type pollen from +/+ homozygous plants, indicates intergametophytic competition over nutrients among the pollen grains developing in the same anther, but the mechanism by which it occurs is unknown.

Rf3 is a nuclear gene that restores fertility of maize CMS-S cytoplasm, whose expression is post-meiotic: CMS-S *Rf3*/*rf3* plants are semi-sterile, segregating normal and sterile grains at 1 : 1 ratio (Laughnan and Gabay 1983). *Rf* genes may be regarded as suppressors of the CMS (cytoplasmic male sterility) phenotype, since they produce no heritable change in the mtDNA, where the mutation responsible for cytoplasmic male sterility resides. Studying naturally occurring restorers of CMS cytoplasm and those produced by spontaneous mutations in experimental studies, a large number (more than 30) of independent *Rf* genes have been identified, that map in different chromosomal positions and are all capable of restoring fertility (Gabay-Laughnan 1997).

A large number of nuclear male sterile mutants, as already pointed out, has been described in many plant species. These mutants can be grouped in three classes (Kaul 1988): functional, structural and sporogenous. Functional mutants produce viable pollen but have defects in flower structure that prevent pollen release; structural mutants have defects in stamens so that pollen is not produced; the flowers of sporogenous mutants are morphologically normal, but they don't produce functional pollen. Only the third category of mutants can provide specific information about the process of pollen development. Sporogenous male sterility includes: pre-meiotic male sterile mutations, that occur in tapetal cells, preventing the development of microsporocytes into microspores (Chaudhury et al. 1994; Loukides et al. 1995); meiotic mutants, characterised by defects in different stages of the meiotic processes, producing aberrant microspore tetrads (Albertsen and Phillips 1981; Peirson et al. 1996; He et al. 1996); post-meiotic mutants. The latter are expressed during microsporogenesis, up to the first mitosis, or after the completion of this event (Albertsen and Phillips 1981; Sari-Gorla et al. 1996).

Almost all male sterile genes described to date are expressed before microspore mitosis and act sporophytically, suggesting that the sporophytic gene contribution to this process is substantial; for instance, in tomato, it apparently involves 45 different genes (Gorman and McCormick 1997). On the contrary, mutations that act gametophytically and that alter nuclear division pattern, nuclear migration and grain differentiation are scarcely represented in higher plants. Correns (1900) reported data suggesting the presence of two gametophytically expressed pollen lethals in *Oenothera*. *Bcp1*, a gene required for male fertility in *Arabidopsis*, has a diploid/haploid mode of action (Xu et al. 1995). In the same species a male gametophytic gene was identified, *sidecar pollen*, which is required for the normal cell division pattern during pollen development; pollen carrying the mutant allele exhibits an extra cell, that has the cell identity of a vegetative cell and is produced prior to any asymmetric microspore mitosis (Chen and McCormick 1996).

In order to identify genes controlling pollen development and acting gametophytically, we carried out a program of transposon insertion mutagenesis. Two male sterile mutants of this type have been isolated: *gaMS-1* (*gametophytic male sterile-1*) (Sari-Gorla et al. 1996), and *gaMS-2* (*gametophytic male sterile-2*) (Sari-Gorla et al., 1997a). Both the mutants demonstrated gametophytic expression, since plants heterozygous for the mutant allele segregated 1 : 1 normal to sterile pollen grains, and their selfed progeny segregated 1 : 1 fertile to semi-sterile plants. The same plants exhibited a normal phenotype and were fully female-fertile.

The cytological characterization of *gaMS-1* and *gaMS-2*, was performed on single anthers from heterozygous plants, that were collected at different developmental stages, from meiosis to pollen shedding. It revealed that the mutant microspores developed normally up to the completion of the first mitosis; at this early stage, no differences between normal and mutant binucleate microspores were observable, but shortly afterwards an appreciable amount of starch appeared in the binucleate normal grains unlike the mutants grains. During the following phases, the normal grains proceeded in the maturation process, increasing in size and undergoing the second pollen mitosis, yielding a well differentiated engorged pollen grain with one vegetative nucleus and two sperm nuclei. On the contrary, *gaMS-1* mutant grains did not further increase in size, re-

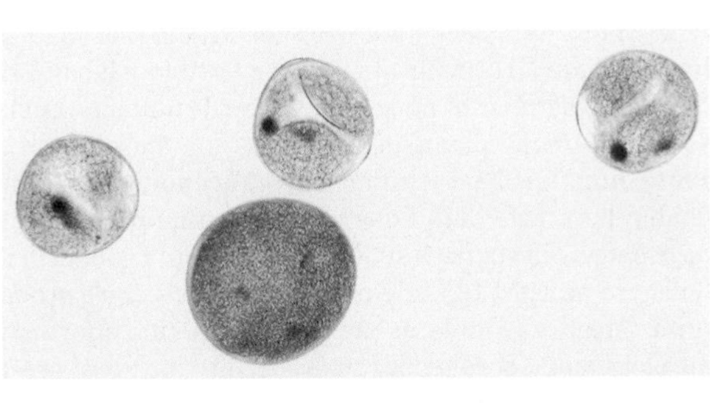

Fig. 16.1. Pollen from GaMS-1/gaMS-1 plants at anthesis

maining until anthesis at a stage of maturation corresponding to that of a binucleate stage (Fig. 16.1). Thus, the function that appears to be affected by the mutation is the capability of producing the developmental switch marked by the first microspore mitosis. Also *gaMS-2* grains did not accumulate starch, their two nuclei did not differentiate in a vegetative and a generative, but both undertook other division cycles in a completely unregulated way, so that a common developmental pattern for all mutant grains was not detectable and, at anthesis, the *gaMS-2* pollen grains exhibited several different phenotypes, from two undifferentiated nuclei, to multinucleate, or completely free of clearly observable nuclear material (Fig. 16.2). On the basis of these observations, *GaMS-2* is thought to be involved in the mechanism of commitment of vegetative and generative nuclei and perhaps in the regulation of cell division.

First pollen mitosis appears to be a critical point in commitment to the gametophytic pathway (McCormick 1993). Results obtained in many different species indicate that there does appear to be a genuine developmental switch occurring in microspore gene expression around the time of microspore mitosis (Hamilton and Mascarenhas 1997). The observation that the sporophytic male sterile mutants are expressed before mitosis, whereas the gametophytically acting alleles are expressed after this event, suggests that after the first pollen mitosis the role of the sporophytic genome in directing pollen development is reduced.

With regard to the mechanism at the basis of the cell fate determination, a model has been proposed (Eady et al. 1995). It assumes the presence of a factor (*gametophytic factor*) that, due to partition of the grain cytoplasm at the first mitosis, could act either activating the genes of the vegetative nucleus, or repressing gene expression in the generative cell. It is worth noting that in the developing *Drosophila* nervous system homeodomain containing proteins have been characterized, that localize asymmetrically during cell division and are exclusively partitioned to one of the daughter cell (Hirata et al. 1995; Knobloch et al. 1995). In *Arabidopsis*, six gametophytic mutants (*gemini*) have been isolated which produce twinned or more symmetrically divided pollen at maturity (Twell et al 1997); *GEMINI* genes could encode polarity determinants required for nuclear migration and the determination of the spindle axis parallel to the microspore wall.

Whether the *GaMS-1* and *GaMS-2* genes have haploid or haplo-diploid type of expression is presently not known : they could be expressed also in plant tissues other than pollen, but at present this cannot be verified, since heterozygous plants are pheno-

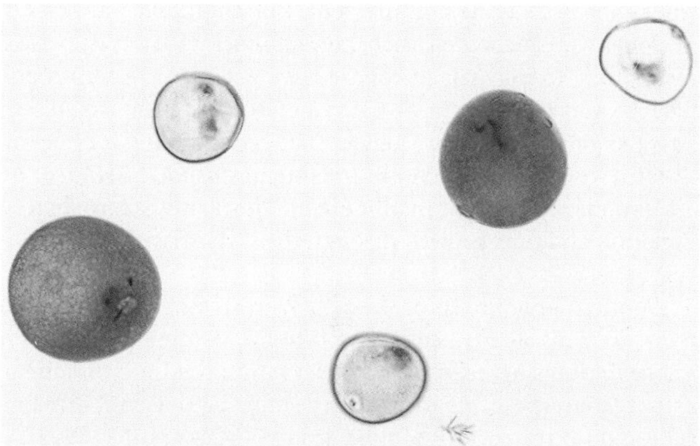

Fig. 16.2. Pollen from GaMS-2/gaMS-2 plants at anthesis

typically normal and homozygous recessive plants cannot be produced, since the pollen grains carrying the mutant allele are sterile. Insight into their possible functional role, at both pollen and plant level, will be provided by isolation of the genes. Two strategies for cloning ms are to-date the most promising: positional cloning and transposon tagging.

The first approach has been used by Gorman et al. (1996) to isolate the *ms-14* gene of tomato, the function of which has not yet been established. For map-based cloning, first molecular markers are identified flanking and closely linked to the *ms* trait. These markers then serve as a startng point for a chromosome walk in a large insert genomic library, such as a YAC (yeast artificial chromosome) or BAC (bacterial artificial chromosome) library. Once a megaclone containing the flanking markers is identified, unique sequences internal to the flanking markers are used as additional molecular markers to further delimit the region containing the gene. The smallest fragment containing the ms gene is used for transformation and the transgenic plants are analyzed for restoration of fertility.

Transposon tagging has been used to isolate a male sterile gene in maize (Albertsen 1993) and one in *Arabidopsis* (Aarts et al 1993). In the first case the *Ac-Ds* transposon system was used; the putative protein encoded by this gene shows some homology (33%) to strictosidine synthase, an enzyme involved in the indole alkaloid pattern. Thus it is not clear which is the real function of this gene in pollen development. The *En-1* (*Spm*) maize transposable element system was used to transposon-tag and isolate the *Arabidopsis* male sterile gene; the sequence of the predicted protein bears some similarity to a short region of an ORF in the mitochondrial genome, but the significance of the data is also ambiguous. A similar approach, using T-DNA induced mutations, was used for the isolation of MS5, a specific male sterile of *Arabidopsis* (Glover et al., 1988). MS5 gene product shows interesting homologies with a protein of the synaptonemal complex and with a subunit of a cyclin-depentent kinase.

16.2.2
From DNA to its Function

Table 16.1 lists the genes published to date that have been shown to be pollen-specific or preferentially expressed in pollen. Many of the genes whose isolation was based on their gametophytic mode of expression are expected to be important for pollen viability.

Considering the genes listed in the Table, three large groups emerge. The first one includes genes showing homology to wall-degrading enzymes. Maize, tobacco and tomato clones were identified, that show similarity to pectate lyase. Several cDNA clones corresponding to polygalacturonase have been isolated from pollen of maize, *Oenothera*, tobacco and *Brassica*. In *Brassica* a cDNA has been isolated that shows sequence similarity to pectin esterase. The fact that genes coding for enzymes of this type have been isolated from many cDNA libraries suggests that mRNAs for pectin-degrading enzymes are abundantly present in all pollens This is not surprising, since the processes of germination and tube growth through the stylar tissues require both wall degradation and synthesis.

Another large group of the pollen-expressed proteins are cytoskeleton-related proteins, such as actin, profilin, α-tubulin, β-tubulin and actin depolymerizing factor. These findings agree with the observation that microfilaments of the actin contractile system are responsible for the very active cytoplasmic streaming in the growing pollen tube and for the control of tube growth (Mascarenhas 1993).

Table 16.1. Pollen-specific cDNA clones isolated from flowering plants

Species	Clone name	Putative function	References
Alfalfa	P73	polygalacturonase	Qiu & Erickson 1996
Ambrosia artemisifolia	Ambra 1	pectate lyase	Rafner et al. 1991
Arabidopsis	TUA1	α-tubulin	Carpenter et al. 1992
	PFN4	profilin	Christensen et al. 1996
	ACT4	actin	Mckinney et al. 1995
	At-RAB2	GTP-binding protein	Moore et al. 1997
Betula verrucosa	Betv 1	disease resistance	Breiteneder et al. 1989
			Swoboda et al. 1995
Brassica campestris	Bcp1	?	Theerakulpisut et al. 1991
Brassica napus	Bp4	?	Albani et al. 1990
	Bp10	ascorbate oxidase	Albani et al. 1992
	Bp19	pectine esterase	Albani et al. 1991
	Sta-44	polygalacturonase	Robert et al. 1993
	I3	oleosin	Roberts et al. 1993
Gossypium hirsutum	G9	polygalacturonase	John & Petersen 1994
Lolium perenne	Lolp 1A	allergen	Singh et al. 1991
	Lolp 1B	amylase inhibitor	Singh et al. 1991
Lycopersicon esculentum	LAT51	ascorbate oxidase	Ursin et al. 1989
	LAT52	Kunitz trypsin inhibitor	Twell et al. 1991
	LAT56	pectate lyase	Wing et al. 1989
	LAT59	pectate lyase	McCormick 1991
Nicotiana tabacum	NTP303	ascorbate oxidase	Weterings et al. 1992
	TP10	pectate lyase	Rogers et al. 1992
	NPG1	polygalacturonase	Tebbut et al. 1994
	G10	pectate lyase	Rogers et al. 1992
	Tac25	actin	Thangavelu et al. 1993
	NTM19	?	Oldenhof et al. 1996
Oenohera organensis	P1		Mascarenhas 1992
	P2	polygalacturonase	Brown & Crouch 1990
Olive	Ole e 1	allergens	Villalba et al. 1994
Pearl millet	Mpt1	ascorbate oxidase	Ferrario, this lab
Petunia hybrida	chi A	chalcone isomerase	van Tunen et al. 1990
Petunia inflata	PPE1	pectin esterase	Mu et al. 1994
	PRK1	receptor-like kinase	Lee et al. 1996
Poa pratensis	KBG31	?	Silvanovich et al. 1991
Rice	PS1	?	Zou et al. 1994
Rye-grass	Lol p	allergens	Singh et al. 1991
Ragweed	Amb a	pectate lyase, allergen	Rafnar et al. 1991
Sorghum bicolor	MSb8	Kunitz trypsin inhibitor	Pè et al. 1994
	MSb2.1	ascorbate oxidase	Pè et al. 1994
Sunflower	SF3	DNA binding protein	Baltz et al. 1992
	SF17	leucin-rich protein	Reddy et al. 1995
Tradescantia paludosa	Tpc44	ascorbate oxidase	Stinson et al. 1987
Zea mays	Zm13	Kunitz trypsin inhibitor	Hanson et al. 1989
	Zm58	pectate lyase	Turcich et al. 1993
	PGI	polygalacturonase	Niogret et al. 1991
	3C12	polygalacturonase	Allen & Lonsdale 1993
	ZmPRO	profilin	Staiger et al. 1993
	ZmABP	ADF-like protein	Lopez et al.1996

Finally, several of the highly expressed proteins in pollen turned out to be also human allergens. In several cases, their function in pollen is not yet known, but some of them have been shown to be pectate lyase or profilin.

In most cases the putative function of the isolated clones has been deduced from sequence homology to known genes; in a few cases, however, involvement of the gene in pollen development has been directly demonstrated.

The promoter of the tomato gene *LAT52* was fused to the reporter gene *gusA*. Studies on transgenic plants carrying this construct indicated that the GUS activity was under the control of the haploid genome, since in the heterozygous plant only half of the pollen showed GUS activity (Twell et al. 1990) ; inactivation of this gene by antisense technology revealed that its expression is required for fertility: the suppression of the expression of the gene leads to abnormal pollen hydration, from which abnormally shaped, sterile grains are produced (Muschietti et al. 1994). Thus, although the exact role of this protein is still indeterminate, it is clear that it plays a critical role in pollen maturation and/or germination.

PRK1, a receptor-like kinase of *Petunia inflata*, was revealed to be essential for post-meiotic development of pollen (Lee et al. 1996). The function of the gene was studied by transformation of *P. inflata* plants with a construct containing an antisense PRK1 cDNA under a pollen-specific promoter. Aborted pollen grains showed that their outer wall, the exine, was essentially normal, but their cytoplasm did not contain nuclei or any other organelles except for starch-like granules; the microspore of the transgenic plants developed normally until the uninucleate stages; however in the subsequent stages half of the microspores completed mitosis and developed into normal binucleate pollen, while the other half initially remained uninucleate and subsequently lost their nuclei. These results suggest that PKR1 plays an essential role in a signal transduction pathway that mediates post-meiotic development of microspores.

One particularly interesting pollen gene has been isolated from a cDNA library made from sunflower pollen (Baltz et al. 1992) This gene, named *SF3*, has zinc-finger domains that correspond to a so-called LIM motif present in several animal metal-binding cysteine-rich proteins, that have been identified as regulatory proteins This suggests that *SF3* might be involved in the regulation of late pollen genes.

Finally, it is worth pointing out that sequence conservation among genes of taxonomically distant species could be an important clue for the identification of genes involved in important physiological processes. This could be the case of *MSb2.1*, a gene of *Sorghum bicolor*, isolated by heterologous hybridization with the *Tradescantia* pollen-specific clone *Tpc44* (Pè et al. 1994). Similar sequences were revealed by Southern blot hybridization in mature pollen of maize and rice. Using *MSb2.1* as heterologous probe for the screening of a cDNA library from pollen of pearl millet, three different pollen specific-genes were isolated (*MPt1*, *MPt3* and *MPt4*); they share an homology of 67% with a family of RNA isolated from *Brassica napus* (*Bp10*) and of 65% with tobacco gene *NPT303*. Transcripts homologous to these genes are also present in pollen from tomato (*LAT51*). The fact that similar sequences have been conserved in species that are taxonomically very distant, such as cereals, *Nicotiana*, *Brassica*, *Tradescantia*, tomato, suggests their not marginal role in the control of pollen development and/or function, even though the function of the relating gene products is still to be determined.

16.3 Pollen Function

Information about genes specifically or preferentially expressed during germination and pollen tube growth is quite scant. This is mainly due to technical problems: the isolation and the analysis of transcripts in germinated grains and in particular in pollen tubes *per se* is in general quite difficult and for some species practically impossible. Furthermore, the demonstration that a gene product present in pollen tubes is the result of gene expression after tube germination is not a trivial task. Thus, most of the information available on this class of genes has been indirectly obtained on the basis of their expression timing or of similarity to other known genes. Many of the mutants affecting pollen function seem to be involved in pollen-pistil interaction. In this respect, incompatibility phenomena will not be considered here, since this topic will be discussed in details in other parts of this book.

16.3.1 From Characters to Genes

Due to the difficulties of detecting gametophytic mutants for pollen function, a useful strategy has been based on the analysis of sporophytic mutants, which are also expressed in the gametophyte. In the analysis of a large collection of defective endosperm (*de*) mutants of maize, Ottaviano et al. (1988) found that 65% of these mutants were also expressed in the gametophyte. Precise analysis of the distorted segregation from the Mendelian ratios permitted the demonstration of gametophytic gene expression and the identification of the phase in which the genes are expressed. One class of *de-ga* mutants affects pollen development processes, while a second class affects the pollen tube growth rate. Of the same type are lethal embryo mutations in *Arabidopsis*, expressed during tube growth (Meinke 1982; 1985). However, the biological function of these genes remains unknown.

Several *Arabidopsis* mutants have been characterized in which genes under sporophytic control alter the interaction of the pollen with the stigma (Chaudury 1993). Although these mutants produce abundant pollen grains, they fail to germinate *in vivo*, generally because of pollen-pistil recognition defects, but pollen could be induced to grow pollen tubes by the addition of wild type pollen, indicating that diffusible products required for an early step in pollen stigma recognition are missing in these mutants. A mutant of this type, *pop1* (Preuss et al. 1993), has been described in *Arabidopsis*, that has a reduced amount of tryphin; in this mutant in vivo pollen germination is impaired, and callose is apparently formed in stigma cells that contact mutant pollen, indicating a role of tryphin in pollen-stigma interaction.

Recent evidence indicates that some flavonols are essential for pollen function. A lack of chalcone-synthase (CHS) activity, an enzyme catalysing the initial step of flavonoid biosynthesis, has a pleiotropic effect in maize and petunia: it disrupts not only flavonoid synthesis, but also pollen function. In both maize and petunia, CHS-deficient mutants are self-sterile; pollen is apparently normal, but is unable to germinate and grow, apparently because mutant pistils are unable to sustain pollen growth. However, CHS-deficient pistils set seed when pollinated with wild-type pollen, and the mutant pollen is partially functional on wild-type stigmas (Coe et al. 1981; van der Meer et al. 1992). The same phenomenon was observed in transgenic petunia plants with suppressed *CHS* gene expression: pollen with greatly reduced function both *in vitro* and *in vivo* is partially restored on wild-type stigmas (Taylor and Jorgensen 1992). Kaempferol was identified as a pollen germination-inducing constituent in wild-type petunia stigma extracts (Mo et al. 1992). Adding micromolar quantities of this compound to the germination medium or to the stigma is sufficient to restore mutant pollen germination and tube growth *in vitro* and seed set *in vivo*. In tobacco, flavonols (quercetin, kaempferol, myricetin), when added to the germination medium, strongly promote germination and tube growth of *in vitro*-matured pollen (Ylstra et al. 1992). These results indicate that some type of flavonoid is essential for normal pollen function, and can be provided either by the pollen or by the stigma. The fact that the compound is effective at very low dosages (micromolar range) and its specificity suggest that flavonols act as signal molecules.

A category of genes directly controlling pollen-pistil interaction is represented by the so-called *Gametophyte factors*, the presence of which has been reported in many species, such as maize (Nelson 1952; 1994), lima bean (Allard 1963), barley (Tabata 1961), rice (Iwata et al. 1964) and *Oenothera* (Harte 1969). These genes strongly affect pollen tube growth: pollen bearing the recessive *ga* allele has a great disadvantage when competing with pollen carrying the dominant *Ga* allele in stylar tissue with the same dominant allele, but not in *ga/ga* styles. In maize, some *Ga* factors have been extensively studied using both *Ga* and *ga* pollen labelled with ^{32}P (House and Nelson 1964): *Ga* pollen germinated and pollen tubes grew into the stylar tissues, but their growth became progressively slower than that of *Ga* pollen, and eventually ceased before reaching the ovules. Nothing is known about the function of these genes, that appear to be organized in gene families: for instance, in maize at least nine *Ga* factors have been detected, that map in different chromosomes (Bianchi and Lorenzoni 1975). On the other hand, these genes have not been the object of study in recent years, when more powerful molecular approaches could offer the opportunity for their isolation and characterization.

Many pollen traits that are important components of pollen fitness, such as tube growth rate or tolerance to some stresses, are controlled by many genes. Thus, the strategies for identifying the underlying genetic architecture are based on quantitative genetic methodologies. To detect and localise in maize chromosomes the factors involved in the control of these types of pollen characters (early growth *in vitro*, germination and early growth *in vivo*, late tube growth *in vivo*, pollen-pistil interaction), linkage analysis between the expression of each character and molecular markers (RFLPs) was performed (Sari-Gorla et al. 1992; 1994; 1995). The analysis is based on the following rationale. The segregating population to be analyzed is produced starting from two parental lines divergent for the trait of interest and polymorphic for molecular markers. Each individual plant of the resulting F_2 generation is evaluated for the expression of the trait and is typed for the molecular allele present at each locus. If a marker is closely linked to a factor controlling the trait, the association between the phenotypic value of the trait and the allelic composition at the marker locus will be significant, while no correlation will be detected in the case of markers located in a different position with respect to the involved genes. When a densely saturated map of markers is available, most of the factors (QTLs: quantitative trait loci) controlling the trait and segregating in the population can be detected and localized on chromosomes.

The study was carried out in a population of recombinant inbred lines characterized for about 200 RFLPs. Even though the nature of the estimate is only indirect, our findings strongly confirm the complex basis of the genetic system that regulates pollen function (Fig. 16.3). In particular, four QTLs, accounting for about 40% of the phenotypic variability of *in vitro* pollen tube growth, were also significantly associated with *in*

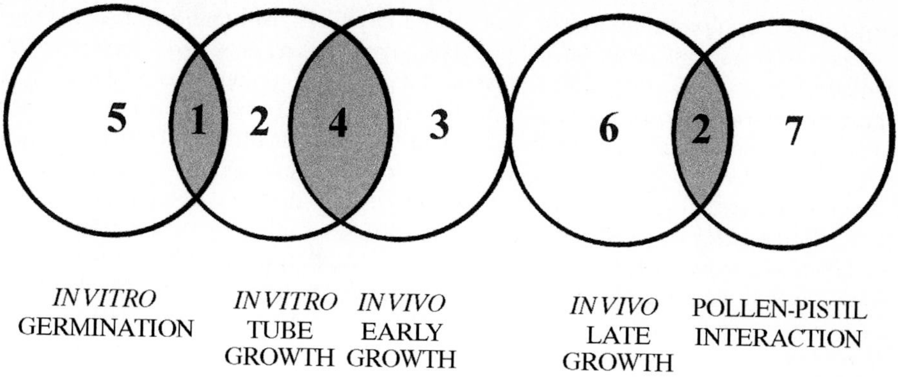

Fig. 16.3. Association between molecular markers and different pollen functions. The number of specific or common loci is indicated

vivo germination and early tube growth, whereas all the identified chromosomal regions carrying QTLs for *in vivo* later tube growth rate within the style appeared to be specific for this trait, and only one was shared between *in vitro* germination and early growth. Bearing in mind that the growth of maize pollen tubes in artificial medium is limited to 1-2 mm compared with *in vivo* growth, that can be up to 20–30 cm, our results suggest the existence of different sets of genes acting during the various stages of the fertilisation processes. A first group of genes affects germination rate evaluated *in vitro*; a second set controls the early phases of tube lengthening, and thus is responsible for variation of *in vitro* tube elongation and *in vivo* early growth; other genes regulate pollen tube growth when the tube is well established into the style, up to fertilisation. Finally, the replication of the experiment using two different genotypes as female plants allowed the detection of a set of QTLs controlling pollen-pistil interaction effects, most of which did not coincide with those involved in the control of pollen competitive ability *per se*.

A long-term objective could be the isolation by positional cloning of the genes identified in this way. Before the advent of QTL era, it was assumed that complex traits were determined by a very large number of genes of relatively small and equal effect; QTL analysis by molecular markers has revealed that the effects of these loci are not equal, but often a substantial portion of the genetic variation of a population can be explained by a few QTLs of moderately large effects. In fact, we have identified single markers whose variation account for up to 26% of the character variability, even though their effect could be modified by genotype × environment (G × E) interaction. Adopting a model based on least square interval mapping, we were able to estimate G × E effects (Sari-Gorla et al. 1997b). Two different categories of QTLs were detected: those whose expression is modulated by the environmental effects and those that are independent of them, and that are likely "major genes", on which the variation of the character mainly depends; these could be the candidate genes for map based gene isolation.

16.2.2
From DNA to its Function

Many late genes isolated from mature pollen grains could, due to their putative function, have a more important role in pollen function than in grain development. Many allergens, such as glycoprotein, could be involved in pollen-pistil interaction signalling. Callase, wall-degrading enzymes and cytoskeletal proteins are expected to be important in the processes of germination and particularly of tube growth. Bet v1 proteins, a group of isoforms that constitutes the major pollen allergen of the white birch (*Betula verrucosa*), are encoded by a complex multigene family which includes a subset of genes transcriptionally activated in the presence of microbial pathogens. The analysis of the temporal and spatial expression pattern of *Bet v1* genes during development of birch male inflorescences revealed that they are specifically expressed in late bicellular and mature pollen, while no transcripts were detected in sporophytic anther tissues of any developmental stage. (Swoboda et al., 1995). The authors suggest a possible role of Bet v1 protein during pollen germination on the stigma, protecting the female reproductive tissues from infection by pathogens.

As already stated, information regarding single genes expressed specifically or preferentially after pollen germination, during pollen tube growth, is very scant. A clear case of transcripts that were demonstrated to be present also in the pollen tube is that of *Tp44* from *Tradescantia*, which were detected in pollen tubes, up to one hour after *in vitro* germination (Stinson et al. 1987). From a pollen tube cDNA library of *Petunia inflata* two clones were isolated (Mu et al. 1994a; 1994b): *PRK1*, an already mentioned pollen receptor-kinase, and *PPE1*, which exhibited sequence similarity to pectin esterase. During pollen development their mRNAs were first detected at the stage of mono or binucleate microspores, reaching the highest level in mature pollen, and remaining at a high level in *in vitro*-germinated pollen tubes.

The localization of transcripts of the pollen-specific gene *NPT303* from tobacco has been determined by in situ hybridization (Weterings et al. 1995); *NTP303* transcripts were first detectable at the mid-binucleate stage of pollen development and persisted even after the pollen tube had reached the ovary. The majority of transcripts was localized in the vegetative cell, and was predominantly present in the apex of the pollen tube. Since the protein appears to be localized on the membranes, *NTP303* could be involved in maintaining their integrity, or in the adherence of membrane to the wall.

NTF4 is a tobacco gene encoding for a MAPkinase; it reveals a peculiar profile of expression: transcripts are detected only after the first pollen mitosis and are present during pollen maturation, but show very low levels in mature grain, and finally increase in germinating pollen (Heberle-Bors, pers. comm.). Thus the gene appears to be involved in some of the early events during pollen germination.

A cDNA clone coding for tobacco profilin, which has been recently identified as a potent allergen in pollen and regulates actin-binding protein, was isolated from a tobacco pollen cDNA library by antibody screening (Mittermann et al. 1995). Profilin was detected in different somatic tissues, but its expression was 50–100 fold higher in mature pollen: no profilin could be detected up to early binucleate stages, while a very high increase occurred at a late stage of pollen development; it was expressed at high levels in mature and *in vitro* germinated pollen and pollen tubes, particularly in their tips. This data suggests that profilin may fulfil a specialized function before and during pollen germination and growth, probably as a microfilament precursor. In *Arabidopsis*, four member of the profilin multigene family have been cloned, sequenced and analyzed with regard to their expression profile (Christensen et al. 1996). The results demonstrated that these four genes fall into two groups: one group (*pfn1* and *pfn2*) is expressed in all the organs of the plant and the other group (*pfn3* and *pfn4*) in floral tissues only; in particular, expression of *pfn4* is restricted to mature and germinating pollen grains.

We used pearl millet with the purpose of isolating genes specifically or preferentially expressed during pollen tube growth, to be used as probes on other cereals, since millet pollen, unlike the pollen of most cereals, is able to elongate *in vitro* at about the same extent as *in vivo*.

A cDNA library from *in vitro* germinated pollen tubes was screened using labelled cDNAs enriched in sequences highly expressed in pollen tubes, by means of subtractive strategies. About one hundred cDNAs were isolated and the preliminary expression pattern of their corresponding genes was determined by "reverse" Northern blot analysis. On the basis of their strong hybridization signals with total labelled cDNA from pollen tubes, 55 cDNAs were selected for further analysis.

The partial sequence at the 5′ end was determined for all the 55 cDNAs. For the majority of the clones no obvious homology to known genes was found, although a number of these cDNAs showed homology to *Arabidopsis* or rice ESTs. One clone showed high homology (more than 84%) to the barley, maize and rice calmodulin genes. Although this gene is expressed in other plant tissues besides pollen, this finding is interesting, providing a link between the expression of genes coding calcium-binding proteins and the data correlating the maintenance of a calcium ion gradient along the pollen tube and the tube growth (Malhò and Trewavas 1996). Another clone showed homology (more than 50%) to a dyneine gene of *Clamydomonas*, coding for a motor protein involved in cytoplasmic streaming. Preliminary data suggests that the expression of this gene is confined to the male gametophyte.

16.4 Concluding Remarks

The developmental pathway followed by the male gametophyte is a process involving precise control over cell division and gene expression by the haploid genome. About 20,000 gametophytic genes have been detected, at least 2000 of which are specifically expressed in the mature pollen grain. Thus it is not surprising that a large number of different mutants has been described in many plant species.

So far, just a very small fraction of these genes has been characterized, and the function of most of them is still unknown. The strategies for discovering the role of the huge number of pollen genes that have or can be isolated could be: i) the isolation of mutant genes, by positional cloning or gene tagging; ii) the determination of the role of the pollen-specific clones by knock-out and function analysis in transgenic plants. The task should be facilitated by the fact that genes controlling the complex reproductive system, on the basis of evolutionary reasons, are expected to be highly conserved. In fact, mutations in genes impairing fertility have a low probability of being transmitted. A large amount of evidence now supports these ideas.

A crucial point to be elucidated is the precise mechanism by which the fate of the generative and the vegetative nuclei is determined. In fact, this phenomenon has huge biological significance, since it is at the basis of every differentiative event in higher organisms. In this frame, the "pollen system", due to its simple structure, is a very suitable material to carry out these types of studies.

Acknowledgements
The research in M. Sari-Gorla lab was partially supported by EC grant BIO4 CT97 2312.

References

Aarts MGM, Dirkse WG, Stlekema WJ, Pereira A (1993) Transposon tagging of a male sterility gene in *Arabidopsis*. Nature 363:715–717

Albani D, Altosaar I, Arnison PG, Fabijanski SF (1991) A gene showing sequence similarity to pectin esterase is specifically expressed in developing pollen of *Brassica napus*: Sequences in its 5′ flanking region are conserved in other pollen-specific promoters. Plant Mol Biol 16:501–513

Albani D, Robert LS, Donaldson PA, Altosaar I, Arnison PG, Fabijanski SF (1990) Characterization of a pollen-specific gene family from *Brassica napus* which is activated during early microspore development. Plant Mol Biol 15:605–622

Albani D, Sardana R, Robert LS, Altosaar I, Arnison PG, Fabijanski SF (1992) A *Brassica napus* gene family which shows sequence similarity to ascorbate oxidase is expressed in developing pollen. Molecular characterization and analysis of promoter activity in transgenic tobacco plants. Plant J 2:331–342

Albertsen MC, Fox TW, Trinell MR (1993) Tagging, cloning and characterizing a male fertility gene in maize. Am J Bot Abstr 80:16

Albertsen MC, Phillips RL (1981) Developmental cytology of 13 genetic male sterile loci in maize. Can J Genet Cytol 23:195–208

Allard RW (1963) An additional Gametophyte factor in the lima bean. Zuchter 33:212–216

Allen RJ, LonsdaleDM (1993) Molecular characterization of one of the maize polygalacturonase gene family members which are expressed during late pollen development. Plant J 3:261–271

Baltz R, Domon C, Pillay DTN, Steimmetz A (1992) Characterization of a pollen-specific cDNAfrom sunflower encoding a zinc finger protein. Plant J 2:713721

Bedinger PA, Broadwater AH, Conway JD, Hardeman KJ, Lonkides CA, Prata RTN, Rubinstain AL (1994) Molecular studies of pollen development in maize. In: Stephenson AG, Kao T (eds) Pollen-pistil interaction and pollen tube growth. Curr Topics in Plant Physiol 12:1–14

Bianchi A, Lorenzoni C (1975) Gametophytic factors in *Zea mays*. In: Mulcahy DL (ed) Gamete Competition in Plants and Animals, Elsevier, Amsterdam, pp 257–264

Bishop CJ, McGowan LJ (1953) The role of the vegetative nucleus in pollen tube growth and in the division of the generative nucleus in *Tradescantia paludosa*. Am J Bot 40:658–659

Breiteneder H, Pettenburger K, Bito A, Valento R, Kraft D, Rumpold H, Scheiner O, Breitenbach M (1989) The gene coding for the major birch pollen allergen Bet v I is highly homologous to a pea disease resistance response gene. EMBO J 8:1935–1938

Brown SM, Crouch ML (1990) Characterization of a gene family abundantly expressed in *Oenothera organensis* pollen that shows seuqnce similatity to polygalacturonase. Plant Cell 2:263–274

Carpenter JL, Ploense SE, Snustad DP, Silflow C (1992) Preferential expression of an α-tubulin gene of *Arabidopsis* in pollen. Plant Cell 4:557–571

Chang MT, Neuffer MG (1989) Maize microsporogenesis. Genome 32:232–244

Chaudhury AM (1993) Nuclear genes controlling male fertility. Plant Cell 5:1277–1283

Chaudhury AM, Lavithis M, Taylor PE, Craig S, Singh MB, Signer ER, Knox RB, Dennis ES (1994) Genetic control of male fertility in *Arabidopsis thaliana*: structural analysis of premeiotic developmental mutants. Sex Plant Reprod 7:17–28

Chen YS, McCormick S (1996) sidecar pollen, an *Arabidopsis thaliana* male gametophytic mutant with aberrant cell division during pollen development. Development 122:3243–3253

Christensen HEM, Ramachandran S, Tan CT, Surana U, Dong CH, Chua NH (1996) *Arabidopsis* profilins are functionally similar to yeast profilins: identification of a vascular bundle-specific profilin and a pollen-specific profilin. Plant J 10:269–279

Coe EH, McCormick SM, Modena SA (1981) White pollen in maize. J Hered 72:318–320

Correns K (1900) Über den Einfluss, welchen die Zahl der zur Bestäubung verwendeten Pollenkörner auf die Nachkommenschaft hat. Ber Bot Ges 18:422–435

Eady C, Lindsey K, Twell D (1995) The significance of microspore division and division symmetry for vegetative cell-specific transcription and generative cell differentiation. Plant Cell 7:65–74

Glover J, Grelon M, Craig S, Chaudhury A, Dennis E (1998) Cloning and characterization of MS5 from *Arabidopsis*: a gene critical in male meiosis. Plant J 15:345–356

Gorman SW, Banasiak D, Fairley C, McCormick S (1996) A 610-kb YAC clone harbors 7 cM of tomato (Lycopersicom esculentum) DNA that includes the *male sterile-14* gene and a hot spot for recombination. Mol Gen Genet 251:52–59

Gabay-Laughnan S (1997) Late reversion events can mimic imprinting of restorer-of-fertility genes in *CMS-S* maize. Maydica 42:163–172

Gorman SW, McCormick S (1997) Male sterility in tomato. Crit Rev Plant Sci 16:31–53

Hamilton DA, Mascarenhas JP (1997) Gene expression during pollen development. In: Shivanna KR, Sawhney VK (eds) Pollen Biotechnology for Crop Production and Improvement. Cambridge University Press, Cambridge, pp 40–58

Hanson DD, Hamilton DA, Travis JL, Bashe DM, Mascarenhas JP (1989) Characterization of a pollen-specific cDNA clone from *Zea mays* and its expression. Plant Cell 1:173–179

Harte C (1969) Gonenkonkrenz bei *Oenothera* unter dem Einfluss eines gametofitisch-wirksamen Gensin der ernsten Koppelungsgruppen sowie ein Modell fur die Untersuchung verzweigter Koppelungsgruppen. Theor Appl Genet 39:163–178
He C, Tirlapur U, Cresti M, Peja M, Crone DE, Mascarenhas JP (1996) An *Arabidopsis* mutant showing aberrations in male meiosis. Sex Plant Reprod 9:54–57
Hirata J, Nakagoshi H, Nabeshima Y, Matsuzaki F (1995) Asymmetric segregation of the homeodomain protein Prospero during *Drosophila* development. Nature 377:627–630
Horner HT, Palmer RG (1995) Mechanism of genetic male sterility. Crop Sci 35:1527–1535
House LR, Nelson OE (1958) Tracer study of pollen tube growth in cross-sterile maize. J Hered 49:18–21
Iwata N, Nagamatsu T, Omura T (1964) Abnormal segregation of *waxy* and *apiculus* coloration by a gametophyte gene belonging to the first linkage group in rice. Japan J Breed 14:33–39
John ME, Petersen MW (1994) Cotton (*Gossypium hirsutum* L) pollen-specific polygalacturonase mRNA: tissue and temporal specificity of its promoter in transgenic tobacco. Plant Mol Biol 26: 1989–1993
Kaul MLH (1988) Male sterility in higher plants. Springer-Verlag, Berlin, Hidelberg, New York
Kindiger B, Beckett TA, Coe EH (1991) Differential effects of specific chromosomal deficiencies on the development of the maize pollen grain. Genome 32:579–594
Knoblich K, Jan LJ, Nung Jan Y (1995) Asymmetric segregation of Numb and Prospero during cell division. Nature 377:624–627
Laughnan JR, Gabay SJ (1983) Cytoplasmic male sterility in maize. Annu Rev Genet 117:27–48
Lee HS, Karunanandaa B, McCubbin A, Gilroy S, Kao T (1996) PRK1, a receptor-like kinase of *Petunia inflata*, is essential for post-meiotic development of pollen. Plant J 9:613–624
Lopez I, Anthony RG, Maciver SK, Jiang CJ, Khan S, Weeds AG, Hussey PJ (1966) Pollen specific expression of maize genes encoding actin depolymerizing factor-like proteins. Proc Natl Acad Sci USA 93: 7415–7420
Loukides CA, Broadwater AH, Bedinger PA (1995) Two new male sterile mutants of *Zea mays* (Poaceae) with abnormal tapetal cell morphology. Am J Bot 82:1017–1023
Malhò R, Trewavas J (1996) Localized apical increases of cytosolic free Calcium control pollen tube orientation. Plant Cell 8:1935–1949
Mangelsdorf PC (1932) Mechanical separation of gametes in maize. J Hered 23:288–295
Mascarenhas JP (1990) Gene activity during pollen development. Ann Rev Plant Physiol Plant Mol Biol 41:317–338
Mascarenhas JP (1992) Pollen gene expression: molecular evidence. Int Rev Cytol 140:3–18
Mascarenhas JP (1993) Molecular mechanisms of pollen tube growth and differentiation. Plant Cell 5: 1303–1314
McCormick S (1991) Molecular analysis of male gametogenesis in plants. Trends Genet 7:289–303
McCormick S (1993) Male gametophyte development. Plant cell 5:1265–1275
Mckinney EC, Aali N, Traut A, Feldmann KA, Belostotsky DA, Mcdowelll JM, Meagher RB (1995) Sequence-based identification of T-DNA insertion mutations in *Arabidopsis*: actin mutants *act2-1* and *act 4-1*. Plant J 8:613–622
Meinke DW (1982) Embryo-lethal mutants of *Arabidopsis thaliana*. Evidence for gametophytic expression of the mutant genes. Theor Appl Genet 63:381–386
Meinke DW (1985) Embryo-lethal mutants of *Arabidopsis thaliana*: analysis of mutants with a wide range of lethal phases. Theor Appl Genet 69:543–552
Mittermann I, Swoboda I, Pierson E, Eller N, Kraft D, Valenta R, Heberle-Bors E (1995) Molecular cloning and characterization of profilin from tobacco (*Nicotiana tabacum*): increase profilin expression during pollen maturation. Plant Mol Biol 27:137–146
Mo Y, Nagel C, Taylor LP (1992) Biochemical complementation of chalcone synthase mutants defines a role for flavonols in functional pollen. Proc Nat Acad Sci 89:7213–7217
Moore I, Diefenthal T, Zarsky V, Schell J, Palme K (1997) A homolog of the mammalian GTPase Rab2 is present in *Arabidopsis* and is expressed predominantly in pollen grains and seedlings. Proc Natl Acad Sci USA 94:762–767
Mu JH, Lee HS, Kao T (1994b) Characterization of a pollen-expressed receptor-like kinase gene of *Petunia inflata* and the activity of its encoded kinase. Plant Cell 6:709–721
Mu JH, Stains JP, Kao T (1994a) Characterization of a pollen-expressed gene encoding a putative pectin esterase of *Petunia inflata*. Plant Mol Biol 25:539–544
Muschietti J, Dirks L, Vancanneyt G, McCormick S (1994) LAT52 protein is essential for tomato pollen development: pollen expressing antisense LAT52 RNA hydrates and germinates abnormally and cannot achieve fertilisation. Plant J 6:321–338
Nelson OE (1952) Non reciprocal cross sterility in maize. Genetics 37:101–124
Nelson OE (1994) The Gametophyte factors of maize. In: Freeling M, Walbott V (eds) The Maize Handbook, Springer-Verlag, New York, pp 496–502
Niogret MF, Dubald M, Mandaron P, Mache R. (1991) Characterization of pollen polygalacturonase encoded by several cDNA clones in maize. Plant Mol Biol 17:1155–1164
Oldenhof MT, De Groot PFM, Visser JH, Schrauwen J, Wullems G (1996) Isolation and characterization of a microspore-specific gene from tobacco. Plant Mol Biol 31:213–225

Ottaviano E, Petroni D, Pè ME (1988) Gametophytic expression of genes controlling endosperm development in maize. Theor Appl Genet 75:252–258

Pè ME, Frova C, Colombo L, Binelli G, Mastroianni N, Tarchini R, Visioli G (1994) Molecular cloning of genes expressed in pollen of *Sorghum bicolor*. Maydica 39:107–113

Peirson BN, Owen HA, Feldmann KA, Makaroff CA (1996) Characterisation of three male sterile mutants of *Arabidopsis thaliana* exhibiting alterations in meiosis. Sex Plant Reprod 9:1–16

Preuss D, Lemieux B, Yen G, Davis RW (1993) A conditional sterile mutation eliminates surface components from *Arabidopsis* pollen and disrupts cell signalling during fertilization. Genes Develop 7:974–985

Qiu X, Erickson L (1996) A pollen-specific polygalacturonase-like cDNA from alfalfa. Sex Plant Reprod 9:123–124

Rafnar T, Griffith LJ, Kuo M, Bond JF, Rogers BL, Klupper DG (1991) Cloning of Amb a I (antigen E), the major allergen family of short ragweed pollen. J Biol Chem 266:1229–1236

Reddy JT, Dudareva N, Evrard JL, Krauter RR, Steinmetz A, Pillay DTN (1995) A polleen-specific gene from sunflower encodes a member of the leucin-rich-repeat protein superfamily. Plant Sci 111: 81–93

Robert LS, Allard S, Gerster JL, Cass L, Simmonds J (1993) Isolation and characterization of a polygalacturonase gene highly expressed in *Brassica napus* pollen. Plant Mol Biol 23:1273–1278

Roberts MR, Robson F, Foster GD, Draper J, Scott RC (1993) A *Brassica napus* mRNA expressed specifically in developing microspores. Plant Mol Biol 17:295–299

Rogers HJ, Harvey A, Lonsdale DM (1992) Isolation and characterization of a tobacco gene with homology to pectate lyase which is specifically expressed during microsporogenesis. Plant Mol Biol 20: 493–502

Sari-Gorla M, Binelli G, Pè ME, Villa M (1995) Detection of genetic factors controlling pollen-style interaction in maize. Heredity 74:62–69

Sari-Gorla M, Calinski T, Kaczmareck Z, Krajewski P (1997b) Detection of QTL × environment interaction in maize by a least squares interval mapping method. Heredity 78:146–157

Sari-Gorla M, Ferrario S, Villa M, Pè ME (1996) *gaMS-1*, a gametophytic expressed male sterile mutant of maize. Sex Plant Reprod 9:216–220

Sari-Gorla M, Gatti E, Villa M, Pè ME (1997a) A multi-nucleate male sterile mutant of maize with gametophytic expression. Sex. Plant Reprod. 10:22–26

Sari-Gorla M, Pè ME, Mulcahy DL, Ottaviano E (1992) Genetic dissection of pollen competitive ability in maize. Heredity 69:423–430

Sari-Gorla M, Pè ME, Rossini L (1994) Detection of QTLs controlling pollen germination and growth in maize. Heredity 72:332–335

Silvanovich A, Astwood J, Zhang L, Olsen E, Kisill F, Sehon A, Mohapatra S, Hill R (1991) Nucleotide sequence analysis of three cDNA coding for Poa p IX isoallergens of Kentucky bluegrass pollen. J Biol Chem 266:1204–1210

Singh MB, Hough T, Theerakulpisut P, Avjioglu A, Davis S, Smith PM, Taylor P, Simpson RJ, Ward LD, McCluskey J, Puy R, Knox RB (1991) Isolation of cDNA encoding a newly identified major allergenic protein of rye-grass pollen: intracellular targeting to the amyloplast. Proc Natl Acad Sci USA 88: 1384–1388

Staiger CJ, Goodbody KC, Hussey PJ, Valenta R, Brobak B, Lloyd CW (1993) The profilin multigene family of maize: differential expression of three isoforms. Plant J 4:631–641

Stinson JR, Eisemberg AJ, Willing RP, Pè ME, Hanson DD, Mascarenhas JP (1987) Genes expressed in the male gametophyte of flowering plants and their isolation. Plant Physiol 83:442–447

Swoboda I, Dang TCH, Heberle-Bors E, Vicente O (1995) Expression of Bet v I, the major birch pollen allergen, during anther development: an in situ hybridization study. Propoplasma 187:103–110

Tabata M (1961) Studies of a gametophyte factor in barley. Japan J Genet 36:157–167

Taylor LP, Jorgensen R (1992) Conditional male fertility in chalcone synthase-deficient *Petunia*. J Hered 83:11–17

Tebbut SJ, Rogers HJ, Lonsdale DM (1994) Characterization of a tobacco gene encoding a pollen specific polygalacturonase. Plant Mol Biol 25:283–297

Thangavelu M, Belostotsky D, Bevan MW, Flavell RB, Rogers HJ, Lonsdale DM (1993) Partial characterization of the *Nicotiana tabacum* actin gene family: evidence for pollen-specific expression of one of the gene family members. Mol Gen Genet 240:290–295

Theerakulpisut P, Xu H, Singh MB, Pettitt JM, Knox RB (1991) Isolation and developmental expression of *Bcp1*, an anther specific cDNA clone in *Brassica campestris*. Plant Cell 3:1073–1084

Touraev A, Lezin F, Heberle-Bors E, Vicente O (1995) Maintenance of gametophytic development after symmetrical division in tobacco microspore culture. Sex Plant Reprod 8:70–76

Turcich MP, Hamilton DA, Mascarenhas JP (1993) Isolation and characterization of pollen-specific maize genes with sequence homology to ragweed allergens and pectate lyases. Plant Mol Biol 23: 1061–1065

Twell D (1995) Diphtheria toxin-mediated cell ablation in developing pollen: vegetative cell ablation blocks generative cell migration. Protoplasma 187:144–154

Twell D, Park S, Howden R (1997) Division asimmetry (*gemini*) mutants and the control of pollen cell fate. Proc 5th Int Congr of Plant Mol Biol, Singapore, 44

Twell D, Yamaguchi J, McCormick S (1990) Pollen specific gene expression in transgenic plants: coordinate regulation of two different tomato gene promoters during microsporogenesis. Development 109:705–713

Twell D, Yamaguchi J, Wing R, Ushiba J, McCormick S (1991) Promoter analysis of genes that are coordinately expressed during pollen development reveals pollen-specific enhancer sequences and shared regulatory elements. Gene Devel 5:496–507

Ursin VM, Yamaguchi J, McCormik S (1989) Gametophytic and sporophytic expression of anther-specific genes in developing tomato anthers. Plant Cell 1:727–736

van der Meer IM, Stam ME, van Tunen AJ, Mol JNM, Stultje AR (1992) Antisense inhibition of flavonoid biosynthesis in petunia anthers results in male sterility. Plant Cell 4:253–262

van Tunen AJ, Mur LA, Brouns GS, Rienstra JD, Koes RE, Mol JNM (1990) Pollen- and anther-specific promoters from petunia: tandem promoter regulation of the *chiA* gene. Plant Cell 2:393–401

Villalba M, Batanero E, Monsalve RI, Gonzales de la Pena MA, Lahoz C, Rodrigues R (1994) Cloning and expression of ole e I, the major allergen from olive tree pollen. J Biol Chem 269:15217–15222

Weterings K, Reijnen W, van Aarssen R, Kortstee A, Spljkers J, van Herpen M, Schrauwen J, Wullems G (1992) Characterization of a pollen specific cDNA clone from *Nicotiana tabacum* expressed during microgametogenesis and germination. Plant Mol Biol 18:1101–1111

Weterings K, Reijnen W, Wijn G, van de Heuvel K, Appeldoorn N, De Kort G, van Herpen M, Schrauwen J, Wullems G (1995) Molecular characterization of the pollen-specific genomic clone *NTPg303* and in situ localization of expression. Sex Plant Reprod 8:11–17

Wing RA, Yamaguchi J, Larabell SK, Ursin VM, McCormik S (1989) Molecular and genetic characterization of two pollen-expressed genes that have sequence similarity to pectate lyases of the planr pathogen *Erwinia*. Plant Mol Biol 14:17–28

Xu H, Knox RB, Taylor PE, Singh MB (1995) Bcp1, a gene required for male fertility in *Arabidopsis*. Proc Nat Acad Sci USA 92:2106–2110

Ylstra B, Touraev A, Benito Moreno RM, Stoger E, van Tunen A, Viciente A, Mol JNM, Heberle-Bors E (1992) Flavonols stimulate development, germination and tube growth of tobacco pollen. Plant Physiol 100:902–907

Zou JT, Zhan XY, Wu HM, Wang H, Cheung AY (1994) Characterization of a rice pollen-specific gene and its expression. Amer J Bot 81:552–561

CHAPTER 17

The Use of the Vibrating Probe Technique to Study Steady Extracellular Currents During Pollen Germination and Tube Growth

A. M. Shipley [1] · J. A. Feijó [2,3]

[1] Applicable Electronics, P.O. BOX 589, Forestdale, MA 02644, U.S.A
telephone: +1.508.833.5042 fax: +1.508.833.1544 e-mail: amshipley@capecod.net
[2] Departamento Biologia Vegetal, Faculdade de Ciências Lisboa, Campo Grande, C2, P-1700 Lisboa, Portugal
telephone: +351.1.7500069 fax: +351.1.7500048 e-mail: jose.feijo@fc.ul.pt
[3] Author for correspondance

17.1 Introduction

The origins of the vibrating probe technique go back to the early part of the twentieth century (Zisman 1932). To the best of our knowledge, the first biological application of the vibrating probe technique was done in the middle of this century (Bluh and Scott 1950) and later on was used to investigate fluxes on skeletal muscle fibers (Davies 1966). The application to biological systems was, however, re-discovered when a major development of the vibrating probe system was done by the addition of a lock-in amplifier (Jaffe and Nucitelli 1974). This modification increased the system sensitivity many fold and provided a method of measuring steady, low magnitude extracellular currents at the surface of biological membranes without damage to the specimen with what has become known as the **vibrating voltage probe** (sometimes referred to as the "wire-probe", since it uses a metal electrode as a probe). The technique provided a practical method of showing total electrical currents in a number of biological systems and triggered a very fecund period of research on currents generated by a number of different biological systems (Jaffe 1981, Jaffe 1985, DeLoof 1986). The voltage probe was additionally refined by computerisation methods as well as simultaneous, two-dimensional measurement capability utilising small, insulated wire microelectrodes (Scheffey 1988). These and other improvements were introduced to cope with specific experimental needs or to overcome general pitfalls of the technique. They have transformed the vibrating voltage probe to a reliable and reproducible technique, making its way to a number of applications in several fields of biology (reviews in Nuccitelli 1986, 1990). Materials researchers have also found these techniques to be applicable to the study of corrosion on metal surfaces (Isaacs 1986, Isaacs et al. 1991) with practical consequences in quality control assaying and new product development.

An important limitation of the voltage probe is the analysis of the actual ionic composition of the currents. When mapping currents around a specimen in control media, the only way to have some hint about the different ionic components was to perform ion substitution experiments and add chemical inhibitors of channel and pump activity to determine the individual ion contributions to the net current detected. This laid down the grounds for the next major improvement of the vibrating probe technique, which was to come from the utilisation of ion-selective microelectrodes instead of the original metal electrodes. These are glass micropipettes with a column of LIX (Liquid Ion Exchanger) in the tip (Jaffe and Levy 1987). The LIX contains an ion-specific ionophore, and the selectivity and efficiency of these probes depends mostly on the characteristics of the specific ionophore utilised. A system was later developed that employed a com-

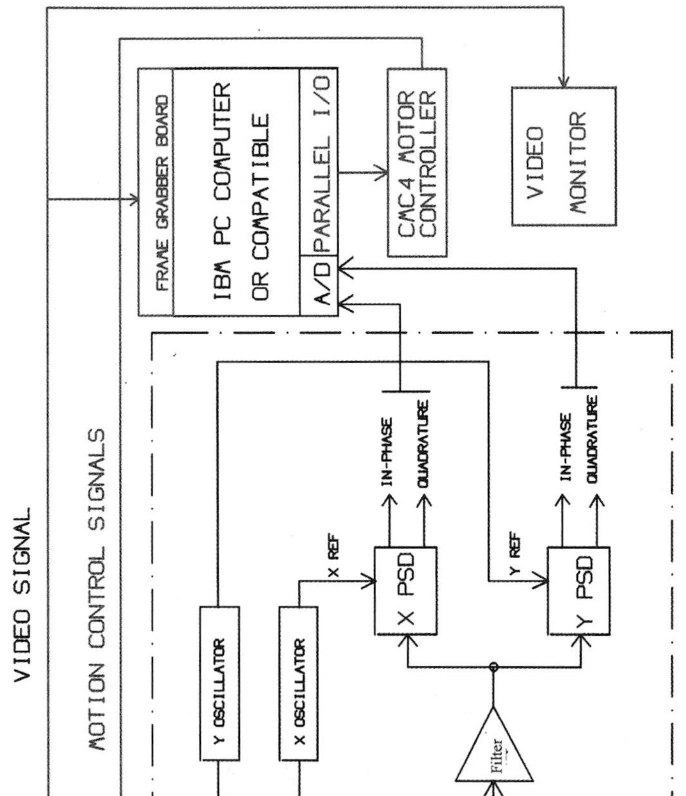

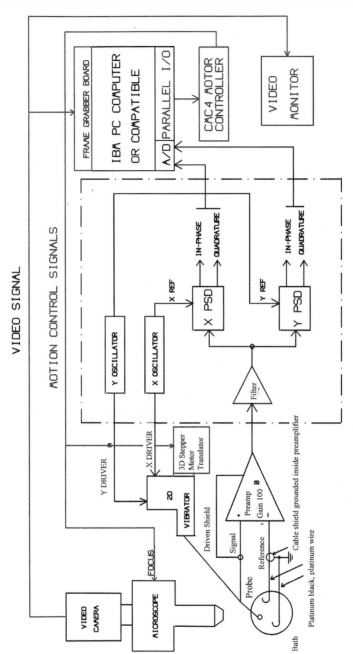

Fig. 17.1. Diagram of a voltage sensitive vibrating probe. The probe is vibrated by a 2D vibrator, which is moved by a 3D positioner driven by stepper motors. The head stage also contains a pre-amplifier. The return electrode(s) is made of platinum black, and fed into the preamplifier. The signal is then filtered and attacks a lock-in amplifier (LIA). The LIA incorporates two oscillators which both drive the piezoelectric elements of the vibrator and serve as reference to the phase sensitive detectors (PSD). The resulting in-phase and quadrature signals are taken by a A/D board and processed by the software. The software also integrates the video input of CCD camera coupled to the microscope. This allows the direct visualisation and control of the probe and may be used to superimpose current vectors at the measure points (see fig. 4)

puter controlled repetitive electrode movement protocol to measure microgradients of specific ions in aqueous media (Kühtreiber and Jaffe 1990). This second set of methods is generally referred to as the **ion-selective vibrating probe** or, in short, the vibrating ion probe.

Briefly, this method was developed to measure the individual ions that carry the net electric current as measured by the vibrating voltage probe. Since this technique complements the vibrating voltage probe, both techniques have sometimes been applied, enabling theoretical confirmation of the expected results in each separated one (Marcus and Shipley 1994). Most biological applications imply the focus on a limited set of individual ions (e.g. calcium, protons or potassium) which are related to the more important physiological and developmental processes.

Current technology is still based on the conceptual framework described above. However, application of technological advances in electronics, computers, micromanipulators and control systems have pushed the envelope of use for these kind of noninvasive, self-referencing, micro-investigation techniques to a level where reliability, automation and user-friendly software are readily usable in a relevant number of biological systems. In this paper we review the basic principles, methods and limitations behind the use of these different types of vibrating probes, with special emphasis on their application to the study of pollen germination and tube growth.

17.2 The Vibrating Voltage Probe

The vibrating voltage probe is an insulated, fine diameter, sharpened wire with an exposed tip in the order of microns (see Fig. 17.2). Commercially available stainless steel or platinum/iridium alloy microelectrodes (Martin Bak, Micro Probe Inc., P.O. Box 87, Clarksburg, MD 20871, USA) are most commonly used. The exposed tips are electroplated in a platinum chloride solution to deposit platinum black. Platinization of the metal tip of the electrode makes it appear black due to the tiny hair-like strands of platinum which are smaller than the wavelength of visible light. This platinum blacking makes the tip highly capacitive as a result of the final, large surface area at the tip. These tips typically have an access resistance between 10k and 100k ohms and a capacitance value from about 3 to 30 nF, which depends on the final platinum blacked tip diameter, typically between 5 and 50 micrometer. Maximum size is determined by the desired spatial resolution necessary for the particular sites of interest on the specimen being measured. If this tip is vibrated (typically over one probe diameter) in an aqueous medium and in the presence of an electric field, it will detect a voltage drop across its vibration excursion distance. Presently, platinum blacked minimum tip sizes are at the 5 micrometer diameter level when using the commercially available electrodes, but it's possible to make electrodes with sharper tip profiles to increase the spatial resolution.

The basic principle of how a vibrating voltage probe works is perhaps better explained by simple electrical analogies. One possibility would be to consider that the specimen to be measured acts as a voltage source and the liquid medium as a resistor. The vibrating voltage probe would then be the wiper arm of a potentiometer having one end connected to a voltage source (specimen) and the other end connected to circuit ground. As the wiper arm moves back and forth along the resistive element (medium), a change in voltage occurs from one end of the movement to the other. That change occurs at the same rate as the motion of the wiper arm. Since the probe is vibrated using a piezoelectric wafer driven by a sine wave oscillator, the signal on the probe, in the presence of an electrical field, is a sine wave having the same frequency as the vibration driver signal and an amplitude proportional to the voltage gradient across the resistive element (medium). Thus the vibrating voltage probe detects an AC signal developed by the motion of a conductor in a DC voltage gradient. Typical vibration frequencies range

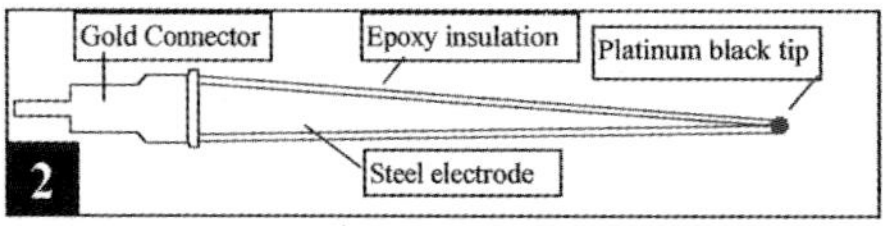

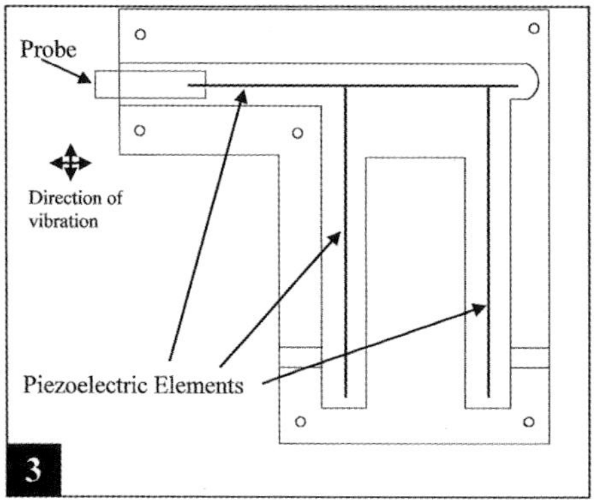

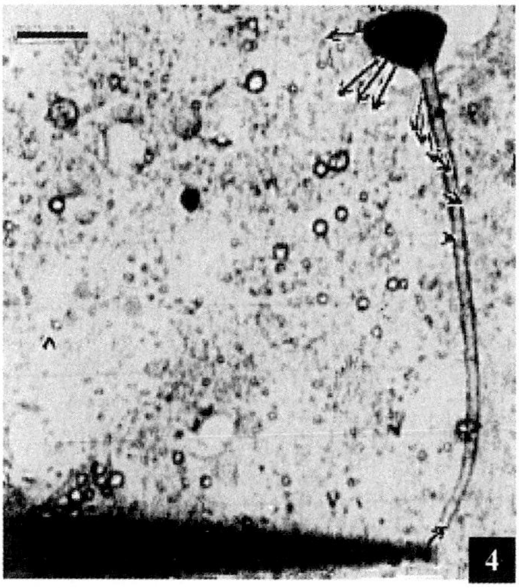

Fig. 17.2. Diagram of a typical wire-probe. A steel electrode is parylene-coated with an exposed tip of about 5 µm. The tip is then electrochemically covered first with gold, and then with a ball of platinum black up to a diameter of about 20–30 µm. The probe is wired to the vibrator through a gold pin connector

Fig. 17.3. Diagram of a typical "π" shaped 2D vibrator. In this configuration three piezoelectric elements are mechanically glued to form an orthogonal axis. The bases of the two arms of the "π" are attached to the frame. These two elements are vibrated with the same frequency to generate one direction of movement, and the top element is vibrated with a different frequency to generate the other. The probe is then attached to the end of the top element and will move in both directions with different frequencies

Fig. 17.4. Typical image of the total electric currents generated by a growing pollen tube as measured by the above set-up. The pollen grain displays an outward current, while the tube has an inward current. The tip has also an inward current but always with more intensity. The length and direction of the represented vectors were obtained by integration of 2 seconds and are imposed by the software on the top of the CCD captured image (the bar on the upper left corner represents 1 $\mu Amp \cdot cm^{-2}$). The image is then taken directly from the video monitor. Also on the field is the probe, which appears blurred due to the vibration. The arrowhead inside the dashed box is the background reference, showing that the detected currents are not due to anisotropic ion movements or thermal convection. The air bubbles on the background are due to the presence of silicone grease used to attach the pollen grains to the dish bottom

from 100 Hz to 1 kHz. Therefore when detecting a component of an electric field, the electrode senses a potential difference from one end of the vibration excursion to the other. If an ion channel or pump generates a flow of ions perpendicular to the membrane surface and there's enough spatial separation between the current's source and sink, vibrating the probe at a position perpendicular to the cell's membrane and with its excursion extreme as close as possible to the membrane surface, will allow the net electrical current generated by the membrane to be measured (see Scheffey 1986 and Ferrier and Lucas 1986 for theoretical details). While producing a sensitivity close to the level of the electronic noise (Scheffey 1988) the vibrating voltage probe is capable of accurately measuring very low biocurrents, at a level unattainable by any other method, especially when compared to stationary electrode methods. In the latest versions, in which the signals are digitally expressed on top of digitally acquired images, it provides an excellent time and space resolution of many important membrane phenomena (see Fig. 17.4). However the vibration frequency utilized (hundreds of Hertz) does induce a strong agitation of the medium close to the specimen (Jaffe and Nuccitelli 1974), destroying stationary gradients, and this aspect must be taken into consideration, especially when looking at phenomena in which these gradients may be of some physiological significance.

17.3 The Vibrating Ion-Selective Probe

Ion-selective microelectrodes made with commercially available ionophores have been available for many years (Amman 1986) (see Fig. 17.6). Measuring gradients in aqueous media was possible in some situations by simply moving the measurement electrode from one place to another and recording the differences in potential on the ion-selective electrode. This method is possible provided there is a sufficient signal to noise ratio to measure any meaningful change. Yet, at the level of small fluxes from single cells, electrode drift and noise limit the application of these stationary electrodes to a few situations (see Kochian et al. 1992).

Kühtreiber and Jaffe (1990) designed a computer controlled repetitive positioning system to avoid manual manipulation of an ion-selective microelectrode from one point to another. It was a logical extension of the vibrating voltage probe technique which measures the net electric field as a function of ionic currents in the extracellular media. This new method allowed for direct measurement of the specific ions that carry the current. In most biological systems, this technique can avoid the use of ion substitution as a means to isolate the components of the net current. A limitation to this stems from the fact that the signal to noise ratio is defined by the measured specific ion gradient compared to the total background concentration. Thus, in some situations, to measure a specific ion, the background concentration of that ion of interest must be lowered in order to attain a sufficiently good signal-to-noise ratio. Since ion activity coefficients change due to ion substitution, and important biological effects can occur, this limitation of the ion probe must be considered when designing low ionic concentration media (Vaughn-Jones and Aickin 1987). Experience shows that in most cases this is not, however, a major limitation and the ion vibrating probe was shown to be useful in a number of systems (see reviews by Smith et al. 1994 and Smith 1995).

Ion-selective electrodes develop DC voltage potentials across a liquid membrane which are proportional to the concentration of the particular ion being measured and can be calculated as Nernst potentials, i.e., 56 mV per decade of concentration gradient for a monovalent anion, and half of that for a bivalent ion (Amman 1986). These electrode potentials result from the difference in specific ion concentration between the measured solution and the fixed internal concentration of the salt bridge in the microelectrode that ensures connection to an electronic amplifier.

Ion-selective electrodes are typically made using a 100 mM concentration of the ion to be measured as an electrolyte in the microelectrode. The electrode Nernst potentials in typical biological saline are generally around -300 to $+300$ millivolts DC depending on the ionophore used and the experimental solution concentration of the specific ion to be measured. The manufacturer's (Fluka, Geneva, Switzerland or WPI, Sarasota, FL, USA) specified response slopes of their commercially available ionophores are generally so reliable that experience shows that when Nernst values of the electrode differ from the manufacturer's specified voltage range per decade of concentration, it is usually an indication that the calibration solutions are not correct or the ionophore is old and, or contaminated.

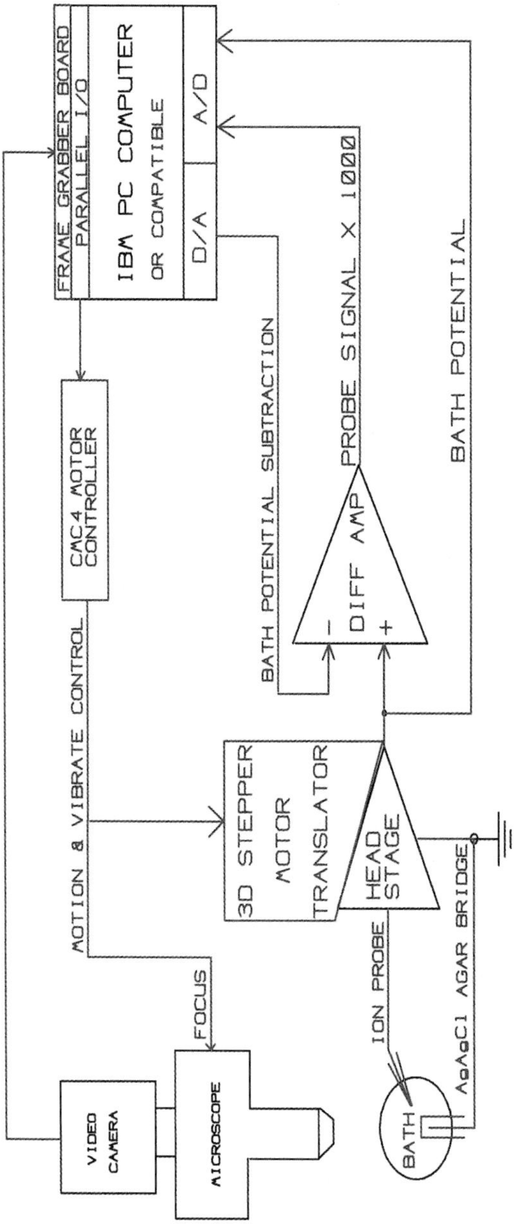

Fig. 17.5. Diagram of an ion-vibrating probe system. The probe is directly connected to the head stage by a silver wire. The head stage is moved by a 3D micro-positioner driven by stepper motors. Both vibration and positioning of the electrode is mechanically driven by these motors. The return current goes by a Ag·AgCl bridge, and the signal feeds a differential amplifier. The use of an A/D board allows direct DC coupling and bath concentration subtraction through the controlling software. The software can also integrate input from a CCD camera for direct monitoring of the probe operation. The microscope, head stage and the mechanical translator must all be inside a closed Faraday cage for noise isolation

When using this extracellular measurement technique, the ion-selective electrode potentials change in the order of microvolts over a distance of 10–30 micrometers from biological membrane surfaces in a bathing medium. However, these microvolt changes, must be measured on top of the constant voltage offset caused by the electrode's Nernst potential. For example, assuming an ion-selective electrode potential of 50 mV in a given medium. Adding an electronic gain factor of 1000 in order to see the microvolt

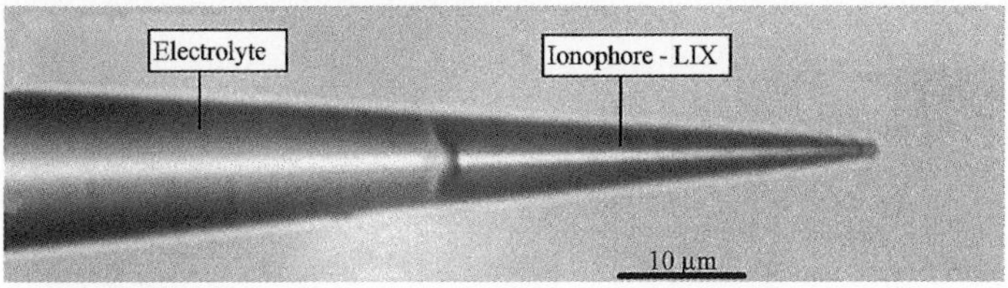

Fig. 17.6. Microphotograph of a typical ion-probe. A 1.5 mm glass pipette is two-step pulled to form a shank of about 1.0 mm and a tip length of about 3 mm. The final tip diameter should be between 1 and 4 µm and will define the ultimate spatial resolution of the probe. The pipette is then sylanised, back filled with a syringe with the appropriated electrolyte and front filled to form a column of about 10–30 µm of the desired ionophore cocktail

changes will also amplify the Nernst potential and saturate an analog to digital converter that has a bipolar 10 VDC input range (see Fig. 17.5). Therefore, the Nernst potential must be cancelled by an appropriate and stable DC offset voltage. The gain factor of 1000 between the microelectrode and the A/D converter appears appropriate in such a system. Using a 16 bit resolution AD board, the bipolar dynamic range of the system is 0.3 µV to 10 mV, referred to the input. Quantizing noise is less than the thermal or Johnson noise typical of these kinds of electrodes. This system resolution has been shown to be appropriate for most biological systems (Kühtreiber and Jaffe 1990, Smith et al 1994).

To measure an ion gradient, the microelectrode is compared to itself at two different positions in the experimental solution, typically 10 micrometers apart. One point as close to the membrane as possible, the other point away from the membrane. The vibration frequency (and consequently the time resolution) is much slower than that of the vibrating voltage probe. Generally speaking this limitation is imposed by (1) the disruption by mechanical stirring of stationary gradients (typically in a fraction of a second) and (2) the stabilisation of the microelectrode's Nernst potential (typically in the millisecond range). Positioning and vibration of the microelectrode is usually done with a computer-controlled, stepper motor driven, micropositioning system (see Fig. 17.5).

Experience has shown that 0.3 to 0.5 Hz is the range where most of the steady or slowly changing gradients in many specimens should be measured (Küthreiber and Jaffe 1990). The H^+ ionophores allow for faster measurement rates, which can be increased up to 0.75 Hz reliably (Demarest and Morgan 1995). The inherent limitations of these low frequencies are time resolutions of no better than about 2–4 seconds per data point. This excludes detection of fast transients such as action potentials, but is adequate for many steady, or slowly changing currents in developmental processes to be followed in time with enough accuracy.

17.3.1 Limitations and Artefacts of the Vibrating Ion-Selective Probe

Despite having been introduced in its present form nearly a decade ago (Küthreiber and Jaffe 1990) some of its technical aspects are still in development. There has not been a standardised software platform and some aspects concerning data generation and acquisition as well as calibration vary significantly from study to study. Most papers published so far use custom made electronics and different ways of acquiring and treating

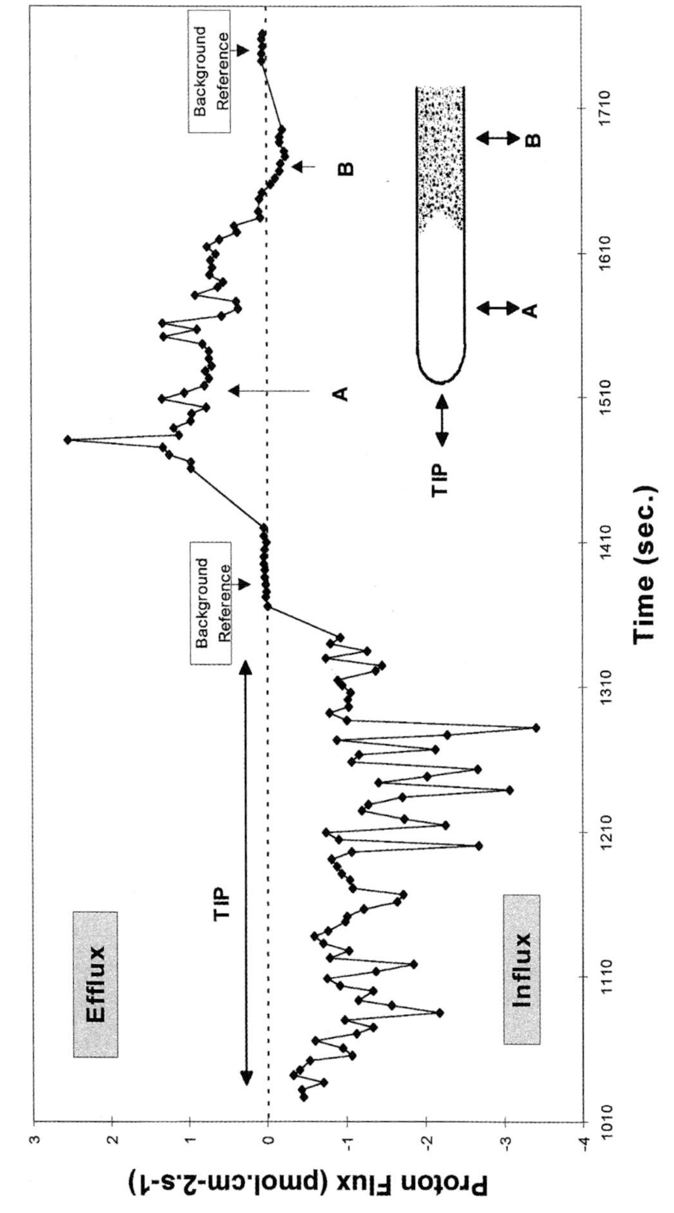

Fig. 17.7. Typical trace of a time-course experiment with the ion-probe. In this specific case a pollen tube has been measured with a proton specific probe. Firstly the probe was moved along with the tip as the tube was growing. As long as the steps are small enough this procedure does not affect significantly the current values. Then the direction of the vibration was changed and the probe stays in the same position as the tube grows (see insert). In all cases the probe was vibrated across 10 μm, with the extreme as close and orthogonal as possible to the membrane surface. Following the conventional direction of current, a negative potential means an influx and a positive potential means an efflux. While the tip displays a clear oscillating influx, the lateral membranes of the clear cap have an efflux of protons. The sub-apical membrane also displays an influx of protons, but not so obvious. Between measures, reference values are made in points 300–400 μm far from any membrane to assure that the potentials detected are not derived from solution gradients or thermal convection

data. This lack of standardisation introduces an extra source of variability, which makes quantitative comparison between different systems very difficult, if not impossible. In this section we analyse some major issues which must be addressed in the future in order to improve this technology for more common standard laboratory use.

Accurate determination of ion-selective electrode efficiency is the most important problem to address. By definition, the efficiency of an ion-selective electrode is the slope of its experimentally determined potential versus the log ion concentration curve, measured at the ion concentration of interest, and expressed as a percentage of the ideal Nernstian behaviour (see Jaffe and Levy 1987, Kühtreiber and Jaffe 1990, Smith et al. 1994).

An attempt to measure an ion-selective vibrating electrode efficiency can be made by creating an artificial gradient of the relevant ion. The source can be a glass micropipette with tip diameter in the order of 20 micrometer, filled with 0.5% agar and containing a high concentration of the respective ion compared to the bath. This arrangement produces an ion gradient (outward from the pipette into the bath) which is difficult to accurately define. Therefore certain limitations must be understood. The most important is the shape and size of the pipette source and convection currents, either by evaporation of the bath or mechanical vibrations (aside from moving the electrode), which will disturb the bath and cause stirring of the artificial gradient. Accurate positioning of the ion-selective electrode relative to the artificial source is also crucial in order to measure the decay of the gradient with distance from the source. The accurate determination of ion-selective electrode efficiency is perhaps, by necessity, going to wait until a known ion microgradient can be created reliably, whereas the reported method of calibration is a good alternative in the absence of a standard, provided that absolute quantitative figures are not required.

On the instrumental side, one of the major problems, especially when trying to characterise time course changes, is that the initial versions of the hardware/software schemes for the vibrating ion-selective probe technique have relied on an AC coupling arrangement (electrode signal connected to the electronics via a capacitor; see electronic schematic in Smith et al. 1994). This method works well for the vibrating voltage probe, but at lower frequencies it has serious limitations. The major one is set by the high and low pass filters in the analog amplifier since AC coupling distorts the signal as a function of charge and discharge time (Kühtreiber and Jaffe 1990, Kochian et al. 1992).The frequency domain which can be used constitutes another limitation in the AC coupled design. Many ion-specific measurements have been made with an input stage that acts as a high pass filter with a cut-off frequency of 0.32 Hz (Kühtreiber and Jaffe 1990 and Smith et al. 1994). This approach may lose a substantial part of the desired signal if the electrode's vibration is slower than 1 Hz. In the case of magnesium ionophores, which require minutes to settle at a new voltage value, the AC coupled design is not useable at all. Also, ion-selective electrode efficiencies at microvolt levels cannot be accurately measured due to the electronic constraints of the AC design. The system is not able to measure voltage drift of the electrode over time in the microvolt domain. The system does, however, permit DC coupled measurement at unity gain to record bath millivolt potentials for Nernst calibrations, but the microvolt information is partially lost through the AC section of the electronics. As a corollary AC coupling is a source of a number of potentially important artefacts and, although it was used in the first prototypes of the ion vibrating probe, should be avoided whenever possible.

17.4
The Vibrating Probe and Plants

Most plant cells have significant activities of ionic transport through the plasma membrane, either due to the activity of electrogenic proton-ATPases or through active ion channels for, namely, potassium, calcium and chloride. These activities have often been related to several developmental processes, namely in apical growing cells or organ extension (Weisenseel and Kicherer 1981; Harold and Caldwell 1990). Generally speaking the presence of a cell wall and an osmotic active vacuole implies the existence of sophisticated ion movement mechanisms and exchanges. The pivotal role of these movements seems to be played by membrane proton translocating ATPases (H^+-ATPase). These pumps display significant electrogenic activity and are responsible, in most plant cells, for the acidification of the apoplast and the vacuole content. They are also functionally different, with two major classes, one located in the plasma membrane and extruding protons to the cell wall ("P" type) and another one secreting protons to the vacuole ("V" type). Like most cells, plants have a relatively neutral cytosol, with high potassium and low calcium concentrations (see review in Feijó et al. 1995). These conditions favour the appearance of significant membrane fluxes, especially involving protons, calcium and potassium (bicarbonate may also induce large fluxes, namely in the filamentous algae with high photosynthetic rates – see Harold and Caldwell 1990). While the asymmetric nature of the distribution of ions across membranes must generate electrical potentials across membranes, little is known about it in plant cells. The reason for this is simple and lies in the fact that the cell wall constitutes a formidable obstacle to the application of standard electrophysiology methods, namely intracellular recording and, even more complex, patch-clamp, which implies a complete removal of the cell wall by enzymatic hydrolysis. Not only are gigaohm seals difficult to obtain, but the true physiological nature of the results may be questioned as protoplasts are known to develop great stress towards the formation of a new cell wall. Recent sophisticated methods have evolved using laser perforation, but still with limited success (DeBoer et al. 1994). Being a non-invasive method, the vibrating probe may be thus a suitable approach provided that a large enough signal to noise ratio may be obtainable. Such is the case for most plant cells which may grow in minimal ionic media and promote strong membrane fluxes. A survey of the literature shows a significant body of information in different systems. Advances have been made in a number of brown algae but only higher plant systems will be covered (for other systems see reviews by Weisenseel and Kicherer 1981, DeLoof 1986, Harold and Caldwell 1990).

At the organ level, systems as diverse as somatic embryos of *Daucus carota* (Rathore et al. 1988) or leaves of *Graptopetalum* (Hush et al. 1991) were successfully investigated concerning polarity establishment with the voltage vibrating probe. Yet the organ which proved to be best suited for experimental study with the vibrating probe were roots. They display enormous currents, grow on minimal media (some in nearly distilled water) and have an inherent polarised growth. The vibrating voltage probe was used to investigate the presence of currents (Weisenseel at al. 1979, Miller et al. 1985), root development and its ionic basis (Miller and Gow 1988, 1989) gravitropism (Weisenseel et al. 1992) and different wounding responses (Miller et al. 1988, Hush et al. 1992). To some extent most of these authors managed to infer some of the ionic basis of the total electric currents by performing substitution/inhibition experiments, or complementing the profiles with the use of stationary ion specific microelectrodes. In all described cases, protons seem to display a significant physiological role.

With the emergence of the calcium specific vibrating probe (Kühtreiber and Jaffe 1990) a number of cellular systems were investigated. Plant systems (maize roots and maize suspension cells) were actually used as model systems to successfully ascertain the applicability of ionophore cocktails, namely for protons and potassium (Kochian et al. 1992). Root hairs have subsequently been investigated for calcium fluxes during normal development (Schiefelbein et al. 1992, Felle and Hepler 1997), or in leguminous roots under the influence of the nod factor (Allen et al. 1994, Cardenas et al. submitted).

Calcium fluxes in the root apex and root hairs were also studied under the influence of aluminium in toxic concentrations (Huang et al. 1992, Jones et al. 1995) (see note at the end).

Last, but certainly not least, pollen tubes and their intrinsic ionic currents have been closely related to the development of the vibrating probe in its present incarnations. Growing lily pollen tubes were the first cell system to be seriously investigated with a voltage vibrating probe (Weisenseel et al. 1975). Tobacco pollen tubes were also one of the systems used to demonstrate the potential power of the calcium specific vibrating probe (Kühtreiber and Jaffe 1990).

Large electrical currents (up to 300 $\text{pAmp}\cdot\text{cm}^{-2}$) were shown to traverse pollen tubes, with the current source being the grain and the current sink being the tube (Weisenseel et al. 1975). Since pollen can be germinated in very low ionic media (for instance in most cases calcium can be lowered down to about 50 μM) this allows for excellent signal to noise ratios. Furthermore, pollen germinates in a couple of hours, allowing completion of several experiments per day, and it's an actively growing cell (one of the fastest growing cells in nature). On the other hand, the major experimental obstacle lies in the fact that these isolated cells are not easy to immobilise, especially in the presence of a moving object like a vibrating probe. While this problem is not serious for the slow moving ion probe, which allows most pollen to settle and remain immobile, with the voltage probe pollen must be attached in some way to a solid support. In the first approaches Weisenseel et al. (1975) and Weisenseel and Jaffe (1976) used a cellophane membrane over the pollen tubes, and measured the current over the membrane. While electrically the cellophane membrane is an acceptable barrier, the irregular shape of the pollen grain and tube induces spatial deformations on the electrical field formed, which cannot be accurately described over the physical distance imposed by the cellophane thickness. Therefore their qualitative and quantitative description of the electric field generated by pollen tubes was not totally accurate (Feijó et al. 1994, 1995). A typical result obtained with an updated two-dimensional probe is shown in Fig. 17.4. In this case, pollen grains were attached to the substrate with silicone grease, allowing direct access to the perpendicular plane of the cell and much better accuracy in resolving the functional domains of the membrane. The current source is clearly located on the grain, and all the current flows into the tube. However the tip was shown to always drive larger currents than the rest of the tube, a feature not observable over a membrane covered with cellophane, due to field spatial distortion. Furthermore absolute values of the current are also one order of magnitude higher than previously described. The ionic basis of these currents was first investigated by substitution experiments (Weisenseel and Jaffe 1976) and important contributions were assigned to an outward proton current in the grain and a inward potassium current in the tube. The use of a calcium specific probe showed, however, in apparent contradiction with these results, that the calcium influx at the tip alone should provide most of the tube inward current (Kühtreiber and Jaffe 1990). This result was later confirmed, as well as a relationship established between the extracellular calcium flux and the concentration of intracellular free calcium on the tip focused gradient that exists in growing pollen tubes (Pierson et al. 1994). Calcium influx at the tip was circumstantially related to the possible presence of stretch activated channels on the tip since gadolinium inhibited the apical specific current (Feijó et al. 1994, 1995 and in preparation). A direct relation between the tip electrical current and the growth rate (Feijó et al. 1994 and in preparation), later confirmed to be mainly due to calcium (Pierson et al. 1996), also corroborated the relationship between calcium and the growth process. Other ion fluxes have been analysed, and Fig. 17.7 displays a typical plot of the proton pattern of a long (>1 mm) growing pollen tube. In these cells, the growth process becomes distinctly oscillatory (Pierson et al. 1996) and it could be shown that the oscillations on growth are paralleled by oscillations on the extracellular calcium fluxes and the intracellular calcium concentration. Holdaway-Clarke et al. shown that the oscillations in growth are paralleled by oscillations in the extracellular calcium fluxes and the intracellular calcium concentration (Holdaway-Clarke et al. 1997 and chapter by Feijó JA in this book). This periodic behaviour is also perfectly obvious in the influx of protons displayed by the tip (see Fig. 17.7). Mapping of the apical flanks in the clear cap zone showed a reversal of the flux direction, creating a local short-circuit of protons which is reflected on the intracellular pH and may be of functional significance in the process of polarity establishment and growth. Lower parts of the tube showed more reversal of the proton flux, but at ranges not comparable to the ones displayed in the tip, and eventually related to respiration. Since both extracellular and intracellular calcium are related to the growth rate, this parameter may serve as an experimental synchronising clock, allowing precise phase determination to be performed (Holdaway-Clarke et al. 1997). Quite surprisingly, the results show that the extracellular flux is delayed by about 13 seconds after the intracellular calcium elevation. This unexpected result contradicts previous quantitative assessments that seemed to relate the calcium influx and the cytosolic free calcium through the activity of membrane channels (Pierson et al. 1994). If confirmed in other systems and experimental conditions, an even more profound consequence could come from this finding. A possible explanation for this result, for the time being, is the role of the cell wall as a calcium and proton buffer (see Holdaway-Clarke et al. 1997 for discussion and models). This would call for a full re-evaluation of all the data acquired with the vibrating probe in plant cells and, although it could open new avenues for the study of cell wall physiology, would certainly diminish the experimental potential of the vibrating probe to look at membrane phenomenon. Further confirmatory work is, therefore, urgent to validate these conclusions.

17.5 The Vibrating Probe Technology

Much of the current vibrating probe technology has been developed under the direction of Lionel Jaffe in the National Vibrating Probe Facility of the Marine Biological Laboratory at Woods Hole (which recently evolved to the BioCurrents Research Center). Development of the technology still takes place in Woods Hole and equipment diffusion is a part of the objectives of the Center. However, since most of the effort is done around the established equipment in the prototype form, other commercial alternatives, with their own research and development, have appeared to answer the demand on these systems both for biological and material sciences. Most of that which is described in the following sections will focus on these commercial systems.

Although the initial prototypes of the vibrating probe were significantly simpler (especially the voltage probe, which was built around a vibrator, a lock-in amplifier and a chart recorder!) most of the present configurations of vibrating voltage probes and vibrating ion-selective probes utilise a common platform made up of the following devices (Figs. 17.1 and 17.5): a central control computer (usually an IBM compatible PC), 16 bit analog to digital converter board (e.g. CIO-DAS1602/16, Computer Boards Inc., Mansfield, MA, USA), video color framegrabber board (e.g. ComputerEyes/PCI, Digital Vision Inc., Dedham, MA, USA), color or black and white CCD video camera, a microscope (inverted or upright) , vibration isolated table (with solid metal Faraday cage for the ion-selective probe) and some software platform to control motion and acquire data. In the first prototypes of the ion-selective probe the motion was accomplished using piezoelectric devices; however, in all present configurations, the probe is moved from point to point as well as stepped from one position to another by means of a three dimensional, stepper motor driven micromanipulator. The stepper motors are usually driven by a four-axis (one axis to focus microscope) microstepping motor controller unit. In the above configuration, the three dimensional motorized micromanipulator and motor controller combined can have as low as a 50 nm/step rate providing sub-micrometer accuracy. Since these techniques employ a move-wait-measure protocol, stepper motors have been found to be quite suitable for these applications.

17.5.1
The SVET (Scanning Vibrating Electrode Technique)

This commercially available system was originally designed in 1989. The SVET system is based upon the design of Scheffey (1988). Presently, this system consists of two identical units (one dimensional vibrating voltage probe systems) merged into one (see Fig. 17.1). The vibrator linkage is simultaneously driven by two different signals from two separate sine wave oscillators with adjustable frequency ranges of 100 Hz to 1.0 kHz. This moves the vibrating probe in two separate directions which outlines a square pattern of vibration. On one possible mechanical linkage, an "Inverted T" type with two PZT5H piezoelectric bimorphs (Vernitron Inc., Bedford, OH, USA) vibrates in one vertical direction and one horizontal direction for investigating flat samples such as skin or corroding metal coupons. Another type of linkage called a "pi" linkage (see Fig. 17.3) is used to vibrate in two horizontal directions for other specimen types (Scheffey 1988). Though the "pi" linkage is also capable of being used in a vertical configuration, the "T" linkage was specifically designed for scanning corrosion samples due to their size and the need to scan large areas. It may, however, also be used for epithelial studies (Marcus and Shipley 1994; Nagel et al. 1997). Besides these specific cases, the "pi" configuration is the one that fits most of the applications for biological systems.

The low noise preamplifier is AC coupled to eliminate electrode drift. It operates in the 100 Hz to 1 kHz frequency domain, which permits AC coupling due to the nature of the wire microelectrode detection characteristics. It also has a F.E.T. input stage to give it a very high input resistance (see Scheffey 1988 for preamplifier specifications). The preamplifier unit amplifies the signal by a factor of 100. The signal is then fed into another amplifier in the main chassis that filters out signals above 10 kHz. To lock-in the signal, it is then fed to phase sensitive detectors (PSDs) along with a reference signal from the oscillators driving the vibrating electrode. If the electrode's signal frequency is the same frequency as the reference signal from the oscillator driving the electrode's vibration then it is detected. The system has also phase adjustments to compensate for small phase lags. These lags between the electrode's signal and the reference signal from the oscillator are mostly due to mechanical vibrations of the electrode and small electronic phase shifts. There are two phase sensitive detectors (PSD) in the system each having an **in-phase** and **quadrature** output. The **in-phase** PSD output is intended to produce a signal (DC voltage representing peak to peak electrode signal) when properly phased, while detecting an electric field. The **quadrature** PSD output is intended to be zero or minimum when an electric field is detected. If the system is properly phased a quadrature PSD output indicates an artefact. Total system gain is 50,000 (1 µV peak to peak at the electrode yields 50 mV DC at the PSD output).

17.5.2
The SIET (Scanning Ion-Selective Electrode Technique)

With most ion-selective electrodes, movement frequencies above 0.1 Hz are needed to reject noise from electrode drift. To measure microvolt changes and determine ion-selective electrode efficiency at such low frequency, the electrode should be directly connected to the system electronics. An improved vibrating ion-selective probe amplifier was designed in 1995 by Applicable Electronics that DC couples the ion-selective electrode signal to the electronics without saturating the system's dynamic range. There are a number of important advantages in using DC coupling (see discussion above § 3.1). In short, in the low frequencies necessary to measure ion gradients the hardware does not limit or distort the microvolt gradient signal. Actually, there is no low end frequency limit to the time interval of measurement. This allows for measurement of the microvolt DC drift of the ion-selective electrode over time as well as the response time of the ionophore being used.

To perform an efficient DC-coupling, the Nernst potential of the ion-selective electrode must be subtracted from the gradient's microvolt signal to keep the electronics from saturating. Therefore, the Nernst potential must be cancelled with a DC offset voltage of equal and opposite magnitude. An easy solution to this is to sample the electrode's Nernst potential via an Analog to Digital converter then send this voltage to a Digital to Analog converter for output back to the ion-selective electrode amplifier (see Fig. 17.5). By using a digital sample and hold scheme, the Nernst potential on the ion-selective electrode can be subtracted from the changing microvolt gradient signal by means of a differential amplifier. The plus (+) input of the differential amplifier being the changing microvolt ion gradient signal and the minus (−) input being the ion-selective electrode's Nernst potential. This allows auto-zeroing of the electronics to keep the signal output within the dynamic range of the A to D board. This subtraction voltage is changed on software command (typically once after the ion-selective electrode is placed in the experimental bath) rather than by an offset potentiometer, which adds one more control to keep track of accurately. The computer can then record the Nernst offset value for future reference.

Combined with the ASET software, the best frequency of stepping can be operator programmed for the ion electrode and the particular ionophore being used and the intensity of the gradient being measured.

17.5.3
The Software: ASET

The ASET (Automated Scanning Electrode Techniques) software (Science Wares, 30 Sandwich Rd., Falmouth MA, 0253 and Applicable Electronics) was written in Visual C++ for WINDOWS. This software package provides a powerful interface to control these non-invasive techniques. Programmable grid and vector scans are available as well as the ability to measure multiple ions simultaneously. The SVET (vibrating voltage electrode) provides two dimensional information with a maximum speed of 50 milliseconds per measurement point¡. The SIET (ion-selective electrode) software provide programmable move, wait, and measure time intervals. Two separate ion-selective electrodes can be measured simultaneously. It is also possible to control two separate measurement systems to allow alternate recording of the net field with the SVET followed by specific ions making up that field with the SIET. To do this, the software is able to control two separate motion control systems in the same program environment.

Typically, the software measures and records the ion-selective probe signal(s) without any elaborate averaging routines, although on-line averaging of data is also programmable. The important point about this technique is to minimise data collection time and maximise system sensitivity. Data filtering is best done after data is collected utilising commercially available software (worksheet or math interfaces) written for such purposes and many data analysis software packages can be easily adapted for that purpose.

Routines for video frame storage and graphic data overlays are also available. The software has the advantage of automatic data collection via computer without the need for an on-line operator if the system being investigated allows for such. Three dimensional ion-selective probe scans are also available which eliminate the need for keeping the probe vibration angle normal to the surface of the measured sample. However, as in the case of growing pollen tubes, routines are available to "chase" the growing pollen tube tip with the ion-selective electrode at different angles and step rates depending on growth rates.

An additional development has been the design of a **vibrating polarographic probe.** This system utilizes a current amplifier head stage attached to the ion-selective probe main amplifier. This system will accommodate commercially available Clark type and cathode type dissolved oxygen microelectrodes (Diamond General Corp., Ann Arbor, MI, USA) as well as gold plated stainless steel and platinum/iridium electrodes as used with the vibrating voltage probe. The ASET software controls this technique called the SPET (Scanning Polarographic Electrode Technique). In addition to previously mentioned modes of operation routines are also available for vertical scanning of the probe as used in the case of dissolved oxygen profiles in mud sediments. The system will measure small changes in dissolved oxygen in aqueous media. Electrode polarisation is provided by hardware adjustment or by D/A control via a PC computer. Other polarographic electrodes can also be used with the system. Experience has shown that frequencies of measurement greater than 1 Hz are possible with this additional technique due to the dynamics of the oxygen reduction reaction which takes place at the gold/solution interface.

17.6 Conclusions

It is now over 20 years since the vibrating probe technology was re-discovered in its biological applications. Yet, looking at general textbooks and trying to discern its true impact, one would probably underestimate its potential since it has failed to generate a body of literature one cannot by-pass in the general context of biology. Some of the reasons for this apparent lack of popularity may stem from a generalised belief that a real hard technical background and profound knowledge on electronics, computer programming or biophysics is needed. Although basic knowledge in these areas may prove to be useful, methods, designs and software packages are described that effectively bridge the gap between the prototype development and the user-friendly utilisation, requiring no more than the comprehension of the physical principles described. Other criticisms have to do with the kind of data produced by the vibrating probe. As an electrophysiological approach it cannot be compared to other techniques, such as patch-clamp or intracellular electrodes with respect to temporal and spatial resolution or quantitative accuracy. In plant cells in general, and pollen grains and tubes, in particular, there is fecund ground for this technique to impose itself as a true standard since there is little use for traditional electrophysiological approaches in normal living conditions. The vibrating probe is perhaps the most non-invasive, non-interactive technique available for studying living systems provided that a good enough signal-to noise ratio is attainable in good experimental conditions. It provides real-time, beautifully clear results for a number of highly significant membrane phenomena. Such is the case for pollen tubes, roots and root hairs and possibly many other systems yet to be described, which are natural dipoles driving significant fluxes and growing in less demanding ionic millieu. In particular the ion-selective technique needs standardisation of methods, calibration and electrode efficiency, which are crucial for true quantitative use and comparison of results. These developments will provide more accurate data collection which may establish this technique as a major tool for membrane research in cell biology and development.

Acknowledgements
We thank C. Scheffey, W. Nagel and R. Malhó for critical reading of the manuscript, Joe Kunkel for stimulating discussion and Sofia Cordeiro for technical assistance. JAF has been funded by the Fundação Luso-Americana para o Desenvolvimento (FLAD) and Fundação Calouste Gulbenkian (FCG). Fig. 4 was obtained during a short stay by J. A. F. in the NVPF at the MBL in Woods Hole, for which Dr. Lionel Jaffe is gratefully acknowledged.

References

Allen NS, Bennet MN, Cox DN, Shipley A, Ehrhardt DW, Long SR (1994) Effects of nod factors on alfalfa root hair Ca^{++} and H^+ currents and on cytoskeletal behavior. In: Daniels MJ et al (eds) Advances in Molecular Genetics of Plant-Microbe Interactions, vol. 3, Kluwer Acad Pub, pp 107–113

Amman D (1986) Ion-selective microelectrodes: Principles, design and application. Springer-Verlag, Berlin Heidelberg New York

Bluh O, Scott BIH (1950) Vibrating probe electrometer for the measurement of bioelectric potentials. Rev Sci Inst 10:867–868

Davies WP (1966) Membrane potential and resistance of perfused skeletal muscle fibers with control of membrane current. Fed Proc 25:332 (abstract 801)

DeBoer AH, Van Duijn B, Giesberg P, Wegner L, Obermeyer G, Kohler K, Linz KW (1994) Laser microsurgery: a versatile tool in plant (electro) physiology. Protoplasma 178:1–10

DeLoof A (1986) The electric dimension of cells: the cell as a miniature electrophoresis chamber. Int Rev Cytol 104:251–351

Demarest JR, Morgan JLM (1995). Effect of pH buffers on proton secretion from gastric oxyntic cells measured with vibrating ion-selective microelectrodes. Biol Bull 189:197–198

Feijó JA, Shipley AM, Jaffe LF (1994) Spatial and temporal patterns of electric and ionic currents around in vitro germinating pollen of lilly: a vibrating probe study. In: Heberle-Bohrs E, Vicente O (eds) Frontiers on Sexual Plant Reproduction. University of Viena. Viena, p 40

Feijó JA, Malhó RM, Obermeyer G (1995) Ion dynamics and its possible role during in vitro pollen germination and tube growth. Protoplasma 187:155–167

Felle HH, Hepler PK (1997) The cytosolic Ca^{2+} concentration gradient of Sinapis alba root hairs as revealed by Ca^{2+}-selective microelectrode tests and fura-dextran ratio imaging. Plant Physiol 114: 39–45

Ferrier JM, Lucas WJ (1986) Theory of Ion transport and the vibrating probe. In: Nuccitelli R (ed) Ionic Currents in Development. Alan Liss, New York, pp 45–51

Harold FM, Caldwell JH (1990) Tips and currents: Electrobiology of apical growth. In: Heath TH (ed) Tip Growth in Plant and Fungal Cells. Academic Press, NY, pp 59–89

Holdaway-Clarke TL, Feijó JA, Hackett GA, Kunkel JG, Hepler PK (1997) Pollen tube growth and the intracellular-cytosolic calcium gradient oscillate in phase while extracellular calcium influx is delayed. Plant Cell 9:1999–2010

Huang JW, Grunes DL, Kochian LV (1992) Aluminum effects on the kinetics of calcium uptake into cells of the wheat root apex. Planta 188:414–421
Hush JM, Overall RL, Newman IA (1991) A calcium influx precedes organogenesis in Graptopetalum. Plant, Cell Environ 14:657–665
Hush JM, Newman IA, Overall RL (1992) Utilization of the vibrating probe and ion-selective microelectrode techniques to investigate electrophysiological responses to wounding in pea roots. J Exp Bot 43:1251–1257
Isaacs HS (1986) Applications of current measurement over corroiding metallic surfaces. Progr Clin Res 210:37–44
Isaacs HS, Davenport AJ, Shipley A (1991) The electrochemical response of steel to addition of dissolved cerium. J Electrochem Soc 138:390–393
Jaffe LF (1981) The role of ionic currents in establishing developmental pattern. Phil Trans R Soc Lond B 295:553–566
Jaffe LF (1985) Extracellular current measurements with a vibrating probe. TINS 8:517–521
Jaffe LF, Levy S (1987) Calcium gradients measured with a vibrating calcium-selective electrode. IEEE (Inst Elect Elect Eng Med Biol Soc) Conf 9:779–781
Jaffe LF, Nucitelli R (1974). An ultrasensitive vibrating probe for measuring steady extracellular current. J Cell Biol 63:614–628
Jones DL, Shaff JE, Kochian LV (1995) Role of calcium and other ions in directing root hair tip growth in *Limnobium stonoloniferum*. Planta. 197:672–680
Kochian LV, Shaff JE, Kühtreiber WM, Jaffe LF, Lucas WJ (1992) Use of an extracellular, ion-selective, vibrating microelectrode system for the quantification of K^+, H^+, and Ca^{2+} fluxes in maize roots and maize suspension cells. Planta 188:601–610
Kühtreiber WM, Jaffe LF (1990) Detection of extracellular calcium gradients with a calcium-specific vibrating electrode. J Cell Biol 110:1565–1573
Marcus DC, Shipley AM (1994) Potassium secretion by vestibular dark cell epithelium demonstrated by vibrating probe. Biophysical J 66:1939–1942
Miller AL, Gow NAR (1988) Correlation between profile of ion-current and root development. Physiol Plantarum 75:102–108
Miller AL, Gow NAR (1989) Correlation between root-genrated ionic currents, pH, fusicoccin, indoleacetic acid and growth of the primary root of *Zea mays*. Plant Physiol 89:1198–1206
Miller AL, Raven JA, Sprent JI, Weisenseel MH (1985) Endogenous ion currents traverse growing roots and root hairs of Trifolium repens. Plant Cell Environment 9:79–83
Miller AL, Shand E, Gow NAR (1988) Ion currents associated with root tips, emerging laterals and induced wound sites in *Nicotiana tabacum*: spatial relationship proposed between resulting electrical fields and phytophtoran zoospore infection. Plant Cell Environ 11:21–25
Nagel W, Somieski P, Shipley AM (1998) Mitochondia-rich cells and voltage-activated chloride current in toad skin epithelium: Analysis with the scanning vibrating electrode technique. J Memb Biol (in press)
Nucitelli R (ed)(1986) Ionic Currents in Development. Alan Liss, New York
Nucitelli R (1990) Vibrating probe technique or studies of ion transport. In: Foskett JK, Grinstein S (eds) Noninvasive Techniques in Cell Biology. Wiley-Liss, NewYork, pp 273–310
Pierson ES, Miller DD, Callaham DA, Shipley AM, Rivers BA, Cresti M, Hepler PK (1994) Pollen tube growth is coupled to the extracellular calcium ion flux and the intracellular calcium gradient: Effect of BAPTA-type buffers and hypertonic media. Plant Cell 6:1815–1828
Pierson ES, Miller D, Callaham D, Van Aken J, Hackett G, Hepler PK (1996). Tip-localized calcium entry fluctuates during pollen tube growth. Dev Biol 174:160–173
Rathore KS, Hodges TK, Robinson KR (1988) Ionic basis of currents in somatic embryos of Daucus carota. Planta. 175:280–289
Scheffey C (1986) Electric fields and the vibrating probe for the uninitiated. In: Nucitelli R (ed) Ionic Currents in Development. Alan Liss, New York, pp xxv–xxxvii
Scheffey C (1988) Two approaches to construction of vibrating probes for electrical field measurements. Rev Sci Inst 59:787–792
Scheffey C, Foskett JK, Machen TE (1983) Localization of ionic pathways in the teleost opercular membrane by extracellular recording with a vibrating probe. J Membrane Biol 75:193–203
Schiefelbein JW, Shipley A, Rowse P (1992). Calcium influx at the tip of growing root-hair cells of *Arabidopsis thaliana*. Planta 187:455–459
Smith PJS (1995) The non-invasive probes – tools for measuring transmembrane ion flux. Nature 378: 645–646
Smith PJS, Sanger R, Jaffe L (1994) The Vibrating Ca^{++} electrode: A new technique for detecting plasma membrane regions of Ca^{++} influx and efflux. Meth Cell Bio 40:115–134
Vaughn-Jones RD, Aickin CC (1987) Ion-selective microelectrodes. In: Gray P, Standen N, Whittaker M (eds) Microelectrode Techniques for Cell Physiology. Company of Biologists, Cambridge (UK), pp 137–167
Weisenseel MH, Becker HF, Ehlgotz JG (1992) Growth, gravitropism, and endogenous ion currents of cress roots (*Lepidium sativum* L.). Plant Physiol 100:16–25

Weisenseel MH, Dorn A, Jaffe LF (1979) Natural H^+ currents traverse growing roots and root hairs of barley (*Hordeum vulgare* L.). Plant Physiol. 64:512–518
Weisenseel MH, Jaffe LF (1976) The major growth current through lily pollen tubes enters as K^+ and leaves as H^+. Planta 133:1–7
Weisenseel MH, Kicherer RM (1981) Ionic currents as control mechanism in cytomorphogenesis. In: Kermeyer O (ed) Cytomorphogenesis in Plants, vol. 8. Springer-Verlag, Berlin, Heidelberg, New York, pp 379–399
Weisenseel MH, Nucitelli R, Jaffe LF (1975) Large electrical currents traverse growing pollen tubes. J Cell Biol 66:556–567
Zissman WA (1932) A new method of measuring contact potential differences in metals. Rev Sci Inst 7: 367–370

CHAPTER 18

The Role of Calcium and Associated Proteins in Tip Growth and Orientation

R. Malhó

Dept of Plant Biology, FCL, Block C2, University of Lisbon, Campo Grande, Portugal
e-mail: r.malho@fc.ul.pt
telephone: 351 1 7500069
fax: 351 1 7500048

18.1 $[Ca^{2+}]_c$ in Pollen Tube Growth and Reorientation

The primary function of the pollen tube is to convey the sperm cells to the embryo sac. Pollen germination is usually rapid and pollen tubes can grow at rates close to 1 $cm \cdot hr^{-1}$ and over distances as long as 50 cm. During their progress through the style, ovary, micropyle and finally through the nucellus, pollen tubes ***undergo many necessary changes in direction*** which result from signalling by poorly-understood mechanical, electrical and chemical cues. Their behaviour appears strikingly purposive! But the specific morphogenetic and polarising mechanisms which provide the pollen tube with these extraordinary attributes remain very much a mystery. Without them, however, fertilisation and the subsequent crucial events which lead to seed formation would never happen.

In pollen tubes, growth is tightly associated with a tip-focused gradient of $[Ca^{2+}]_c$ that drops from 2 mM to 200 nM over the first 10 to 20 μm from the tip (Figure 18.1A; Pierson et al., 1996). Dissipation of this gradient either by raising (Franklin-Tong et al., 1993; Malhó et al., 1994) or lowering (Pierson et al., 1994) $[Ca^{2+}]_c$ results in the inhibition of tube growth (Figure 18.1B). When the pollen tubes recover and resume growth, they do so in a random direction. However, when the cells recover, **the new site of growth direction is preceded by an increase in Ca^{2+} in the same region** (e.g. figure 18.15 of Pierson et al., 1996 and figure 18.1 of Malhó et al., 1995). The data gathered so far thus indicates that before reorientation takes place, the new site of growth always experiences an increase of Ca^{2+}, either from a resting level to the generation of a new gradient (in the case of a recovering pollen tube) or from a high level to an even higher level (in the case of a growing tube). To the best of my knowledge, there is no data showing that in a growing tube a localized decrease in Ca^{2+} promotes bending towards the zone of less Ca^{2+}. In fact, our data with Diazo-2 (a caged Ca^{2+} chelator) shows exactly the opposite (Malhó and Trewavas, 1996).

The gradient results from an apical influx of Ca^{2+} through putative Ca^{2+} channels (Figure 18.1C) as suggested by experiments with the Mn^{2+} quenching technique (Malhó et al., 1995) and with a Ca^{2+}-sensitive probe (see Feijó, *this book*). Several lines of evidence, including patch-clamp experiments in fungal hyphae (Garrill et al., 1993) suggest these channels to be stretch-activated. However, it must be noted that definitive presence of these channels is still lacking. In the shaft of the tube Ca^{2+}-ATPases pump the Ca^{2+} out of the cytosol (see Obermeyer, *this book*). Recovery is concurrent with the formation of a tip gradient of Ca^{2+} channel activity (Figure 18.1D). This activity induces a localized hot spot of $[Ca^{2+}]_c$ that predicts the new apex and the new direction of

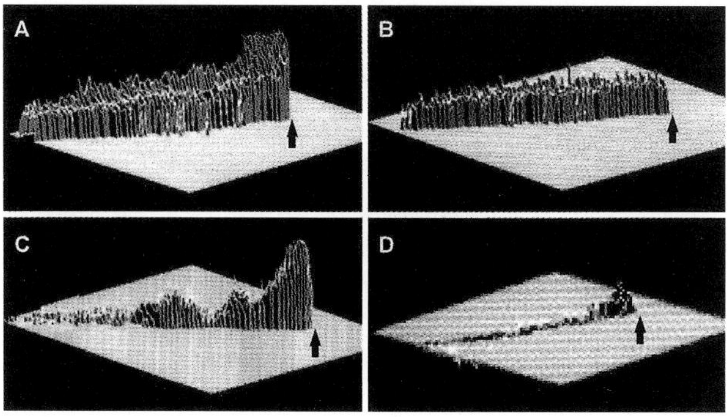

Fig. 18.1. Imaging $[Ca^{2+}]_c$ and putative Ca^{2+} channel activity with Indo-1 and the Mn^{2+} quenching technique in growing pollen tubes of *Agapanthus umbellatus* using wide-field microscopy. Each graph represents a Ymod plot of fluorescence intensity from the first 50 µm of one pollen tube. In each case the tip of the cell is indicated by an arrow. **(A)** Ratio imaging of Indo-1 in a growing tube revealing the tip-focused $[Ca^{2+}]_c$ gradient. **(B)** Ratio imaging of Indo-1 in a non growing tube. The tip-focused $[Ca^{2+}]_c$ gradient was replaced by a uniform distribution. **(C)** Putative Ca^{2+} channel activity in a growing tube revealed after Mn^{2+} quenching of Indo-1. The activity is spatially correlated with the tip-focused $[Ca^{2+}]_c$ gradient. **(D)** Similar to C but in a swelling tube immediately before growth resumes. The emergence of channel activity is concomitant with the reestablishment of the $[Ca^{2+}]_c$ gradient

growth. Thus, the growth of pollen tubes is strongly polarized and is underpinned by the spatial separation of channels and pumps. However, the gradient can also be modified, either by the localized photolysis of caged Ca^{2+} on one side of the apical dome or by the application of ionophores or channel blockers via a microelectrode applied extracellularly near the tip (Malhó and Trewavas, 1996). Different clusters of Ca^{2+} channels in the pollen tube tip may separately regulate orientation and growth. In these cases and others, the tube bends toward the zone of higher $[Ca^{2+}]_c$, which seems to be maintained by Ca^{2+}-induced Ca^{2+}-release (CICR) while the tube is reorienting.

18.1.1 $[Ca^{2+}]_c$ Oscillations and $[Ca^{2+}]_c$ Waves

The gradient of $[Ca^{2+}]_c$ at the pollen tube tip oscillates quantitatively with a frequency of several minutes and tube growth can exhibit a similar oscillatory variation (Pierson et al., 1996). Oscillations add a new dimension to $[Ca^{2+}]_c$ signaling because the frequencies of the resulting digital signals can be sensed, thereby providing a rich source of complex information to the cell. Oscillations may originate from a continual filling and emptying of intracellular stores of $[Ca^{2+}]_c$. In animal cells, the endoplasmic reticulum is the primary store and is described as acting like a capacitor; hence the term 'capacitative Ca^{2+} signalling' to describe $[Ca^{2+}]_c$ oscillations (Berridge, 1995). The pollen tube tip contains smooth endoplasmic reticulum and it is possible that the endoplasmic reticulum performs an equivalent capacitative function here (see §3.3).

An interesting feature of the $[Ca^{2+}]_c$ oscillations in pollen tubes is that they are confined to the apical region and mechanisms must exist to ensure this. Some hypotheses, not mutually exclusive, are (1) negative feed-back by excessive Ca^{2+}, (2) inhibition of $Ins(1,4,5)P_3$ receptors and (3) regulation by protein kinases (see sections below). In addition, oscillations of $[Ca^{2+}]_c$ may be coupled to a noise-induced stochastic opening of

Ca^{2+} channels. In the presence of weak signals, transduction thresholds may then be periodically approached but not exceeded. Stochastic resonance results when noise increases yet further. More channels will open, thus enabling the critical signaling threshold to be stochastically and frequently crossed. Protein kinase cascades that regulate channel opening may also be used to amplify stochastic resonance. In this way, the perception of very weak signals – such as reorienting stimulus – in a noisy environment is made possible.

In animal cells, Ca^{2+} waves are known to regulate meiosis, mitosis, fertilization and secretion (Bock and Ackrill, 1995), and the documentation suggests that Ca^{2+} may regulate these important processes in plants as well. In pollen tubes, Ca^{2+} waves were observed either after the *in vitro* addition of self-incompatibility glycoproteins (Franklin-Tong et al., 1993) or induced after localized photolysis of caged compounds (Franklin-Tong et al., 1996; Malhó and Trewavas, 1996). These Ca^{2+} waves, which travelled at speeds similar to slow Ca^{2+} waves in other organisms (Jaffe, 1993), were observed to trigger the reorientation of growth in *Agapanthus* pollen tubes and the cessation of tube growth in poppy as they moved toward the tip. The Ca^{2+} wave is clearly carrying information, which may be transduced via an IP_3-based relay (see §3.1). However, it must be noted that these waves were only observed under artificial conditions and one can question their significance for *in vivo* growth. Nevertheless, they do convey a physiological response which is already an indication of their importance.

18.1.2
Ca^{2+} Measurements in Tip Growing Cells; the Limits of Technology

18.1.2.1
Imaging Ca^{2+} Dynamics with Ca^{2+}-Specific Dyes

Ca^{2+} research has benefited largely from the appearance of sensitive fluorescent dyes (Grynkiewicz et al., 1985). These dyes, usually described as single- (SW) or dual-wavelength (DW), can be imaged by a variety of microscopy techniques (Read et al., 1992) providing measurements of intracellular Ca^{2+} concentration and/or dynamics in living cells. SW-dyes (like Fluo-3 and Calcium Green-1) have one major excitation/emission peak at which the fluorescence emitted is proportional to the ion concentration thus being suitable to measure relative changes in cytosolic free calcium ($[Ca^{2+}]_c$); SW dyes are excited by visible light which accounts for their popularity among laser confocal (CSLM) users. However, due to intracellular asymmetries in viscosity, dye concentration and other problems like photobleaching (all affecting the emission intensity of the dye) SW dyes cannot be used for absolute quantification of $[Ca^{2+}]_c$. For that purpose, there are UV excited DW-dyes (such as Fura-2 and Indo-1) which exhibit spectral shifts upon ion binding and can be used for more precise ratiometric quantifications of $[Ca^{2+}]_c$. An alternative procedure to do ratio is the simultaneous use of 2 dyes, e.g. a Ca^{2+}-insensitive dye (acting as a volume indicator) which could be ratioed against a Ca^{2+}-sensitive dye. Unfortunately, this is another technique prone to many artefacts and whose efficiency varies greatly with the biological system. We tried this technique using Calcium Green-1 against Fura-Red and the typical tip-focused gradient was not observed (Figure 18.2C). However, the bleaching/leaking rate of the two dyes was not the same so a proper ratio/calibration could not be obtained and the technique was abandoned. More recently (unpublished data) we have tried a dye mixture of Oregon Green BAPTA-1 and SNARF-1. These 2 dyes have a more stable and similar behaviour in the cytosol so a suit-

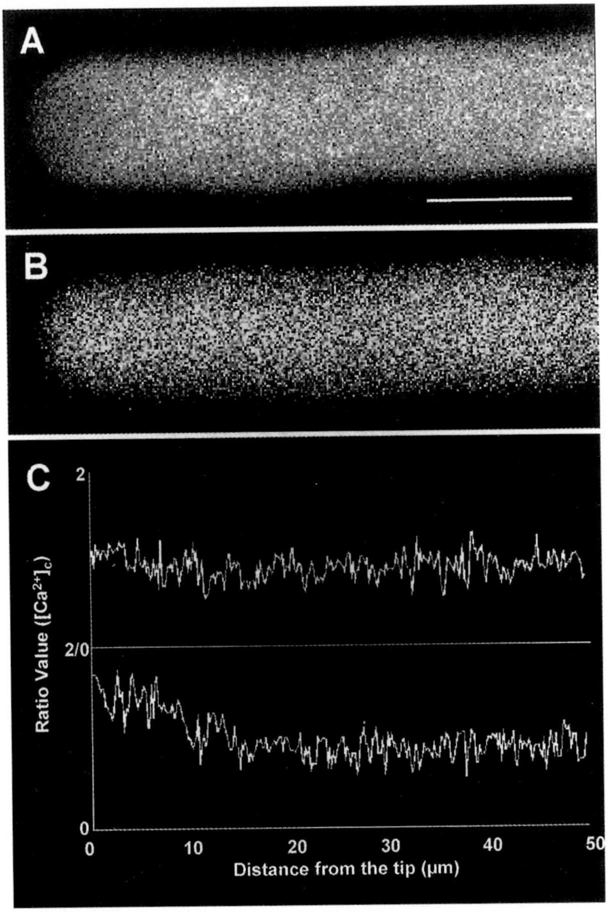

Fig. 18.2. Imaging Ca^{2+} dyes and $[Ca^{2+}]_c$ in growing pollen tubes of *Agapanthus umbellatus* using CSLM. Bar = 10 μm. **(A–B)** Unprocessed image of dye distribution in tubes loaded with Calcium-Green 1 free acid **(A)** and Calcium Green 1 dextran **(B)**. The signal emitted by the "clear cap" is about half the signal emitted by the shank of the tube. **(C)** Ratio plots of $[Ca^{2+}]_c$ in the first 50 μm of tubes loaded with Oregon Green Bapta 1/SNARF-1 (upper part) or Calcium Green-1/Fura-red (lower part). Ratio values (corresponding to $[Ca^{2+}]_c$) are given in a 0-2 scale. Each trace represents the average pixel intensity of 5 midlines transects down the length of the apical 50 μm of one pollen tube

able ratio can be obtained. The cells were grown under the conditions described by Parton et al. (1997) where a pronounced pH gradient was not observed; thus SNARF-1 can be used as volume indicator (Read et al., 1992). A slightly oscillating higher ratio was detected in the tube tip but of reduced steepness (Figure 18.2C). This may result from the lower dynamic range of Oregon Green when compared with similar dyes such as Fluo-3.

Ratio methods should be used whenever a quantification of $[Ca^{2+}]_c$ is essential. However, this is not always the case and SW dyes have some advantages over DW dyes which make them useful reporters of changes in $[Ca^{2+}]_c$: (1) **they provide a much higher quantum yield** hence reduced cytotoxicity. Since a good signal/noise ratio is needed to make proper measurements, one has to load the cells with high concentration of dyes (~1–50 μM) which may act as buffers. The buffering effect of the dyes may not be enough to perturb large vital processes such as the $[Ca^{2+}]_c$ gradient in pollen tubes, but enough to mask small $[Ca^{2+}]_c$ changes. Blumenfeld et al. (1992) analyzed this possibility and found that free fura-2 diminished the amplitude of $[Ca^{2+}]_c$ transients, increased the Ca^{2+} diffusion rate and the cells time response without totally compromising its viability. The use of dextran coupled dyes may partially overcome this problem since they have a much lower mobility. However, if very high concentrations are used the opposite effect may happen with a decrease in diffusion rate; in the particular case of pollen tubes this could result in a steeper gradient. Since SW-dyes are brighter, less dye is needed inside the cell which means less photodamage and higher experimental temporal resolution; time courses of 5–10 sec interval can be routinely achieved when Calcium Green-1 is the single dye introduced in the cell (Malhó and Trewavas, 1996). Reducing the dye's concentration also means reducing the viscosity inside the microinjection needle (the most usual loading method). This may significantly increase the easiness of the microinjection procedure, something that many researchers (the author at the top of the list) will consider a blessing; **(2) higher stability and reduced photobleaching,** allowing more time for experimentation. The use of dextran-conjugates is particularly meaningful in this case since they show less compartmentation into organelles and no leakage from the cell allowing measurements up to several hours; (3) **higher K_d, meaning sensitivity to a wider range of ion concentration** (indicators have a detectable response in the concentration range from approximately $0.1 \times K_d$ to $10 \times K_d$). This is particularly relevant in tip growing cells with a $[Ca^{2+}]_c$ gradient where the high values in the apex may saturate the detector; **(4) excitation with visible light** which permits the simultaneous use of UV-photoactivated caged-probes. One of the major drawbacks of SW-dyes, the fact that a calibration is difficult to achieve, can be ameliorated by rationing the signal emitted by the cell at 2 different (but close) time intervals. The obtained ratio is an accurate indication of changes in ion concentration experienced by the cell in that period of time. In many cases this is probably the most meaningful information, since even with ratio dyes a perfect calibration is impossible to achieve. In a recent paper (Malhó and Trewavas, 1996) we used this SW-dye/caged-probe technology coupled to CSLM to study the effect of localized increases in $[Ca^{2+}]_c$ on pollen tube reorientation. With this "pseudo-ratio" approach, we reported that increasing $[Ca^{2+}]_c$ by flash photolysis of caged-Ca^{2+} in the left or right side of the apical dome promotes reorientation of the growth axis towards the side of Ca^{2+} release. Similar experiments were made in root hairs by Bibikova et al. (1997) but using a ratio method – Calcium Green-2 against rhodamine. Interestingly, the data obtained with this ratio procedure was quite similar to our work with Calcium Green-1. Despite the fact that non-ratio dyes preclude the visualization of the true Ca^{2+} distribution, both sets of data revealed (1) localized increases in Ca^{2+} of approximately the same magnitude upon cell reorientation, (2) similar duration of the Ca^{2+} transients recorded and (3) the existence of a maximum level to which the Ca^{2+} gradient can be manipulated without causing growth disruption.

18.1.2.2
Confocal or Wide-Field Detectors?

Apical growing cells are an extremely interesting model for studying spatial ion and molecular signaling. However, they do pose particular problems when it comes to imaging. These cells often have asymmetric organelle distribution and geometrical forms that results in uneven dye distribution and stresses the importance of using DW dyes and CSLM optical sectioning for ion imaging. However, the simultaneous use of the 2 techniques has not been easy. CSLM with UV lasers (needed to excite DW-dyes) only recently became available and are extremely expensive making their acquisition prohibitive to most labs. Furthermore, many biological systems do not tolerate well the high dose of UV radiation to which they are exposed during imaging. Thus DW-dyes have been mainly used with wide-field detectors, while in CSLM SW-dyes have been the dyes of choice.

In pollen tubes, imaging of DW dyes with wide-field detectors revealed a tip-focused Ca^{2+} gradient (Malhó et al., 1995; Miller et al., 1992). However, when SW dyes were used, the gradient was not visible. In fact, the signal emitted by the apical tip (also known as the clear cap), was usually lower than the signal emitted by the shank of the tube (Figure 18.2A-B; Pierson et al., 1996). Why is this? One of the suggestions was that this could arise from reduced path length, although the thin 2 μm slice we have used would only make this a problem in the first 0–1 μm of the image, not the first 10–15 μm where the gradient is observed. Furthermore, the lower signal is often coincident with the clear cap (it has the shape of an inverted cone) which cannot be explained by differences in path length. Other hypotheses associated with the technical characteristics of the CSLM are also unlikely. Misrepresentation due to spherical and chromatic aberrations is only a problem when imaging planes are buried deep in the sample (Opas, 1997) which is clearly not the case in single cell systems. Photodamage and dye saturation are two other problems in CSLM that can be easily controlled and do not account for this apparent discrepancy. CSLM have a lower optical transfer efficiency (due to the large number of filters and dichroic mirrors in the light path) than wide field systems (Sandinson et al., 1995), but the quantum yield of the dyes excited by visible light is far higher than their counterparts' used with UV. Our own experience with Indo-1, showed that the necessary concentration forproper imaging is 10 times the one used with Calcium Green-1 in the CSLM (~ 1–2 μM). Thus, it is unlikely that dye buffering is extensive to the point of masking a steep gradient (over an order of magnitude). On the other hand, plant cells have high autofluorescence so there is a hazard to using very low dye concentrations because the signal/noise ratio may get too low. A more simple explanation is that the amount of cytosol in the clear cap is lower due to the high density of exocytotic vesicles. Another aspect, which is rarely mentioned, is the fact that the large number of vesicles (and thus membranes) could confer on the cytosol particular characteristics and therefore change the dye's emission properties. Blumenfeld et al. (1992) compared a biophysical and an experimental model and reported that $[Ca^{2+}]_c$ just under the membrane surfaces cannot be accurately recorded with fura-2.

Recent advances in technology have made it possible to use ratiometric techniques in CSLM which do not preclude the simultaneous use of caged-probes. Calcium Green/ Texas-Red 70 kDa dextran conjugate is a new probe for Ca^{2+} imaging which does not need UV light. Instead, the 2 fluorophores of the dye can be excited simultaneously with a Kr-Ar laser at 488 nm and 568 nm respectively. However, the use of this dye has so far been impaired by the difficulties of pressure injecting apical growing cells with sufficient amounts of dye to image without perturbing growth. Furthermore, Bibikova et al. (1997) were unable to obtain high quality ratiometric images using this dye. This may be due to the fact that *in vitro* tests have revealed an extremely low dynamic range for this dye (only about 3-fold increase in fluorescence intensity). Since most of the dyes show a further reduction of their dynamic range once inside the cells, the usefulness of this dextran conjugate may be also quite reduced. In that case we should use other techniques as long as the right controls are performed and the data are carefully and critically analyzed. But it is also important to realize that even the use of a ratiometric procedure is not free of errors. In a recent study (Parton et al., 1997) using confocal ratio imaging, it was shown that if the collected signal falls below a certain limit, the ratio no longer exhibits a linear response and artefactual gradients may appear. These authors also reported that numerical data should be extracted from unprocessed images rather than from ratio images otherwise statistical inherent noise may arise in distorted values.

These considerations became recently more complicated with the observation of a pulsating Ca^{2+} hot spot in the apex of pollen tubes using Calcium Green-dextran (Messerli and Robinson, 1997). However, the images were collected with the confocal aperture completely open. In these conditions (and with the objective used), the optical section is >10 μm thick. Unfortunately, the authors do not say how many cells were analyzed, whether the hot spot was observed in all of them and whether it was still recorded with thin optical sections. Still, this paper raises an interesting question. Does the imaged $[Ca^{2+}]_c$ distribution change with the type of dye? If so, then what dye (if any) reveals the true $[Ca^{2+}]_c$ distribution? Studies on Ca^{2+} imaging with wide-field detectors in pollen tubes of different species and with different dyes (Indo-1 and Calcium Green free acid; Fura-2 dextran and free acid) revealed similar results (Malhó et al., 1994; Pierson et al., 1996): even with wide-field detectors, the dye fluorescence in the apical region is lower and only after ratio is the gradient visible. Similarly, it was found that AM-esters, free acid or dextran-conjugated dyes all revealed the same pH distribution in apical growing cells when CSLM was used (Parton et al., 1997; Fricker et al., 1997). The slight changes detected between these different examples can be due either to the biological sample or to the calibration procedure. A third hypothesis, which could help to explain the Calcium Green-dextran data, is that the dextran may somehow alter the dye properties. Dextran conjugates are generally assumed to be better because of their superior localization and retention within the cytosol. But they also have a lower dynamic range and a careful analysis of dextran versus non-dextran dyes is still lacking.

To the best of our knowledge there is no convincing imaging showing a tip-focused $[Ca^{2+}]_c$ gradient in apical growing cells using thin optical sections (≤2 μm), even with ratiometric procedures. A possible explanation may reside in the technical aspects of the 2 detectors. In wide-field systems, light from out-of-focus regions of the cell is also processed as in a 2D system. In CSLM images, only light coming from a median region of the cell is captured. What are the consequences of this fact? In pollen tubes it is known that the high $[Ca^{2+}]_c$ in the tip is maintained by influx through plasma membrane channels (Malhó et al., 1995); the ion concentration is therefore likely to be higher in the cell periphery decreasing exponentially towards the centre. If this is so, the magnitude of the gradient will reach its lowest intensity in the median sections of the cell, precisely the ones analyzed in CSLM.

It is clear that more work has to be done in order to fully understand the spatial distribution of $[Ca^{2+}]_c$. In the particular example of apical growing cells, it would be very interesting to compare, in the same species and using ratio dyes, the distribution of $[Ca^{2+}]_c$ in confocal versus wide-field systems, perhaps even with the help of deconvolution methods.

18.1.2.3
Ca^{2+} in Organelles

The high levels of relatively-immobile cytoplasmic Ca^{2+} buffers limit the range of Ca^{2+} diffusion to within a few microns, so transmission of the signal to different cell regions is achieved either by (1) diffusional distribution of the Ca^{2+} mobilizing second messenger or (2) regenerative Ca^{2+} release. Both mechanisms depend on the role of organelles as intracellular stores. Therefore, the subcellular distribution of intracellular Ca^{2+} stores can determine the spatial organization of the $[Ca^{2+}]_c$ signal (Thomas et al., 1997). This is particularly relevant in pollen tubes where the tip-high $[Ca^{2+}]_c$ spatially overlaps with the clear cap, a cellular region where large organelles are almost absent.

The ability of mitochondria to accumulate Ca^{2+} has been known for many years. Mitochondria have been shown to tune out sustained $[Ca^{2+}]_c$ signals (Hajnózcky et al., 1995) and can thus interpret the apical oscillations as well as "confine" them to that cell region. A synchronization of Ca^{2+} waves by mitochondria has also been demonstrated (Jouaville et al., 1995). In addition, Ca^{2+}-induced waves were initiated primarily in the nuclear/rough endoplasmic reticulum cellular locale, but not in the apical region, suggesting sites of particular sensitivity and/or differential regulation. This hypothesis finds support in our data (see § 2.1) which indicates that $Ins(1,4,5)P_3$ stores, probably located in the endoplasmic reticulum, participate in the transduction of signals to the shank of the tube. However, it must be noted that most of the evidence we have so far comes from indirect measurements. This is clearly an area that requires further attention before more significant conclusions can be drawn.

18.2 The Role of Calmodulin

Calmodulin is a small but highly conserved protein, ubiquitous in eukaryotic cells, and is considered to be the primary decoder of Ca^{2+} information (Vogel, 1994). Barley aleurone cells have Ca^{2+}/calmodulin as central elements in the control of secretory activity (Gilroy, 1996; Schuurink et al., 1996) and developmental pathways can be restored by microinjection of calmodulin into phytochrome-defective cells (Neuhaus et al., 1993). Love et al. (1997) recently reported calmodulin involvement in the process of axis fixation and formation in developing zygotes of *Fucus*, and in pollen tubes, it is known that calmodulin antagonists inhibit growth and dissipate the $[Ca^{2+}]_c$ gradient (Obermeyer and Weisenseel, 1991).

18.2.1 Imaging Calmodulin in Growing Cells

If calmodulin is involved in decoding the tip-focused $[Ca^{2+}]_c$ gradient, then a spatial distribution to match the tip-focused $[Ca^{2+}]_c$ gradient might be expected. Haußer et al. (1984) reported a tip-to-base calmodulin gradient in pollen tubes but the very-early methods employed are nowadays regarded as dubious. In a recent work (Moutinho et al., 1998) we imaged fluorescently labelled calmodulin (FITC-CaM) pressure microinjected into pollen tubes. Interestingly, we found that **calmodulin distributes evenly throughout the cell and does not form a gradient** (Figure 18.3A). If, however, the concentration of introduced exogenous calmodulin was raised, a V-shaped collar could be observed behind the apical region (Figure 18.3B). This specific type of localisation was observed only in growing cells and was abolished upon tube growth arrest. Imaging of FITC-dextran over similar periods of time and irradiation showed that photobleaching or pH changes are unlikely to explain this fact. It can be argued that this particular distribution is the result of a high concentration of exogenous calmodulin, leading to the formation of a pattern which is otherwise absent. Nevertheless, it is still worth recording the ability of an active calmodulin to associate with particular regions of the cell. Calmodulin functions in growing pollen tubes are necessarily dependent on the localisation of its targets and this data points most likely to some of those.

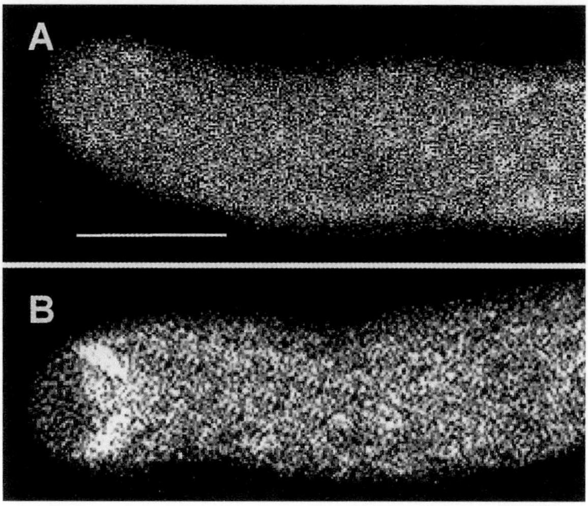

Fig. 18.3. Confocal imaging of growing pollen tubes previously loaded with FITC-CaM. Estimated concentration: 10–20 µM **(A)** and 30–50 µM **(B)**. Bar = 10 µm

These results contrast with other reports for calmodulin localisation either in pollen tubes or other tip growing cells (Hauber et al., 1984) which show a tip-associated accumulation of the protein, rather than a more diffuse pattern. In yeast it was observed that calmodulin concentrates at regions of cell growth (Brockerhoff and Davis, 1992; Sun et al., 1992) but these studies used immunological techniques which lead to cell death. In pollen tubes, the use of similar methods also resulted in a gradient of calmodulin (Vidali and Hepler, 1997) suggesting some type of artefact.

18.2.2 Distribution of Calmodulin mRNA

In eukaryotic cells certain mRNAs are spatially restricted to discrete regions within the cell, and this specific localisation appears to be a widely used mechanism for establishing gradients of proteins (Wilhelm and Vale, 1993; Bouget et al., 1996). The distribution of calmodulin mRNA was also analysed by microinjection of fluorescently labelled mRNA (Moutinho et al., 1998). It was found that upon microinjection the growth rates rapidly resumed and this was accompanied by a decrease in the levels of FITC-mRNA in the apex; the regions further back showed a uniform distribution. Such distribution is similar to the FITC-CaM suggesting a certain degree of co-localisation between the protein and its mRNA. Similar results were obtained in *Fucus* zygotes regarding the distribution of actin mRNA (Bouget et al., 1996).

18.2.3 Regulation of Tip Growth by Calmodulin

The pollen tube actomyosin system regulates the strong cytoplasmic streaming that translocates organelles during polarised growth. Although this system was generally believed to be more dense in the extreme apical region, the first study of actin distribution in living pollen tubes (Miller et al., 1996) showed a different picture. While numerous actin microfilaments arranged axially where present in the sub-apical region, the tip itself appeared essentially devoid of these elements. Moreover, the microfilament distribution observed in the sub-apical region of both freeze-substituted (Lancelle et al., 1997) and living pollen tubes resembles the V-shaped collar reported for calmodulin. This suggests a direct interaction between calmodulin and actin. The interaction of calmodulin with proteins from the cytoskeleton is very well documented in animal cells, specially in the case of myosin light-chain kinase (James et al., 1995). In yeasts, Myop2, an unconventional myosin required for polarised growth is a target of calmodulin at sites of cell growth (Brockerhoff et al., 1994). Several myosins (I, II and V) have already been identified in pollen tubes by immunological methods (Miller et al., 1995; Tang et al., 1989; Yokota et al., 1995). Myosins I and V are calmodulin-binding proteins at low $[Ca^{2+}]_c$ (Cheney and Mooseker, 1992). Therefore, high levels of cytosolic Ca^{2+}, like those occurring in the tip of growing pollen tubes, will cause the release of calmodulin and consequent inhibition of motility. Similarly, the disruption of the actin cytoskeleton in the extreme apex by high $[Ca^{2+}]_c$ concentration (Kohno and Shimmen, 1987) may imply lower levels of mRNA in that region. In polarised *Fucus* zygotes, the distribution of poly(A)$^+$RNA has been shown to require intact microfilaments (Bouget et al., 1995).

The involvement of calmodulin and calmodulin-binding peptides in actin-dependent processes has already been suggested for the differentiation of tracheary elements in *Zinnia* cells (Kobayashi and Fukuda, 1994) and during the polarised growth of *Neurospora crassa* (Capelli et al., 1997). Our data supports the view that calmodulin is an essential element for growth probably by interaction with cytoskeletal elements. Exactly which enzymatic or structural proteins are activated by the Ca^{2+}/calmodulin complex in the growing tube is yet unknown.

Another possible function for sub-apical calmodulin is the regulation of Ca^{2+} stores. In pollen tubes, $[Ca^{2+}]_c$ drops a few microns from about 1–3 μM in the apex to 200–220 nM in the sub apical region. Thus, a tight control between Ca^{2+} stores and pumps in this later region must exist. Ainger et al. (1997) have recently shown calmodulin regulation of Ca^{2+} stores in the phototransduction of *Drosophila*, and modulation of ion channel and pump activity by plant calmodulin is also known (Askerlund, 1997; Bethke and Jones, 1994).

18.2.4
A Gradient of Bound Calmodulin?

An even distribution of calmodulin does not necessarily mean an even functionality. Calmodulin activity is dependent on the cooperative binding of Ca^{2+} ions, therefore on $[Ca^{2+}]_c$. It is predictable that the high levels of $[Ca^{2+}]_c$ found in the tube apex create a tip-focused zonation of bound calmodulin, coincident with the $[Ca^{2+}]_c$ gradient. This could explain the higher tip fluorescence of apical growing cells treated with fluphenazine, an antagonist that binds calmodulin in a Ca^{2+}-dependent manner (Hauber et al., 1984). Using fluorescence anisotropy imaging, Gough and Taylor (1993) showed that while the protein itself is uniformly distributed, calmodulin binding is elevated in the tails of highly polarised and migrating fibroblasts during wound healing. The V-shaped collar of calmodulin interaction in pollen tubes may be due to the presence of specific targets acting like a sink for the available FITC-CaM, and hence reducing its mobility relatively to the endogenous isoform. Such targets could well be actin-binding calmodulin-regulated polypetides. One of these peptides has already been shown to regulate polarised growth in *Neurospora crassa* (Capelli et al., 1997). Further insights on the role of calmodulin in polarised growth depend on the identification of the potential calmodulin targets and the extent of calmodulin binding to them.

18.3
The Role of IP3 and Protein Kinases in Tube Growth and Reorientation

It is clear from the data gathered so far that pollen tubes can respond to a variety of stimulus when it comes to changing their growth direction. However, we must distinguish those which promote a consistent and predictable change in growth axis from those which act simply by disrupting the polarity axis, even if transiently. *In vivo*, it is clear that the former are the ones which guide the tube to the micropyle.

18.3.1
The Role of Ins(1,4,5)P_3 in Guidance Mechanisms

We have previously shown (Malhó et al., 1994, 1995; Malhó and Trewavas, 1996) that reorientation is preceded by an elevation in $[Ca^{2+}]_c$ resulting from Ca^{2+} channel activity in the pollen tube plasma membrane. However, it is unlikely that the recorded elevation is due only to an influx of extracellular Ca^{2+}. It is plausible to assume that the initial influx of Ca^{2+} could result in subsequent releases from internal stores, a mechanism common to different cell types (Fewtrell, 1993).

Ins(1,4,5)P_3 functions as a cytosolic second messenger in animal cells by stimulating Ca^{2+} release from intracellular stores (Berridge, 1993). It was also shown that a functional relation might exist between polarization and stimulation of the inositol phosphate turnover (Audigier et al., 1988) and this in turn could mobilize internal Ca^{2+} stores. We found that the artificial increase in the levels of Ins(1,4,5)P_3 resulted in a transient increase in $[Ca^{2+}]_c$ and reorientation of the cell growth axis (Figure 18.4A–C; Malhó, 1998). This elevation could be inhibited by perfusion of the cells with heparin (Figure 18.4D), a specific blocker of the intracellular Ins(1,4,5)P_3 receptor (Kobayashi et al., 1988) whose efficiency in plants has already been demonstrated on red beet microsomes (Brosnan and Sanders, 1990) and poppy pollen tubes (Franklin-Tong et al., 1996).

Release of high concentrations of Ins(1,4,5)P_3 ($\sim$0.5–1 μM), particularly in the apical region, resulted in sustained $[Ca^{2+}]_c$ elevation and tip bursting. A simple elevation of $[Ca^{2+}]_c$ cannot account for this effect as the same does not happen with photolysis of caged Ca^{2+} (Malhó and Trewavas, 1996). Franklin-Tong et al. (1996) observed that the levels of Ins(1,4,5)P_3 remain elevated after photolysis which can result in prolonged inhibition of PIC and disturbance of cell homeostasis.

In our experiments, heparin incompletely inhibited the Ca^{2+} elevation triggered by Ins(1,4,5)P_3. A likely explanation for this is that the concentration of heparin loaded into the cells was not enough to completely block all the Ins(1,4,5)P_3 receptors. Unfortunately, this hypothesis can not be fully tested due to the difficulties in estimating the intracellular concentration of heparin and its inhibitory effects on growth altogether. Heparin also inhibited the electrical field-induced reorientation. Imaging of Ca^{2+} in these circumstances showed that the typical elevation throughout the shank of the tube no longer occurred (Malhó, 1998). The data suggests that following initial Ca^{2+} influx, the signal transduction pathway involved in reorientation includes further Ca^{2+} release from Ins(1,4,5)P_3-dependent stores. Such Ins(1,4,5)P_3-induced-Ca^{2+}-release provides a means of amplifying stimulus-induced Ca^{2+} changes both temporally and spatially, and it has been suggested as generating oscillations in cytosolic Ca^{2+} (Schroeder and Hagiwara, 1990).

It has been shown that the inositol phosphate cascade could be activated by localized cathode-facing membrane depolarization (Audigier et al., 1988). However, we have already demonstrated that inclusion of the Ca^{2+} channel blocker $LaCl_3$ in the culture medium prevents reorientation (Malhó et al., 1994), which agrees with the signal being triggered by an initial Ca^{2+} influx. In this sense, activation of plasma membrane Ca^{2+} channels may not be the only event needed for reorientation but the primary target for a reorientation stimulus. Our hypothesis finds support in the experiments of Petersen and Berridge (1994) who showed that injection of heparin completely blocked responses to flash photolysis of caged Ins(1,4,5)P_3 but had no apparent effect on Ca^{2+} entry.

Fig. 18.4 see color plate at the end of the book.

Ins(1,4,5)P_3 binding does not seem to be required for the activation of Ca^{2+} entry, but instead for transducing the signal to the body of the tube. Among the possible targets for the phosphoinositide pathway, actin microfilaments and actin-binding proteins are particularly important. Profilin, a protein particularly abundant in pollen, forms complexes with ADP-actin and promotes its phosphorylation to ATP-actin, a form which polymerizes faster (McCurdy and Williamson, 1991). These complexes are dissociated by PIP2 (Drøbak et al., 1994) so a decrease in this molecule could change the physical state of the cytoskeletal network. Profilin also binds to phospholipase C, inhibiting the function of the enzyme which can be a way of regulating the levels of PIP_2 (Drøbak et al., 1994). The Ca^{2+} released by Ins(1,4,5)P_3 can, in turn, activate gelsolin, a protein that dissolves the actin gel (McCurdy and Williamson, 1991). Ins(1,4,5)P_3 receptor-like proteins were also shown to be linked to actin filaments (Fujimoto et al., 1995) and it is possible that a part of the self-incompatibility mechanism described by Franklin-Tong et al. (1996) involves Ins(1,4,5)P_3-induced changes in the cytoskeleton.

18.3.2 Activity of a Ca^{2+}-dependent Protein Kinase Enzyme in Pollen Tubes

Phosphorylation cascades are known to be part of the signaling aspects of plant development and physiology, particularly cell motility (Linden and Kreimer, 1995). In animal cells, PKC is a vital element in these pathways; it is a known regulator of Ca^{2+}-dependent exocytosis (Burgoyne and Morgan, 1993), and it was shown to regulate Ca^{2+} influx by modification of channel activity (Petersen and Berridge, 1995), both processes being essential for cell reorientation. In plants, PKC is not well characterized but several recent reports have shown the presence of a functional homologue (Nanmori et al., 1994; Subramanian et al., 1997).

In recent years, many papers have reported the existence of a wide range of protein kinases in plant tissues. One report (Van der Hoeven et al., 1996) detected a protein kinase from plasma membrane activated by Ca^{2+} and cis-unsaturated fatty acids, that phosphorylates a synthetic peptide with a motif that can be a target for both PKC and CDPK. Elliot and Kokke (1987) also purified a PKC-type enzyme with a dependence on µM concentrations of free Ca^{2+}; the properties of this enzyme include phospholipid activation and reaction with antibodies raised against the regulatory domain of bovine brain PKC. Similar results were obtained by different groups (Abo-El-Saad and Wu, 1985; Nanmori et al., 1994; Subramanian et al., 1997) and in a recent work (Moutinho et al., 1998a) we obtained data which suggests that growing pollen tubes present a tip-high gradient of kinase activity. The activity is mainly associated with the plasma membrane

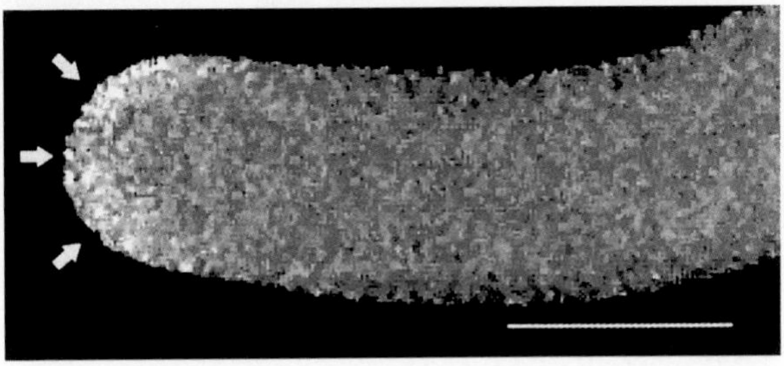

Fig. 18.5. Activity of a Protein Kinase in a growing pollen tube visualized by ratio imaging of Bodipy Fl Bisindolylmaleimide using confocal microscopy. Growing tubes present a higher kinase activity in the extreme apical region (plasma membrane and cortical cytosol). Bar = 10 μm

of the apical region and cytosolic regions underneath (Figure 18.5). When kinase activity was analyzed in growing reorienting tubes the data indicated that the side toward which the cell bent showed a higher activity (Malhó et al., 1998). The data thus suggests that an asymmetric activity of kinases within the growing dome is involved in the reorientation mechanisms. Whether this asymmetric activity precedes reorientation, cannot be said at this stage, but considering the data gathered with different inhibitors and activators (Moutinho et al., 1998a), we believe this may be so.

18.3.3 Does Ca^{2+} Capacitative Entry Control Apical Growth?

In many cell types, the emptying of the intracellular calcium stores by inositol trisphosphates opens a plasma membrane Ca^{2+} influx pathway, also named capacitative Ca^{2+} entry (Berridge, 1995). It is now widely accepted that the tip focused gradient of $[Ca^{2+}]_c$ present in growing pollen tubes results from a Ca^{2+} influx through plasma membrane channels but how these channels are regulated is unknown. An attractive but speculative hypothesis is that the channels, or a population of channels, have their activity regulated by the degree of emptiness of the intracellular Ca^{2+} stores, most likely localized in the endoplasmic reticulum. This hypothesis fits both our data and the effects of BAPTA buffers in pollen tubes (Pierson et al., 1994). BAPTA buffers not only chelate Ca^{2+} but also bind to the Ins(1,4,5)P_3 receptor and inhibit Ins(1,4,5)P_3-induced Ca^{2+} mobilization (Luttrell, 1994).

Our results on the activity of the protein kinase enzyme (Moutinho et al., 1998a) also support this model, which is consistent with data obtained in other cell systems (Petersen and Berridge, 1994) where a low level activation of PKC could potentiate Ca^{2+} influx, apparently by blocking the inactivation of the Ca^{2+} influx. It also explains the continued elevation in $[Ca^{2+}]_c$ observed when caged Ca^{2+} was released in one side of the apex (Malhó and Trewavas, 1996). Alternatively, PKC may not be directly associated with the entry of Ca^{2+} but with the release of stored Ca^{2+} (Fine et al., 1994). In any case, this data puts forward protein kinases as strong candidates for the regulation of the apical $[Ca^{2+}]_c$ oscillations, an hypothesis by Thomas et al. (1996).

18.4 Exocytosis and Apical Growth

Much of the research done in apical growing plant cells suggests, sometimes in a subliminal way, that Ca^{2+} is the regulator of a, not yet formulated, unifying theory for tip growth and orientation. The data recently obtained in root hairs (Bibikova et al., 1997;

Wymer et al., 1997), and growing lobes of the algae *Micrasterias* (Holzinger et al., 1995), suggests that tip growth and orientation may exist without being associated with the presence of a continuous tip-focused $[Ca^{2+}]_c$ gradient (Malhó, 1998a).

Tip growth relies on the localized exocytosis of Golgi vesicles, which depends not only on Ca^{2+} but also on the presence of annexins. Annexins are Ca^{2+}-dependent, phospholipid-binding proteins known to mediate calcium-dependent fusion between biological membranes; evidence has been gathered for their involvement in membrane trafficking, membrane channel activity, and phospholipid metabolism (Moss, 1997). In animal cells the Ca^{2+} requirement can be reduced by the presence of cis-unsaturated fatty acids (such as arachidonic acid) so vesicle fusion could be achieved without a sustained increase in $[Ca^{2+}]_c$. The breakdown of lipids, could be controlled by other proteins (e.g. GTP-binding proteins, kinases, phosphatases, etc) or even by Ca^{2+}. Trotter et al. (1995) showed that relocation of annexins from the cytosol to the membranes started when $[Ca^{2+}]_c$ was elevated to 300 nM and, at a concentration of 800 nM, were no longer detected in the cytosol reaching a maximum level in the membrane fraction. These $[Ca^{2+}]_c$ values are very close to those we determined in *A. umbellatus* pollen tubes with the Ca^{2+} indicator Indo-1 (Malhó et al., 1994, 1995) for the body of the tube and tip region (180–220 and 700–1000 nM respectively). Although highly speculative, it is still interesting to note that under such conditions, annexins would be mainly associated with the plasma membrane in the apical region, whereas in the cytosol in regions further back. In pollen tubes and fern rhizoids, annexins were found to localize mainly in the apical region (Blackbourn and Battey, 1993; Clark et al., 1995) and van der Hoeven et al. (1996) demonstrated binding of an annexin-like peptide to a PKC isoform. In addition, Lin et al. (1996) reported a high concentration of Rho GTPase in the cortical region of the pollen tube apex. Rho proteins (GTP binding proteins) are known to be PKC regulators (Nagata and Hall, 1996) and to participate in the maintenance of cell wall integrity (Drgonová et al., 1996). All this evidence thus points to a tight regulation of exocytosis (and possibly deposition of new wall material) by the activity of PKC and dependence on the local concentration of $[Ca^{2+}]_c$. Golgi vesicles in the extreme apical region could not be "delivered" to the plasma membrane but have random movements and fuse with the plasma membrane whenever they enter into contact with it. Asymmetric fusion would depend only on the conditions existent on the plasma membrane (activity of channels, concentration of ions, phospholipids and proteins) without necessarily involving a cytoskeletal rearrangement. This would occur at a later stage as a result of the interactions of actin with phosphoinositides, actin-binding proteins and protein kinases.

During the course of evolution, different tip growth mechanisms may have developed one (or more) of these control mechanisms for exocytosis. In any case, Ca^{2+} may act as an important factor in localizing growth but not be the only determinant – a sort of common element between different cell strategies to generate tip growth. In favour of this assumption is the fact that, among cell types showing a tip-focused $[Ca^{2+}]_c$ gradient, there seems to exist a correlation between growth rates and steepness of the gradient, but only *within* the same cell type (Pierson et al., 1996; Wymer et al., 1997). For example, if the absolute values presented for $[Ca^{2+}]_c$ are to be trusted, pollen tubes growing about one order of magnitude faster than root hairs show approximately the same gradient steepness ($\sim$800 μM in the first 15–20 μm). Thus, it seems unlikely that the rate of Ca^{2+} influx through plasma membrane channels is dictated solely by the speed of growth. Instead, such experimental evidence suggests that **(1)** Ca^{2+} entry is regulated by the extent to which the intracellular Ca^{2+} stores are filled – capacitative Ca^{2+} entry, (2) metabolic processes in different cell types present different sensitivities to $[Ca^{2+}]_c$, and (3) more important than the actual $[Ca^{2+}]_c$ value is the physiological function to which it is associated.

18.5 Future Prospects

Although the use of ion-sensitive dyes is one of the most powerful techniques available to study molecular and ion dynamics, it is still fraught with numerous limitations preventing us from going further in data analysis. Bibikova et al. (1997) found that short time sequences (< 1 min) caused root hair damage, presumably due to photoirradiation of the 2 dyes used for ratio measurements (Calcium green-2 and rhodamine). Therefore, new, brighter, and non toxic ratioable dyes – such as cameleon 2 (Miyawaki et al., 1997) – will probably soon start to be used in plant cell research. Due to the spatial segregation of cellular components, tip growing cells are also ideal models for studies with the vibrating microprobe which should complement internal Ca^{2+} measurements (see Feijó, *in this book*). For example, it would be very interesting to compare the Ca^{2+} influx rates between pollen tubes and root hairs and establish if there is a correlation with gradient steepness.

Further dissection of the Ca^{2+} signalling cascade is also of fundamental importance. In all cases examined so far, the cellular response to a reorienting stimulus seems to involve changes in $[Ca^{2+}]_c$ but our data suggests a picture of much higher complexity. (1) localized Ca^{2+} channel activity and membrane potential changes, (2) localized changes in $[Ca^{2+}]_c$ possibly regulated by protein kinase, and (3) activation of the inositol phosphate cascade, all seem to be necessary for localized exocytosis and reorientation. Biochemical and molecular tools are now available to identify the plant counterparts of the animal signalling systems. Rho GTPases, protein kinases, calmodulin, cAMP, and many others will certainly receive a great deal of attention in the next few years. The study of polarity-altered mutants should also provide enlightening insights in this area. In flowering plants, the only cases described where the guidance mechanisms were somehow disrupted correspond to genetic lesions in the female tissue (Wilhelmi and Preuss, 1997). The *Arabidopsis* rhd2 mutant shows altered $[Ca^{2+}]_c$ distribution in root hairs (Wymer et al., 1997) but growth is concomitantly impaired. Other known mutants (rhd3 and tip1) also seem to be defective in a gene encoding some product involved in tip growth (Galway et al., 1997; Schiefelbein et al., 1993). Identification (and full characterization) of a mutant(s) affecting orientation but not growth, would certainly be of fundamental importance to our understanding of polarity mechanisms. The fact that such mutant has not yet been found could mean that growth and orientation cannot be dissociated, and that we have to understand them as a two-component system with a co-ordinated action.

These observations should naturally be extended to other tip growing cells namely rhizoids and fungal hyphae where numerous questions are still unanswered. Is there a $[Ca^{2+}]_c$ gradient present in all hyphae? If so, how do we interpret the high $[Ca^{2+}]_c$ levels in the tip and the presence of a spitzenkörper? Will the results fit in a unifying theory for apical growth?

Acknowledgements
The author wishes to acknowledge Tony Trewavas, Ana Moutinho and José Feijó for their support and helpful discussions. Research in the author's lab is supported by a PRAXIS XXI grant (2/2.1/BIA/401/94) and a FCT grant (PBICT/P/BIA/2068/95).

References

Abo-El-Saad, M, Wu, R (1995) A rice membrane calcium-dependent protein kinase is induced by gibberellin. Plant Physiol 108:787–793

Ainger K, Avossa D, Morgan F, Hill SJ, Barry C, Barbarese E, Carson JH (1993) Transport and localisation of exogenous myelin basic protein mRNA microinjected into oligodendrocytes. J Cell Biol 123:431–441

Arnon A, Cook B, Montell C, Selinger Z, Minke B (1997) Calmodulin regulation of calcium stores in phototransduction of *Drosophila*. Science 275:1119–1121

Askerlund P (1997) Calmodulin-stimulated Ca^{2+}-ATPases in the vacuolar and plasma membranes in cauliflower. Plant Physiol 114:999–1007

Berridge MJ (1995) Capacitative calcium entry. Biochem J 312:1–11

Bethke PC, Jones RL (1994) Ca^{2+}-Calmodulin modulates ion channel activity in storage protein vacuoles of barley aleurone cells. Plant Cell 6:277–285

Bibikova TN, Zhigilei A, Gilroy S (1997) Root hair growth in *Arabidopsis thaliana* is directed by calcium and an endogenous polarity. Planta 203:495–505

Blackbourn HD, Battey NH (1993) Annexin-mediated secretory vesicle aggregation in plants. Physiol Plant 89:27–32

Blumenfeld H, Zablow L, Sabatini B (1992) Evaluation of cellular mechanisms for modulation of calcium transients using a mathematical model for fura-2 Ca^{2+} imaging in *Aplysia* sensory neurons. Biophys J 63:1146–1164

Bock GR, Ackrill K (1995) Calcium Waves, Gradients and Oscillations. (Chichester, UK: John Wiley and Sons)

Boitano S, Dirksen ER, Sanderson MJ (1992) Intercellular propagation of calcium waves mediated by inositol triphosphate. Science 258:292–295

Bouget F-Y, Gerttula S, Quatrano RS (1995) Spatial redistribution of poly(A)$^{+}$ RNA during polarisation of the *Fucus* zygote is dependent upon microfilaments. Dev Biol 171:258–261

Bouget F-Y, Gerttula S, Shaw S, Quatrano RS (1996) Localisation of actin mRNA during the establishment of cell polarity and early cell divisions in *Fucus* embryos. Plant Cell 8:189–201

Brockerhoff SE, Davis TN (1992) Calmodulin concentrates at regions of cell growth in *Saccharomyces cerevisiae*. J Cell Biol 118:619–629

Brockerhoff SE, Stevens RC, Davis T (1994) The unconventional myosin Myop2 is a calmodulin target at sites of cell growth in *Saccharomyces cerevisiae*. J Cell Biol 124:315–323

Burgoyne, RD, Morgan, A (1993) Regulated exocytosis. Biochem J 293:305–316

Bush DS (1993) Regulation of cytosolic calcium in plants. Plant Physiol 103:7–13

Capelli N, Barja F, Van Tuinen D, Monnat J, Turian G, Ortega Prez R (1997) Purification of a 47 kDa calmodulin-binding polypeptide as an actin-binding protein from *Neurospora crassa*. FEMS Microb Lett 147:215–220

Cheney RE, Mooseker MS (1992) Unconventional myosins. Curr Op Cell Biol 4:27–35

Clark GB, Turnwald S, Tirlapur UK, Haas CJ, von der Mark K, Roux SJ, Scheurlein R (1995) Polar distribution of annexin-like proteins during phytochrome-mediated initiation and growth of rhizoids in the ferns *Dryopteris* and *Anemia*. Planta 197:376–384

Dawson AP (1997) Calcium signalling: How do IP_3 receptors work? Curr Biol 7:R544–RR547

Doris FP, Steer MW (1996) Effects of fixatives and permeabilisation buffers on pollen tubes: implications for localisation of actin microfilaments using phalloidin staining. Protoplasma 195:25–36

Drgonová J, Drgon T, Tanaka K, Kollar R, Chen GC, Ford RA, Chan CSM, Takai Y, Cabib E (1996) Rho1p, a yeast protein at the interface between cell polarization and morphogenesis. Science 272:277–279

Elliot, DC, Kokke YS (1987) Partial purification and properties of a protein kinase C type enzyme from plants. Phytochemistry 26:2929–2935

Estruch JJ, Kadwell S, Merlin E, Crossland L (1994) Cloning and characterisation of a maize pollen-specific calcium-dependent calmodulin-independent protein kinase. Proc Natl Acad Sci 91:8837–8841

Fine BP, Marques ES, Hansen KA (1994) Calcium-activated sodium and chloride fluxes modulate platelet volume: role of Ca^{2+} stores. Amer. J. Physiol. 36:C1435–C1441

Franklin-Tong VE, Drøbak BK, Allan AC, Watkins PAC, Trewavas AJ (1996). Growth of pollen tubes of *Papaver rhoeas* is regulated by a slow moving calcium wave propagated by inositol triphosphate. Plant Cell 8:1305–1321

Franklin-Tong VE, Ride JP, Read ND, Trewavas AJ, Franklin CH (1993) The self-incompatibility response in *Papaver rhoeas* is mediated by cytosolic free calcium. Plant J 4:163–177

Fricker MD, White N, and Obermeyer G (1997) pH gradients are not associated with tip growth in pollen tubes of *Lilium longiflorum*. J Cell Sci 110:1729–1740

Galway ME, Heckman JW, Schiefelbein JW (1997) Growth and ultrastructure of *Arabidopsis* root hairs: the rhd3 mutation alters vacuole enlargement and tip growth. Planta 201:209–218

Garrill A, Jackson SL, Lew RR, Heath IB (1993) Ion channel activity and tip growth: tip-localized stretch-activated channels generate an essential Ca^{2+} gradient in the oomycete *Saprolegnia ferax*. Eur J Cell Biol 60:358–365

Gilroy S (1996) Signal transduction in barley aleurone protoplasts is calcium dependent and independent. Plant Cell 8:2193–2209

Gilroy S, Read ND, Trewavas AJ (1990) Elevation of cytoplasmic calcium by caged calcium or caged inositol triphosphate initiates stomatal closure. Nature 346:769–771
Gough AH, Taylor DL (1993) Fluorescence anisotropy imaging microscopy maps calmodulin binding during cellular contraction and locomotion. J Cell Biol 121:1095–1107
Grynkiewicz G, Poenie M, Tsien RY (1985) A new generation of Ca^{2+} indicators with greatly improved fluorescence properties. J Biol Chem 260:3440–3450
Hajnóczky G, Robb-Gaspers LD, Seitz MB, Thomas AP (1995) Decoding of cytosolic calcium oscillations in the mitochondria. Cell 82:415–424
Harper JF, Sussman MR, Schaller GE, Putnam-Evans C, Charbonneau H, Harmon AC (1991) A calcium-dependent protein kinase with a regulatory domain similar to calmodulin. Science 252:951–954
Haußer I, Herth W, Reiss H-D (1984) Calmodulin in tip-growing plant cells, visualised by fluorescing calmodulin-binding phenothiazines. Planta 162:33–39
Holzinger A, Callaham DA, Hepler PK, Meindl U (1995) Free calcium in Micrasterias: local gradients are not detected in growing lobes. Eur J Cell Biol 67:363–371
Jaffe LF (1993) Classes and mechanisms of calcium waves. Cell Calcium 14:736–745
James P, Vorherr T, Carafoli E (1995) Calmodulin-binding domains: just two faced or multi-faceted? Trends Biochem Sci 20:38–42
Jouaville L, Ichas F, Holmuhamedov E, Camacho P, Lechleiter JD (1995) Synchronization of calcium waves by mitochondrial substrates in *Xenopus*. Nature 377:438–441
Kobayashi H, Fukuda H (1994) Involvement of calmodulin and calmodulin-binding peptides in the differentiation of tracheary elements in *Zinnia* cells. Planta 194:388–394
Kohno T, Shimmen T (1987) Ca^{2+}-induced fragmentation of actin filaments in pollen tubes. Protoplasma 141:177–179
Lancelle SA, Cresti M, Hepler PK (1997) Growth inhibition and recovery in freeze-substituted *Lilium longiflorum* pollen tubes: structural effects of caffeine. Protoplasma 196:21–33
Lechletter J, Girard S, Peralta E, Clapham D (1991) Spiral calcium wave propagation and annihilation in *Xenopus laevis* oocytes. Science 252:123–126
Legué V, Blancaflor E, Wymer C, Perbal G, Fantin D, Gilroy S (1997) Cytoplasmic free Ca^{2+} in *Arabidopsis* roots changes in response to touch but not gravity. Plant Physiol 114:789–800
Lin Y, Wang Y, Zhu J-K, Yang Z (1996) Localization of a Rho GTPase implies a role in tip growth and movement of the generative cell in pollen tubes. Plant Cell 8:293–303
Linden, L, Kreimer, G (1995) Calcium modulates rapid protein phosphorylation/dephosphorylation in isolated eyespot apparatuses of the green alga *Spermatozopis similis*. Planta 197:343–351
Lipp P, Niggli E (1993) Ratiometric confocal Ca^{2+}-measurements with visible wavelength indicators. Cell Calcium 14:359–372
Love J, Brownlee C, Trewavas AJ (1997) Calmodulin dynamics during photopolarisation in *Fucus serratus* zygotes. Plant Physiol in press
Luttrell BM (1994) Cellular actions of inositol phosphates and other natural calcium and magnesium chelators. Cellular Signaling 6:355–362
Malhó R, (1998a) Expanding tip growth theory. Trends Pl Sci 3:40–42
Malhó R (1998) Role of 1,4,5-inositol triphosphate-induced Ca^{2+} release in pollen tube orientation. Sex Plant Reprod, in press
Malhó R, Read ND, Pais MS, Trewavas AJ (1994) Role of cytosolic free calcium in the reorientation of pollen tube growth. Plant J 5:331–341
Malhó R, Read ND, Trewavas AJ, Pais MS (1995) Calcium channel activity during pollen tube growth and reorientation. Plant Cell 7:1173–1184
Malhó R, Trewavas AJ (1996) Localized apical increases of cytosolic free calcium control pollen tube orientation. Plant Cell 8:1935–1949
Mascarenhas JP (1993) Molecular mechanisms of pollen tube growth and differentiation. Plant Cell 5: 1303–1314
Messerli M, Robinson KR (1997) Tip localized Ca^{2+} pulses are coincident with peak pulsatile growth wrates in pollen tubes of *Lilium longiflorum*. J Cell Sci 110:1269–1278
Miller DB, Callaham DA, Gross DJ, Hepler PK (1992) Free Ca^{2+} gradient in growing pollen tubes of *Lilium*. J Cell Sci 101:7–12
Miller DD, Lancelle SA, Hepler PK (1996) Actin filaments do not form a dense meshwork in *Lilium longiflorum* pollen tube tips. Protoplasma 195:123–132
Miller DD, Scordilis SP, Hepler PK (1995) Identification and localisation of three classes of myosins in pollen tubes of *Lilium longiflorum* and *Nicotiana alata*. J Cell Sci 108:2549–2563
Miyawaki A, Llopis J, Heim R, McCaffery JM, Adams JA, Ikura M, Tsien RY (1997) Fluorescent indicators for Ca^{2+} based on green fluorescent proteins and calmodulin. Nature 388:882–887
Moss (1997) Annexins. Trends in Cell Biol 7:87–89
Moutinho, A, Love, J, Trewavas, AJ, Malhó, R (1998) Distribution of calmodulin protein and mRNA in growing pollen tubes. Sex Plant Reprod 11:131–139
Moutinho A, Trewaves AJ, Malhó R (1998a) Relocation of a Ca^{2+}-dependent protein kinase activity during pollen tube reorientation. Plant Cell 10:1499–1510
Nagata K-i, Hall A (1996) The Rho GTPase regulates protein kinase activity. Bioessays 18:529–531

Nanmori, T, Taguchi, W, Kinugasa, M, Oji, Y, Sahara, S, Fukami, Y, Kikkawa, U (1994) Purification and characterization of protein kinase C from a higher plant, *Brassica* campestris L. Biochem Biophys Res Commun 203:311–318
Neuhaus G, Bowler C, Kern R, Chua N-H (1993) Calcium/Calmodulin dependent and independent phytochrome signal transduction pathways. Cell 73:937–952
Obermeyer G, Weisenseel MH (1991). Calcium channel blocker and calmodulin antagonists affect the gradient of free calcium ions in lily pollen tubes. Eur J Cell Biol 2:319–327
Opas M (1997) Measurement of intracellular pH and pCa with a confocal microscope. Trends in Cell Biol 7:75–80
Parton R, Fischer S, Malhó R, Papasouliotis O, Jelitto T, Leonard T, Read ND (1997) Pronounced cytoplasmic pH gradients are not required for tip growth in plant and fungal cells. J Cell Sci 110: 1187–1198
Petersen CCH, Berridge MJ (1995) G-protein regulation of capacitative calcium entry may be mediated by protein kinases A and C in *Xenopus oocytes*. Biochem J 307:663–668
Pierson ES, Miller DD, Callaham DA, Shipley AM, Rivers BA, Cresti M, Hepler PK (1994) Pollen tube growth is coupled to the extracellular calcium ion flux and the intracellular calcium gradient: Effect of BAPTA-type buffers and hypertonic media. Plant Cell 6:1815–1828
Pierson ES, Miller DD, Callaham DA, van Aken J, Hackett G, Hepler PK (1996) Tip-localized calcium entry fluctuates during pollen tube growth. Develop Biol 174, 160–173
Polya GM, Micucci V (1985) Interaction of wheat germ Ca^{2+}-dependent protein kinases with calmodulin antagonists and polyamines. Plant Physiol 79:968–972
Putnam-Evans C, Harmon AC, Palevitz BA, Fechheimer M, Cormier MJ (1989) Calcium-dependent protein kinase is localised with F-actin in plant cells. Cell Mot Cyt 12:12–22
Rathore KS, Cork RJ, Robinson KR (1991) A cytoplasmic gradient of Ca^{2+} is correlated with the growth of Lily pollen tubes. Develop Biol 148:612–619
Read ND, Allan W, Knight H, Knight M, Malhó R, Russell A, Shacklock P, Trewavas AJ (1992) Imaging and measurement of cytosolic free calcium in plant and fungal cells. J Microsc 165:586–616
Roberts DM, Harmon AC (1992) Calcium-modulated proteins: targets of intracellular calcium signals in higher plants. Ann Rev Plant Physiol Plant Mol Biol 43:375–414
Sandinson DR, Williams RM, Wells, KS, Strickler J, Webb WW (1995) Quantitative fluorescence confocal scanning laser microscopy (CSLM). In: Pawley JB (ed) Handbook of Biological Confocal Microscopy. Plenum Press, pp 39–53
Schiefelbein J, Galway M, Masucci J, Ford S (1993) Pollen tube and root-hair tip growth is disrupted in a mutant of *Arabidopsis thaliana*. Plant Physiol 103:979–985
Schuurink RC, Chan PV, Jones RL (1996) Modulation of calmodulin mRNA and protein levels in barley aleurone. Plant Physiol 111:371–380
Subramanian R, Després C, Brisson N (1997) A functional homolog of mammalian protein kinase C participates in the elicitor induced defense response in potato. Plant Cell 9:653–664
Sun G-H, Ohya Y, Anraku Y (1992) Yeast calmodulin localises to sites of cell growth. Protoplasma 166: 110–113
Tang X, Hepler PK, Scordilis SP (1989) Immunochemical and immunocytochemical identification of a myosin heavy chain polypeptide in *Nicotiana* pollen tubes. J Cell Sci 92:569–574
Taylor AR, Manison NFH, Fernandez C, Wood J, Brownlee C (1996) Spatial organization of calcium signalling involved in volume control in the *Fucus* rhizoid. Plant Cell 8:2015–2031
Thomas AP, Bird GSJ, Hajnóczky G, Robb-Gaspers LD, Putney JW, Jr (1996) Spatial and temporal aspects of cellular calcium signalling. FASEB J 10:1505–1517
Trewavas AJ, Malhó R (1997) Signal perception and transduction: the origin of the phenotype. Plant Cell 7:1181–1195
Trotter PJ, Orchard MA, Walker JH (1995) Ca^{2+} concentration during binding determines the manner in which annexin V binds to membranes. Biochem. J. 308:591–598
Van der Hoeven, PCJ, Siderius, M, Korthout, HAAJ, Drabkin, AV, De Boer, AH (1996) A calcium and free fatty acid-modulated protein kinase as putative effector of the fusicoccin 14-3-3 receptor. Plant Physiol 111:857–865
Vidali L, Hepler PK (1997) Characterisation and localisation of profilin in pollen grains and tubes of *Lilium longiflorum*. Cell Mot Cyt 36:323–338
Vogel HJ (1994) Calmodulin – a versatile calcium mediator protein. Biochem Cell Biol 72:357–376
Webb AAR, McAinsh MR, Taylor JE, Hetherington AM (1996) Calcium ions as intracellular second messengers in higher plants. Adv Bot Res 22:45–96
Wilhelm JE, Vale RD (1993) RNA on the move: the mRNA localisation pathway. J Cell Biol 123:269–274
Wilhelmi LK, Preuss D (1997) Blazing new trails – pollen tube guidance in flowering plants. Plant Physiol 113:307–312
Woods NM, Cuthbertson KSR, Cobbold PH (1986) Repetitive transient rises in cytoplasmic free calcium in hormone stimulated hepatocytes. Nature 319:600–602
Wymer CL, Bibikova TN, Gilroy S (1997) Cytoplasmic free calcium distributions during the development of root hairs of *Arabidopsis thaliana*. Plant J 12:427–439
Yokota E, McDonald AR, Liu B, Shimmen T, Palevitz BA (1995) Localisation of a 170 kDa myosin heavy chain in plant cells. Protoplasma 185:178–187

CHAPTER 19

Measuring Ion Channel Activity During Polar Growth of Pollen Tubes

F. Armstrong · R. Benkert · F.-W. Bentrup · G. Obermeyer*

Institut für Pflanzenphysiologie, Universität Salzburg, Hellbrunnerstr. 34, 5020 Salzburg, Austria
e-mail: gerhard.obermeyer@sbg.ac.at
telephone: +43-(0)662-8044-5556
fax: +43-(0)662-8044-619
* corresponding author

Abstract. Ion transporters like H^+ ATPases, K^+ channels, and Ca^{2+} channels probably play an essential role during pollen tube tip growth by mediating ion fluxes that are involved in regulation of tube growth, e.g. Ca^{2+} conducting channels localized in the tube tip plasma membrane allow a Ca^{2+} influx that is responsible for the growth speed and the growth direction. However, no Ca^{2+} channels have been identified in the plasma membrane of pollen grains and tubes, so far; secondly, use of conventional patch-clamp techniques requires enzymatic removal of the cell wall to gain access to the protoplast. This approach, in turn, causes loss of the cell polarity. We, therefore, used laser microsurgery to isolate protoplasts from the very tip of pollen tubes while the tube's polarity is maintained. Using the patch-clamp technique it is then possible to investigate the ion channels in this tip protoplast.

19.1 Pollen and Plasma Membrane Ion Transporter

Pollen tubes represent a superb model system for the study of various cellular processes which are involved in tip growth. Many of these processes have been subject to intensive research, e.g. organization of the cytoskeleton (Cai et al. 1997, Pierson and Cresti 1992), pollen-pistil-interaction (Cheung 1996, Hiscock et al. 1996), transport and fusion of secretory vesicles (Battey and Blackbourn 1993), and tremendous progress has been made during the last years. Nevertheless, some other actions of pollen during germination like uptake of ion, water, and nutrients are not yet well understood, and investigation of these subjects may reveal new aspects of the complex process of tip growth (Obermeyer and Bentrup 1996).

There may be two states in the life of a pollen grain where ion transporters become obviously important. Firstly, during activation of the pollen grain the rehydrating pollen has to establish its cytoplasmic ionic components and a membrane potential has to be built up. In particular, the cytoplasmic concentrations of H^+ and Ca^{2+} have to be well adjusted. All these processes need the activity of specific ion transporters, e.g. Ca^{2+} ATPases and Ca^{2+} channels may be involved in the regulation of $[Ca^{2+}]_{cyt}$. Additionally, before any visible sign of pollen tube emergence, a cationic inwardly directed current was observed with the vibrating electrode indicating the presumptive site of pollen tube outgrowth (Weisenseel et al. 1975). This cation influx, presumably mediated by ion channels, is only possible when V_m is more negative than the electrochemical equilibrium potential of the specific ion carrying this influx. Therefore, at least a plasma membrane (PM) H^+ ATPase 'energizing' the plasma membrane and a cation channel are involved in this endogenous current (Obermeyer and Blatt 1995).

Secondly, ion transporters play an important role in the state of pollen tube growth when Ca^{2+} enters the very tip of the tube possibly via Ca^{2+} channels localized in the tip plasma membrane (Obermeyer and Weisenseel 1991, Malhó et al. 1995, Pierson et al. 1995). In other regions of the growing tube a K^+ influx has been detected (Weisenseel and Jaffe 1976). For both ion fluxes the responsible ion channels have not yet been identified.

A more detailled survey of the possible functions of ion channels during pollen activation and tube growth was written earlier (Feijó et al. 1995, Obermeyer and Bentrup 1996). In this chapter we would like to present some recent results on ion channels measured using the combination of laser microsurgery and patch-clamp technique. This is the first report on ion channels in the very tip of pollen tubes and our data on ion channel activity in the tube tip plasma membrane are still incomplete. We will discuss the application of laser microsurgery to localized isolation of protoplasts from polar growing organisms and focus on problems arising during application of both methods.

19.2 Release of Tip Protoplasts With Laser Microsurgery

A suitable method for investigation of ion channels is the patch-clamp technique which allows measurements of the activity of a single channel molecule or of entire channel populations depending on the configuration of the patch-clamp method. However, electrical recording of ion channel activity is restricted to accessable membranes onto which a patch electrode can be 'sealed' tightly. Therefore, before application of the patch-clamp technique protoplasts have to be prepared from plant cells. As a result of forming a spherical protoplast the cell loses all information of its previously established polarity. Combination of the patch-clamp technique with laser microsurgery may basically solve this problem. Using an UV-laser a tiny hole is cut into the cell wall of a partially plasmolyzed cell, and a small area of the plasma membrane then extrudes as the external osmotic potential is increased (DeBoer et al. 1994, Kurkdjian et al. 1993, Taylor and Brownlee 1992). Finally, a patch pipette is sealed onto the released sub-protoplast and ion channels present in this patch of plasma membrane may be investigated.

In the following, we will describe experimental protocols we have used to release tip protoplasts from lily pollen tubes. These protocols may be used as a guideline to release protoplasts from other plant cells using laser microsurgery. But special care has to be taken in adjusting the osmolalities of the bath media to maintain the structural and functional polarity of these cells during plasmolysis and protoplast release.

19.2.1 Typical Perfusion Protocol

To release a protoplast from the tip pollen tubes have to be plasmolyzed first. A measuring chamber containing pollen tubes from *Lilium longiflorum* Thunb. was perfused sequentially with 2 ml of pollen growth medium (**Med B**) containing 10% sucrose, 100 mg/l boric acid, 1 mM KCl, 0.1 mM $CaCl_2$, pH 5.6, 2 ml of **Med 500** containing (in mM): 384 mannitol, 50 KCl, 1 $CaCl_2$, 1 $MgCl_2$, 10 Hepes/BTP pH 7.2, a gradient of Med 500 to Med 1100, and finally with 2 ml **Med 1100** containing (in mM): 984 mannitol, 50 KCl, 1 $CaCl_2$, 1 $MgCl_2$, 10 Hepes/BTP pH 7.2. Med 1100 also contained 0.1% Calcofluor White (CFW) to stain the cell wall. A hole was then shot in the cell wall with the UV-laser, and the chamber was perfused carefully with a gradient of Med 1100 to Med 500. During this perfusion the pollen tube recovered and a part of the plasma membrane protruded through the hole in the cell wall.

Additionally, pollen tubes exposed to various perfusion protocols as indicated, were then fixed in 2.5% glutaraldhyde in Med 1100 for 20 min, washed and incubated in 2% OsO_4 for 120 min at room temperature. Dehydration and embedding were performed according to standard protocols (Reiss and Herth 1979, Hall 1991). Pollen tube sections were observed with a Phillips EM 400T electron microscope at 60–80 kV.

19.2.2
Laser Microsurgery Protocol

Pollen grains from 1 anther of *Lilium longiflorum* were incubated for 2–4 h in 6 ml Med B. Pollen tubes were selected and carefully transferred into perspex chambers with a cover slip at the bottom which was coated with poly-L-lysine (1 mg ml^{-1}) to fix the pollen tubes. These chambers allowing a rapid exchange of the surrounding bath medium were mounted on a motor-driven stage (Märzhäuser, Wetzlar, Germany) of an inverted microscope (Zeiss Axiovert 135) equipped with quartz optics. A set-up for laser microsurgery was built according to Weber and Greulich (1992) and De Boer et al. (1994). A nitrogen laser (VSL 337 ND, Laser Science Inc., USA) was coupled into the epifluorescence light path of the microscope using a beam expander and a beam steerer. The laser light had a wavelength of 337.1 nm, 0.1 nm band width, >250 µJ pulse energy, and beam divergence was below 0.3 mrad. The laser was triggered by a computer and pulse duration was 3 ns. The laser beam was finally focussed through the microscope objectives (Ultrafluar $\times 100$ or Ultrafluar $\times 40$, Zeiss) almost to its diffraction limit. In the focal plane the laser beam had a diameter of 0.5–1 µm depending on the objective used. With both objectives the energy density in the focussed beam was high enough to cut biological material including cell walls. Laser microsurgery experiments were performed under video control using a video camera (PCO, Kelheim, Germany), a monitor, a video recorder, and a video printer for rapid documentation (all from Sony, Vienna, Austria).

19.2.3
Patch-Clamp Measurements

All electrical recordings were made at room temperature (20–22 °C) and patch-clamp experiments were performed according to the method introduced by Hamill et al. (1981). Patch-pipettes were pulled from thick-walled borosilicate glass capillaries (Bio-Products, Hofheim, Germany) on a two-stage pipette puller (List Electronics, Darmstadt, Germany). After filling with Med 500 patch pipettes were dipped into SigmaCote (Sigma, Vienna, Austria) to reduce pipette capacitance. Pipettes were used immediately after pulling without polishing and usually had a resistance of 10–50 MΩ when filled with Med 500. The reference electrode was also filled with Med 500 to minimize the pipette offset voltage. Channel events were recorded with the List EPC-7 amplifier in the cell-attached configuration of the patch-clamp technique. Data from channel recordings were stored on a digital tape recorder and finally digitized at a sampling rate of 2 kHz after filtering at 500 Hz using an A/D converter (DigiData 1200, Axon Instruments, Foster City, USA) and the pClamp software (vers. 6, Axon Instruments).

Negative voltages applied to the pipette (V_{pip}) caused a depolarization of the plasma membrane and outward currents, meaning cation fluxes from the cytoplasm into the external medium, were presented as downward deflections of negative current.

19.3 Results and Discussion

19.3.1 Plasmolysis of Pollen Tubes and Laser Microsurgery

When pollen tubes were plasmolyzed carefully, they did not lose their polarity which may be noticed at the maintenance of the clear cap during plasmolysis and at the immediate recovery of cytoplasmic streaming and growth after increasing the osmolality of the external bath solution again. A typical plasmolyzing (solution perfusion) protocol is shown in Fig. 19.1. Aliquots of 50 µl were taken every 20 s from the middle of the perfusion chamber and the osmolality was measured. The perfusion rate varied between 1–2 ml min^{-1}. Using these conditions for exchanging the external media, the polarity of the tube was maintained in over 90% of the experiments ($n > 100$). A video sequence of a pollen tube which has been treated according to the perfusion protocol is presented in Fig. 19.2. This tube did not lose its polarity: a clear cap could be observed during the entire experiment. Although the cytoplasmic streaming stopped at an osmolality between 800 and 1100 mosmol kg^{-1}, it showed immediate recovery after perfusion with Med 500.

The time point $t = 0$ min marks the beginning of the perfusion protocol and after 5 min incipient plasmolysis was observed which was completed after 20 min. At $t = 22.5$ min a hole was shot into the cell wall (arrows). A CFW fluorescence image is shown as the last image of the video sequence and the hole may be clearly seen as a lack of fluorescence at the very tip. After laser microsurgery the osmolality of the medium was decreased by perfusion with an inversed gradient (Med 1100 to Med 500) and recovery of the pollen tube can be observed. In this experiment the cytoplasm of the tube divided into two parts during the recovery but still moved forward. After 30 min a small part of the tip plasma membrane protruded through the previously cut hole in the cell wall.

To compare the behavior of individual pollen tubes during the perfusion protocol aliquots of perfusion media were taken at times when a certain event happened and the osmolality was measured. The relationship between osmolality and pollen tube behavior is also presented in Fig. 19.1. Pollen tubes started to plasmolyze at 645 ± 89 mosmol kg^{-1}, recovery started at 1052 ± 145 mosmol kg^{-1}, and tip protoplasts were released at 631 ± 144 mosmol kg^{-1}. For laser microsurgery the osmolality of the medium has to be at least 1000 mosmol kg^{-1}. Otherwise, the tube protoplast still had a turgor pressure which would cause the protoplast to extend suddenly and finally to burst when a hole was cut into the cell wall. This indicates that the osmolality of the solutions inside and outside the cell wall was not always the same during plasmolysis. The cell wall may therefore work as a molecular sieve prohibiting rapid exchanges of all solutes. For the release of an intact protoplast it was therefore very important to equilibrate the osmostic pressure across the cell wall before cutting a hole with the laser. Incidentally, the pollen tube turgor pressure estimated from osmolality values measured at incipient plasmolysis yields turgor pressures of approximately 8 bar. However, recent measurements using a turgor pressure probe have revealed much lower values for turgor pressures (1–4 bar, Benkert et al. 1997), thus indicating that incipient plasmolysis does not reflect the actual turgor pressure of growing pollen tubes.

The released protoplasts were assayed for remaining cell wall rests and as shown in the last image of Fig. 19.2, obviously no cellulose covered the tip protoplast. Protoplasts were also incubated in 0.03% coriphosphine to visualize pectin rests adhering to the

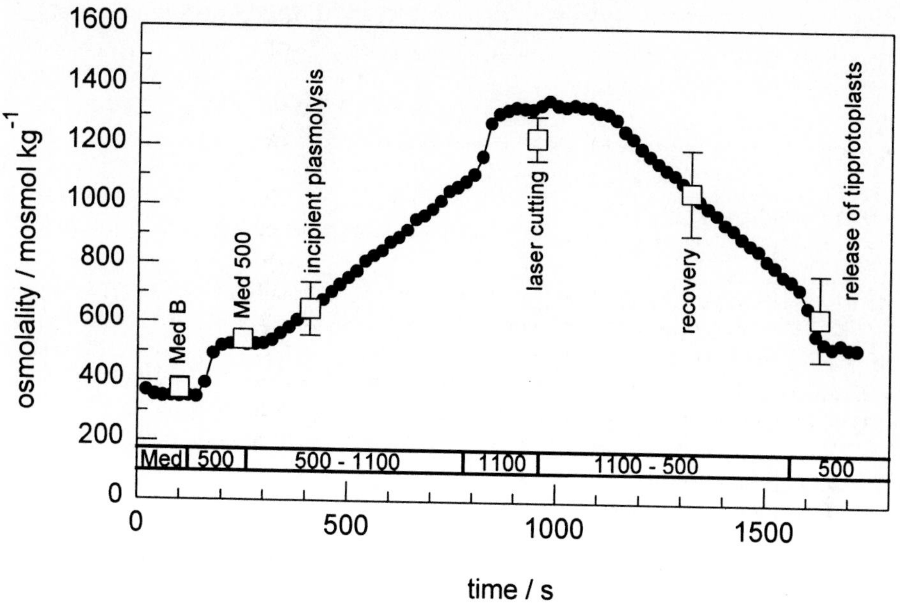

Fig. 19.1. Perfusion protocol. The osmolality of 50 µl aliquots taken every 20 s from the perfusion chamber was measured (filled circles). The vertical lines in the bar at the x-axis indicate the time at which the media were changed. The open squares present the average osmolality of the bath solution ± S.D. at different events during the perfusion. n = 18 (*Med B*), 16 (*Med 500*), 22 (*incipient plasmolysis*), 18 (*laser cutting*), 16 (*recovery*), and 7 (*release of tip protoplast*)

protoplast's surface (Weiss et al. 1988). But again, no pectin could be detected on the tip protoplasts (data not shown).

To underline the importance of a gentle osmotic treatment of pollen tubes in order to maintain their polarity during tip protoplast release, we tried different perfusion protocols: a step gradient, a fast continuous, and a slow continuous gradient (Fig. 19.3). There were no visible differences between the slow and the fast continuous gradient but when applying a step gradient the polarity of the pollen tubes was disturbed. Larger organelles entered the clear cap which eventually disappeared totally after a few minutes, and cytoplasmic streaming stopped immediately. To verify these light microscopic observations, pollen tubes exposed to a step and to a continuous gradient respectively, were fixed for electron microscopy. Figure 19.4 A shows the tip of a control pollen tube growing in Med B before exposure to hyperosmolar media. Note the different staining of secretory vesicles in the pollen tubes that were incubated in mannitol during fixation (Fig. 19.4B and C). Pollen tubes plasmolyzed during exposure to a fast or slow continuous gradient maintained the polar organization of the tip cytoplasm (Fig. 19.4B). Secretory vesicles can be found in the tip cytoplasm only while ER, mitochondria, and larger organelles are excluded from the tip. In contrast, pollen tubes exposed to a step gradient lost their polar organelle distribution (Fig. 19.4C). There are still secretory vesicles in the tip, but also liposomes and mitochondria entered the tip region. These results are also important for investigating the effect of growth medium osmolality on tube growth and related phenomena because fast changes in osmolality will not only cause a change in pollen tube turgor pressure but will irreversibly change the tube polarity.

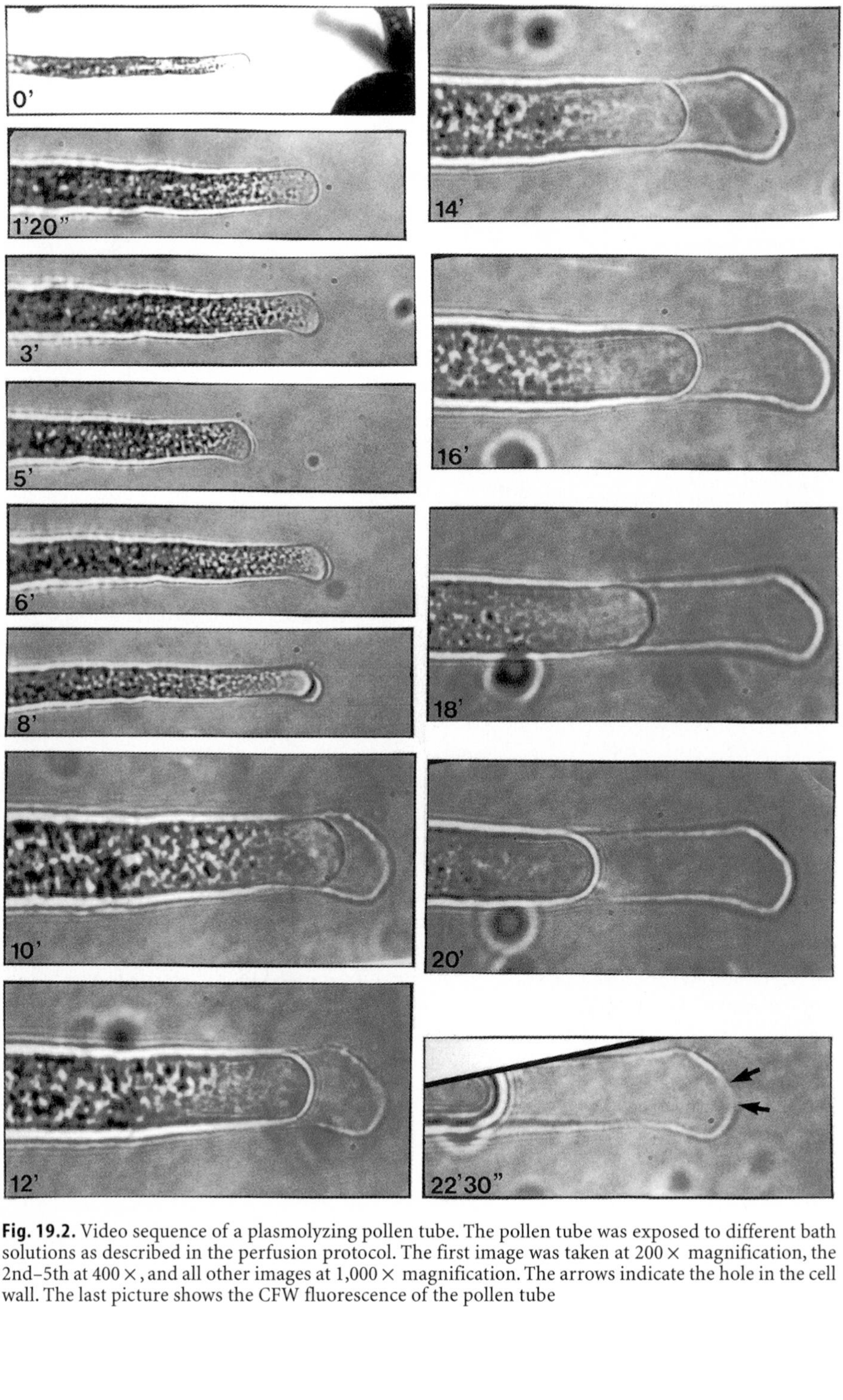

Fig. 19.2. Video sequence of a plasmolyzing pollen tube. The pollen tube was exposed to different bath solutions as described in the perfusion protocol. The first image was taken at 200 × magnification, the 2nd–5th at 400 ×, and all other images at 1,000 × magnification. The arrows indicate the hole in the cell wall. The last picture shows the CFW fluorescence of the pollen tube

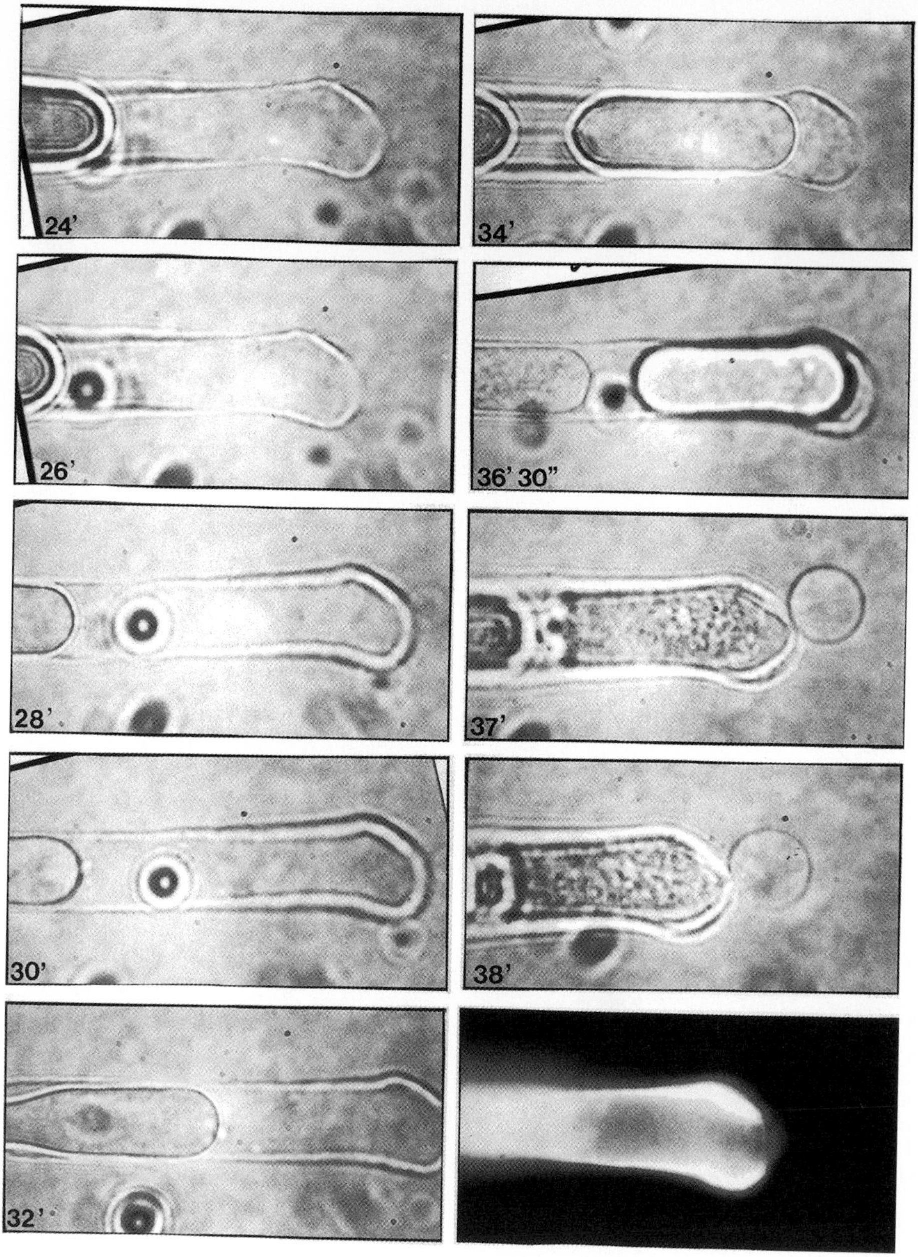

Fig. 19.2. (Continued)

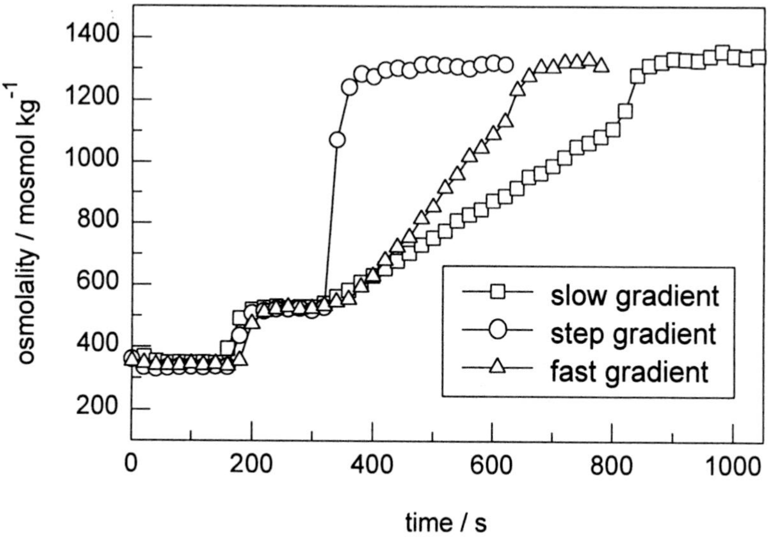

Fig. 19.3. Step and continuous gradient perfusion protocol. Three different perfusion protocols were applied to growing pollen tubes. Every 20 s aliquots of 50 µl were withdrawn from the measuring chamber to determine the osmolality

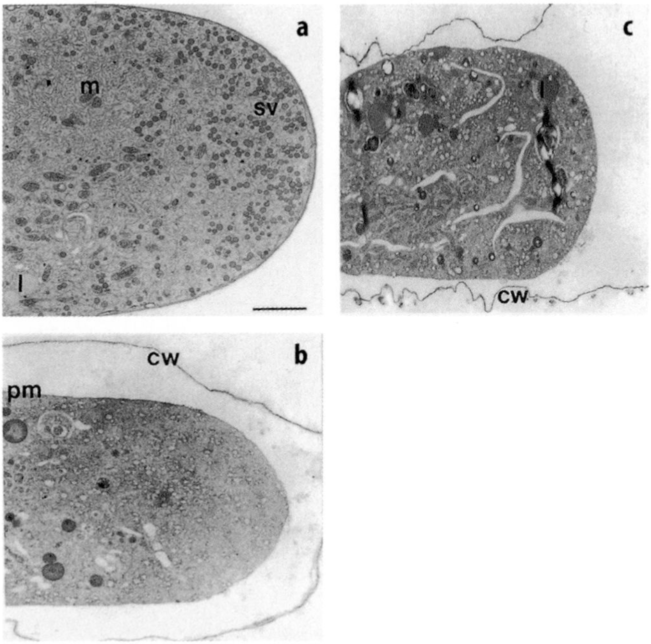

Fig. 19.4. Effect of perfusion protocol on pollen tube polarity. Plasmolyzed pollen tubes were fixed with glutaralhedyde and OsO_4, dehydrated with increasing ethanol concentrations, and finally embedded in Epon using standard protocols for electron microscopy. **A.** Control pollen tube not exposed to any gradient. **B.** Pollen tube plasmolyzed gently with a fast continuous gradient (Med 500 to Med 1100). **C.** Pollen tube plasmolyzed with a step gradient of mannitol. Cell wall (*cw*), mitochondria (*m*), liposome (*l*), secretory vesicle (*sv*). Bar = 2.5 µm, × 4,000 magnification

19.3.2 Ion Channels in the Pollen Tube Tip

When a tip protoplast has been released successfully a patch pipette has to be 'sealed' onto the plasma membrane of the tip protoplast. This means that the contact between the protoplast plasma membrane and the glass of the pipette has to be as tight as possible so that a resistance in the Giga-Ohm (GΩ) range is measured when small voltage pulses are applied to the pipette. Despite the fact that the tip protoplast surface was obviously clean of cell wall rests it was still difficult to achieve GΩ seals. In most trials the seal resistance was between 400 and 700 MΩ only. This effect has also been observed in other protoplasts released by laser microsurgery. The seal rate in these experiments was

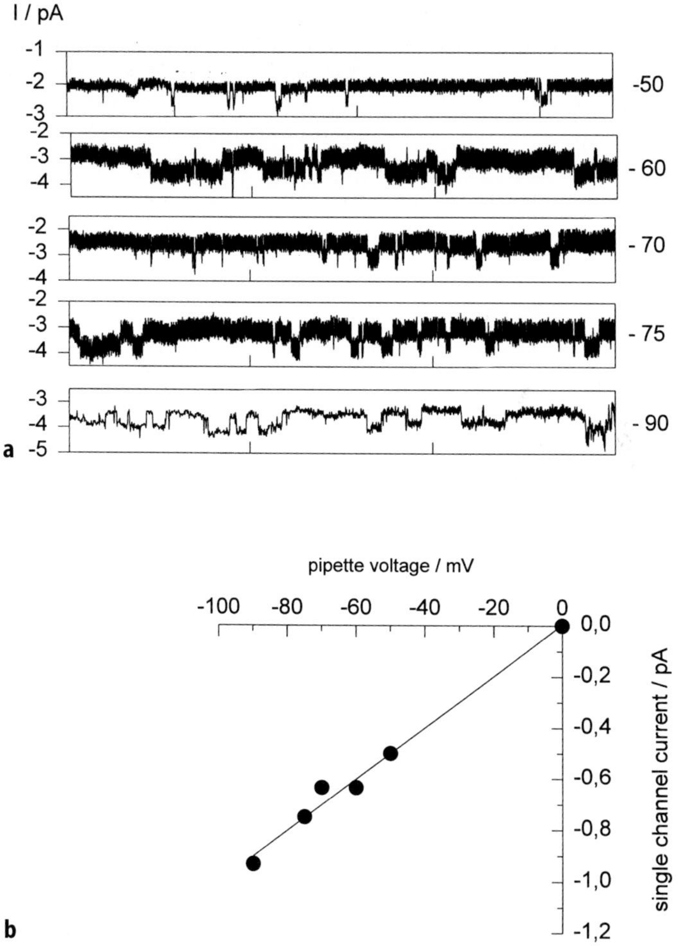

Fig. 19.5. Single channel recording from a tip protoplast. Channel currents were measured in the cell-attached configuration of the patch-clamp technique. Bath and pipette solution contained (in mM) 500 mannitol, 50 KCl, 5 $CaCl_2$, 10 Hepes/Mes, pH 7.2. **a.** Single channel currents at different pipette voltages as indicated. Current traces of a length of 15 s are shown. **b.** Current-voltage (IV) plot of data shown in **a.** The slope of the regression line is 10 pS

much lower than obtained with protoplasts prepared by enzymatic digestion of the cell wall (De Boer et al. 1994, Kurt Köhler and Bert De Boer, pers. comm., and own unpubl. results). A high seal resistance of at least 1 GΩ is important for detecting single ion channel events in membrane patches. Otherwise small, spontaneous changes in seal resistance would cause currents in the range of single channel currents thus mimicking opening and closing of ion channels.

In Fig. 19.5 ion channel recordings obtained from a tip protoplast in the cell-attached configuration are presented. In this experiment the seal resistance was 5 GΩ and the observed channels had a conductance of 10 pS. According to the sign convention these channels will allow a cation efflux when the plasma membrane is depolarized. All recorded channels were almost similar to an outward rectifying K^+ channel already observed in pollen grain protoplasts from lily pollen (Obermeyer and Kolb 1993). So far, no channels allowing specifically a Ca^{2+} influx have been detected in tip protoplasts of pollen tubes. It is conceivable that calcium ions enter the tip cytoplasm via ion channels exhibiting a low cation selectivity. Such a Ca^{2+} influx pathway has been observed in guard cells (Fairley-Grenot and Assmann 1992).

In addition to single channel events, spontaneous increases and decreases in electrical noise were observed, perhaps indicating the opening and closing of low conductivity ion channels which cannot be resolved by the patch-clamp method.

19.4 Conclusion

Laser microsurgery is a suitable tool for gaining local access to the plasma membrane in pollen tubes without changing the tubes' polarity. The isolated parts of the plasma membrane still showed ion channel activity when assayed with the patch-clamp technique. Surprisingly, these laser-isolated protoplasts are difficult to seal onto a patch-pipette despite their cell wall-free surface. But there may be other factors determining the seal rate of plant protoplasts. Laser microsurgical release of protoplasts has been well established in growing *Fucus* rhizoids (Taylor et al. 1997) and approximately 80% of the released protoplasts could be sealed to a patch pipette. But other tip growing cells including root hairs and pollen tubes are more difficult to investigate probably because of their high growth rates. *Fucus* rhizoids grow very slowly (1–10 µm h^{-1}) compared to pollen tubes (10 µm min^{-1}) and root hair cells (1 µm min^{-1}). Secretion of cell wall material must be very efficient in these fast growing cells and during the gentle release of tip protoplasts the secretion machinery may not be affected and thus prevents an easy 'sealing' with the patch pipette. Therefore, factors inhibiting secretion may increase the seal rate in fast growing cells and future research will focus on this aspect.

Acknowledgements
We thank Dr. Lian Jaensch (Flinders University, South Australia) for critical reading of the manuscript and Dr. Margit Höftberger for technical assistance with the electron microscopy. This project was financially supported by the Austrian Fond zur Förderung der Wissenschaftlichen Forschung (FWF, P09666). RB and FA were supported by postdoctoral fellowships from the Deutsche Forschungsgemeinschaft and the British Royal Society, respectively.

References

Battey NH, Blackbourn HD (1993) The control of exocytosis in plants. New Phytol 125:307–338

Benkert R, Obermeyer G, Bentrup F-W (1997) Turgor pressure of growing pollen tubes. Protoplasma 198:1–8

Cai G, Moscatelli A, Cresti M (1997) Cytoskeletal organisation and pollen tube growth. Trends Plant Sci 2:86–91

Cheung A (1996) Pollen-pistil interactions during pollen tube growth. Trends Plant Sci 1:45–51

De Boer AH, Van Duijn B, Giesberg P, Wegner L, Obermeyer G, Köhler K, Linz K (1994) Lasermicrosurgery: A versatile tool in plant (electro)-physiology. Protoplasma 178:1–10

Fairley-Grenot KA, Assmann S (1992) Permeation of Ca^{2+} through K^+ channels in the plasma membrane of *Vicia faba* guard cells. J Membr Biol 128:103–113

Feijó JA, Malhó R, Obermeyer G (1995) Ion dynamics and its possible role during in vitro germination and tube growth. Protoplasma 187:155–167

Hall JL (1991) Electronmicroscopy of plant cells. Academic Press, London

Hamill OP, Marty A, Neher E, Sakmann B, Sigworth FJ (1981) Improved patch-clamp technique for high-resolution current from cells and cell-free membrane patches. Pfluegers Arch 391:85–100

Hiscock SJ, Kües U, Dickinson HG (1996) Molecular mechanisms of self-incompatibility in flowering plants and fungi – different means to the same end. Trends Cell Biol 6:421–428

Kurkdjian A, Leitz G, Manigault P, Harim A, Greulich O (1993) Non-enzymatic accesss to the plasma membrane of *Medicago* root hairs by laser microsurgery. J Cell Sci 105:263–268

Malhó R, Read ND, Trewavas AJ, Pais MS (1995) Calcium channel activity during pollen tube growth and reorientation. Plant Cell 7:1173–1184

Obermeyer G , Bentrup F-W (1996) Regulation of polar growth and morphogenesis. Prog Bot 57:54–67

Obermeyer G, Blatt MR (1995) Electrical properties of intact pollen grains of *Lilium longiflorum*: Characteristics of the non-germinating pollen grain. J Exp Bot 46:803–813

Obermeyer G, Kolb H-A (1993) K^+ channels in the plasma membrane of lily pollen protoplasts. Bot Acta 106:26–31

Obermeyer G, Weisenseel MH (1991) Calcium channel blocker and calmodulin antagonists affect the gradient of free calcium ions in lily pollen tubes. Eur J Cell Biol 56:319–327

Pierson ES, Cresti M (1992) Cytoskeleton and cytoplasmic organization of pollen and pollen tubes. Int Rev Cytol 140:73–128

Pierson ES, Miller DD, Callaham DA, Shipley AM, Rivers BA, Cresti M, Hepler PK (1995) Pollen tube growth is coupled to the extracellular calcium ion flux and the intracellular calcium gradient: Effect of BAPTA-type buffers and hypertonic media. Plant Cell 6:1815–1828

Reiss H-D, Herth W (1979) Calcium ionophore A23187 affects localized wall secretion in the tip region of pollen tubes of *Lilium longiflorum*. Planta 145:225–232

Taylor A, Brownlee C (1992) Localized patch-clamping of a polarized plant cell. Plant Physiol 99: 1686–1688

Taylor A, Mannison N, Brownlee C (1997) Regulation of channel activity underlying cell volume and polarity signals in *Fucus*. J Exp Bot 48:579–588

Weber G, Greulich KO (1992) Manipulation of cells, organelles, and genomes by laser microsurgery and optical trap. J Microscopy 167:127–151

Weisenseel MH, Jaffe LF (1976) The major growth current through lily pollen tubes enters as K^+ and leaves as H^+. Planta 133:1–7

Weisenseel MH, Nuccitelli R, Jaffe LF (1975) Large electrical currents transverse growing pollen tubes. J Cell Biol 66:556–567

Weiss KG, Polito VS, Labovitch JM (1988) Microfluometry of pectic materials in the dehiscence zone of almond (*Prunus dulcus*) fruits. J Histochem Cytochem 36:1037–1041

CHAPTER 20

The Rheological Properties of the Pollen Tube Cell Wall

A. Geitmann

PCM, Wageningen Agricultural University, Arboretumlaan 4, 6703 BD Wageningen, The Netherlands
e-mail: anja.geitmann@guest.pcm.wau.nl
telephone: +31-317-484864
fax: +31-317-485005

Abstract. Pollen tubes are very fast, polarly growing cells which produce a high amount of cell wall. The architecture of the pollen tube cell wall, its formation and rheological properties are discussed with regard to plant cell growth. The phenomenon of pulsating pollen tubes which show hoop-shaped depositions of pectin along their longitudinal axis represents a particular model for the investigation of plant cell growth since the equilibrium between cell wall force and internal turgor must be extremely well controlled in these cells. The putative biological function of this unique mode of growth is discussed.

20.1 Introduction

20.1.1 Rheology

Rheology is "a science dealing with the deformation and flow of matter" (Webster's Dictionary 1985). The rheology of the plant cell wall has been studied intensively because the balance between cell wall force and the cell turgor represents a major controlling factor in plant cell growth and because plant cells sustain external tension by the presence of a rigid cell wall (Showalter 1993). Two important cell wall characteristics are plasticity (capacity to be irreversibly deformed) and elasticity (capacity to be reversibly deformed). To date most investigations concerning the rheological properties of particular plant cell walls are carried out using entire organs or blocks of multicellular tissues such as stem segments. Standard experiments include the creep test consisting in a load extensiometer in which tissue segments are fixed between clamps and exposed to a stretching force.

In the case of a single cell a similar determination of the rheological properties of the cell wall poses considerable problems for obvious reasons: any direct mechanical handling of the cell is apt to influence the cell or more likely to destroy it. Therefore a different approach has to be chosen. In the present chapter some aspects of the rheological properties of the pollen tube cell wall are elucidated by interpreting various observations obtained during studies of pollen tube growth *in vitro*.

20.1.2 The Pollen Tube

Pollen tubes are plant cells with particular characteristics – they show extreme growth rates and they exhibit tip growth. The mechanism regulating their polar elongation has been studied intensively and yet various aspects remain obscure. The common feature of tip growing cells, such as fungal hyphae, rhizoids and root hairs, is their polar organization which allows the cells to grow in one direction only. In these cells multidirectional growth occurs only in the sense that bifurcation can give rise to two growing tube tips, a common feature in fungal hyphae, whereas pollen tubes divide only rarely in most plant species. Frey-Wissling (1959) defined tip growth as "locally restricted enlargement of a cell that usually has a tubular shape". This particular case of cell morphogenesis was described in a similar way by Green (1969) and Sievers and Schnepf (1981) as an expansion which is confined to a roughly hemispherical, dome shaped growth zone, which produces a cylinder at its base. In consistence with this concept, elongation growth in pollen tubes is confined to the apical end of the cell, the growth zone (Rosen et al. 1964; Cresti et al. 1977). In the basal part of the cell only thickening of the wall and other secondary processes take place. Consequently, tip growing cells are highly polarized showing morphological and physiological gradients concerning vesicle fusion, cellulose and callose synthesis, transport processes, and electric currents. Cell wall precursors for formation of the primary cell wall, i.e. the outer layer, are transported from their site of production, the dictyosomes, to the growing end of the cell and are released only here to contribute to the elongation of the cell. It is still not well understood, how in such a system the cell diameter is controlled over a long period of growth. Likewise, the maintenance of wall thickness and the control of elongation rate both of which are strictly correlated with the regulation of cell diameter (Schnepf 1986) warrant further investigation.

Apart from conveying stability during pollen tube elongation the cell wall fulfills a variety of other functions, such as adhesion, recognition, interaction with the transmitting tissue, and selective uptake of substances which is partially controlled by the porosity of the walls. These topics as well as the connection of the cell wall with the membrane cytoskeleton are covered in various reviews (for reviews on the cell wall in general see McCann and Roberts 1994; Gibeaut and Carpita 1994; Reuzeau and Pont-Lezica 1995. The pollen tube cell wall has been reviewed recently by Li et al. 1997).

20.2 The Plant Cell Wall – Rheology and Expansion in Growing Cells

The rheological properties of the plant cell wall play a crucial role in cell growth. Essential properties are stretch resistance, which is provided by the load bearing cellulose/xyloglucan/hemicellulose network, and compression resistance, which includes the pectic polysaccharide network. In addition, structural proteins are likely to play a role in cell wall resistance (Cosgrove 1993a; McCann and Roberts 1994). Cell enlargement involves a stretching of the "old" wall already present, as well as synthesis of new wall material (reviewed by Setterfield and Bayley 1961; Wilson 1964; Cleland 1971; Ray et al. 1972; Preston 1974; Burström 1979; Taiz 1984; Masuda 1990; Ray 1992; Boyer 1992; Cosgrove 1993a, 1993b). Therefore plant cell growth is governed by the ability of the cell walls to extend elastically and plastically which in turn depends on the chemical composition of the cell wall polymers, on their crosslinking with each other, on the orientation of microfibrils, on the pH, on the presence of ions, such as Ca^{2+} and Mg^{2+} and on the presence of putative lubricants, such as arabinogalactan proteins (Willats and Knox 1996) and extensins (McQueen-Mason and Cosgrove 1995). *In vivo* and *in vitro* these characteristics can be influenced by the action of growth hormones and by the growth conditions.

Growing cell walls were compared with the behavior of viscoelastic liquids (Cosgrove 1993a). Among the physiological processes occurring basically simultaneously during steady growth, stress relaxation of the wall was claimed to be the primary event (Ray et al. 1972; Preston 1974). This relaxation subsequently causes a decrease in water potential, water uptake, increase in volume and extension of the cell wall. Concomitantly synthesis and liberation of wall material take place (Ray et al. 1972; Burström 1979). Inclusion of newly synthesized wall material occurs either by apposition, i.e. deposition of the new polymers between the plasma membrane and the old wall. Alternatively, new polymers can be inserted between old polymers, termed intercalation or intussusception. Both, this constant metabolic input (Taiz 1984) as well as turgor pressure (Ray 1992) represent necessary requirements for wall loosening. According to Ray (1992) loosening can be attributed to the breaking of crosslinks which permits a sliding of the wall polymers against each other, thus allowing stretching. The properties of the cell wall, its formation, relaxation, stretching and deformation play a central role in pollen tube growth, which is why this aspect is treated with particular attention.

20.3 Pollen Tube Cell Wall

20.3.1 Formation and Structure

Brewbaker and Kwack (1964) stated that pollen tube elongation is almost exclusively a process of cell wall synthesis, involving the production of cellulose, callose, pectic substances, and minor amounts of protein. According to these authors little or no increase in cytoplasm occurs during elongation. The pollen tube wall is a typical plant cell wall in as far as it consists of several layers which correspond to the primary and secondary walls of other plant cells (for reviews on the pollen tube cell wall see Heslop-Harrison 1971, 1987; Knox 1984; Mascarenhas 1993; van Amstel et al. 1994; Harris et al. 1984; Steer and Steer 1989; Derksen et al. 1995, Li et al. 1997). The outer fibrillar layer is present around the entire tube and consists mainly of pectins, which are synthesized within the cell. The middle layer which is not always clearly distinguishable from the outer layer contains cellulose. The innermost layer is composed of callose and appears rather electron translucent and homogeneous in the electron microscope. The inner callosic layer is seemingly absent at the apical end of the pollen tube and becomes thicker towards the distal part. In older parts of the pollen tube the inner callosic layer is prominent and distinct from the fibrillar outer layer, since it does not stain after Thiéry treatment (Rougier et al., 1973). Its texture seems to be random in tobacco pollen tubes (Kroh and Knuiman 1982) and its callosic contents has been demonstrated by immunolabel (Meikle et al. 1991; Geitmann et al. 1995a). Compared to other plant cells pollen tubes contain unusually high amounts of callose (Rae et al. 1985, Read et al. 1996). Callose is supposed to be synthesized and deposited at the surface of the plasma membrane by callose synthases which are localized in the membrane (Gibeaut and Carpita 1994). It was suggested that callose is synthesized by complexes which initially produce cellulose fibrils at the tip of the pollen tube (Van Amstel 1994). Recently two glycoproteins have been identified which might be involved in the correct deposition of callose in the pollen tube cell wall (Capková et al. 1987, 1994, 1997). Depositions of visible amounts of callose start about 10–30 μm back from the pollen tube tip. The apical part of the pollen tube is usually devoid of callose except for pollen tubes which are arrested for example due to incompatibility reaction (Geitmann, this volume, and references therein). However, immunolabel evidenced recently that callose seems actually to be present also in the tip and in the secretory vesicles of normally growing pollen tubes, albeit in the form of molecules of low degree of polymerization. This might be the reason why the tip localized callose has not been detected by aniline blue, the standard dye for callose (Hasegawa et al. 1996).

The outer layer of the pollen tube cell wall is present around the entire tube including the tip. It appears fibrillar in the electron microscope and is strongly electron scattering after Thiéry treatment indicating its high content of polysaccharides other than callose. The main components in this wall layer are cellulose and pectins. The deposition of the fibrils does not seem to be controlled by the direction of the microtubular system underlying the wall as is the case in other plant cells (Emons 1989).

Immunogold labeling has shown that pectins are present almost exclusively in the outer layer of the pollen tube cell wall (Kroh and Knuiman 1982; Geitmann et al. 1995a, 1995b; Li et al. 1995b; Jauh and Lord 1996; Geitmann 1997). Pectins have been postulated to play an important role in cell wall hydration, filtering, adhesion between cells, and wall plasticity during growth (Jarvis 1984; Baron-Epel et al. 1988; Levy and Staehelin 1992; Carpita and Gibeaut 1993). They are composed of homogalacturonan, rhamnogalacturonan I and II, oligosaccharide side chains and arabinogalactan. The degree and pattern of the methyl-esterification of the homogalacturonan domains has influence on the physical state of the cell wall matrix. The presence of Ca^{2+} causes dimerization of acidic pectins to form a relatively rigid pectate gel, whereas methyl-esterified pectins are rather water soluble (Grant et al. 1973; Rees 1969). The subcellular route of pectin synthesis is believed to start with first polymerization steps in the ER, however, immunogold label for pectins indicated that the assembly of the pectin backbone up to a high degree of polymerization takes place in the cis and medial cisternae of the dictyosomes (Bowles and Northcote 1976; Delmer and Stone 1988; Levy and Staehelin 1992). Nearly simultaneously to the backbone assembly methyl-esterification of the carboxyl groups by pectin methyl transferases takes place (Moore and Staehelin 1988; Staehelin et al. 1991; Zhang and Staehelin 1992; Liners et al. 1994). Dictyosome derived vesicles transport the pectins from their site of production to the plasma membrane, probably with the help of the cytoskeletal system (Delmer and Stone 1988). During this transport final synthesis steps, such as additional methyl-esterification, take place (Vian and Roland 1991; Fincher and Stone 1981; Carpita and Gibeaut 1993). In pollen tubes these highly esterified pectins are transported in secretory vesicles to the pollen tube tip, where these fuse with the plasma membrane and release their contents, forming the outer pollen tube layer (Geitmann et al. 1995a). Once outside, the esterified pectins are subjected to de-esterification by pectin methyl-esterases present in the transmitting tissue or in the pollen tube wall proper. Therefore the concentration of methylesterified pectins is highest at the pollen tube apex and decreases along the pollen tube axis towards the pollen grain as shown by immunofluorescence label [Li et al. 1994; Jauh and Lord 1996 (The latter publication has to be considered with caution, since apparently the wrong secondary antibodies were used)].

The differences in pectin esterification along the pollen tube axis result in a gradient in cell wall rigidity. The apical pectin is highly methyl-esterified and therefore not Ca^{2+} dimerized, which is expected to render the cell wall rather deformable at this part of the tube. In contrast, the distal part of the tube cell wall contains de-esterified pectins which in the presence of Ca^{2+} dimerizes to the more rigid pectate. The assumption that cross-linking between cell wall polymers involves Ca^{2+}, as already suggested by Brewbaker and Kwack (1963), is supported by the demonstration of the abundant presence of this ion in the pollen tube cell wall using antimonate precipitation (Geitmann 1997).

However, the difference in wall plasticity between the apical and the distal parts of the pollen tube is likely to be a consequence not only of the differences in pectin configuration but also of the presence of callose. As was described earlier a visible callosic inner lining of the pollen tube cell wall is absent from the apical 20 to 30 µm of the tube. Experiments have shown that this zone coincides with the part of the pollen tube that yields upon alteration of growth conditions or application of toxic substances. Application of N-ethylmaleimide (inhibitor of ATPase activity) or cytochalasin D (inhibitor of actin polymerization) causes among other effects the swelling of the apical part of the pollen tube, whereas the diameter of the distal tube is maintained (Geitmann et al. 1996b; Geitmann 1997). Similar swellings were reported after application of cadmium of mercury ions (Sawidis and Reiss 1995). The opposite was observed after application of monensin (ionophore), which results in formation of a thinner tube at the apical tip while the older parts of the tube maintain the diameter (Geitmann 1997). These observations indicate that there is indeed a strong gradient of cell wall plasticity along the longitudinal axis of a growing pollen tube with higher deformability being located at the apical 20 to 30 µm of the cell. This gradient seems to depend on the degree of cross-linking between the pectic polymers as well as on the presence of callose.

The characteristics of a plant cell wall do not only depend on the composition of polysaccharide components, but also on the presence of proteins. Immunolabel evidenced the presence of arabinogalactan proteins in the outer and middle layer of the pollen tube cell wall (Li et al. 1992, 1995b; Jauh and Lord 1996). The distribution of these proteins in the cell wall as well as their possible function in pollen tube growth is discussed by Li et al. (1997). The same review also covers the presence of other wall-bound glycoproteins as well as the putative role of extensin-like proteins in the pollen tube cell wall (Rubinstein et al. 1995a, 1995b). These and other proteins might be involved in adhesion events and interaction with the stylar transmitting tissue (Cheung et al. 1995).

It has to be mentioned that also the cytoskeleton contributes to the stability of plant cells. However, in pollen tubes the elements of the cytoskeleton are mainly oriented longitudinally which indicates that they probably do not contribute to a great degree to the radial stability of the tube. The stability of the apical growth zone might however be conveyed partially by cytoskeletal elements (for a recent review on the pollen tube cytoskeleton see Li et al. 1997 and references therein).

20.3.2 Boron and the Pollen Tube Cell Wall

As early as 1932 Schmucker observed that pollen germination and pollen tube growth of many species could be considerably improved by adding boric acid to the culture medium (Schmucker et al. 1932; Schmucker 1934). It was found that the optimal concentration of boric acid was between 0.001% to 0.02%, depending on the species (Johri and Vasil 1961; Stanley and Loewus 1964; Vasil 1964; Brewbaker and Kwack 1964); higher concentrations were observed to be inhibitory for pollen germination. The effect of boron on the cellular level is still far from clear. Various hypotheses suggested involvement of boron in sugar transfer (Fähnrich 1964), effects on plasmalemma bound ATPase (Obermeyer et al. 1996) and reductase activities, phenol oxidation, Ca^{2+} distribution and others (Loomis and Durst 1992; review by Goldbach 1997). As far as the cell wall is concerned, boron was proposed to be involved in cell wall synthesis and formation at the level of the Golgi apparatus as well as liberation of secretory vesicles (Kouchi and Kumazawa 1976). Teasdale and Richards (1990) suggested that boron might contribute to the strength of cell walls during the expansion phase of cell growth through its participation in the reversible formation of a carbohydrate gel around cellulose microfibrils. Yamauchi et al. (1986) as well as Hu and Brown (1994) proposed boron to be related to the pectic fraction of the cell wall influencing the amount of bound Ca^{2+} thus interfering with extensibility. Kobayashi et al. (1996) and Matoh et al. (1996) showed that boron forms crosslinks between rhamnogalacturonan II chains. Such a complex has recently been demonstrated to occur in the cell walls of lily pollen tubes (Matoh as cited in Li 1997). These crosslinks are supposed to allow pH-controlled cell expansion (Ghanadan et al. 1995).

During the experiments on *in vitro* growing pollen tubes many of the tubes burst upon removal of boron (Schmucker 1934; Stanley and Lichtenberg 1963; Dickinson 1978; Polster et al. 1992; Obermeyer et al. 1996; Geitmann 1997). Since it can be assumed that internal turgor was not affected by the removal of boron (Findeklee and Goldbach 1996), the bursting events confirm the hypothesis, that the lack of boron renders the cell wall more unstable, possibly because of the reduction of borate ester crosslinks between the cell wall polymers as suggested by Goldbach (1997).

In contrast to fungal hyphae in which bursting mostly occurs in the apical part but not at the very tip of the cell (B Heath, personal communication) pollen tubes being deprived of boron burst and release the cytoplasmic contents at the very tip. Since the pollen tube is basically a single compartment, the internal pressure exerted on the surface should be equal at all points. Apparently the pollen tube tip represents the weakest point of the cell surface and therefore it is the first part to break upon a disturbance in the balance between turgor and cell wall resistance. This observation confirms the presence of a gradient of cell wall stability along the pollen tube axis with the most unstable part being located at the tip.

20.4 Pulsating Growth in Tip Growing Cells

Video observation of *in vitro* growing pollen tubes and fungal hyphae has evidenced that tip growth is not a steady process, but that growth rates are subject to continuous fluctuations (Tang et al. 1992; Plyushch et al. 1995; Pierson et al. 1995, Geitmann et al. 1996a; Derksen 1996; Li et al. 1997; Holdaway-Clarke et al. 1997; Geitmann 1997; Geitmann and Cresti 1998). One can distinguish two types of fluctuations. One of them has been described as sinusoidal oscillations of the growth rate with amplitudes in the range of two- to fourtimes the basic rate and periods in the range of seconds. This mode of growth was observed in many fungal hyphae (Kaminskyj et al. 1992; López-Franco et al. 1994; Bracker et al. 1995) and the pollen tubes of *Aloë zebrina* (Tang et al. 1992) and *Lilium longiflorum* (Pierson et al. 1995; Holdaway-Clarke et al. 1997). In contrast, the other class of fluctuations shows spike-shaped oscillations of the growth rate and was described as pulsatory growth: In pollen tubes of *Petunia hybrida* and *Nicotiana tabacum* slow growth is interrupted in intervals with the length of several minutes by pulse-like elongations lasting for about 5–20 sec (Pierson et al. 1995, Geitmann et al. 1996a; Geitmann 1997; Geitmann and Cresti 1998) (Fig. 20.1). Similar observations were made in pollen tubes of *Gasteria verrucosa,* which show a higher pulsing frequency (Plyushch et al. 1995). Pierson et al. (1995) report the increase of growth rate during these pulses to be about twelvefold, Geitmann et al. (1996a) observed alterations of growth rate up to 50fold.

Oscillatory growth in fungal hyphae were proposed to be caused by fluctuations in the liberation of secretory vesicles (López-Franco et al. 1994; Bracker et al. 1995) or to be

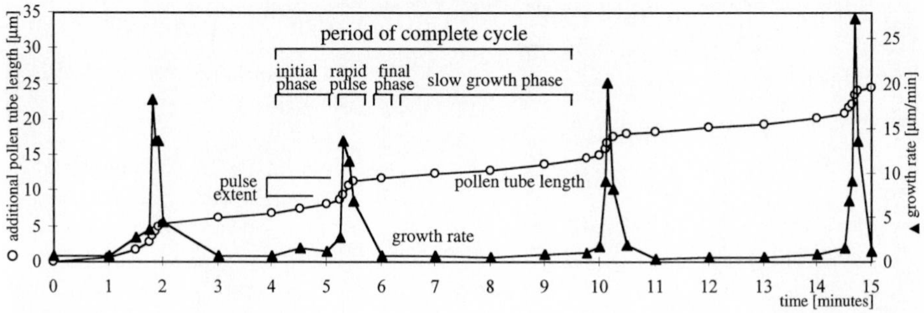

Fig. 20.1. Length and growth rate of pulsating pollen tube of *Petunia hybrida*. It is noteworthy that the rapid phase of expansion is preceded by an initial intermediate phase, which in many cases is clearly distinguishable from the main pulse (arrow)

controlled by the cytoskeleton accompanied by changes in turgor pressure (Kaminskyj et al. 1992). In *Lilium* pollen tubes Ca^{2+} measurements indicated that fluctuations in Ca^{2+} influx at the apical tip might be correlated to the alterations of growth rate in oscillatory growth (Pierson et al. 1996; Holdaway-Clarke et al. 1997). A putative mechanism based on fluctuations of cell wall plasticity, vesicle liberation and cytosolic Ca^{2+} concentration is proposed by Holdaway-Clarke et al. (1997).

Even though pulsatory growth in *Nicotiana tabacum* and *Petunia hybrida* shows fluctuations of the growth rate on a much higher scale the principle mechanism might be similar to the one of oscillatory growth. It was observed that inhibitors of Ca^{2+} channels prevent these pollen tubes from pulsing (Geitmann 1997; Geitmann and Cresti 1998) which indicates the participation of regulated Ca^{2+} influx in the control of the pulses. A two-step model for the events occurring during pulsating growth was proposed (Geitmann 1997; Fig. 20.2). This model is based on the concept, that oscillations of the Ca^{2+} influx, which are controlled by a feedback mechanism, trigger pollen tubes to elongate in a pulse-like manner, similar to plant cell systems with Ca^{2+}-triggered secretion (Plattner 1989; Baydoun and Northcote 1980, 1981). A model for the regulation of pulsatory pollen tube growth based on Ca^{2+} flux had already been proposed by Derksen (1996). The author suggested that Ca^{2+} induced vesicle fusion leads to a thickened wall and thus causes a decrease in expansion rate. The opposite is proposed in our model (Geitmann 1997): the released secretory material causes a thickening of the cell wall but also its simultaneous relaxation, thus triggering rapid turgor driven expansion of the tube tip.

The participation of the expansion component of the pulse in this model seems obvious considering the speed of elongation achieved during the rapid phase of the pulse. Furthermore it has been observed that in the case of bursting of a pulsating pollen tube, this event always occurs upon completion of a pulse, thus indicating that at this point of time the pollen tube cell wall is thinned and/or less stable, which could be explained by previous stretching (Geitmann et al. 1996a; Geitmann 1997).

The fact that the pulse-like elongations do not only consist of rapid turgor driven expansions of the apical cell wall, but that they comprise an initial phase of sudden vesicle release is reasonable considering the fact that within 5 to 20 sec the tip is elongated by up to 2 µm. In pollen tubes the zone involved in the growth process is very limited, Rosen et al. (1964) estimated it to be confined to the apical 3 to 5 µm in steadily growing

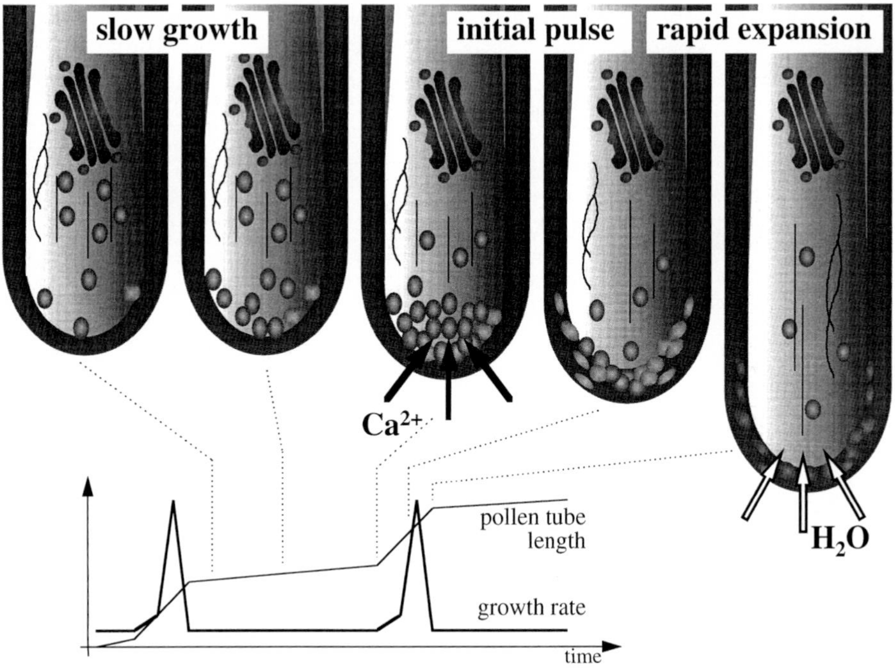

Fig. 20.2. Hypothetical model for the sequence of events during one cycle of pulsating growth. During slow growth the rate of vesicle fusion is rather low. Possibly governed by stretching of the membrane and/or the underlying membrane cytoskeleton, Ca^{2+} channel located at the pollen tube apex open, thus causing a sudden Ca^{2+} influx. The elevated cytosolic Ca^{2+} at the tip triggers a rapid release of secretory vesicles, giving rise to the initial phase of the pulse. The low rigidity of the newly secreted cell wall material causes the cellular water potential to increase, thus leading to a water influx which in turn provokes an expansion of the apical cell wall. This turgor driven expansion represents the rapid phase of the pulse

lily pollen tubes. Assuming that in *Nicotiana* and *Petunia* tubes not more than the apical 10 to 20 µm of the tube are involved in the stretching (probably much less), each pulse would contribute at least an additional 10 to 20% of apical cell surface. It has to be considered, that simultaneously to the wall also the plasma membrane at the tip would have to be stretched to the same degree. However, it is known that the elastic limit of lipid membranes is as low as 2–3% (Kwok and Evans 1981; Wolfe and Steponkus 1981; Wolfe et al. 1986). It is therefore reasonable to propose that during the initial pulse phase a release of secretory vesicles has to take place, which provides cell wall material as well as plasma membrane surface necessary for the subsequent expansion. This is also supported by the observation of an initial wall thickening before the pulses in *Gasteria* pollen tubes (Plyushch et al. 1995), which might represent wall material released prior to the rapid expansions.

20.4.1 The Cell Wall of Pulsating Pollen Tubes

Pierson et al. (1995) were the first to speculate that pulsating growth might be correlated with the appearance of periodic ringshaped depositions of pectins and arabinogalactan proteins in the cell walls of these pollen tubes after immunofluorescence or immunogold labeling (Li et al. 1992, 1994, 1995a; Geitmann et al. 1995b; Geitmann 1997). These periodic ring-shaped or hooped depositions of pectins and of arabinogalactan proteins seemed too prominent and too regular in their distance, to be purely accidental irregularities. The spatial distance of 4–6 μm between these rings correlates with the spatial distance between two pulses in the pollen tubes of these species. This however does not exclude, that these hoops might be artifacts resulting from the *in vitro* cultivation of pollen. The counter-evidence was provided by Li et al. (1995a), who found the same phenomenon in *in situ* grown pollen tubes. Furthermore Li and coworkers were able to demonstrate that the ring-shaped depositions of pectin in the pollen tube cell wall correlate with phases of slow growth, whereas the less densely labeled parts of the cell wall correspond to the pulse-like expansions. This was concluded from experiments during which growth was reduced by changing the osmotic value of the medium or by application of caffeine and thus inducing artificial rings in pollen tubes of *Lilium* which usually do not show pulsating growth or pectin rings (Li et al. 1996).

The ultrastructural investigation of pulsating pollen tubes showed that the accumulations of pectin label were localized at fibrillar structures in the inner zone of the outer cell wall layer (Geitmann et al. 1995b). It is noteworthy that the thickness of the entire cell wall in the zones of higher pectin density was identical to the areas in between. In conclusion, the cell wall does not actually thicken during slow growth, instead it seems as if only its composition and/or density changes. These results are consistent with the observation of pulsating pollen tubes in brightfield microscopy, which do not show any significant changes in cell diameter along the pollen tube axis.

In general, during plant cell growth all of the processes, deposition of precursor material, wall loosening, water uptake and stretching of the cell wall, are closely interrelated and occur simultaneously (Setterfield and Bayley 1961; Ricard and Noat 1986). Pulsating pollen tubes, however, repeatedly undergo changes in their growth rate, probably involving constant alterations in the equilibrium between turgor and cell wall. Therefore in these cells this balance must be particularly delicate and extremely well controlled, since expansions occur on a time scale of seconds and in dimensions of μm. For this reason pollen tubes exhibiting pulsating growth represent an excellent model for the investigation of cell wall expansion and cell elongation in plant cells. The equilibrium between cell wall and turgor in pulsating pollen tubes has been studied applying various agents which interfere either with the properties of the cell wall or with the internal turgor (Geitmann 1997). In the following paragraphs the findings are discussed with respect to their significance for plant cell growth.

20.4.2
Auxin and Pulsating Pollen Tube Growth

Auxin is a plant growth hormone which is supposed to have a relaxing effect on the cell wall. Determination of the optimal amount of auxin for pollen tube growth resulted in a concentration around 10 μM which enhanced pollen tube growth and/or germination in *Nicotiana tabacum* and *Petunia hybrida* (Geitmann 1997). This is consistent with previous observations on pollen of other plant species (Konar 1958; Iwanami 1980; Dhawan and Malik 1981; Li 1990) and on other plant tissues in general (Brummell and Hall 1987; Thimann and Biradivolu 1994). Auxin concentrations above 100 μM inhibited pollen tube growth and/or germination (Sondheimer and Linskens 1974; McLeod 1975; Dhingra and Varghese 1985; Dhingra 1990; Geitmann 1997). In general, the physiological effect of auxin can be divided into two phases. Whereas the long term effect, which is necessary for sustained auxin-induced growth, depends on increased protein and cell wall synthesis (Evans 1974; Nishitani and Masuda 1982), the immediate, short term effect is manifested in the cell wall. Auxin causes a loosening or relaxation of the cell wall, probably mediated by protons (Evans 1974; Masuda 1990), thus allowing the hydrostatic pressure to stretch the wall and cause cell elongation under simultaneous uptake of water (Cleland 1967a, 1967b). It is important to note that the effect of the hormone is based on a change in the cell wall properties and not on an alteration in hydraulic conductivity (Cosgrove and Cleland 1983). It was suggested that auxin causes an increase of cell wall plasticity (Galston and Purves 1960; Galston and Davies 1969; Kutschera and Schopfer 1986) and lowers the adjustable yield threshold of the cell wall (Okamoto et al. 1990; Nakahori et al. 1991; for a review see Masuda 1990). However, the picture is still not completely clear, since Hohl and Schopfer (1992, 1995) demonstrated that the hormone affects the amount of hysteresis in viscoelastic wall extensibility. In the experiments on pulsating pollen tubes the capacity of auxin to cause cell wall loosening without influencing the osmotic pressure of the cytoplasm (Cleland 1959, 1971; Boyer and Wu 1978; Cosgrove and Cleland 1983) was utilized.

External application of auxin considerably affected the behavior of growing pollen tubes of *Nicotiana tabacum* and *Petunia hybrida*. Both increase in pulse frequency and in slow growth rate indicate that the properties of the cell wall play a role in the control of pulsating growth. The degree of crosslinking in the cell wall seems to be essential for the frequency and the slow growth rate. The balance between turgor and mechanical resistance of the cell wall, which is influenced by external addition of auxin (Green and Cummins 1974) seems to play a key role in the control of pulse frequency. This is supported by the finding that addition of auxin induced pulsating growth in steadily growing pollen tubes. Apparently, the hormone caused a disturbance in the balance between turgor and cell wall, which initiated oscillations. This together with the fact that the presence of internal pressure is a prerequisite for auxin-induced wall loosening (Cleland 1967a, 1967b; Hohl et al. 1995), provides support for the previous suggestion that the pulse-like elongations are driven by the internal hydrostatic pressure and not exclusively by intussusception of cell wall material (Geitmann et al. 1996a). On the other hand auxin had no significant effect on pulse amplitude proper. This allows the conclusion that the material expanded during the pulse is newly secreted and was therefore not exposed to the hormone prior to its expansion. This confirms the two-step model claiming a release of secretory material to be required during the initial phase of the pulse.

It is noteworthy that the response of growing pollen tubes upon auxin addition occurs as fast as one to three mins after addition of the hormone. Except for a few examples (e.g. Sweeney and Thimann 1938; Preston 1974 and refs therein) in most experiments the earliest increase of growth rate observed in plant tissues treated with auxin was reported to occur after 7 to 10 mins (Brummell and Hall 1987; Evans and Vesper 1980). The fact that pollen tubes react as fast as one min after hormone treatment indicates, that the lag time observed in most of the other plant systems might be due to the limited speed of penetration of auxin through the tissue, a limiting factor not relevant for pollen tubes *in vitro*, since these are in direct contact with the surrounding medium.

Another open question concerning the effect of auxin was elucidated by application of the hormone to pollen tubes in which the intracellular transport of secretory vesicles as well as growth had been inhibited with monensin or brefeldin A (Geitmann 1997).

The absence of cell elongation in these living but not growing cells upon the application of auxin indicates that stretching of the plant cell wall due to auxin does not only depend on internal hydrostatic pressure (Cleland 1959, 1967b, 1971; Hohl et al. 1995), but also on an active vesicular traffic as proposed by Phillips et al. (1988) and Morré (1994).

20.4.3 Pectin Methyl-Esterase and Pollen Tube Growth

Application of pectin methyl-esterase on pulsating pollen tubes resulted in a considerable reduction of pulse frequency (Geitmann 1997). It is conceivable that the rigidifying effect of the enzyme (Stoddart and Northcote 1967; Binet 1976) is manifested in particular at the apical part of the pollen tube, since under normal conditions this region of the cell wall shows the highest plasticity. As a result pollen tube growth slowed down, and in some cases even stopped completely, which is in correspondence with the observations of Roggen and Stanley (1969). The finding that the pulse frequency in pulsating pollen tubes decreased due to the enzyme indicates that the properties of the cell wall have a regulating function in the control of the pulses. On the other hand, the lack of a significant effect of the enzyme on pulse amplitude proper confirms the conclusions drawn from the experiments involving auxin, i.e. that rapid expansion seems to involve mainly newly released cell wall material, which was not exposed to agents affecting the cell wall properties.

20.4.4 Boron and Pulsating Pollen Tubes

It has been mentioned previously that rapid removal of boron from the growth medium caused pollen tubes to burst. If however boron was removed slowly, bursting was avoided and the pollen tubes continued growing, albeit, in the case of pulsating tubes, with reduced growth rates during slow growth phase.

It seems therefore that on the one hand, the lack of boron causes the cell wall to rigidify, since pulse frequency is decreased similar to the effect of pectin methyl-esterase. However, unlike the effect of the enzyme the cell wall does not actually become stronger (as is the case after pectin methyl-esterase application due to pectate formation), but it becomes brittle and bursts easily. The lack of boron therefore seems to lower the plasticity of the cell wall and at the same time to destabilize its architecture. The mechanism of the influence of boron on the pollen tube cell wall in these experiments is far from clear. A possible explanation for the effect might be the interaction of boron with pectin polymers. Alternatively, the effect of boron on membrane H^+-ATPases, which has been observed in lily pollen tubes (Obermeyer et al. 1996) might be of significance in this case. The removal of boron might cause a reduction of H^+-ATPase activity, which in turn provokes a rigidification of the cell wall due to its alkalinization.

20.4.5 The Effect of Ruthenium Red on the Pollen Tube Cell Wall

Ruthenium red is supposed to inhibit Ca^{2+} sequestration by mitochondria (Picton and Steer 1985, Bernath and Vizi 1987; Hayat 1993), but it also has a fixating effect on the plant cell wall, due to pectin binding by salt linkages with anionic groups (Hayat 1993). Any immediate effect should be due to the effect of the dye on the cell wall, since the penetration of ruthenium red into intact cells, which would be a necessary prerequisite for its action on Ca^{2+} sequestration, is rather slow (Baldwin 1973; Gupta et al. 1988).

Application of ruthenium red either caused a complete arrest of pollen tube growth (Picton and Steer 1985, Geitmann 1997) or a reduction of slow growth rate and pulse frequency in pulsating pollen tubes, without inhibiting the pulses proper (Geitmann 1997). In the presence of ruthenium red the initial phase of the pulse-like expansions involves a "breaking" of the apical cell wall but the cell does not burst. Pulses still proceed normally except for a lamellar ring of rigid cell wall left behind the apical tip. This suggests that the cell wall is thickened and/or rigidified by ruthenium red in such a way that during a pulse it cannot expand smoothly anymore. A similar observation was made by van Amstel et al. (1994), who described the resuming of growth by temporarily arrested pollen tubes as "breaking through the pectin layer". Since in the presence of ruthenium red the pollen tube does not burst, unlike during removal of boron, one might presume that only the very outer part of the cell wall is actually rigidified by the dye and subsequently "breaks" upon the onset of a pulse, whereas the inner part of the outer cell wall layer still possesses enough plasticity to be able to yield.

20.5 Turgor and Tip Growth

Plant cell growth requires the presence of intracellular turgor pressure as a driving force (Cleland 1971; Ray et al. 1972; Preston 1974; Taiz 1984; Boyer 1992; Cosgrove 1993a). More correctly the driving force is embodied by a difference in water potentials between the inside and outside of the cells, since expansion is due to water uptake (Burström 1979). This was shown by application of media having high or low osmotic values, which affected growth of root hairs (Schröter and Sievers 1971) and fungal hyphae (Robertson 1958; Park and Robinson 1966). Similar observations were made in pollen tubes (Li et al. 1996). Polar growth involves a delicate balance between the mechanical resistance of the cell wall at the tip and the internal turgor, between deposition of new wall material and stretching of the present cell wall (Schnepf 1986; Steer and Steer 1989; Heath 1990; Wessels 1990). Various mechanisms have been proposed to control this particular mode of plant cell growth. On one hand it was suggested that apical growth represents simply a turgor driven stretching of the relatively weakest part of the wall, the apex, accompanied by continuous deposition of wall material. On the other hand a cytoskeletal control of the tip shape and the seemingly ameboid like movements was proposed (Steer and Steer 1989; Steer 1990). Whether or not this latter hypothesis can be confirmed, in any case both microfilaments and microtubules are likely to be in some way involved in the control of the tip shape in pollen tubes (Derksen and Emons 1990).

Inhibition of pollen tube growth by changes of the osmotic value of the medium (Li et al. 1996) indicate however that intracellular turgor is a necessary prerequisite for growth, even though pollen tubes seem to be able to regulate turgor very well upon changing of the external conditions (Benkert et al. 1997). The authors showed that growth rate is not linearly correlated with the osmotic value of the external medium. In the case of pulsating pollen tubes changes in the osmotic value of the medium influenced the pulse frequency but not the pulses proper (Geitmann 1997). This might indicate that the onset of the pulses is somehow controlled by the turgor, but not the pulse-like elongation proper. However, since changes in osmolarity were induced by alterations of the sucrose concentration, the direct effect of sucrose on the pollen tube metabolism cannot be ruled out in these experiments.

It would certainly be of interest to monitor the cell turgor in pulsating pollen tubes in order to determine its role in this particular growth mechanism. However, technical problems render turgor measurements difficult in *Petunia hybrida* or *Nicotiana tabacum* pollen tubes (P Findeklee, personal communication). Nevertheless there is evidence for the turgor not being the central oscillating factor in pulsating pollen tubes. The simultaneous observation of branches deriving from a bifurcated pollen tube demonstrated that pulse frequencies are never identical, which provides evidence for the independent growth behavior of the two tips (Geitmann 1997). The pulsations are therefore unlikely to be caused by fluctuations of the global cell turgor, since identical frequencies would have been expected in this case. However, it has to be considered that the pollen tube cytoplasm is rather viscous and contains high amounts of organelles and stabilizing structures, which might in some way allow local turgor differences.

Based on these results, it is concluded that turgor is a necessary prerequisite for pollen tube growth in general; in pulsating pollen tubes it influences the slow growth rate, but not the amplitude of the pulses. It is therefore suggested that turgor is the necessary driving force for the pulses, but that it is unlikely to be responsible for the fluctuations by way of pressure oscillations (as proposed by Kaminskyj et al. (1992) for fungal hyphae).

20.6 The Biological Function of Pulsating Growth

The origin and significance of pulsating growth and the resulting ring-like structures in the pollen tube cell wall have been subject to speculation. Since pulsating growth gives rise to the hoop-shaped cell wall depositions, this particular mode of growth and the resulting cell wall architecture are symptoms of the same phenomenon. If the occurrence of pulsating pollen tube growth had a biological significance, the question arises, which of the two symptoms – pulses or rings – is the significant part, and which represents the necessary prerequisite or inevitable consequence. Two scenarios can be visualized:

a) The ring-shaped architecture provides stability for the pollen tube
A comparison of species showed that the annular cell wall pattern occurs almost exclusively in plant species having a solid style (Li et al. 1994, 1995a). In these plants the pollen tube is forced to grow intercellularly between the cells of the transmitting tissue, whereas in plants with hollow style the pollen tube grows inside a stylar canal. It is likely, that the intercellular growth in solid styles requires higher resistance of the pollen tube, compared to growth through the stylar canal cavity. The tube has to withstand the pressure of the cells of the surrounding transmitting tissue and the firmness of their intercellular matrix. The hoop-shaped depositions found in the pollen tube cell wall contain mainly pectins which are compression resistant when Ca^{2+} dimerized. This indicates that the rings might have a stabilizing function for the cylindrical pollen tube architecture. In this scenario the pulsatory mode of growth would be the necessary mechanism for the production of these reinforcing rings.

b) The pulsing allows better penetration
Alternatively, the pulses proper might allow the pollen tubes to better enter the transmitting tissue. Rapid pulse-like elongations might enable the tube to penetrate through the dense extracellular matrix, very much like the principle of a sledge hammer. Furthermore one can speculate that the rapid pulses might elicit the liberation of pectin methyl-esterase or other wall degrading enzymes from the cells of the

transmitting tissue. This way the pulses would serve as a trigger for the auto-degradation of the extracellular matrix of the transmitting tissue. In this scenario the pulsating mode of growth would be the significant aspect of the phenomenon, whereas the ring-shaped deposits in the cell wall would represent a secondary consequence.

Whatever the primary aspect might be, in any case the coincidence of pulsating growth with the presence of a transmitting tissue in the respective pistils leads to the assumption, that both phenomena developed parallel, that perhaps a coevolution took place. Further investigations of different species might provide more information for a well-founded hypothesis regarding this issue.

Alternatively it has to be taken into consideration that pulsating growth might simply be a phenomenon without any functional significance. One could argue that pollen tube growth relies on a functioning feedback mechanism to ensure rapid steady growth. This feedback mechanism might be easily disturbed in some species (e.g. *Nicotiana tabacum*, *Petunia hybrida*), whereas it would be more resistant to external influences in other species. Disturbance of the feedback mechanism would induce oscillations in the entire regulation mechanism resulting in pulsating growth. This hypothesis is favored by the fact that in the *in vitro* situation the overall growth rate of pulsating pollen tubes is lower than the growth rate of steadily growing pollen tubes from the same flower. Therefore the pulsating behavior seems to represent a negative criterion causing evolutionary disadvantage for the respective pollen tubes. Upon pollination of a flower growth speed is a strong selection factor, since many pollen tubes are "competing" for fertilization of few egg cells. Following this line of argumentation there would be no reason for the evolutionary "function" of the phenomenon. However, the *in vitro* situation does not at all correspond to the *in situ* conditions, since in the latter the transmitting tissue might be a considerable obstacle to pollen tube growth, which could possibly be overcome by the pulsating mode. Monitoring of *in situ* growth rates would be necessary to answer this question.

20.7 Summary

The overview on various observations on pollen tube growth and behavior allows us to draw a first picture of the rheological properties of this particular plant cell. It could be shown that along the longitudinal axis of the cylindrical pollen tube there exists a gradient of cell wall plasticity, the apical tip being rather plastic and deformable, whereas the distal part of the tube shows considerable stability towards external influences. This gradient it due to the different distribution of callose in the cell wall as well as to the varying degree of methyl-esterification and dimerization of the pectin polymers. Externally applied agents interfering with the crosslinking of the cell wall influence the growth behavior of these normally extremely fast growing cells. It was demonstrated that the equilibrium between turgor and cell wall force is one of the crucial factors determining pollen tube growth. This balance has to be very well regulated especially in the case of pulsating pollen tubes. The two symptoms, pulsatory growth and hoop-shaped pectin depositions, were described for pollen tubes of *Nicotiana tabacum* and *Petunia hybrida* and a putative model was established to explain the sequence of events occurring during this particular mode of growth.

The success of pollen tube growth is on a first-come first-serve basis: Fertilization and offspring are granted only to those tubes which are the first to arrive at the ovary.

Consequently, fast elongation involving the rapid formation of cell wall is the main purpose of growing pollen tubes which is why it is probably not pretentious to claim that these cells provide an excellent model for the investigation of cell wall formation and architecture in relation to cell growth. The observation of pulsatory growth might indicate that during evolution highly sophisticated regulatory mechanisms developed to serve the purpose of rapid elongation, successful penetration and delivery of sperm cells.

Acknowledgements

The author would like to thank M. Cresti, Department of Environmental Biology, Siena, for the opportunity to carry out the research presented here. Furthermore the collaboration with Y.Q. Li concerning some of the work is acknowledged.

References

Baldwin KM (1973) A study of electrical uncoupling using ruthenium red. J Cell Biol 59:15a

Baron-Epel O, Gharyal PK, Schindler M (1988) Pectins as mediators of wall porosity in soybean cells. Planta 175:389–395

Baydoun E A-H, Northcote DH (1980) Measurement and characteristics of fusion of isolated membrane fractions from maize root tips. J Cell Sci 45:169–186

Baydoun E A-H, Northcote DH (1981) The extraction from maize (*Zea mays*) root cells of membrane-bound protein with Ca^{2+}-ATPase activity and its possible role in membrane fusion *in vitro*. Biochem J 193:781–792

Benkert R, Obermeyer G, Bentrup FW (1997) The turgor pressure of growing lily pollen tubes. Protoplasma 198:1–8

Bernath S, Vizi ES (1987) Inhibitory effect of ionized free intracellular calcium enhanced by ruthenium red and m-chloro-carbonylcyanide phenylhydrazone on the evoked release of acetylcholine. Biochem Pharmacol 36:3683

Binet P (1976) Succulence et activité pectinesterase chez *Aster trifolium* L. Physiologie Végétale 14: 283–295

Bowles DJ, Northcote DH (1976) The size and distribution of polysaccharides during their synthesis within the membrane system of maize root cells. Planta 128:101–106

Boyer JS (1992) Walls, water and solute in plant growth. Curr. Topics Plant Biochem Physiol 11:1–17

Boyer JS, Wu G (1978) Auxin increases the hydraulic conductivity of auxin-sensitive hypocotyl tissue. Planta 139:227–237

Bracker CE, López-Franco R, Bartnicki-Garcia S, Murphy DM, Howard RJ (1995) Satellite Spitzenkörper, pulsed growth, and the determination of cell shape in growing hyphal tips of fungi. Proc Royal Micr Soc 30:17

Brewbaker JL, Kwack BH (1963) The essential role of calcium ion in pollen germination and pollen tube growth. Amer J Bot 50:859–865

Brewbaker JL, Kwack BH (1964) The calcium ion and substances influencing pollen growth. In: Linskens HF (ed) Pollen physiology and fertilization. Elsevier North-Holland, Amsterdam, 145–151

Brummell DA, Hall JL (1987) Rapid cellular responses to auxin and the regulation of growth. Plant Cell Environ 10:523–543

Burström HG (1979) In search of a plant growth paradigm. Amer J Bot 66:98–104

Capkova V, Fidlerova A, VanAmstel T, Croes AF, Mata C, Schrauwen JAM, Wullems GJ, Tupý J (1997) Role of N-glycosylation of 66 and 69 kDa glycoproteins in wall formation during pollen tube growth *in vitro*. Eur J Cell Biol 72:282–285

Capková V, Hrabetova E, Tupý J (1987) Protein changes in tobacco pollen culture: a newly synthesized protein related to pollen tube growth. J Plant Physiol 130:308–314

Capková V, Zbrozek J, Tupý J (1994) Protein synthesis in tobacco pollen tubes: preferential synthesis of cell wall 69-kDa and 66-kDa glycoproteins. Sex Plant Reprod 7:57–66

Carpita NC, Gibeaut DM (1993) Structural models of primary cell walls in flowering plants: consistency of molecular structure with the physical properties of the walls during growth. Plant Journal 3: 1–30

Cheung AY, Wang H, Wu H-M (1995) A floral transmitting tissue-specific glycoprotein attracts pollen tubes and stimulates their growth. Cell 82:383–393

Cleland R (1959) Effect of osmotic concentration on auxin-action and on irreversible and reversible expansion of the Avena coleoptile. Physiol Plant 12:809–825

Cleland R (1967a) Extensibility of isolated cell walls: measurement and changes during cell elongation. Planta 74:197–209
Cleland R (1967b) A dual role of turgor pressure in auxin-induced cell elongation in Avena coleoptiles. Planta 77:182–191
Cleland R (1971) Cell wall extension. Ann Rev Plant Physiol 22:197–226
Cosgrove DJ (1993a) Wall extensibility: its nature, measurement and relationship to plant cell growth. New Phytol 124:1–23
Cosgrove DJ (1993b) Water uptake by growing cells: an assessment of the controlling roles of wall relaxation, solute uptake, and hydraulic conductance. Int J Plant Sci 154:10–21
Cosgrove DJ, Cleland RE (1983) Osmotic properties of pea internodes in relation to growth and auxin action. Plant Physiol 72:332–338
Cresti M, Ciampolini F, Pacini E, Ramulu K Sree, Devreux M, Laneri U (1977) Ultrastructural aspects of pollen tube growth inhibition after gamma irradiation in *Lycopersicum peruvianum*. Theor Appl Genet 49:297–303
Delmer DP, Stone BA (1988) Biosynthesis of plant cell walls. In: Preiss J (ed) The biochemistry of plants, Vol 14 Carbohydrates. Academic Press, Inc, San Diego, California, 373–420
Derksen J (1996) Pollen tubes: a model system for plant cell growth. Bot Acta 109:341–345
Derksen J, Emons AMER (1990) Microtubules in tip growth systems. In: Heath IB (ed) Tip growth in plant and fungal cells. Academic Press, San Diego, 147–181
Derksen J, Rutten T, VanAmstel T, DeWin A, Doris F, Steer M (1995) Regulation of pollen tube growth. Acta Bot Neerl 44:93–119
Dhawan AK, Malik CP (1981) Effect of growth regulators and light on pollen germination and pollen tube growth in *Pinus roxburghii* Sarg. Ann Bot 47:239–248
Dhingra HR (1990) Effect of growth substances and salts on in vitro germination and tube growth of *Datura metel* L. pollen. Indian J Plant Physiol 1:76–79
Dhingra HR, Varghese TM (1985) Effect of growth regulators on the *in vitro* germination and tube growth of maize (*Zea mays* L.) pollen from plants raised under sodium chloride salinity. New Phytol 100:563–569
Dickinson DB (1978) Influence of boron and pentaerythriol concentrations on germination and tube growth of *Lilium longiflorum* pollen. J Amer Soc Horticult Sci 103:413–416
Emons AM (1989) Helicoidal microfibril depositioning in a tip-growing cell and microtubule alignment during tip morphogenesis: a dry-cleaving and freeze-substitution study. Can J Bot 67:2401–2408
Evans ML (1974) Rapid responses to plant hormones. Ann Rev Plant Physiol 25:195–223
Evans ML, Vesper MJ (1980) An improved method for detecting auxin-induced hydrogen ion efflux from corn coleoptile segments. Plant Physiol 66:561–565
Fincher GB, Stone BA (1981) Metabolism of noncellulosic polysaccharides. Encycl Plant Physiol New Ser 13B 68–132
Findeklee P, Goldbach HE (1996) Rapid effects of boron deficiency on cell wall elasticity modulus in *Cucurbita pepo* roots. Bot Acta 109:1–3
Frey-Wissling A (1959) Die pflanzliche Zellwand. Springer, Berlin Göttingen Heidelberg
Fähnrich P (1964) Untersuchungen über den Einfluß des Bors bei der Pollenkeimung und beim Pollenschlauchwachstum. In: Linskens HF (ed) Pollen physiology and fertilization. Elsevier North-Holland, Amsterdam, 120–127
Galston AW, Davis PJ (1969) Hormonal regulation in higher plants. Science 163:1288–1297
Galston AW, Purves WK (1960) The mechanism of action of auxin. Ann Rev Plant Physiol 11:239–279
Geitmann A (1997) Growth and formation of the cell wall in pollen tubes of *Nicotiana tabacum* and *Petunia hybrida*. Hänsel-Hohenhausen, Egelsbach Frankfurt Washington
Geitmann A, Cresti M (1998) Ca^{2+} channels control the rapid expansions in pulsating growth of pollen tubes. J Plant Phsyiol, in press
Geitmann A, Hudak J, Vennigerholz F, Walles B (1995a) Immunogold localization of pectin and callose in pollen grains and pollen tubes of *Brugmansia suaveolens* – implications for the self-incompatibility reaction. J Plant Physiol 147:225–234
Geitmann A, Li YQ, Cresti M (1995b) Ultrastructural immunolocalization of periodic pectin depositions in the cell wall of *Nicotiana tabacum* pollen tubes. Protoplasma 187:168–171
Geitmann A, Li YQ, Cresti M (1996a) The role of the cytoskeleton and dictyosome activity in the pulsatory growth of *Nicotiana tabacum* and *Petunia hybrida*. Bot Acta 109:102–109
Geitmann A, Wojciechowicz K, Cresti M (1996b) Inhibition of intracellular pectin transport in pollen tubes by Monensin, Brefeldin A and Cytochalasin D. Bot Acta, 109:373–381
Ghanadan H, Loomis WD, Durst RW (1995) A possible borate-apiose cross-link in plant cell walls. In: Zarra I, Revilla G (eds) Abstracts of the 7th Cell Wall Meeting. Santiago de Compostela, Spain, p 2
Gibeaut DM, Carpita NC (1994) Biosynthesis of plant cell wall polysaccharides. FASEB J 8:904–915
Goldbach HE (1997) A critical review on current hypotheses concerning the role of boron in higher plants: suggestions for further reserach and methodological requirements. J Trace Micropr Techn, in press

Grant GT, Morris ER, Rees DA, Smith PJC, Thom D (1973) Biological interactions between polysaccharides and divalent cations: the egg-box model. FEBS Letters 32:195–198

Green PB (1969) Cell morphogenesis. Ann Rev Plant Physiol 20:365–394

Green PB, Cummins WR (1974) Growth rate and turgor pressure. Plant Physiol 54:863–869

Gupta MP, Innes IR, Dhalla NS (1988) Responses of contractile function to ruthenium red in rat heart. Amer J Physio 225:H1413

Harris PJ, Anderson MA, Bacic A, Clarke AE (1984) Cell-cell recognition in plants with special reference to the pollen-stigma interaction. Ox Surv Plant Mol Cell Biol 1:161–203

Hasegawa Y, Nakamura S, Nakamura N (1996) Immunocytochemical localization of callose in the germinated pollen of *Camellia japonica.* Protoplasma 194:133–139

Hayat MA (1993) Stains and cytochemical methods. Plenum Press, New York, London

Heath IB (1990) The roles of actin in tip growth of fungi. Int Rev Cytol 123:95–127

Heslop-Harrison J (1971) The pollen wall: structure and development. In: Heslop-Harrison J (ed) Pollen: Development and Physiology. Butterworths, London, 75–98

Heslop-Harrison J (1987) Pollen germination and pollen-tube growth. Int Rev Cytol 107:1–78

Hohl M, Greiner H, Schopfer P (1995) The cryptic-growth response of maize coleoptiles and its relationship to H_2O_2-dependent cell wall stiffening. Phys Plant 94:491–498

Hohl M, Schopfer P (1992) Physical extensibility of maize coleoptile cell walls: apparent plastic extensibility is due to elastic hysteresis. Planta 187:498–504

Hohl M, Schopfer P (1995) Rheological analysis of viscoelastic cell wall changes in maize coleoptiles as affected by auxin and osmotic stress. Physiol Plant 94:499–505

Holdaway-Clarke TL, Feijó JA, Hackett GR, Kunkel JG, Hepler PK (1997) Pollen tube growth and the intracellular cytosolic calcium gradient oscillate in phase while extracellular calcium influx is delayed. Plant Cell 9:1999–2010

Hu H, Brown P (1994) Localization of boron in cell walls of squash and tobacco and its association with pectin. Evidence for a stuctural role of boron in the cell wall. Plant Physiol 105:681–689

Iwanami Y (1980) Stimulation of pollen tube growth *in vitro* by dicarboxylic acids. Protoplasma 102: 111–115

Jarvis MC (1984) Structure and properties of pectin gels in plant cell walls. Plant Cell Environ 7: 153–164

Jauh GY, Lord EM (1996) Localization of pectins and arabinogalactan-proteins in lily (*Lilium longiflorum* L.) pollen tube and style, and their possible roles in pollination. Planta 199:251–261

Johri BM, Vasil IK (1961) Physiology of pollen. Bot Rev 27:325–381

Kaminskyj SGW, Garrill A, Heath IB (1992) The relation between turgor and tip growth in *Saprolegnia ferax*: Turgor is necessary, but not sufficient to explain apical extension rates. Exp Mycol 16:64–75

Knox RB (1984) Pollen-pistil interactions. In: Linskens HF, Heslop-Harrison J (eds) Cellular interactions. Springer Verlag, 508–608

Kobayashi M, Matoh T, Azuma J (1996) Two chains of rhamnogalacturonan II are cross-linked by borate-diol ester bonds in higher plant cell walls. Plant Physiol 110:1017–1020

Konar RN (1958) Effect of IAA and kinetin on pollen tubes of *Pinus roxburghii* Sar. Curr Sci 6:216–217

Kouchi H, Kumazawa K (1976) Anatomical responses of root tips to boron deficiency III. Effect of boron deficiency on sub-cellular structure of root tips, particularly on morphology of cell wall and its related organelles. Soil Sci Plant Nutr 22:53–71

Kroh M, Knuiman B (1982) Ultrastructure of cell wall and plugs of tobacco pollen tubes after chemical extraction of polysaccharides. Planta 154:241–250

Kutschera U, Schopfer P (1986) Effect of auxin and abscisic acid on cell wall extensibility in maize coleoptiles. Planta 167:527–535

Kwok R, Evans E (1981) Thermoelasticity of large lecithin bilayer vesicles. Biophys J 35:637–652

Levy S, Staehelin LA (1992) Synthesis, assembly and function of plant cell wall macromolecules. Curr Op Cell Biol 4:856–862

Li X (1990) Study on relation of plant pollen germination and phytohormones. Acta Sci Nat Univ Norma Human 13:248–252

Li Y-Q, Bruun L, Pierson ES, Cresti M (1992) Periodic deposition of arabinogalactan epitopes in the cell wall of pollen tubes of *Nicotiana tabacum* L. Planta 188:532–538

Li Y-Q, Fang C, Faleri C, Ciampolini F, Linskens HF, Cresti M (1995a) Presumed phylogenetic basis of the correlation of pectin deposition pattern in pollen tube walls and the stylar structure of angiosperms. Proc Kon Ned Akad v Wetensch 98:39–44

Li YQ, Chen F, Linskens HF, Cresti M (1994) Distribution of unesterified and esterified pectins in cell walls of pollen tubes of flowering plants. Sex Plant Reprod 7:145–152

Li YQ, Faleri C, Geitmann A, Zhang HQ, Cresti M (1995b) Immunogold localization of arabinogalactan proteins, unesterified and esterified pectins in pollen grains and pollen tubes of *Nicotiana tabacum* L. Protoplasma 189:26–36

Li YQ, Moscatelli A, Cai G, Cresti M (1997) Functional interactions among cytoskeleton, membranes, and cell wall in the pollen tube of flowering plants. Int Rev Cytol 176:133–199

Li YQ, Zhang HQ, Pierson ES, Huang FY, Linskens HF, Hepler PK, Cresti M (1996) Enforced growth-rate fluctuation causes pectin ring formation in the cell wall of *Lilium longiflorum* pollen tubes. Planta 200:41–49
Liners F, Gaspar T, VanCutsem P (1994) Acetyl- and methyl-esterification of pectins of friable and compact sugar-beet calli: consequences for intercellular adhesion. Planta 192:545–556
Loomis WD, Durst RW (1992) Chemistry and biology of boron. BioFactors 3:229–239
López-Franco R, Bartnicki-Garcia S, Bracker CE (1994) Pulsed growth of fungal hyphal tips. Proc Natl Acad Sci USA 91:12228–12232
Mascarenhas JP (1993) Molecular mechanisms of pollen tube growth and differentiation. Plant Cell 5: 1303–1314
Masuda Y (1990) Auxin-induced cell elongation and cell wall changes. Bot. Mag. Tokyo 103:345–370
Matoh T, Kawaguchi S, Kobayashi M (1996) Ubiquity of a borate-rhamnogalacturonan II complex in the cell walls of higher plants. Plant Cell Physiol 37:636–640
McCann MC, Roberts K (1994) Changes in cell wall architecture during cell elongation. J Exp Bot 45: 1683–1691
McLeod KA (1975) The control of growth of tomato pollen. Ann Bot 39:591–596
McQueen-Mason SJ, Cosgrove DJ (1995) Expansion mode of action on cell walls. Plant Physiol 107: 87–100
Meikle PJ, Bonig I, Hoogenraad NJ, Clarke AE, Stone BA (1991) The location of (1–3)-β-glucans in the walls of pollen tubes of *Nicotiana alata* using a (1–3)-β-glucan-specific monoclonal antibody. Planta 185:1–8
Moore PJ, Staehelin LA (1988) Immunogold localization of the cell-wall-matrix polysaccharides rhamnogalacturonan I and xyloglucan during cell expansion and cytokinesis in *Trifolium pratense* L.; implication for secretory pathways. Planta 174:433–445
Morré DJ (1994) Physical membrane displacement: reconstitution in a cell-free system and relationship to cell growth. Protoplasma 180:3–13
Nakahori K, Katou K, Okamoto H (1991) Auxin changes both the extensibility and the yield threshold of the cell wall of Vigna hypocotyls. Plant Cell Physiol 32:121–129
Nishitani K, Masuda Y (1982) Roles of auxin and gibberellic acid in growth and maturation of epicotyls of *Vigna angularis*: cell wall changes. Physiol Plant 56:38–45
Obermeyer G, Kriechbaumer R, Strasser D, Maschessnig A, Bentrup F-W (1996) Boric acid stimulates the plasma membrane H^+-ATPase of ungerminated lily pollen grains. Physiologia Plantarum 98: 281–290
Okamoto H, Miwa C, Masuda T, Nakahori K, Katou K (1990) Effects of auxin and anoxia on the cell wall yield threshold determined by negative pressure jumps in segments of cowpea hypocotyl. Plant Cell Physiol 31:783–788
Park D, Robinson PM (1966) Internal pressure of hyphal tips of fungi, and its significance in morphogenesis. Ann Bot 30:426–439
Phillips DG, Preshaw C, Steer MW (1988) Dictyosome vesicle production and plasma membrane turnover in auxin-stimulated outer epidermal cells of coleoptile segments from *Avena sativa* L. Protoplasma 145:59–65
Picton JM, Steer MW (1985) The effects of ruthenium red, lanthanum, fluorescein isothiocyanate and trifluoperazine on vesicle transport, vesicle fusion and tip extension in pollen tubes. Planta 163: 20–26
Pierson ES, Li YQ, Zhang HQ, Willemse MTM, Linskens HF, Cresti M (1995) Pulsatory growth of pollen tubes: investigation of a possible relationship with the periodic distribution of cell wall components. Acta Bot Neerl 44:121–128
Pierson ES, Miller DD, Callaham DA, Van Aken J, Hackett G, Hepler PK (1996) Tip-localized calcium entry fluctuates during pollen tube growth. Develop Biol 174:160–173
Plattner H (1989) Regulation of membrane fusion during exocytosis. Int Rev Cytol 119:197–286
Plyushch TA, Willemse MTM, Franssen-Verheijen MAW, Reinders MC (1995) Structural aspects of *in vitro* pollen tube growth and micropylar penetration in Gasteria verrucosa (Mill.) H. Duval and *Lilium longiflorum* Thunb. Protoplasma 187:13–21
Polster J, Schwenk M, Bengsch E (1992) The role of boron, silicon and nucleic bases on pollen tube growth of *Lilium longiflorum*. Z Naturf 47c:102–108
Preston RD (1974) The physical biology of plant cell walls. Chapman and Hall, London
Rae AL, Harris PJ, Bacic A, Clarke AE (1985) Composition of the cell walls of *Nicotiana alata* Link et Otto pollen tubes. Planta 166:128–133
Ray PM (1992) Mechanisms of wall loosening for cell growth. Curr Top Plant Biochem Physiol 11:18–41
Ray PM, Green PB, Cleland R (1972) Role of turgor in plant cell growth. Nature 239:163–164
Read SM, Li H, Doblin M, Turner A, Ferguson C, Bacic A (1996) Deposition of callose: The main metabolic preoccupation of the male gametophyte. Abstracts of the 14th International Congress of Sexual Plant Reproduction. Lorne, Australia, p 41
Rees DA (1969) Structure, conformation, and mechanism in the formation of polysaccharide gels and networks. Adv Carboh Chem Biochem 24:267–332

Reuzeau C, Pont-Lezica RF (1995) Comparing plant and animal extracellular matrix-cytoskeleton connections – are they alike? Protoplasma 186:113–121
Ricard J, Noat G (1986) Electrostatic effects and the dynamics of enzyme reactions at the surface of plant cells. 1. A Theory of the ionic control of a complex multi-enzyme system. Eur J Biochem 155: 183–190
Robertson NF (1958) Observations of the effect of water on the hyphal apices of *Fusarium oxysporum*. Ann Bot 22:159–173
Roggen HPJR, Stanley RG (1969) Cell-wall-hydrolysing enzymes in wall formation as measured by pollen tube extension. Planta 84:295–303
Rosen WG, Gawlik SR, Dashek WV, Siegesmund KA (1964) Fine structure and cytochemistry of *Lilium* pollen tubes. Amer J Bot 51:61–71
Rougier M, Vian B, Gallant D, Roland JC (1973) Aspects cytochimiques de l'étude ultrastructurale des polysaccharide végétaux. Annèe biol 12:43–75
Rubinstein AL, Broadwater AH, Lowrey KB, Bedinger PA (1995) Pex1, a pollen-specific gene with an extensin-like domain. PNAS 92:3086–3090
Rubinstein AL, Màrquez J, Suèrez-Cervera M, Bedinger PA (1995) Extensin-like glycoproteins in the maize pollen tube wall. Plant Cell 7:2211–2225
Sawidis T, Reiss H-D (1995) Effects of heavy metals on pollen tube growth and ultrastructure. Protoplasma 185:113–122
Schmucker T (1934) Über den Einfluß von Borsäure auf Pflanzen, insbesondere keimende Pollenkörner. Planta 23:264–283
Schmucker T, Hartmann M, VonLaue M, Neuberg C, Rosenheim A, Volmer M (1932) Bor als physiologisch entscheidendes Element. Naturwissenschaften 20:839
Schnepf E (1986) Cellular polarity. Ann Rev Plant Physiol 37:23–47
Schröter K, Sievers A (1971) Wirkung der Turgorreduktion auf den Golgi-Apparat und die Bildung der Zellwand bei Wurzelhaaren. Protoplasma 72:203–211
Setterfield G, Bayley ST (1961) Structure and physiology of cell walls. Ann Rev Plant Physiol 12:35–62
Showalter AM (1993) Structure and function of plant cell wall properties. Plant Cell 5:9–23
Sievers A, Schnepf E (1981) Morphogenesis and polarity of tubular cells with tip growth. In: Kiermayer O (ed) Cell Biology Monographs Cytomorphogenesis in Plants. pp 265–299
Sondheimer E, Linskens HF (1974) Control of in vitro germination and tube extension of Petunia hybrida pollen. Proc Koninkl Nederl Akad Wetensch Amsterdam 77:116–124
Staehelin LA, Giddings TH, Levy S, Lynch MA, Moore PJ, Swords KMM (1991) Organisation of the secretory pathway of cell wall glycoproteins and complex polysaccharides in plant cells. In: Hawes CR, Coleman JOP, Evans DE (eds) Endocytosis, exocytosis and vesicle traffic in plants. University Press, Cambridge, UK, pp 183–198
Stanley RG, Lichtenberg EA (1963) The effect of various boron compounds on in vitro germination of pollen. Physiologia Plantarum 16:337–346
Stanley RG, Loewus FA (1964) Boron and myo-inositol in pollen pectin biosynthesis. In: Linskens HF (ed) Pollen physiology and fertilization. Elsevier North-Holland, Amsterdam, pp 128–136
Steer MW (1990) Role of actin in tip growth. In: Heath IB (ed) Tip growth in plant and fungal cells. Academic Press, San Diego, pp 119–145
Steer MW, Steer JM (1989) Pollen tube tip growth. New Phytol 111:323–358
Stoddart RW, Northcote DJ (1967) Metabolic relationships of the isolated fractions of the pectic substances of actively growing sycamore cells. Biochem J 105:45–59
Sweeney BM, Thimann KV (1938) The effect of auxins on protoplasmic streaming. J Gen Physiol 21: 439–463
Taiz L (1984) Plant cell expansion: Regulation of cell wall mechanical properties. Ann Rev Plant Physiol 35:585–657
Tang XW, Liu GQ, Yang Y, Zheng WL, Wu BC, Nie DT (1992) Quantitative measurement of pollen tube growth and particle movement. Acta Bot Sinica 34:893–898
Teasdale RD, Richards DK (1990) Boron deficiency in cultured pine cells: quantitative studies of the interaction with Ca^{2+} and Mg^{2+}. Plant Physiol 93:1071–1077
Thimann KV, Biradivolu R (1994) Actin and the elongation of plant cells. II. The role of divalent ions. Protoplasma 183:5–9
van Amstel ANM, Knuiman B, Derksen J (1994) The ontogeny of the pollen tube wall. In: Nijmegen T (ed) Construction of plant cell walls. Koninklijke Bibliotheek, Den Haag, pp 65–100
Vasil IK (1964) Effect of boron on pollen germination and pollen tube growth. In: Linskens HF (ed) Pollen physiology and fertilization. Elsevier North-Holland, Amsterdam, pp 107–119
Vian B, Roland JC (1991) Affinodetection of the sites of formation and of the further distribution of polygalacturonans and native cellulose in growing plant cells. Biol Cell 71:43–55
Webster (1985) Webster's Ninth New Collegiate Dictionary. Merriam-Webster Inc, Springfield, Massachusetts, USA
Wessels JGH (1990) Role of cell wall architecture in fungal tip growth generation. In: Heath IB (ed) Tip growth in plant and fungal cells. Academic Press, San Diego, pp 1–29

Willats WGT, Knox JP (1996) A role for arabinogalactan-proteins in plant cell expansion: evidence from studies on the interaction of β-glucosyl Yariv reagent with seedlings of *Arabidopsis thaliana*. Plant J 9:919–925
Wilson K (1964) The growth of plant cell walls. Int Rev Cytol 17:1–49
Wolfe J, Dowgert M, Steponkus P (1986) Mechanical study of the deformation and rupture of the plasma membranes of protoplasts during osmotic expansions. Membr Biol 93:63–74
Wolfe J, Steponkus P (1981) The stress-strain relation of the plasma membrane of isolated plant protoplasts. Biochim Biophys Acta 643:662–668
Yamauchi T, Hara T, Sonoda Y (1986) Distribution of calcium and boron in the pectin fraction of tomato leaf cell wall. Plant Cell Physiol 27:729–732
Zhang GF, Staehelin LA (1992) Functional compartmentation of the golgi apparatus of plant cells. Plant Physiol 99:1070–1083

CHAPTER 21

Actin Filament- and Microtubule-Based Motor Systems: Their Concerted Action During Pollen Tube Growth

A. Moscatelli · G. Cai · M. Cresti

Dipartimento di Biologia Ambientale, Università degli Studi di Siena, Via P. A. Mattioli 4, 53100 Siena, Italia
e-mail: moscatelli@unisi.it
telephone: 577-29 88 56
fax: 577-29 88 60

Abstract. Cytoskeleton and cytoskeletal motor proteins play a major role in pollen germination and growth. Both actin filaments and microtubules are believed to be involved in generating different cytoplasmic movements which allow the cytoplasmic streaming and the polar organelle distribution during the pollen tube elongation. Myosins as well as kinesins/dyneins related polypeptides have been identified as microfilament and microtubule motor proteins respectively, but information is still needed in order to understand the relative contributions of these two motor systems. This paper draws attention to their possible cooperation during pollen tube germination and growth.

21.1 The Male Gametophyte During Fertilization

The pollen grain represents the male gametophyte of higher plants. At the anthesis the pollen grain consists of a cell having its own nucleus (vegetative nucleus) and comprises at its inner two sperm cells (trinucleated pollen) or their progenitor, the generative cell (binucleated pollen).

Having arrived on the surface of a compatible stigma it germinates producing a cylindracal structure, the pollen tube, which grows through the style transmitting tissue and function as a biological channel for the migration of sperm cells to the embryo sac. The pollen tube growth is a critical point for successful fertilization and depends from several factors including the maintainance of a precise cytoplasmic organization and the integrity of the cytoskeletal apparatus.

21.2 Cytoplasmic Organization

The cytoskeletal apparatus, namely actin filaments (AFs) and microtubules (MTs), is essential for the movement of a wide range of components within plant cells. The cytoplasmic streaming represents the most prominent form of these intracellular movements in plants and depends on the interaction of organelles with stationary AFs (Williamson 1993, Shimmen and Yokota 1994). This kind of transport is important for the movement of organelles especially in elongated cells such as differentiated somatic cells, root hairs and pollen tubes. However, other cytoplasmic transportation events such as organelle positioning and trafficking should also be considered within pollen tubes such as the slow movement of the male germ unit (MGU) (vegetative nucleus and generative cell/sperm cells) particularly important for the fertilization process (for review see Pierson and Cresti 1992).

The pollen tube shows a specialized growth depending on the fusion of Golgi-derived secretory vesicles (SVs), carrying cell wall material and new plasma membrane, to a restricted apical region of the tube (Derksen et al. 1995). The rapid and bidirectional cytoplasmic streaming observed within the pollen tube is necessary for the transport and accumulation of SVs at the tip. The process of vesicle fusion occurs at the very tip, a region described by Iwanami (1956) as "hyaline zone" or "cap block", where the cytoplasmic streaming is inactive and only random Brownian motions of SVs are observed (Derksen et al. 1995). The cytoplasmic streaming, inactivation seems to be an essential condition for the fusion to take place and could be related to the Ca^{2+} tip focused gradient within growing tubes (Hepler 1997).

An excess of SVs is produced for the growth requirement. Part of them do not fuse but are transported backward to the oldest parts of the tube (Derksen et al. 1995). It is generally believed that the excess of plasma membrane incorporated during the fusion process is retrieved and probably recycled by endocytosis (Blackbourn and Jackson 1996). As a matter of fact, vesicles of 50 nm and 300 nm have been observed at the tip, leading to the hypothesis that vesicles with different properties could be present within the pollen tube apex (Cresti and Tiezzi 1990). The polarization of the cytoplasm occurs just after hydration, when the pollen grains round up and rough endoplasmic reticulum, dictyosomes and SVs accumulate at the future germination pore (Fig. 21.1a, b) (Ciampolini et al. 1988). In elongating pollen tubes, a polarized organelle distribution is then observed behind the very tip, being the SVs accumulated at the tip whereas other membranous organelles as well as the MGU are escluded from the apical region (Rosen et al 1964, Sassen 1964, Cresti et al. 1977). As the tube elongates, the oldest parts become isolated by the deposition of callose plugs and both the active cytoplasm and the MGU must be in the younger part of the tube for a successful fertilization. Several factors con-

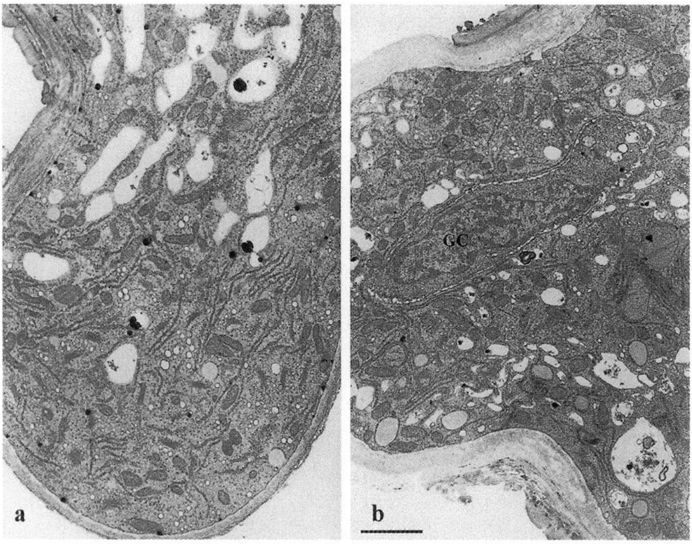

Fig. 21.1a, b. Electron micrographs showing the cytoplasmic organization during early stages of pollen tube emission from *Aloe ciliaris* pollen grains. Most of organelles as well as the generative cell become oriented toward the germinative pore in the vegetative cytoplasm during hydration and early stages of germination

tribute to the polarized growth; among these, the cytoskeletal apparatus is believed to play a central role in maintaining the cytoplasmic streaming and the polarized organization of the cytoplasm (Pierson and Cresti 1992, Derksen et al. 1995). Whereas the spatial organization of the cytoskeletal structures is well characterized, their distinct contributions in generating cytoplasmic movements is still unclear. In animal cells motor proteins are known to be able to convert the chemical energy stored in ATP into mechanical force along AFs and MTs. Two contractile systems are known to generate movements within cells: one based on the AFs-myosin system and the other on MT-kinesins/dyneins. Myosin-like and kinesin-like proteins have been identified and characterized in plant cells and recently also dynein heavy chain-related polypeptides have been found in pollen and pollen tubes (see below).

In this article the concerted action of AFs and MTs and their interaction with cytoskeletal motor proteins in determining cytoplasmic movements during the polarized growth of pollen tube will be discussed.

21.3 The AFs-Myosin System

AFs were first identified in pollen tubes by electron-microscopical studies (Franke et al. 1972). However the first demonstration of AFs in pollen tubes was reported by Condeelis (1974) which showed that fibers from burst pollen tubes of *Amaryllis belladonna* could be decorated with muscle meromyosin. The presence of AFs in pollen tubes was later confirmed by immunogold labelling (Lancelle and Hepler 1989, 1991). The spatial organization of AFs has been investigated using specific markers for actin (Perdue and Parthasarathy 1985, Pierson et al. 1986, Pierson 1988, Tang et al. 1989a, Tiwari and Polito 1990a) after both chemical fixation and permeabilization with DMSO in angiosperms. In chemically- fixed dehydrated pollen of *Helleborus*, actin was shown to be present as spindle-shaped aggregates. During activation and germination, these actin bodies undergo a rapid dispersion and they are no longer visible as the pollen tube emerges. As the cytoplasm leaves the grain and the cytoplasmic streaming starts, AFs appear and converge toward the germination pore (Heslop-Harrison et al. 1986, Tiwari and Polito 1990a). A clear view of AFs organization was observed in lily and tobacco pollen grain and tubes after permeabilization by DMSO: AFs in the vegetative cytoplasm were organized in bundles longitudinally oriented in the oldest part of pollen tubes, whereas they seemed to be short and randomly oriented in the tip region both in chemically-fixed and fresh permeabilized pollen tubes (Heslop-Harrison and Heslop-Harrison 1991, Pierson 1988). AFs at the pollen tube apex have been observed in pollen tube after chemical fixation (Perdue and Parthasarathy 1985, Pierson et al. 1986, Pierson et al 1988) whereas only thin disorganized filaments have been observed in cryo fixed-freeze substituted (CF-FS) pollen tubes (Lancelle et al. 1987, Lancelle and Hepler 1992) and no actin bundles heve been evidentiated at the very tip after microinjection of FITC labelled actin (Miller et al. 1996). This apparently divergent data could be explained by the fact that the AFs in the apical region of CF-FS or fresh pollen tubes are too short and thin to be evidentiated by fluorescent microscopy. A less degree of AFs polymerization in the apical portion of the tube could also be a consequence of the high Ca^{2+} concentration at the very tip (20–30 µM) which could regulate the AFs polymerization through putative actin binding proteins. When the tip focussed Ca^{2+} gradient is disturbed, AFs extend to the very tip region. (Miller et al. 1996). These observations represent only partially the distribution of AFs *in vivo* when the AF polymerization and organization is probably mediated by external stimuli generated by the style transmitting tissue. Actin foci, resembling adhesion plaques of animal cultured cells, are observed in association with the plasma membrane only *in vivo* grown pollen tubes. They are believed to represent putative actin-organizing centers that form in response to the interaction with molecules of the style transmitting tissue (Pierson et al. 1986).

Profilin is an actin binding protein that is believed to modulate AFs organization in response to external stimuli and it has been identified in plant cells, including pollen tubes. It is generally believed that profilin is a regulator of actin polymerization, probably acting as an actin sequestring protein when the barbed ends of AFs are capped. When the barbed ends are free and actin subunits are available, profilin can stimulate the actin polymerization promoting the convertion of the inactive ADP-monomer to active ATP-actin (Staiger et al. 1997). In pollen tubes profilin could be associated with the apical and subapical plasma membrane and bound to inositol phospholipids, which are known to be involved in the signal transduction pathway. External signals could act through inositol phospholipids, and the small RhoGTPase proteins recently identified in pollen tube tip plasma membrane (Lin et al. 1996) to induce the release of profilin, thus increasing the conversion of the ADP-inactive actin to the ATP-active form.

AFs play an essential role in determining the internal pollen polarity before germination by conveying cytoplasmic components, such as SVs and dictyosomes, to the germination pore through the active cytoplasmic streaming. After that, the fusion of SVs at the tube tip allows the plasma membrane to become polarized since specific ion channels are inserted into the apical plasma membrane (Tester 1990). The first indication that AFs could be responsible for the cytoplasmic streaming during pollen tube growth derived from the observation that the AFs affecting drug, cytochalasin D, can disrupt the actin cytoskeleton, thus inhibiting the cytoplasmic streaming, the movement of the MGU and pollen tube growth (Heslop-Harrison and Heslop-Harrison1989a, Tiwari and Polito 1989a). Most of the current information on cytoplasmic streaming derives from studies on the characean algae, in which organelles move along stationary AFs driven by myosin molecules. *In vitro* reconstitution experiments have shown that organelles isolated from *Lilium* pollen tubes move along characean AFs and pollen tube extracts support AF movement, thus suggesting that the cytoplasmic streaming in pollen tube could also be dependent on the presence of actin-based motor proteins on the organelle surface (Kohno T. and Shimmen T. 1988, Kohno et al. 1991). Anti-myosin II antibodies recognized homologous polypeptides in tobacco pollen tubes (Heslop-Harrison et al. 1989b, Tang et al 1989b) and labelled spot-like structures within the vegetative cell after immunofluorescence microscopy. Staining was also observed on the vegetative nucleus and generative cell surface, suggesting that myosin molecules could be responsible for the organelle movement during cytoplasmic streaming and for the MGU transport (Tang et al.1989b, Tirlapur et al. 1995-1996). The association of myosins with membranous organelles has been confirmed in subsequent papers by western blotting analysis with a commercial anti-myosin II antibody that recognized a 174 kDa polypeptide in the microsomal fraction of tobacco pollen tube (Tirlapur et al. 1995). The distribution of myosin was studied during pollen hydration and early stages of germination. Myosins were located in the central area of hydrated pollen grain and later they moved toward the germination pore where the pollen tube emerges (Tirlapur et al. 1995, Tirlapur et al. 1996) thus showing a distribution pattern very similar to that of actin (Heslop-Harrison et al. 1986). This confirmed that myosins could interact with AFs in focussing cytoplasmic organelles, such as dictyosomes and SVs, and the MGU at the germination pore and/or could play a role in the organization of AFs within the emerging tube. During pollen tube growth, a punctate staining along the vegetative cytoplasm is still present and accumulates in the apical region. Immunoelectron microscopical analysis revealed an association with SVs, mitochondria and the endoplasmic reticulum (Tirlapur et al. 1995, Tirlapur et al. 1996). In addition, the antibody recognized epitopes on the surface of the vegetative nucleus and generative cell, both by immunofluoresecence microscopy and immunogold labelling (Tirlapur et al. 1995, Tirlapur et al. 1996).

At present, only one functional myosin of 170 kDa has been purified and biochemically characterized in lily pollen tube. These polypeptides displayed typical properties of myosins such as the ability to bind AFs in an ATP-sensitive manner and to promote the movement of AFs in *in vitro* motility assays (Yokota and Shimmen 1994). Polyclonal antibodies raised against the lily 170 kDa polypeptide stained particles within the vegetative cytoplasm of lily and tobacco pollen tubes, with the fluorescence more concentrated at the very tip. The antibody appeared to recognize a different isoform than the immunoreactive homolog of myosin II since it did not stain muscle myosin in western blotting and did not recognized any epitope on the MGU surface. Interestingly, both the anti-myosin II and the anti-170 kDa antibodies stained particle-like structures were more concentrated at the very tip, where AFs seemed to be loosely organized and the cytoplasmic streaming was likely to be inhibited by the high Ca^{2+} concentration. The presence of different isoforms with similar distribution pattern in the apical region could suggest that they have an overlapping function or they could be simply redundant. Myosins could organize newly synthetized AFs in the region behind the very tip, following a model of nucleation and release of AFs (Yokota et al. 1995).

Other myosin isoforms have been identified on the basis of their reactivity against eterologous antibodies to: myosin I, myosin II and myosin V. These antibodies recognized polypeptides of 125, 205 and 190 kDa, respectively, and gave different staining patterns in *Nicotiana alata* and *Lilium longiflorum* pollen tubes by immunofluorescence microscopy. The anti-myosin I stained large particles and the MGU surface, whereas the anti-myosin II appeared to recognize large particles as well but, only occasionally, the generative cell and/or the vegetative nucleus. The anti-myosin V stained small particles and did not recognize any epitope on the MGU. The presence of distinct myosin isoforms on organelles with different size suggests the presence of specific receptors for myosin isoforms and the use of different mechanisms to regulate myosin function (Miller et al. 1995). Interestingly, none of these antibodies showed an accumulation of the staining at the tube tip as was reported by previous authors (Tirlapur et al. 1994, Yokota et al. 1995). These different observations could be due to different fixation methods, since results reported by Miller et al. (1995) were carried out on CF-FS pollen tubes, whereas other authors have made their observations on chemically-fixed pollen tubes. In addition, two myosin isoforms seemed to cooperate in order to promote the movement of the vegetative nucleus and the generative cell, since epitopes on the surface of the MGU have been recognized by both anti-myosin I (Miller et al. 1995) and anti-myosin II antibodies (Tang et al. 1989b, Tirlapur et al. 1994, Tirlapur et al 1995).

21.4 MTs and MT-Motor Proteins

MTs have been evidentiated in growing pollen tubes and their localization studied by both immunofluorescence microscopy using antibodies against specific tubulin isoforms (Derksen et al 1985, Tiezzi et al. 1986, Del Casino et al. 1993) and at the ultratructural level (Lancelle et al. 1987, Lancelle and Hepler 1991). MTs have been shown to be longitudinally organized along the vegetative cytoplasm, being more concentrated in the cortical than in the cytoplasmic region. Unlike AFs, MTs appeared to be distributed as a single element. In the cortical region, MTs often run parallely to AFs and bridges connecting the two cytoskeletal systems to each other and with the plasma membrane were frequently observed (Pierson et al. 1989, Lancelle and Hepler 1991). MT distribution during pollen tube germination and growth has been studied in order to clarify the dynamic aspects. MTs in ungerminated pollen grains have been observed only in *Pyrus communis* after hydration, probably nucleated by specific sites in the cortical region of the pollen grain (Tiwari and Polito1990b). In short pollen tubes (10–30 µm) MTs appeared as short filaments and occupied half of the tube length. In growing pollen tubes MTs seemed to be more organized in the oldest regions of the tube whereas they were still missing at the apex (Del Casino et al. 1993). In pollen tubes 120 µm long MTs were more intense in the central part of the tube and less stained in the region close to the grain. At the tip only some randomly oriented MT segments were seen. As tubes reached 200 µm in length MTs also appeared to be clearly visible in the apical region.

MTs participate in a variety of cell functions involving movement. Therefore, MTs that operate in different areas and/or physiological states of the cell must have different properties to accomplish specific functions. The way by which MTs originate and maintain their differentiation is essentially based on tubulin isoform composition and/or on the interaction of MTs with specific MT-associated proteins (Matus 1990). As a matter of fact, the etherogeneity of tubulins derive from the presence of multiple tubulin genes and additional variety is obtained by post-translational modifications of both α- and β-tubulin (Hussey et al. 1987, 1988). Some of these post-translational modifications, such as acetylation and tyrosination of the α-tubulin, have been correlated with the dynamic instability of MTs during the cell cycle since detyrosinated (Kreis 1987) and acetylated (Piperno et al. 1987) MTs appeared to be more stable. The process of maturation of MTs during pollen tube growth has been investigated using monoclonal antibodies against two specific tubulin isoforms involved in regulating MT stability: acetylated and tyrosinated tubulin (Del Casino et al. 1993, Astrom et al. 1993). The acetylated tubulin was not observed in pollen tubes grown for three hours, whereas it was observed in MTs within the generative cell in very long tubes grown in presence of taxol. It is known that tyrosinated tubulin is incorporated within newly formed MTs and then it is detyrosinated, thus contributing to the stabilization of the polymer. In short tubes tyrosinated tubulin was observed in the subapical regions whereas in longer tubes it was observed also at the apex. In order to understand the MTs dynamic it would be important to have more information about the mechanism of MT assembly during pollen tube growth. The presence and the localization of putative microtubule organizing centers (MTOCs) in pollen tube has been investigated probing antibodies against two pericentriolar antigens: the 6C6 monoclonal antibody against a pericentriolar component of animal cells (Chevrier et al.1992) and the anti-γ tubulin antibody, which is present in all MTOCs so far identified (Marc 1997).

The 6C6 monoclonal antibody recognized a p-77 polypeptide in pollen tube extracts, which is associated with the plasma membrane after subcellular fractionation. Time course experiments evidentiated that p-77 is already present in cell extracts of dry pollen suggesting that MTOCs could be present within the grain in ungerminated pollen and/or in short tubes. In longer tubes the antibody recognized epitopes in the cortical areas of the apical and subapical regions of the tube by immunofluorescence microscopy, suggesting that putative MTOCs could be present in the new portions of long tubes and MTs could be oriented with their minus end close to the apex (Cai et al. 1996).

The other MTOCs marker, γ tubulin, appeared to be dispersed along cortical MTs for the whole tube length (Palevitz et al. 1994).

All together, data on MTs distribution supports the hypothesis that in long pollen tubes MTs could be nucleated by putative MTOCs at the apex and subsequently translocated to the subapical regions of the tube where they appeared to be more organized. During this transport, they would probably carry at their minus end some γ tubulin molecules deriving from the nucleation step (Marc 1997), which could explaine the dispersed localization of the γ-tubulin along the pollen tube MTs (Palevitz et al. 1994). Furthermore, since tyrosinated tubulin is supposed to be incorporated in newly formed and less stable MTs, also the presence of tyrosinated tubulin at the apex of long tubes is consistent with the evidence that MTs could be nucleated in the apical region of the tube plasma membrane.

A complex microtubular apparatus has been observed within the GC and sperm cells (Lancelle et al. 1987); MTs bundles in GC and sperm cells are organized as a basket under the plasma membrane and often they prolong into the proximal end of the cell forming a tail-like structure (Lancelle and Hepler 1991). It has been previously hypothesized that the tail-like structure of the GC could be reminiscent of the axonemes of motile sperms in lower plants (Tiezzi et al. 1991) and could cooperate in the GC/sperm cell movement along the tube. More recently, evidences supported a mechanism of movement external to GC/sperm cells, principally based on the acto-myosin system (Cai et al. 1996, Cai et al. 1997). The presence of MT bundles within the GC/sperm cells has been correlated to the reshaping process that occurs during the GC/sperm cell movements along the tube and is probably mediated by intermicrotubule sliding (Heslop-Harrison et al . 1988).

The function of the microtubular system during pollen tube growth has been questioned for a long time. Recently, effects of MT affecting drugs have been observed both on the pollen tube growth and on GC/sperm cell migration. In presence of colchicine the growth of lily pollen tubes was not inhibited but the tube diameter was affected, whereas the absence of GC MTs determined that the spindle- shaped GC rounded up and its movement along the tube slowed down. This delay in the GC movement was correlated to the opposition which the rounded GC offers to the viscous vegetative cytoplasm when compared to the spindle-shaped cell (Heslop-Harrison et al. 1988). More recent data, obtained using other MTs-affecting drugs (such as oryzalin, carbetamide and prophan) in *in vitro* cultured pollen tubes of *Nicotiana alata* and *Nicotiana sylvestris* allowed the correlation of the absence of MTs to the loss of the polarized organelle distribution in the vegetative cytoplasm (Joos et al. 1994, 1995). In *in vivo* systems, pollen tubes of *Nicotiana sylvestris,* grown on drug treated stigma, stop growing one hour after germination, suggesting that the pollen-style interaction can regulate the elongation of MT-missing pollen tubes (Joos et al. 1995). Under all the above conditions, as well as at low temperature (Astrom et al 1991, 1995), the movement of the GC was delayed underlying a role of MTs in regulating the GC entrance and movement along the tube. A mechanism of cooperation between MTs and AFs has been proposed according to ultrastructural observations showing a close relationship between the two cytoskeletal systems in the cortical cytoplasm (Pierson et al. 1989, Lancelle and Hepler 1991). Following these observations, a model has been proposed in which the loss of MTs would determine changes in the AFs network and, consequently, an inefficient interaction with the GC surface (Åstrom et al. 1995).

The role of MTs in intracytoplasmic movements other than cytoplasmic streaming in pollen tube has been investigated in order to identify putative MT-motor proteins such as kinesin (Hoyt 1994) and dynein (Holzbaur and Vallee 1994). Two kinesin-related polypeptides were identified in pollen tube of *Nicotiana tabacum* based on their cross-reactivity with a monoclonal antibody (k71s23) directed against the calf brain kinesin heavy chain (Tiezzi et al. 1992). The antibody recognized a doublet of 108 and 100 kDa, which retain some biochemical characteristics of MT-motor proteins in the ability to bind MTs in an AMP-PNP-dependent manner (Tiezzi et al. 1992). Later, A 100 kDa kin-

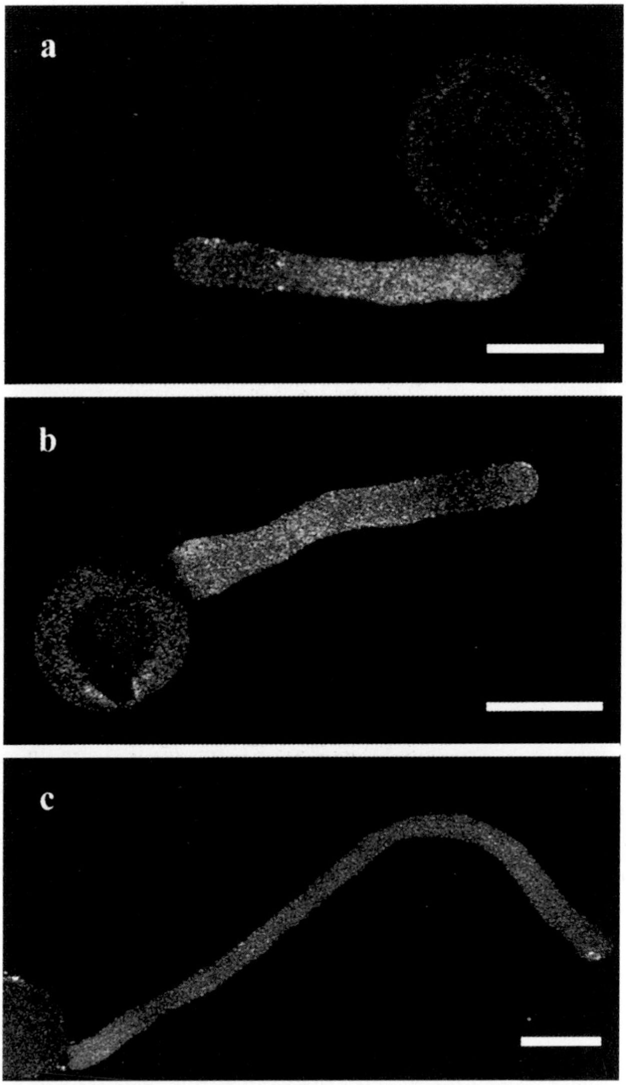

Fig. 21.2a–c. Confocal Laser Scanning Microscope of *Nicotiana tabacum* pollen tubes probed by anti-dynein (Dy-1) antibodies. The Dy-1 antibody recognized spot-like particles whose distribution seemed to be emphasized in the region close to the granule

esin-related polypeptide was purified and biochemically characterized from tobacco pollen tubes; it shows biochemical characteristics typical of kinesin molecules, in the ability to bind to MTs in presence of AMP-PNP and to be released in presence of ATP. Enzymatic activity assays showed that the ATPase activity of the kinesin polypeptide is enhanced by MTs (Cai et al. 1993). Preliminary experiments using immunofluorescence microscopy showed that the antibody gave a punctate staining, more concentrated in the apical region of the tube (Tiezzi et al. 1992). Subsequent studies evidentiated a co-localization of spot-like structures with short MTs in the tip region of long tubes, suggesting that MT-kinesin complexes could contribute to positioning SVs at the apex before their fusion to the plasma membrane (Cai et al. 1993). To support this evidence, a 100 kDa kinesin-related polypeptide was found in association with 80 nm Golgi-derived vesicles in ungerminated *Corylus* pollen (Liu et al. 1994).

Following this hypothesis, SVs could switch from the acto-myosin system to the MT-kinesin complex in the apical part where the cytoplasmic streaming is arrested. Therefore, the complex SVs-kinesin-MTs could represent a check point of SVs separating those that will fuse from others that must be moved backward. In this case, SVs could carry more than one motor on their surface and should be able to regulate their movement on different cytoskeletal systems as reported for polarized epithelial cells (Fath et al. 1994). An example of cooperation between a kinesin like protein and a myosin isoform during polarized growth has also been observed in yeast, where a mutant for a myosin isoform (myo-2) is rescue by a gene for a kinesin-like protein (Lillie and Brown 1994). Colocalization of the kinesin-related polypeptide with short MTs in the apical region of the pollen tube could be in agreement with a mechanism of microtubule nucleation at the tip and subsequent transport of newly formed MTs to the subapical part for their further organization (Cai et al. 1996). The kinesin-related polypeptide could contribute in this case to the MTs translocation process (Cai et al. 1993).

Recently at least five distinct kinesin-like sequences have been identified in the *Arabidopsis* genome (Kat genes) (Mitsui et al. 1993), three of which have been cloned (Kat A-C) (Mitsui et al. 1994). These molecules have their motor domains at the C-terminal region and seem to be closely related to members of the C-terminal motor subfamily of the kinesin superfamily. (Moore and Endow 1996). Antibodies raised against two regions of the motor domain of Kat A recognized a 140 kDa polypeptide in *Arabidopsis* seedlings and a 125 kDa polypeptide in carrot culture cells (Mitsui et al. 1996, Liu et al. 1996). In tobacco pollen tubes the antibody recognized a 110 kDa polypeptide by western blotting analysis and evidentiated a staining within both the vegetative cytoplasm and the generative cell by immunofluorescence microscopy (Liu and Palevitz 1996). Within the cytoplasm of the vegetative cell, a spot-like staining similar to that obtained with the antibody against the calf brain kinesin was observed. Within the generative cell, the staining changed as the cell division progressed. The staining was mostly present on the nuclear surface during the interphase and then the staining moved to the GC MTs. During GC division, the staining was present both in the kinetochore fibers and in the spindle midzone. The authors hypothesized that only the midzone kinesin is active during the cell division whereas the protein localized in the remaining part of the "spindle" represents an inactive form of the protein (Liu and Palevitz 1996).

Recently, other kinesin-like polypeptides have been identified by screening a cDNA library using biotinylated calmodulin affinity. These findings suggested that the function of these new kinesin-like polypeptides could be regulated by Ca^{2+}-calmodulin complexes (Reddy et al. 1996a, b, Wang et al 1996). These kind of molecules have not been identified in pollen tubes yet.

Dynein molecules represent the other known microtubular motors that are able to convert the chemical energy of the ATP into mechanical movement of organelles along MTs. Dynein is a large protein (1–2 million Da) with sedimentation coefficients between 10 and 30 S. Several isoforms of dynein with different sedimentation coefficients were first identified as major constituents of the outer and inner arms of axonemes, where they are responsible for generating sliding forces between adjacent microtubule doublets in order to promote ciliary and flagellar beating (Porter and Johnson 1989). A cytoplasmic dynein with sedimentation coefficient of 20 S was later identified in nerve cells (Paschal et al. 1987) as well as in other cell types and organisms (Holzbaur and Vallee 1994). Recently, heterogeneous isoforms of cytoplasmic dynein in mammalian cultured cells with sedimentation coefficients of 13 S, 15 S and 20 S were localized in different cell compartments and this evidence could account for the multiple dynein functions within cells (Vaisberg et al. 1996).

Furthemore , the presence of dynein heavy chain related polypeptides in pollen tubes of *Nicotiana tabacum* supports the idea that an MT-based motor system could function during pollen tube growth (Moscatelli et al. 1995). The dynein heavy chain related polypeptides were heterogeneous for their sedimentation coefficients since they sedimented in two sets of fractions at 22 S and 12 S. This data suggests that, as well as for the multiple isoforms of mammalian cultured cells, they could have different functions in different cell compartments. The immunolocalization of both these polypeptides was performed using a polyclonal antibody raised against a synthetic peptide reproducing the most conserved p-loop consensus sequence of dynein heavy chains (**GPAGTGKT**) in pollen tubes of *Nicotiana tabacum*. Here the antibody recognized particle-like structures whose distribution appeared to change as the tube elongated (Moscatelli et al. 1998), resembling the organization of MTs in pollen tubes of the same species (Del Casino et al. 1993). In short tubes, when MTs extended in the half region proximal to the grain, the staining did not extend into the apex (Fig. 21.2b), whereas in longer tubes with MTs in the apical region, spot–like structures were visible at the tip as well (Moscatelli et al. 1998).

This data suggested that proteins carrying the two dynein heavy chain related polypeptides could contribute to the movement of membranous organelles along MTs in tobacco pollen tubes. Considering the polar organization of cytoplasmic organelles and the staining pattern of the Dy-1 antibody, we can hypothesize that dynein-related proteins could function in positioning organelles and /or in mediating vesicle trafficking between different membranous compartments. The presence of staining at the tip in long pollen tubes could mean that dynein related proteins cooperate with kinesin related polypeptides in the organization of short, newly formed MTs at the tube apex. The Dy-1 antibody did not recognize epitopes within the generative cell, where crossbridged MTs have been observed for at least three hours after germination. The hypothesis that dynein heavy chain related polypeptides could be present later in dividing generative cells cannot be excluded.

21.5 Concluding Remarks

Data available at present allows us to describe only partially the mechanisms that regulate different movements during pollen tube growth. In the past, attention was mostly given to the study of molecules involved in the most prominent form of intracytoplasmic movement, the cytoplasmic streaming. Some evidence pointed out a fundamental role of the acto-myosin system in determining both the cytoplasmic streaming and the MGU migration (Cai et al. 1997). Other types of cytoplasmic movement such as the trafficking of vesicles between membranous organelles or the organelle positioning were studied to a much lesser extent. The role of MTs and MT-motor proteins has been investigated only in the recent years leading to the identification of several putative MT-motor proteins. The role of these proteins during pollen tube growth is not yet clear since only a role in organelle positioning within the cytoplasm has been hypothesized. More data is needed to understand the cooperation between the two motor systems, their ultrastructural localization, the accessory polypeptides that regulate their function in different steps of tube growth and modulate the motor activity. Further information is also necessary in order to clarify the concerted action of both AFs- and MT-based motor systems in the polarized organization of the cytoplasm, as well as in other events such as endocytosis, which could be important in the uptake of molecules from the style transmitting tissue in "*in vivo*" grown pollen tubes.

References

Asada T, Seonobe S, Shibaoka H (1991) Microtubule translocation in the cytokinetic apparatus of cultured tobacco cells. Nature 350:238–241

Åstrom H, Virtanen I, Raudaskoski M (1991) Cold-stability in the pollen tube cytoskeleton. Protoplasma 160:99–107

Åstrom H (1992) Acetylated α-tubulin in the pollen tube MTs. Cell Biol Int Rep 16:871–881

Åström, H, Sorri, O, Raudaskoski, M (1995) Role of microtubules in the movement of the vegetative nucleus and generative cell in tobacco pollen tubes. Sex Plant Reprod 8:61–69

Blackbourn HD, Jackson AP (1996) Plant Clathrin heavy chain: sequence analysis and restricted localisation in growing pollen tubes. J Cell Sci 109:777–787

Cai G, Bartalesi A, Del Casino C, Moscatelli A, Tiezzi A, Cresti M (1993) The kinesin-immunoreactive homologue from *Nicotiana tabacum* pollen tube: biochemical properties and subcellular localization. Planta 191:496–506

Cai G, Moscatelli A, Del Casino C, Chevrier V, Mazzi M, Tiezzi A, Cresti M (1996) The anti-centrosome mAb 6C6 reacts with a plasma membrane-associated polypeptide of 77-kDa from the *Nicotiana tabacum* pollen tubes. Protoplasma 190:68–78

Cai G, Moscatelli A, Cresti M (1997) Cytoskeletal organization and pollen tube growth. Trends in Plant Science 2:86–91

Chevrier V, Komesli S, Schmit AC, Vantard M, Lambert AM, Job D (1992) A monoclonal antibody, raised against mammalian centrosomes and screened by recognition of plant microtubule organizing centers, identifies a pericentriolar component in different cell types. J Cell Sci 101:823–835

Ciampolini F, Moscatelli A, Cresti M (1988) Ultrastractural features of *Aloe ciliaris* pollen. Sex Plant Reprod 1:88–96

Condeelis JS (1974) The identification of F actin in the pollen tube and protoplast of Amaryllis belladonna. Exp Cell Res 88:434–439

Cresti M, Tiezzi A (1990) Germination and pollen tube formation. In: Blackmore S, Knox RB (eds) Microspores: Evolution and Ontogeny. Academic Press, London, pp 239–263

Cresti M, Pacini E, Ciampolini F, Sarfatti G (1977) Germination and early tube development *in vitro* of *Lycopersicum peruvianum* pollen: ultrastructural features. Planta 136:239–247

Del Casino C, Li Y, Moscatelli A, Scali M, Tiezzi A, Cresti M (1993) Distribution of microtubules during the growth of tobacco pollen tubes. Biol Cell 79:125–132

Derksen J, Rutten T, van Amstel T, de Win A, Doris F, Steer M (1995) Regulation of pollen tube growth. Acta Bot Neerl 44:93–119

Derksen J, Pierson ES, Traas J A (1985) Microtubules in vegetative and generative cells of pollen tubes. Eur J Cell Biol 38:142–148

Fath KR, Trimbur GM, Burgess DR (1994) Molecular motors are differentially distributed on Golgi membranes from polarized epithelial cells. J Cell Biol 126:661–675

Franke WW, Herth H, Van Der Woude WJ, Morré D J (1972) Tubular and filamentous structures in pollen tubes: possible involvement as guide elements in protoplasmic streaming and vectorial migration of secretory vesicles. Planta 105:317–341

Hepler PK (1997) Tip growth in pollen tubes: calcium leads the way. Trends in Plant Science 2:79–80

Heslop-Harrison J, Heslop-Harrison Y, Cresti M, Tiezzi A, Ciampolini F (1986) Actin during pollen germination. J Cell Sci 86:1–8

Heslop-Harrison J, Heslop-Harrison Y (1989a) Actomyosin and movement in the angiosperm pollen tube: an interpretation of some recent results. Sex Plant Reprod 2:199–207
Heslop-Harrison J, Heslop-Harrison Y (1989b) Myosin associated with the surface of organelles, vegetative nuclei and generative cells in angiosperm pollen grains and tubes. J Cell Sci 94:319–325
Heslop-Harrison J, Heslop-Harrison Y, Cresti M, Tiezzi A, Moscatelli A (1988) Cytoskeletal elements, cell shaping and movement in the angiosperm pollen tube. J Cell Sci 91:49–60
Heslop-Harrison J, Heslop-Harrison Y (1991) The actin cytoskeleton in unfixed pollen tubes following microwave-accelerated DMSO permeabilization and TRITC-phalloidin staining. Sex Plant Reprod 4:6–11
Hoyt MA (1994) Cellular roles of kinesin and related proteins. Curr Opin Cell Biol 6:63–68
Holzbaur ELF, Vallee RB (1994) Dyneins: molecular structure and cellular function. Annu Rev Cell Biol 10:339–372
Hussey PJ, Traas JA, Gull K, Lloyd CW (1987) Isolation of cytoskeletons from synchronized cells: the interphase micritubule array utilizes multiple tubulin isotypes. J Cell Sci 88:225–230
Hussey PJ, Lloyd CW, Gull K (1988) Differential and developmental expression of β-tubulins in higher plant cell. J Biol Chem 263:5474–5479
Iwanami Y (1956) Protoplasmic movement in pollen grains and tubes. Phytomorphology 6:288–295
Joos U, Van Aken J, Kristen U (1994) Microtubules are involved in maintaining the cellular polarity in pollen tubes of *Nicotiana sylvestris*. Protoplasma 179:5–15
Joos U, Van Aken J, Kristen U (1995) The anti-microtubule drug carbetamide stops *Nicotiana sylvestris* pollen tube growth in the style. Protoplasma 187:182–191
Khono T, Shimmen T (1987) Ca^{2+}-induced fragmentation of actin filaments in pollen tubes. Protoplasma 141:177–179
Khono T, Okagaki T, Kohama K, Shimmen T (1991) Pollen tube extract supports the movement of actin filaments *in vitro*. Protoplasma 161:75–77
Kohno T, Shimmen T (1988) Accelerated sliding of pollen tube organelles along *Characeae* actin bundles regulated by Ca^{2+}. J Cell Biol 106:1539–1543
Kreis TE (1987) Microtubule containing detyrosinated tubulin are less dynamic. EMBO J 6:2597–2606
Lancelle SA, Hepler PK (1989) Immunogold labelling of actin on sections of freeze substituted plant cells. Protoplasma 150:72–74
Lancelle SA, Hepler PK (1991) Association of actin with cortical microtubules revealed by immunogold localization in Nicotiana pollen tubes. Protoplasma 165:167–172
Lancelle SA, Cresti M, Hepler PK (1987) Ultrastructure of the cytoskeleton in freeze-substituted pollen tubes of Nicotiana alata. Protoplasma 140:141–150
Lancelle SA, Hepler PK (1992) Ultrastructure of freeze-substituted pollen tubes of Lilium longiflorum. Protoplasma 167:215–230
Lillie SH, Brown SS (1994) Immunofluorescence localization of the unconventional myosin, Myo2p, the putative kinesin-related protein Smy1p, to the same region of polarized growth in Saccharomyces cerevisiae. J Cell Biol 125:825–842
Lin Y, Wang Y, Zhu J, Yang Z (1996) Localization of a Rho GTPase implies a role a role in tip growth and movement of the generative cell in pollen tubes. Plant Cell 8:293–303
Liu B, Palevitz BA (1996) Localization of a kinesin-like protein in the generative cells of tobacco. Protoplasma 195:78–89
Liu B, Cyr RJ, Palevitz BA (1996) A kinesin-like protein, katAp, in the cells of *Arabidopsis* and other plants. Plant Cell 8:119–132
Liu GQ, Cai G, Del Casino C, Tiezzi A, Cresti M (1994) Kinesin-related polypeptide is associated with vesicles from *Corylus avellana* pollen. Cell Motil Cytoskeleton 29:155–166
Miller DD, Lancelle SA, Hepler PK (1996) Actin filaments do not form a dense mehwork in *Lilium longiflorum* pollen tube tips. Protoplasma (in press)
Marc J (1997) Microtubule-organizing centres in plants. Trends in Plant Science 2:223–230
Matus A (1990) MT-associated proteins. Curr Opin Cell Biol 21:10–14
Miller DD, Scordilis SP, Hepler PK (1995) Identification and localization of three classes of myosins in pollen tubes of *Lilium longiflorum* and *Nicotiana alata*. J Cell Sci 108:2549–2563
Mitsui H, Yamaguchi-Shinozaki K, Shinozaki K, Nishikawa K, Takahashi H (1993) Identification of a gene family (kat) encoding kinesin-like proteins in *Arabidopsis thaliana* and the characterization of secondary structure of KatA. Mol Gen Genet 238:362–368
Moore DJ, Endow SA (1996) Kinesin proteins: a phylum of motors for microtubule-based motility, BioEssay 18:207–219
Moscatelli A, Del Casino C, Lozzi L, Cai G, Scali M, Tiezzi A, Cresti M (1995) High molecular weight polypeptides related to dynein heavy chains in *Nicotiana tabacum* pollen tubes. J Cell Sci 108:1117–1125
Moscatelli A, Cai G, Ciampolini F, Cresti M (1998) Dynein heavy chain-related polypeptides are associated with organelles in pollen tubes of Nicotiana tabacum. Sex Plant Reprod (in press)
Palevitz BA, Liu B, Joshi C (1994) γ-Tubulin in tobacco pollen tubes: association with generative cell and vegetative MTs. Sex Plant Reprod 7:209–214

Paschal BM, Sheptner HS, Vallee RB (1987) MAP1C is a microtubule-activated ATPase which translocates microtubules *in vitro* and has dynein-like properties. J Cell Biol 105:1273–1282
Perdue TD, Parthasarathy MW (1985) In situ lacalization of F-actin in pollen tubes. Eur J Cell Biol 39: 13–20
Pierson ES (1988) Rhodamine-Phalloidin staining of F-actin in pollen after dimethylsulphoxide permeabilization. Sex Plant Reprod 1:83–87
Pierson ES, Cresti M (1992) Cytoskeleton and cytoplasmic organization of pollen and pollen tubes. Int Rev Cytol 140:73–125
Pierson ES, Derksen J, Traas JA (1986) Organization of microfilaments and microtubules in pollen tubes grown *in vitro* or *in vivo* in various angiosperms. Eur J Cell Biol 41:14–18
Piperno G, LeDizet M, Chang X (1987) Microtubules containing acetylated α-tubulin in mammalian cells in culture. J Cell Biol 104:289–302
Porter EM, Johnson AJ (1989) Dynein structure and function. Annu Rev Cell Biol 5:119–151
Reddy ASN, Narasimhulu SB, Safadi F, Narasimhulu SB, Golovkin M (1996a) A plant kinesin heavy chain is a calmodulin-binding protein. Plant J 10:9–21
Reddy ASN, Safadi F, Narasimhulu SB, Golovkin M, Hu X (1996b) A novel plant calmodulin-binding protein with a kinesin heavy chain motor domain. J Biol Chem 271:7052–7060
Rosen WG, Gawlick SR, Dashek WV, Siegesmund KA (1964) Fine structure and cytochemistry of Lilium pollen tube. Am J Bot 51:61–74
Sassen MMA (1964) Fine structure of Petunia pollen grain and pollen tube. Acta Bot Neerl 13:174–181
Shimmen T, Yokota E (1994) Physiological and biochemical aspects of cytoplasmic streaming. Int Rev Cytol 155:97–140
Shimmen T, Yokota E (1994) Physiological and biochemical aspects of cytoplasmic streaming. Int Rev Cytol 155:97–139
Staiger CJ, Gibbon BC, Kovas DR, Zonia LE (1997) Profilin and actin-depolymerizing factor: modulators of actin organization in plants. Trends in Plant Science 2:275–181
Tang XL, Lancelle SA, Hepler PK (1989a) Fluorescent microscopic localization of actin in pollen tubes: comparison of actin antibody and phalloidin staining. Cell Motil Cytoskel 12:216–224
Tang XL, Hepler PK, Scordilis SP (1989b) Immunochemical and immunocytochemical identification of a myosin heavy chain polypeptide in *Nicotiana* pollen tube. J Cell Sci 92:569–574
Tester M (1990) Plant ion channels: whole cell and single channel studies. New Phytol 114:305–340
Tiezzi A, Pierson ES, Theunis CH, Ciampolini F, Cai G, Bartalesi A, Cresti M (1991) The motile apparatus of sperm cells in angiosperms: correlations with lower plants, gymnosperms and animals. In: Baccetti B (ed) Comparative spermatology 20 years after (Serono Symposia Publications, vol 75). Raven Press, New York, pp 1017–1020
Tiezzi A, Moscatelli A, Cai G, Bartalesi A, Cresti M (1992) An immunoreactive homolog of mammalian kinesin in *Nicotiana tabacum* pollen tubes. Cell Motil Cytoskeleton 21:132–137
Tiezzi A, Cresti M, Ciampolini F (1986) Microtubules in Nicotiana pollen tubes: ultrastructural, immunofluorescence and biochemical data. In: Cresti M, Dallai R (eds) Biology of the Reproduction and Cell Motility in Plants and Animals. Siena Univ Press, Siena, Italy, pp 87–94
Tirlapur UK, Faleri C, Cresti M (1996) Immunoelectron microscopy of myosin associated with the generative cells in pollen tubes of *Nicotiana tabacum* L. Sex Plant Reprod 9:233–237
Tirlapur UK, Cai G, Faleri C, Moscatelli A, Scali M, Del Casino C, Tiezzi A, Cresti M (1995) Confocal imaging and immunogold electron microscopy of changes in distribution of myosin during pollen hydration, germination and pollen tube growth in *Nicotiana tabacum* L. Eur J Cell Biol 67:209–217
Tiwari SC, Polito VS (1990a) An analysis of the role of actin during pollen activation leading to the germination in pear (Pyrus communis L.): Treatment with cytochalasin D. Sex Plant Reprod 3:121–129
Tiwari SC, Polito VS (1990b) The iniziation and organization of MTs in germinating pear (Pyrus communis) pollen. Eur J Cell Biol 53:384–389
Vaisberg EA, Grissom PM, McIntosh JR (1996) Mammalian cells express three distinct dynein heavy chains that are localized to different cytoplasmic organelles. J Cell Biol 133:831–842
Wang W, Takezawa D, Narasimhulu SB, Reddy ASN, Povaiah BW (1996) A novel kinesin-like protein with a calmodulin-binding domain. Plant Mol Biol 31:87–100
Williamson RE (1993) Organelle movements. Annu Rev Plant Physiol Plant Mol Biol 44:181–202
Yokota E, Shimmen T (1994) Isolation and characterization of plant myosin from pollen tubes of lily. Protoplasma 177:153–162
Yokota E, McDonald AR, Liu B, Shimmen T, Palevitz BA (1995) Localization of a 170 kDa myosin heavy chain in plant cells. Protoplasma 185:178–187

CHAPTER 22

The Pollen Tube Oscillator: Towards a Molecular Mechanism of Tip Growth?

J. A. Feijó

Departamento de Biologia Vegetal, Faculdade de Ciências da Universidade de Lisboa, Campo Grande, Ed. C_2, P-1700 Lisboa, Portugal
e-mail:jose.feijo@fc.ul.pt
telephone: +351.1.7500069
fax: +351.1.7500048

22.1 Introduction

Pollen grain germination and tube growth have proven to be suitable models for the study of differentiation and growth at the cellular level. Furthermore they may be considered to be a representative example of the tip growth paradigm, and as such, highlight mechanisms that may be applied to a much larger set of cells with tip growth, including most fungi, a number of plants cells and certain specialised animal cells. Since pollen tubes are fairly easy to maintain and manipulate during *in vivo* experiments, they have become a very popular experimental system. This fact is reflected in the marked increase of the recently published information about the mechanisms that underlie the regulation of its growth (reviews by Mascarenhas 1993; Feijó et al. 1995; Derksen et al. 1996; Taylor and Hepler 1997). Besides other aspects, which are dealt with in more detail in other chapters of this book, one of the issues which emerged recently as more exciting is the fact that pollen tube growth may be characterised in some of its more important characteristics by the existence of pulses, sustained oscillations, or other non-linear temporal and spatial phenomena. In this paper we try to cover these recent advances and their possible impact in the elucidation of the molecular mechanisms that regulate tip growth.

22.2 Oscillations in Pollen

22.2.1 Extracellular Currents

Perhaps the first description of oscillatory behaviour during pollen tube growth was described by Weisenseel et al. (1975). This paper described for the first time the major characteristics of pollen membrane transport and ion physiology: an electric dipole, with the grain working as the source and the tube as the sink of a large cationic current which traverses the tube cytosol. Furthermore, while investigating the temporal patterns of the tube current as growth proceeds, it was shown that when tubes reached a certain size (typically over a 1 mm long) the total inward current in a significant proportion of the tubes changed from a statistically steady current to pulsatile monophasic bursts that may reach 3 to 4 fold the normal basal current. Under the conditions used, the period of these pulses ranged from 30 to 50 seconds. Once this oscillatory behaviour starts it continues until the tube dies, but there is no apparent change in the morphology or dramatic alterations on the mean growth rate when compared to other tubes with steady currents. The ionic nature of these electrical currents was not known but later investigations strongly implied calcium (Jaffe et al. 1975; Kühtreiber and Jaffe 1990) and potassium (Weisenseel and Jaffe 1976) as the main ionic carriers. More recent studies using a specific vibrating ion-probe showed clearly that in fact calcium is one of the major (if not the principal) current carrier during this oscillatory phase (Holdaway-Clarke et al. 1997), but protons are also an important component (Feijó et al. 1998; see also § 4 bellow).

Another oscillatory or, more accurately, pulsatory behaviour has received preliminary evidence during the early germination of *Hemerocallis* and *Lilium* pollen (Feijó et al. 1994). After hydration the pollen tube starts to emerge, and there is a refractory period before the tube starts to grow at the final rate. In this period (roughly when tubes are shorter than 75–100 μm) a series of 3 to 5 pulses of total extracellular current at the tip occurs and seems to establish the onset of the mechanisms that will condition the subsequent tube growth. These pulses were also measured with a calcium probe, and, again, this ion seems to be the main carrier of the current. A distinction should, however be clearly established between these highly localised transitory pulses (which resemble much more the pulsatile behaviour of calcium in animal eggs after fertilisation) and the previously described sustained oscillations, which may proceed for hours with the same pattern.

22.2.2
Growth Rates

A second set of elegant experiments led to the conclusion that not only physiological variables oscillated, but, perhaps more importantly, also growth rates. The first morphological evidence for this was brought about by Li et al. (1992) who showed that arabinogalactans are not present in the cell wall surrounding the tip and the clear cap. Yet, further back in the tube shank their deposition in the cell wall obeyed a ring-like pattern with remarkable periodicity (which later became known as the "Li-rings"). The frequency of these rings was grossly correlated with periodic changes in the growth rate, and this finding was also extended to explain the apparent lack of ring-like structures in species like *Lilium* which don't seem to have growth oscillations as distinct as the ones of *Nicotiana* or *Petunia*, which sometimes reach 24 fold increases in the fast growth cycle (Li et al. 1994; Pierson et al. 1995). Experimentally induced growth fluctuation confirmed this finding, since a similar ring-like pattern of arabinogalactan deposition was obtained in experimentally slowed down pollen tubes of *Lilium* (Li et al. 1996). Under these conditions ring formation occurred during the slow growing phases. This is a surprising result, since it reveals some sort of spatial coupling between the regulatory events on the very tip, where growth takes place, and the tube shank behind the clear cap, where arabinogalactans are incorporated in the cell wall. The nature of this coupling remains totally obscure, and while a molecular "continuum" has been proposed involving the endomembrane system, the cytoskeleton and the cell wall, this concept remains somehow vague (Li et al. 1997). An alternative explanation could lay in the assumption that the cell wall formed on the tip would bear some sort of periodic physical or chemical anisotropy derived from the differences in material deposition when the tubes slow down, that would function as a preferential site of arabinogalactan accumulation. It should be noted however, that since the chemistry of the pollen wall remains largely unexplained, any extrapolations on the regulation of its biogenesis and, vice-versa, of its regulatory role on growth remain largely speculative.

The pulsatory nature of the growth rate was further investigated in *Nicotiana* and *Petunia* by Geitmann and Cresti (1996, 1998). These authors could modify some of the oscillatory characteristics, like period and amplitude, by classical pharmacological approaches using inhibitors of cytoskeleton, Golgi activity and calcium channels. While interfering with the pathways that bring up wall materials to the tip, these approaches would in all likelihood interfere with the growth process itself and dramatically affect any regulatory mechanisms based on the availability of wall precursors. In this context the most interesting and intriguing result comes from the use of 20 mM of colchicine. The role of microtubules in pollen tube growth is still controversial, and while some evidence shows that they play a role in maintaining cellular polarity (Joos et al. 1994) they don't seem to be necessary at all for germination and growth to proceed (Mascarenhas 1966; reviews in Mascarenhas 1975 and Heslop-Harrison 1987). This result was confirmed by Geitmann and Cresti (1996) since the mean growth rate was not affected by colchicine but, quite surprisingly, the amplitude and the period of the growth peaks markedly decreased. Also the tube diameter decreased, showing a very interesting correlation between two structural features (tube diameter and growth rate) and the temporal regulation. While some side effects of colchicine at this concentration cannot be ruled out, to the extent that tip growth regulation seems to be independent of microtubules, the use of colchicine as a non-interacting perturbing agent may definitely prove to be a valuable experimental tool to induce changes in the dynamics of the system without affecting growth.

22.2.3 Cytosolic Calcium

A third oscillatory feature of pollen tube growth which now seems well established was recently demonstrated by Pierson et al. (1996). These authors were the first ones to provide some evidence that the intracellular free calcium levels could also oscillate. In an attempt to correlate these oscillations with growth rate it was found that in *Lilium*, while short pollen tubes showed small chaotic variations around a mean value, longer pollen tubes (>800 μm) possessed a clearly oscillatory growth rate, reaching 3-fold peak-to-peak variations and a period of about 30 seconds. Further analysis with finer methods, revealed a consistent relationship between the cytosolic free calcium and the growth rate, demonstrated by using calcium green (Messerli and Robinson, 1997) or the ratio dye Fura-2 (Holdaway-Clarke et al. 1997). Both groups convincingly proved that the elevations of cytosolic calcium occur at the same time as the growth peaks or, in other words, growth and cytosolic calcium are in phase. While the relationship between the existence of a calcium gradient and growth had already been postulated in a number of papers (Obermeyer and Weisenseel 1991; Rathore et al. 1991; Miller et al. 1992; Malhó et al. 1994; Pierson et al. 1994) this new finding established beyond any doubt a coupling between the growth process and the cytosolic calcium levels.

22.2.4 Different Kinds of Oscillations?

Taken together, these results seem to point out three classes of pollen tube growth: (1) a pulsatile growth, which could occur at the onset of tube growth of most species, and as the normal pattern in the tubes of certain species (e.g. *Nicotiana*, *Petunia*), (2) a stable growth, which should better be described as statistically stable but with random or chaotic small variations around a mean value (e.g. *Lilium* and *Hemerocallis* with tubes <1 mm) and (3) sustained quasi-sinusoidal oscillations, with a period between 30 and 60 sec. (e.g. *Lilium* with tubes >1 mm). This different behaviour was described mostly by directly measuring the growth rates. Analysis of the total extracellular current or ionic fluxes seems to indicate a close correlation of these variations in growth and the influx of cations (namely protons and calcium) and the levels of cytosolic calcium.

Such a variety of dynamic behaviour could suggest basic differences in the molecular mechanisms underlying the growth process. Yet it should be noted that the final outcome, i.e. tip growth with the typical dome shape, is strikingly robust, even in species in which the three types seem to occur sequentially during elongation (e.g. *Lilium*). Furthermore, also the cytoplasmic structure and its underlying polarity seem to be very stable.

In this paper we will argue that these apparent differences in growth behaviour can (or, one should venture, should) in fact be generated by the same molecular mechanism, if one assumes that this mechanism is governed by a set of simple non-linear regulation steps.

22.3 Why are Oscillations Important?

Only a few decades ago, most of the research effort in various branches of the biological sciences was devoted to equilibrium states. When describing a continuous phenomenon, it was (and to a certain extent, still is) a common procedure to represent time as discrete intervals to which statistics are applied in order to obtain stable or "equilibrium" mean values. These are commonly interpreted as reflecting the natural evolution of the system to an equilibrium point. Quite frequently the discrete values of the temporal sequences are set in order to eliminate the annoying random variations of the system, globally rejected as "noise". The situation is now more and more changing to an appreciation of non-equilibrium as a common feature of a number of highly regulated processes, namely the ones that involve symmetry breaking of time and space, a basic characteristic of all developing biological organisms. Underlying this change is a combination of the advent of a new theoretical framework – deterministic chaos and the mathematical tools that were developed to study and display non-linear phenomenon (Baker and Gollub 1990)- and the refining of several methods and techniques that allowed monitoring of biological features at a continuous or nearly-continuous time. Biological rhythms, which are among the most conspicuous properties of living systems, have now been described at the cellular and molecular level (Goldbeter 1996). Such is also the case of pollen tube growth.

Evidence is growing that a number of major highly regulated processes have an intrinsic non-linear temporal and/or spatial structure (Golberger et al. 1990; Goldbeter 1996). The reason for this apparent "uncertainty" in homeostasis is easier to accept if one envisages the molecular and functional complexity of cells: a multitude of genes and molecules organised in complex pathways around complex structural arrays, and a number of regulatory positive and negative feedback loops. Arguments have been produced that the expression of this complexity at the specific level of the transduction pathways could lead to an important role for epigenetic regulation of the phenotype in plants (Trewavas and Malhó, 1997). The obvious consequence of this "diffuse" or "fuzzy" regulation of development and adaptation would be a much higher plasticity on the responses leading to a better fitting and efficacy of reaction to the unpredictable natural stresses; on the other hand they seem to be more stable and with more robust homeostatic mechanisms (West 1985; Haken 1978; Goldbeter and Dupont 1990). The problem remains, though, as to how to synchronise all these networks of information and to get them to cooperate around an emerging spatial pattern.

Such a complex array of interaction is more than likely to produce non-linear phenomena. In fact, from a formal point of view, it only takes (1) two or three independent dynamic variables (molecular concentration, gene expression, channel activity, etc.) and (2) a non-linear term in the equations that describe their interaction (a feedback loop, membrane depolarisation, co-factor or transcription factor binding, etc.) to generate systems with this kind of behaviour (Baker and Gollub 1990). It should be stressed that, while these conditions do not guarantee oscillations, they do make its existence possible. This is a very important point, and it has been long shown, mainly by researchers in ecology and population dynamics, that the same set of equations that describe the interactions in a dynamic system may produce quite dramatically different results with subtle changes in the constants that affect the population parameters (May 1976). In other words, the same regulatory events, which may usually be expressed by a set of differential equations, can produce stable, oscillatory with variable period or chaotic patterns depending on very subtle changes in the system properties. While some systems tend to have an intrinsic homeostasis and, when perturbed or subjected to external stresses, evolve to one of these patterns (the so called dynamic attractors), others have the same probability of working properly in different conditions. This general principle has received confirmation in various cellular and chemical models (Goldbeter, 1996). Oscillations often represent a typical case of attractors, the limit-cycle attractors. The mathematical and physical characterisation of such aspects is beyond the scope of this paper. It should be noted, however, that there is wide agreement between experimentation and theoretical modelling towards accepting the widespread occurrence of oscillations in crucial regulatory pathways as a means to efficiently convey and propagate information within a system. The efficiency of oscillatory phenomena in doing so seems to be closely related to the fact that an oscillation is much more efficient in coordinating spatial and temporal relationships within a system and, through reaction-diffusion mechanisms, generates a number of basic patterns of morphogenesis. As a corollary, oscillating systems seem also to be more robust and respond homeostatically to a more diverse array of perturbations and irregularities (Haken 1978; Goldbeter 1996).

The same could happen with pollen tubes. The evolutionary success of this kind of growth is overwhelming: while the higher plant female gametophyte follows a number of different development algorithms (Haig 1990), the male gametophyte is universally expressed as a growing pollen tube with the same structural organisation. The differences in the growth behaviour could then represent different conditions of dynamic stability of the same underlying molecular mechanisms that ultimately regulate growth. In this respect it should be noted that (1) several authors (Weisenseel et al. 1975; Pierson et al. 1996; Geitmann and Cresti 1996) reported the transition from an apparently stable to an oscillatory growth without major changes in any of the structural and growth parameters; and (2) where continuous temporal recording is possible (e.g. vibrating probe, time lapse video or imaging with resolution of just a few seconds) what has been regarded as stable growth should be better described as chaotic variations around a mean value.

At this point the central question should then be expressed: will we gain insight into the regulatory mechanisms of a living system by looking at its continuous temporal evolution? The answer should be yes if some criteria are met, namely: (1) the system should be looked at with minimal interference in its normal physiology, (2) a maximal number of variables (with a minimum of two) of the system should be studied with continuous recording and (3) the variables studied must have some sort of temporal and spatial correlation. These criteria have different practical implications. Criterion (1) should imply the exclusive use of non-invasive real-time techniques (real-time defined in the sense that the time scale of the measuring technique should be at least one order of magnitude lower than the one of the phenomena being described). Gross manipulation of a system dynamics, like pharmacological approaches, while important as first steps to make experimental decisions, will eventually trigger a number of unpredictable effects at different levels which will be amplified throughout the non-linear regulatory cascade. Criterion (2) should allow the establishment of groups of variables with the same dynamic behaviour and, combined with (3), the decision on what are the variables which are regulatory. Criterion (3) is perhaps the most difficult to achieve, and raises the importance of stable oscillatory patterns. In fact when a system is evolving around a stable monotonic condition there is no possible way of correlating its variables without serious experimental perturbation (e.g. inhibitors, blockers, mutations, etc.). As in quantum mechanics the question always remains as to whether the correlation that may be established is real or the result of the perturbation. On the other extreme, chaotic conditions are also not easy to deal with, since correlation between variables is difficult (when not impossible) to establish without heavy numeric and analytic methods. In the middle situation, sustained oscillations provide the best dynamic pattern to analyse. Correlations are easy to make just by analysis of the period and amplitude and, within certain limits, allows adimensional, qualitative comparison of data from different techniques and different organisation levels. Wave function analysis, like Fourier transforms, power spectra or wavelet analysis, provide clear quantitative data to be compared and correlated. Phase shifts and phase space analysis provide clues about the regulatory sequence of events. In a prototype experimental system, mild perturbations could then be analysed in terms of period change, shifting wave phases or wave power spectrum. These mild perturbations could eventually be stretched until the disruption of the correlation between variables to reveal the functional links. Such a prototype experimental system would allow numeric approaches based on differential equations to be modelled, tuned and eventually predict transitions between attractors and the boundaries of homeostasis.

While the limited knowledge of pollen tube growth mechanisms does not allow unrealistic optimism, it should be noted that pollen tubes oscillate with remarkable periodicity in a number or parameters accessible by continuous recording with non-invasive methods (growth rate with time lapse video, cytosolic ion concentration with various imaging methods and extracellular ionic fluxes with the vibrating probe) and thus represent a potential system in which the three criteria outlined above may be met.

So what do we know about the partners involved, and what is there that should be discovered before a numeric model can be built?

22.4 The Molecular Basis of the Pollen Tube Oscillator

In order to dissect and model the oscillator a number of assumptions must be made in order to simplify the system and try to isolate the crucial regulatory events.

To the extent that the outcome of the oscillatory activity in a pollen tube is reflected by the oscillation of the growth rates, the scrutiny of the molecular basis of the oscillator should be based on the nature of the growth mechanism itself. It is widely accepted that pollen tubes grow by exocytosis of wall precursors contained in Golgi-derived vesicles. Furthermore, although generalising evidence is still lacking, it is also accepted that this growth takes place exclusively in the tip area (reviews in Derksen et al. 1995; Taylor and Hepler 1997). It thus seems reasonable to assume (1) that the molecular mechanisms that shape and control growth exist exclusively in the tip, and their temporal interplay is governed by a feed-back loop(s) that takes about the time of the reported periods. On the other hand, due to the magnitude of the numbers involved, it seems reasonable to assume that (2) the nature of the regulatory events is based on biophysical phenomena or fast chemical or biochemical reactions. These assumptions immediately outrun any organelle or cytoplasmic structure not present in the tip area and regulation by direct gene expression or enzyme induction.

A special mention should be given to the cytoskeleton, especially the actin filaments. Evidence has accumulated that vesicle and organelle streaming in the pollen tube is driven by acto-myosin interactions along the actin bundles (sometimes referred to as actin cables) which are conspicuously present in these cells. Yet the extension of these bundles into the clear cap is still surrounded by controversy. While approaches based on fixed or permeabilised pollen tubes have sometimes provided evidence for such an intrusion to the very tip (review by Derksen et al. 1996; Cai, 1997) a recent report with direct microinjection of rhodamine-phalloidin into living tubes, a procedure much less keen to artefacts and re-distribution of cytoplasm, has demonstrated a different pattern, with the actin bundles flaring out to the cell's membrane in the base of the clear cap (Miller et al. 1996). This finding fits closely with what is known about the fine ultrastructural data obtained by rapid-freeze and freeze-substitution (Tiwari and Polito 1988; Lancelle and Hepler 1988). It thus seems reasonable for the moment to assume that actin bundles, while crucial for the transport of vesicles into the clear cap, do not participate directly in the mechanism of growth. So who are the other actors in the (tip) stage?

In the absence of cytoskeletal or other structural elements, the central issue is then: how do the Golgi vesicles move along the clear cap and how is fusion and exocytosis promoted? Close examination in DIC or phase contrast microscopy of healthy growing tips shows that in the clear cap vesicles and organelles move in a random, quasi-Brownian motion (review in Derksen et al. 1996). This observation led many authors to assume that under these conditions the regulation of the overall pattern would have to depend solely on the physico-chemical forces present in that area of the cytoplasm. Perhaps the first explanations proposed were based on the existence of large polarised ionic currents that would act as an electric field and promote the movement of charged particles according to the electric attracting or repulsive forces of the vesicle membrane proteins (Jaffe et al 1974; Harold and Caldwell, 1990; Weisenseel and Kicherer, 1981). Polarised channels (or polarised channel activation) would serve as membrane spatial marker and would define not only the growth plane, but also would condition the growing shape (Jaffe 1981). An extension of these concepts to membrane electrophoresis lead to the proposal of a regulation mechanism of pollen tube growth based on the spatial and temporal self-organisation of membrane ion carriers and pumps and their spatial segregation on the grounds of putative surface charges (Feijó et al. 1995). Crucial in this and all other models so far proposed (Trewavas and Malhó 1997; Holdaway-Clarke et al. 1997; Geitmann and Cresti, 1988) are the existence of (1) calcium stretch-activated channels on the tip apex and (2) the mechanical interplay between the existence of an internal turgor pressure and the contention and relaxation of the cell wall at the tip apex. With modifications and adaptations, all models accept that there is a breaking mechanical point based either on the yielding of the wall or on the increase in the internal turgor, leading to a stretch of the plasma membrane and consequent opening of the calcium channels. In turn this would allow calcium to flow in and promote calcium activated vesicle fusion mediated by calcium activated proteins, like annexins. From a formal standpoint, three variables (exocytosis, channel activity and wall rheology) and two positive feed-back loops (wall yielding increases channel activity, calcium increase promotes vesicle fusion) exist, allowing the conditions sufficient for oscillations (or chaotic) patterns to occur. Furthermore, while negative feed-back loops are known to permit oscillations, positive feed-back mechanisms are far more important since they constitute a self-amplification mechanism (e.g., cAMP oscillations in *Dyctiostelium*, glycolitic and calcium oscillations; see Goldbeter, 1996). All the models for pollen remain, however, speculative, as a number of intervening parts have not been properly characterised. The existence of calcium channels on the tip has received important indirect and pharmacological evidence (Miller et al. 1992; Malhó et al. 1994, 1995; Feijó et al. 1994; Pierson et al. 1994, 1996). The most important consequence of its activity, i.e. the existence of a tip-focused gradient of cytosolic calcium at the tip that could function as a sink to vesicle fusion, is now well established. Yet, so far, direct confirmation by patch-clamp or molecular isolation of these channels is still lacking. Even less is known about their specificity, a crucial point since many stretch activated channels are relatively unspecific for cations, allowing the influx of not only calcium but also protons and potassium (Yang and Sachs 1989). The mechanical issue is not resolved neither. While recent data on cell wall extension seems to favour the wall yielding hypothesis (see review in Cosgrove 1997) most of these conclusions were drawn from typical pecto-cellulosic vegetative cell walls, quite different from the pollen tube special wall. While direct measurement of the internal turgor revealed very elevated values but a remarkable stability over time (Benkert et al. 1997) experimental manipulations of turgor revealed strong consequences on growth (Li et al. 1996). Last but not least, exocytosis in pollen tubes remains to be characterised. Initially thought of as a membrane flow system (Picton and Steer 1982, see review in Steer and Steer 1989), only recently was convincing evidence of a coated vesicle mediated endocytic pathway described (Derksen et al. 1995). In other model systems exocytosis has proven to be a much more complex phenomena than originally thought, and a number of different models have been described ("kiss and run", "sit-and wait", full fusion, etc.). Basically nothing is known about the docking, priming and fusion steps on pollen tubes (for a recent review on exocytosis see Südhof 1995). Until the dynamics of the exocytic pathway is clearly described, the existence of channels is confirmed and the nature of the mechanical balance between wall and turgor is unravelled, all models so far described should be regarded with caution. It should be noted that so far little effort has been made in the sense of making these models quantitative and thus predictive, so their value is, for most cases, purely phenomenological.

22.5
Case Study: *Lilium* Oscillations

Lilium pollen tube growth represents one of the best studied systems. The recent finding that intracellular calcium is in phase with the growth rate (Messerli and Robinson 1997; Holdaway-Clarke et al. 1997) strongly supports the assumptions discussed above that the calcium influx is mediated by stretch-activated channels. This elevation of the flux when growth peaks up should be reflected in the elevation of the external calcium flux and should be detectable with a calcium vibrating probe. Data from this technique remains, however, more elusive, which justifies some elaboration behind what has been described in Holdaway-Clarke et al. (1997).

As previously described (Weisenseel et al. 1975; Pierson et al 1996) oscillatory behaviour is only triggered in long tubes (typically over 1 mm). Fig. 22.1 shows a representative example of this kind of pattern. 3–4 fold variations are observable in the growth rate and in the tip calcium influx. Period varies from cycle to cycle, but consistently one never finds an overlap between the two parallel oscillations (growth rate and calcium influx). The correlation between them is however suggested by the existence of the exact number of cycles in each during a given temporal series. One possible analytic way of establishing this relationship is to compute the Fourier transform of the specific traces and try to analyse the main components in the wave. Such an analysis is displayed in Figure 22.2.

Since we are dealing with wave forms which are not pure sinusoids, Fourier analysis decomposes them in a number of components with different frequencies (ω). Most of these are however, small components and reflect, besides the irregularities of the process itself, the noise of the collecting technique. Yet in both cases (growth and extracellular calcium flux) a single prominent peak stands as the main component of the wave function associated to the specific sequence. Since these main peaks overlap and have exactly the same frequency (the inverse of the period) one may conclude that both oscillations are strictly correlated and are generated by the same temporal oscillator.

If the correlation between growth and the extracellular calcium flux can be established reliably by Fourier analysis, the same cannot be said about the phase relations between them. Fig. 22.1 clearly shows that they do not overlap in any cycle, which implies that they occur ordered in time. Since this order may reflect the functional (or causal) relationship between them, the analysis of the phase relationships becomes mandatory. Table 22.1 reflects one such attempt. Columns 2, 3 and 4 display the period values for 5 representative sequences, measured by computing the time between successive peaks. Small variations between the VP (extracellular calcium flux) period and GR (growth rate) period are normal, since the time constant for data acquisition in both these parameters is about 4.5 sec. (for detailed methods see Holdaway-Clarke et al. 1997 and Feijó et al. 1998). The estimation of the period deriving from the main component displayed in the Fourier transforms (see Fig. 22.2) should then be a reliable indication of the average period (Column 4) and displayed in all sequences, this value is within the data acquisition time constants. As a rule of thumb, the period obtained for the VP (column 3) is closer to the Fourier main component, and the values are as closer as the number of cycles (column 5) averaged is higher. The first finding is not surprising since image analysis of video captured growing tubes has resolution limitations. Whenever correlation of growth is not at stake, future analysis of period can rely solely on VP measurements, which are much easier and faster to perform. Cross correlation between two simultaneous out of phase oscillations is not a trivial task. There isn't a numerical method that can be said to produce robust results, and all the possible approaches involve a number of assumptions and statistics based decisions. The easiest method is peak-to-peak analysis, and the results are displayed in columns 7 and 8. In this method the time between two consecutive peaks of both series can be computed in two different manners, one measuring the lapse between one VP peak and the consecutive GR peak (column 7), or the other way around (column 8). The assumption is that when one measures the lapse according to the first alternative it is accepted that the calcium influx peak conditions growth or alternatively that growth conditions the calcium influx. The re-

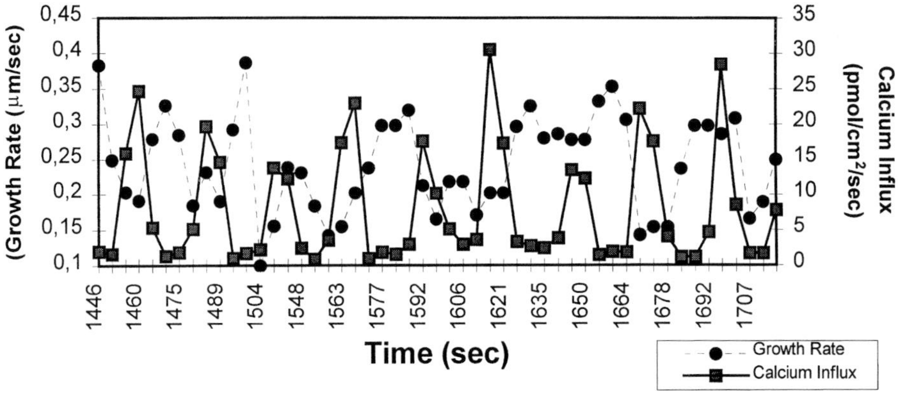

Fig. 22.1. Correlation between the growth rate (dashed line) and the extracellular calcium influx (solid line) at the tip of a growing pollen tube of *Lilium*. The oscillations are clearly not in phase, but the same number of cycles occurs on the interval shown. Under these conditions, causal relationships between the different factors are nearly impossible to establish. In this specific tube, growth rates vary about 5-fold, while calcium oscillates between values 10-fold apart (for methods check Holdaway-Clarke et al. 1997)

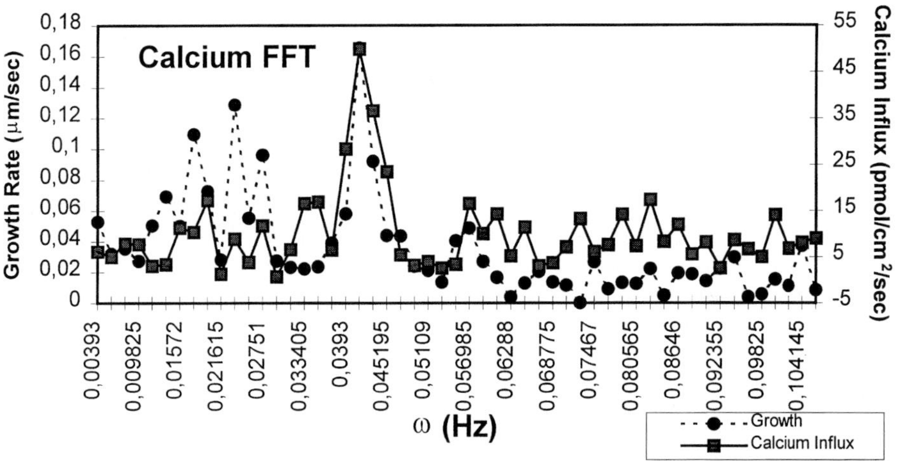

Fig. 22.2. Fourier spectra of a calcium series similar to Figure 22.1. The number of small peaks reflect the irregularities of the waves and the noise of the different techniques used to measure the growth rates and the fluxes. Yet there is a main distinctive peak at both series which perfectly overlaps, with a frequency (ω) of about 0.043 Hz (period of about 25 sec.). This overlapping expresses the fact that the main component of both series is absolutely correlated, and quite likely generated by the same mechanism, i.e. by the same molecular oscillator

sults may then express a more stable temporal relationship in one of these two alternatives, which would favour one of the assumptions, or a similar result for both, which would indicate no temporal relationship. Since a considerable time span in the period occurs in these 5 sequences (24 to 56 sec.) differences between the lags should be clearly expressed if one assumption is more robust than the other. Such is the case, and the

Table 22.1. Summary of the oscillatory characteristic of five representative sequences of growth rates and calcium influx in which Fourier spectra obeys the pattern shown in Figure 22.2 (see text for details). The growth period (GR) was computed assuming a full cycle between two consecutive peaks. Within the SE interval, they all agree with the main component of the Fourier spectrum. Furthermore the period obtained for the calcium flux (VP) is also within the range on the SE. Therefore, bearing in mind the time constants of the measuring techniques (about 4.6 sec. for the VP and 4.2 sec. for GR) the values are quite equivalent and more so, the longer the sequence is, i.e. the more cycles are followed. Within this sample, periods range from 24 to 57 sec., and this finding could not be correlated to growth rate (column 6) or any other characteristic. Cross correlation between sequences was performed by peak-to-peak analysis, assuming that (1) calcium increase induces growth increase (i.e. by measuring the lapse between the calcium peak and a subsequent growth peak- *column 7*) or, inversely, that (2) growth induces calcium increase (i.e. measuring the lapse between a growth peak and the subsequent calcium peak – *column 8*). The crudeness of the method implies that the results should be looked at cautiously, but still it is obvious that both the SE and the span are much smaller when one assumes that calcium follows growth. Note also that, under this assumption, the span and the SE are within the combined time constants for data acquisition of both methods, while under the inverse assumption, i.e. that calcium leads growth, both the span interval and the SE are clearly out of the range, to the extent that the considerable variation in the period in these series (24–57 sec.) is clearly reflected in the delay between the calcium peak and the subsequent growth peak (11–48) while the lapse between each growth peak and the subsequent calcium remains relatively constant (13–18 sec.). This data suggests that there is a much more robust temporal relationship if we consider that growth leads calcium

Series	Period GR (sec ± SE)	Period VP (sec ± SE)	Period Fourier (sec)	Nr. cycles	Gr. Rate (μm/min)	GR after Ca^{2+} (sec ± SE)	Ca^{2+} after GR (sec ± SE)
1	57.2 ± 2.7	56.8 ± 14	56.1	8	17.1	47.8 ± 8.6	14.7 ± 1.0
2	46.1 ± 5.6	50.0 ± 4.2	50.0	7	16.0	37.8 ± 12	17.1 ± 4.8
3	25.3 ± 6.8	26.6 ± 2.2	29.6	9	15.2	11.2 ± 4.9	14.1 ± 2.8
4	26.0 ± 9.0	24.0 ± 5.0	24.2	17	17.4	11.0 ± 9.0	18.0 ± 1.7
5	36.4 ± 6.7	34.6 ± 6.4	38.5	6	20.0	17.0 ± 8.3	13.2 ± 1.2
Average ± SE	38.2 ± 6.16	38.4 ± 6.36	39.68		17.14	24.96 ± 8.56	15.42 ± 2.3
SPAN	26–57	24–57	24–56	6–17	15–20	11–48	13–18

sequence GR inducing or triggering VP (column 8) is clearly favoured since (1) despite the variation in period the lag between these two is always within the range of the data acquisition time constant (13–18 sec.) while (2) under the opposite assumption it expresses clearly the length of the varying period (span 11–48 sec.; difference between the lowest and the highest is 37 sec., a value close to the 32 sec. difference between the lowest and highest value for period). These conclusions are in accordance with a model in which the control factor is the regulation of growth and the extracellular flux of calcium lags by an average of about 15 sec. Peak-to-peak analysis may be considered a crude method but wavelet analysis produced quite a similar result (Holdaway-Clarke et al. 1997). Wavelets are another method of spectral analysis which is becoming increasingly popular since it allows analysis of very short sequences. Yet, despite the sophistication of the numerical methods and the appealing graphical output of this kind of analysis, the assumptions that have to be accepted for the treatment of sequences like the ones in Figure 22.1 and their cross-correlation do not make them much more robust than peak-to-peak analysis, specially when the intervals are as well defined as the ones expressed in Table 22.1 (Benedeto and Frazier 1994). The very concept of causal relationship may not have much meaning in the context of a non-linear self-organised oscillating system since a true understanding of the temporal and spatial relationships in such systems can only come from correct mathematical modelling and analysis of the system dynamics on the phase space (Haken 1978; Goldbeter 1996)(see below and Figs. 22.6 and 22.7).

Fig. 22.3 see color plate at the end of the book.

How then is this delay on the extracellular fluxes explained? A possible explanation was put forward in an elegant model involving the cell wall as a possible buffer of calcium (Holdaway-Clarke et al. 1997). The key point in this model is the fact that de-esterification of pectins on the outside of the tip membrane, and consequent cross-linking is dependent of calcium. In this model the intracellular calcium would then come directly from free calcium in the wall and the extracellular flux would represent the calcium necessary for the pectin cross-linking. While circumstantially supported by the literature, this model still leaves some unanswered questions regarding both the spatial and temporal nature of the extracellular fluxes and their correlation with growth. First, all measures so far taken with calcium vibrating probes, even under conditions where pipettes are built specially thin to increase spatial resolution, locate the source of the influx current in the apex, and the decay of the concentration gradient from that point fits nicely to what could be theoretically expected from a point source (unpublished results). This would be consistent with the activity of calcium channels in that location and indeed that was the interpretation given by Pierson et al. (1994) to the vibrating probe data. On the other hand it is hardly conceivable that the cross-linking occurs strictly on the dome apex, where the wall is especially thin and plastic. Indeed, all observations point out that such mechanical stabilisation reactions should occur on the flanks of the clear cap, an area where the extracellular fluxes decrease by the square of the distance to the apex. Secondly, many sequences, namely the ones which are more irregular (see for example Figure 22.3B in Holdaway-Clarke et al. 1997) show a remarkable similarity between the wave form in both parameters. This apparent synchrony between growth and calcium influx is reflected in many instances in Fourier analysis, where many small peaks are coincident, revealing a very similar wave function. Also wavelet analysis produces very high correlation coefficients for these curves, and the remarkable finding that the shape of the growth curve predicts the shape of the calcium influx curve (J. Kunkel, personal communication). If the cell wall were functioning as a buffer, followed by polymerisation of pectins, a number of chemical and enzyme mediated reactions, each with its own kinetic parameters, would have to take place during a whole cycle. How the combined kinetics of all these steps mimics the exact wave function of the kinetics of growth and makes them so similar is certainly not easy to envisage. Last but not least, the fate of the excess calcium. It should be noted that when tubes change from stable growth rates to pulsatory or oscillatory growth the basal level of calcium influx is the same but extra calcium flows during the peak elevation. Meaning that the calcium needs for growth are sustained by the basal flux, and in oscillating tubes there is a 10-fold increase in the incoming calcium. Since the external medium in *in vitro* conditions does not change significantly during these experiments (external bath calcium is continuously monitored with the vibrating probe) it would be reasonable to accept that the calcium buffering capacity of the cell wall would be in a steady state equilibrium without serious changes. Since there are no dramatic changes in the wall structure the question remains where does the extra calcium go? One possibility would be a localized efflux of calcium somewhere else in the tube, but rigourous scrutiny of currents in the tube and grain never revealed any calcium efflux in tubes above 400 µm (Feijó et al. 1994; Pierson et al. 1994, 1996; Holdaway-Clarke et al. 1997). The other possible explanation relies on the intracellular calcium stores which, being under metabolic control, have certainly much more capacity to suck extra calcium than the limited physico-chemical reactions of the cell wall. The second hypothesis thus implicates extracellular uptake of calcium and all available data points to highly localised presence or activity of the channels in the tip (-Malhó et al. 1995; Pierson et al. 1996).

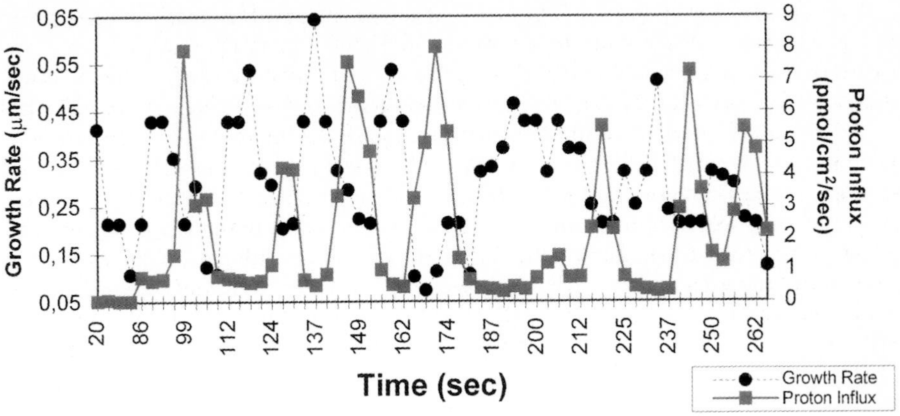

Fig. 22.4. Correlation between the growth rate (dashed line) and the extracellular proton influx (solid line) at the tip of a growing pollen tube of *Lilium*. The plot is remarkably similar to the one above for calcium. In fact, when protons and calcium were measured sequentially in the same tube ($n=3$) the oscillation characteristics and wave functions were quite similar for both ions. In this specific tube the growth rate varied about 6-fold, and the proton flux about 8-fold. It should be noted however that, while the calcium fluxes were always within the same order of magnitude, proton fluxes varied within 3 orders of magnitude, reflecting a much less controlled situation. In a few instances the flux may change direction abruptly, and very strong efflux occurstransiently, usually followed by a strong rebound influx. In most cases the tube bursts a few seconds after this undershoot/overshoot (for methods check Holdaway-Clarke et al. 1997 and Feijó et al. 1998)

From this discussion it stems that, while the model outlined by Holdaway-Clarke et al. (1997) provides a working hypothesis, the delay between the growth rate increase and the extracellular calcium influx still raises many questions which lack satisfactory answers. Clearly new data is needed, either from other species with oscillatory growth, or by manipulation of the growing conditions of *Lilium* in order to generate other pattern and phase shifts that would help to elucidate the functional relationship of these two parameters.

Calcium has received much attention in recent literature, and this has to be understood in the context of the pivotal role that this ion plays in the signal transduction pathways. Simultaneously, calcium low diffusion rates make this ion the preferential choice for the expression of spatial coordination phenomena like oscillations and waves (Jaffe 1991, 1993; Berridge, 1995, Tsien and Tsien 1990). In this respect other ions have remained much more elusive, but their potential for spatial and temporal integration of signals in developing cells through oscillations should not be underlooked. Such is the case for protons. Protons can affect a number of enzymes, metabolic pathways, cytoskeletal proteins and gene expression and accumulated evidence has introduced the concept that protons may act as second messengers in a number of transducing pathways (Felle 1989; Guern et al. 1991). When compared to calcium, however, organisation of structural patterns around pH is much more rare or otherwise involves a much more dynamic steady state. The reason for this stems from the higher mobility of protons and, consequently, the lower capacity to form fixed electrical fields. For the same reasons, cytosolic pH (pH_c) is much more difficult to measure accurately, and some controversy still surrounds the estimates of pH_c.

Two recent reports (Parton et al. 1997; Fricker et al. 1997) came to the conclusion that no pH_c gradient or spatial asymmetry occurs in pollen tubes. Both groups used confocal microscopy as imaging technique. However a recent review on confocal applications in the study of ions has led Opas (1997 and personal communication) to conclude that "in its present incarnation is not best suited for multiwavelength detection of weak fluorescent signals and their fast changes" for which widefield microscopy should be used instead. The reasons for this limitation are directly connected to the comparatively much higher levels of photon excitation/emission necessary for the confocal standard photomultipliers. In contrast, widefield systems use high sensitivity cooled-CCD cameras that can integrate and resolve much weaker signals. As long as these technical intricacies are not sorted out the true nature of pH_c in most systems will still be controversial when measured with the present generation of confocal microscopes.

Quite the opposite is true for the proton specific vibrating probe: pH cocktail LIX's are the ones that have better responses and signal/noise ratio, even in high background concentration (low pH). Use of the proton sensitive probes has recently led to the description of a proton extruding membrane domain located in the flanks of the clear cap (Feijó et al. 1998; see Fig. 22.7 in Shipley and Feijó, this volume). The fluxes involved led to the supposition that some sort of "hot spot" or localised pH domain could also exist on the cytosol. Use of widefield microscopy with a high sensitivity cooled-CCD allowed the concentration of the probe BCECF-Dextran to be lowered to levels in which these hints were confirmed (Feijó et al. 1998 and Fig. 22.3).

The presence of a distinctive alkaline band is spatially correlated with the clear cap and to the cytosol zone where actin bundles were described to flare out to the plasma membrane (Miller et al. 1996).

Investigation of the proton currents in the very tip led to another surprising finding, as protons behave pretty much the same way as calcium ions, i.e., there is a strong influx of protons which oscillates over a 10 fold range (Feijó et al. 1998; Fig. 22.7 in Shipley and

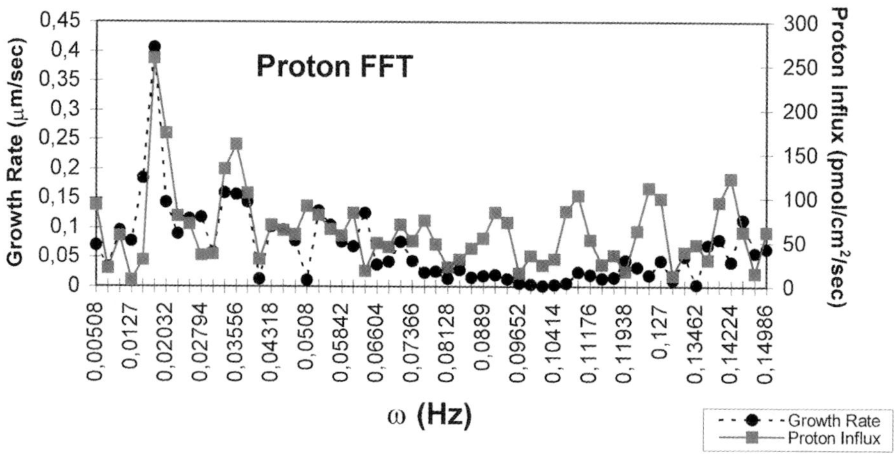

Fig. 22.5. Fourier spectra of a proton series similar to Figure 22.4. The same feature of the calcium series is observed, i.e. the main component of both growth and proton oscillations is absolutely overlapped, in this case with a period of about 60 sec. Since the growth rate is an independent factor, and may be considered as an external clock, this kind of approach clearly shows that the calcium and proton influx and the growth rate are all correlated and conditioned by the same oscillatory mechanism

Table 22.2. Summary of the oscillatory characteristic of four sequences of growth rates and proton influx in which Fourier spectra obeys the pattern shown in Figure 21.4. The general comments about this data are similar to the ones expressed for calcium, since the same kind of correlations seem to hold. The only different figure is the value computed in column 8, for the delay between each growth peak and the subsequent proton peak. While for calcium this value was on average 15 sec., with a span of 13 to 18, for protons this value goes down to about 10 sec., with a span of 8 to 12. To the extent that the difference between these two values exceeds the time resolution, it seems that the delay between growth and protons is slightly smaller than the one between growth and calcium

Series	Period GR (sec ± SE)	Period VP (sec ± SE)	Period Fourier (sec)	Nr. Cycles	Gr. Rate (μm/min)	GR after H^+ (sec ± SE)	H^+ after GR (sec ± SE)
1	21.6 ± 1.6	21.8 ± 1.0	21.6	7	19.04	13.8 ± 2.1	8.0 ± 0.3
2	27.0 ± 2.2	23.5 ± 1.4	24.1	9	16.05	15.3 ± 2.0	8.7 ± 0.8
3	62.0 ± 1.2	63.1 ± 5.2	60.3	6	16.60	50.8 ± 4.4	11.7 ± 1.3
4	25.6 ± 3.1	26.9 ± 3.5	23.0	6	17.40	15.0 ± 8.0	11.1 ± 2.7
Average ± SE	34.1 ± 2.0	33.8 ± 2.8	31.75		17.27	23.7 ± 4.1	9.88 ± 1.3
SPAN	22–62	22–63	22–60	6–9	16–19	13–50	8–12

Feijó, this volume). Contrary to calcium the magnitude of this influx varied over 2–3 orders of magnitude in different pollen tubes, while calcium was always within the same order of magnitude. The qualitative pattern was, however, very regular. The influx in the tip and efflux at the clear cap flanks thus produces a localised short-circuit of protons that must be expressed as an internal electrical current of significant magnitude. The effects of this current can be, as are the ones provoked by the calcium gradient, important in providing enough difference of potential to drive vesicle movement and eventually facilitate membrane fusion.

As could be anticipated these proton currents also correlate well with the oscillations in the growth rate (Fig. 22.4). In fact, individual tubes in which fluxes of protons and calcium could be measured, revealed that the periods of both were quite close. Again Fourier analysis provided strong correlation of the main frequency component, assigning the same oscillator as the temporal source of the tip proton influx and growth rate oscillations (Fig. 22.5). So far, it would seem that everything indicates a similar molecular basis in both calcium and protons influx, and a putative stretch-activated calcium channel could provide the answer since some of these channels are relatively unspecific for cations (Yangs and Sachs, 1989). Closer analysis of the data revealed, however, a subtle difference (Table 22.2). Most of the parameters shown bear great similarities to the values in Table 22.1 for calcium. Accordingly, peak-to-peak analysis also revealed a much more stable and robust relation if one considers that GR (growth rate) induces proton influx (column 8). The value obtained for the delay (average 10 sec.) is however smaller than the one for calcium (15 sec.). Furthermore, sequences in which it was possible to measure both ions in the same tube, always showed this trend (e.g. sequence 1 in Table 22.1 and sequence 3 in Table 22.2; compare columns 8 of both tables). This small variation may thus indicate a difference in the molecular basis of the extracellular fluxes of calcium and protons.

Such a difference would be easy to accommodate in the cell wall model of Holdaway-Clarke et al. (1997) since protons would intervene in a different step of the pectin chemical processing. Again similarities of the wave function between protons and growth, and protons and calcium do however give sufficient grounds for careful examination of this data. Until more detailed analysis of the phase relationships between the extracellular fluxes and growth in different species and experimental and instrumental conditions the question remains as to whether the ion- vibrating probe has became a tool to study wall ion physiology, or whether it still reflects channel activity of the cell membrane.

While the scene is set with the main actors (calcium, protons, cell wall and ion channels) future analysis of the oscillatory behaviour of pollen tubes, especially if mathematical modelling is aimed at, will have to go through the study of the intrinsic dynamic characteristic of the system. One of the most widely used tools for this purpose is phase space analysis. In one of the simplest ways to achieve space phase, each value of an oscillation is plotted against the nest one of the series. In this way, the temporal trend of the system is displayed on a bi-dimensional space which reflects its dynamics in the shape and organisation of the plot. An example of this kind of analysis is shown in Figure 22.6 for calcium and Figure 22.7 for protons. Both plots clearly evolve around what is generally referred to as a limit-cycle attractor, or circular attractor.

Comparison of both plots immediately indicates a common temporal organisation around a similar kind of attractor, which confirms the correlation of these ions in the process. However the plot shape and the limits of dispersion of the sequence values is quite different, a feature that may reflect differences in the mechanism of regulation or in the specific dynamics of the pollen tubes analysed. It is interesting to note that the dynamics of the limit cycle type in all models based on experimental data so far described, indeed shows that these systems admit a non-equilibrium steady state that becomes unstable beyond a critical value of some control parameter; it is in these conditions that sustained oscillations occur, in the form of a limit cycle in the phase space (Goldbeter 1996). In other words steady and oscillatory patterns in pollen tubes may be generated by the same molecular mechanisms, as has been shown in other experimental models which display circular attractors like the ones shown in the Figs. 22.6 and 22.7. Future research will show if that is the case.

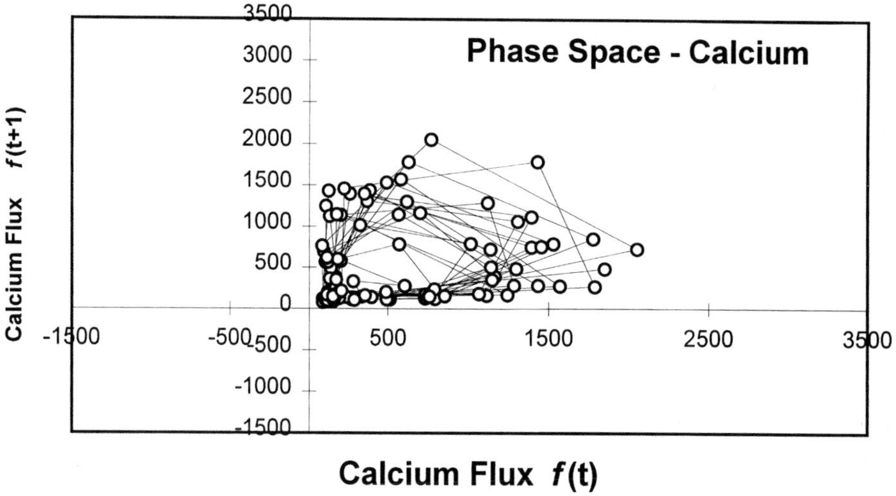

Fig. 22.6. Phase space plot for a calcium sequence. Each point is plotted in the XX' axis against the next in the sequence in the YY' axis. This kind of plot reflects the general dynamics of the system. In this specific case one may see that the system works around loops that converge to what may be characterised as a circulator attractor. The shape and organisation of this attractor is a function of the intrinsic dynamic characteristics of the system, and by shaping its form with different experimental conditions, one may get important insights into the temporal structure of the regulatory events

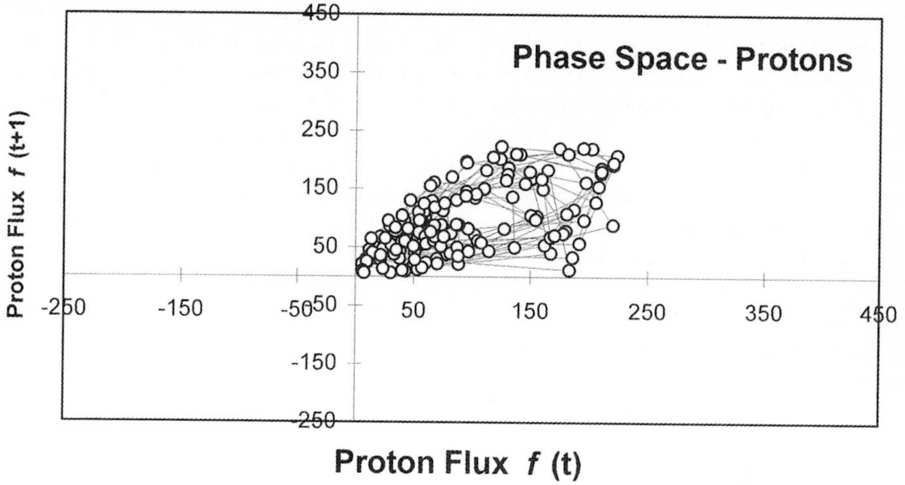

Fig. 22.7. Phase space plot for a proton sequence. Although acquired in a different pollen tube. the phase space trajectory for protons has a very different shape from the one for calcium displayed in Figure 22.6. This fact may reflect either (1) the regulatory dynamics of proton flux are different to the one for calcium or (2) this specific tube is growing with a different temporal pattern of regulation. Systematic collection of sequences for both ions in different growing conditions will generate a collection of patterns reflecting the different temporal regulatory dynamics of the system. This should in turn give important information about the type and form of the system attractors and, consequently, the type and boundaries of homeostasis

22.6 Towards a Molecular Model of Tip Growth?

Evolution preserves robustness and effectiveness. These are clearly the characteristics of the mechanism that makes pollen tubes growth. It is a highly regulated, highly homeostatic mechanism, capable of growing from minimal requirements and enduring differences in the concentration of solutes ranging three orders of magnitude. It faces the mechanical obstacles of the stigma (and other tissues) maintaining the specificity of its target, to deliver the sperms into the ovule. From the above description and discussion this apparent physiological complexity may be based on just a few basic features. The key point is thus the self-organisation of these features around a cycle of putative positive feed-back loops that renders a continuous flow of information around a typical temporal pattern. The participation of conspicuous ions like protons and calcium in this mechanism ensures that these temporal patterns can be easily transmitted over the whole cell and expressed spatially, both by diffusion-reaction mechanisms or through the signalling cascades. Last but not least, it only involves inorganic ions and bio-molecules that appeared very early in evolution. So, is it absurd to admit that the pollen oscillator may be a representative of a class of basic ion oscillators that underlie the spatial and temporal organisation of developing cells? It may not be so, at least for the vast class of cells that expands or differentiates through tip growth. Most of them seem to share some sort of underlying polarity expressed or conditioned by the presence of membrane asymmetries which generate ion gradients, currents and other spatial anisotropies. They all share vesicle fusion mechanisms as a way to deliver molecules to the extracellular surface. Most of them have been characterised as oscillatory. Pollen, however, is one of the simplest and most reliable systems to analyse: it possesses a very simple wall structure, grows at rates which are among the fastest in nature, it is easily accessible and there is a growing body of information which makes it one of the best studied systems in terms of cellular physiology in the plant Kingdom. In this paper we have tried to emphasise that once a molecular mechanism is described and numerically modelled, then it becomes a standard which may be applied to other systems with simple adaptations in the constants that affect the kinetic variables associated to each member of the regulation loop. Future research will show if our knowledge about pollen tube growth is enough to express one such molecular model reliably.

Acknowledgements
Part of the data shown in this paper was obtained during a visiting period in the University of Massachussets at Amherst. Profs. Peter K. Hepler and Joe Kunkel are gratefully acknowledged for the use of their laboratories and equipment and for the helpful discussions. I thank Grant Hackett for some of the growth sequences and Joaquim Oliveira for the endless and motivating discussions about non-linear dynamics and critical reading of the manuscript. Financial support is acknowledged to the Fullbright Program (Scholar Fellowship), Fundação Luso-Americana para o Desenvolvimento, Fundação Calouste Gulbenkian and Universidade de Lisboa, for personal fellowships.

References

Baker GL, Gollub JP (1990) Chaotic Dynamics. Cambridge University Press, NY
Benedeto JJ, Frazier MW (eds) (1994) Wacelets: mathematics and applications. CRC Press, Boca Raton
Benkert R, Obermeyer G, Bentrup FW (1997) The turgor pressure of growing lily pollen tubes. Protoplasma 198:1–8
Berridge MJ (ed) (1995) Ca^{2+} waves gradients and oscillations. CIBA Found Symp, Wiley, Chichester
Cai G, Moscatelli A, Cresti M (1997) Cytoskeletal organization and pollen tube growth. Trends in Plant Sci 2:86–91
Cosgrove D (1997) Relaxation in a high-stress environment: the molecular bases of extensible cell walls and cell enlargement. Plant Cell 9:1031–1041
Derksen J, Rutten T, Amstel TV, Win AD, Doris F, Steer MW (1996) Regulation of pollen tube growth. Acta Bot Neerl 44:93–119
Feijó JA, Hackett G, Kunkel JG, Hepler PK (1998) Extracellular proton fluxes and cytossolic pH analysis reveal that pollen tubes have a constitutive alkaline band on the clear cap and a growth-dependent acidic tip (submitted)
Feijó JA, Malhó RM, Obermeyer G (1995) Ion dynamics and its possible role during in vitro pollen germination and tube growth. Protoplasma 187:155-167
Feijó JA, Shipley AM, Jaffe LF (1994) Spatial and temporal patterns of electric and ionic currents around in vitro germinating pollen of lilly: a vibrating probe study. In: Heberle-Bohrs E, Vicente O (eds) Frontiers on Sexual Plant Reproduction. University of Viena, Viena, p 40
Felle H (1989) pH as a second messenger. In: Boss WF, Morré DJ (eds) Second messengers in plant growth development. Alan R Liss, NY, pp 145–166
Fricker MD, White NS, Obermeyer G (1997) pH gradients are not associated with tip growth in pollen tubes of Lilium longiflorum. J Cell Sci 110:1729–1740
Geitmann A, Cresti M (1996) The role of the cytoskeleton and dyctiosome activity in the pulsatory growth of Nicotiana tabacum and Petunia hybrida pollen tubes. Bot Acta 109:102–109
Geitmann A, Cresti M (1998) Ca^{2+} channels control the rapid expansions in pulsating growth of Petunia hybrida pollen tubes. J Plant Physiol (in press)
Golberger AL, Rigney DR, West BJ (1990) Chaos and Fractals in Human Physiology. Sci Am 262(2): 34–41
Goldbeter A (1996) Biochemical oscillations and cellular rhythms. Cambridge Univ Press, NY
Goldbeter A, Dupont G (1990) Allosteric regulation, cooperativity and biochemical oscillations. Biophys Chem 37:341–353
Guern J, Felle H, Mathieu Y, Kurkdjan A (1991) Regulation of intracellualr pH in plant cells. Int Rev Cytol 127:111–173

Hacken H (1978) Synergetics. Springer-Verlag, Berlin

Haig D (1990) New perspectives on the angiosperm female gametophyte. Bot Rev 56:236–274

Harold FM, Caldweel JH (1990) Tips and currents: electrobiology of apical growth. In: Heath IB (ed) Tip growth in plant and fungal cells. Academic Press, London, pp 59–89

Heslop-Harrison J (1987) Pollen germination and pollen tube growth. Int Rev Cytol 107:1–78

Holdaway-Clarke TL, Feijó JA, Hackett GA, Kunkel JG, Hepler PK (1997) Pollen tube growth and the intracellular-cytosolic calcium gradient oscillate in phase while extracellular calcium influx is delayed. Plant Cell 9:1999-2010

Jaffe LA, Weisenseel MW, Jaffe LF (1975) Calcium accumulations within the growing tips of pollen tubes. J Cell Biol 67:488–492

Jaffe LF (1981) The role of ionic currents in establishing developmental pattern. Philos Trans Soc Lond Biol 295:553–566

Jaffe LF (1991) The path of calcium in cytosolic calcium oscillations: a unifying hypothesis. Proc Natl Acad Sci USA 88:9883–9887

Jaffe LF (1993) Classes and mechanisms of calcium waves. Cell calcium 14:736–745

Jaffe LF, Robinson KR, Nuccitelli R (1974) Local cation entry and self-electrophoresis as an intracellular localisation mechanism. Ann N Y Acad Sci 238:372–389

Joos U, VanAken J, Kristen U (1994) Microtubules are involved in maintaining the cellular polarity in pollen tubes of Nicotiana sylvestris. Protoplasma 179:5–154

Kühtreiber WM, Jaffe LF (1990) Detection of extracellular calcium gradients with a calcium-specific vibrating electrode. J Cell Biol 110:1565–1573

Lancelle, SA, Hepler, PK (1988) Ultrastructure of freeze-substituted pollen tubes of *Lilium longiflorum*. Protoplasma 167:215–230

Li Y-Q, Bruun L, Pierson ES, Cresti M (1992) Periodic deposition of arabinogalactan epitopes in the cell wall of pollen tubes of Nicotiana tabacum L. Planta 188:532–538

Li Y-Q, Chen F, Linskens HF, Cresti M (1994) Distribution of unesterified and esterified pectins in cell walls of pollen tubes of flowering plants. Sex Plant Reprod 7:145–152

Li Y-Q, Moscatelli A, Cai G, Cresti M (1997) Functional Interactions among cytoskeleton, membranes and cell wall in the pollen tube of flowering plants. Int Rev Cytol 176:133–199

Li Y-Q, Zhang H-Q, Pierson ES, Huang FY, Linskens HF, Hepler PK, Cresti M (1996) Enforced growth-rate fluctuation causes pectin ring formation in the cell wall of Lilium longiflorum pollen tubes. Planta 200:41–49

Malhó R, Read ND, Pais MS, Trewavas AJ (1994) Role of cytosolic free calcium in the reorientation of pollen tube growth. Plant J 5:331–341

Malhó RM, Trewavas AJ (1995) Calcium channel actvity during pollen tube growth and reorientation. Plant Cell 7:1173–1184

Mascarenhas JP (1966) Distribution of ionic calcium in the tissues of the gynoecium of Antirrhinum majus. Protoplasma 62:53–58

Mascarenhas JP (1975) The biochemistry of angiosperm pollen development. Bot Rev 41:295–314

Mascarenhas JP (1993) Molecular mechanism of pollen tube growth and differentiation. Plant cell 5: 1303–1314

May R (1976) Simple mathematical models with very complicated dynamics. Nature 261:459–467

Messerli M, Robinson KR (1997) Tip localized Ca^{2+} pulses are coincident with peak pulsatile growth rates in pollen tubes of Lilium longiflorum. J Cell Sci 110:1269–1278

Miller DB, Callaham DA, Gross DJ, Hepler, PK (1992) Free Ca^{2+} gradient in growing pollen tubes of Lilium. J Cell Sci 101:7–12

Miller DB, Lancelle S, Hepler, PK (1996) Actin microfilaments do not form a dense meshwork in Lilium longiflorum. Protoplasma 195:123–132

Obermeyer G, Weisenseel MH (1991) Calcium channel blocker and calmodulin antagonists affect the gradient of free calcium ions in lily pollen tubes. Eur J Cell Biol 56:319–327

Opas M (1997) Measurement of intracellular pH and pCa with a confocal microscope. Trends in Cell Biology 7:75–80

Parton RM, Fischer S, Malhó R, Papasouliotis O, Jelitto T, Leonard T, Read ND (1997) Pronounced cytoplasmic pH gradients are not required for tip growth in plant and fungal cells. J Cell Sci 110: 1187–1198

Picton, JM, Steer, MW (1981) Determination of secretory vesicle production rates by dictyosomes in pollen tubes of *Tradescantia* using cytochalasin D. J Cell Sci 49:261–272

Pierson ES, Li YQ, Zhang HQ, Willemse MTM, Linskens, HF, Cresti, M (1995) Pulsatory growth of pollen tubes: investigation of a possible relationship with the periodic distribution of cell wall components. Acta Bot Neerl 44:121–128

Pierson ES, Miller D, Callaham D, Van Aken J, Hackett G, Hepler PK (1996) Tip-localized calcium entry fluctuates during pollen tube growth. Dev Biol 174:160–173

Pierson ES, Miller DD, Callaham DA, Shipley AM, Rivers BA, Cresti M, Hepler PK (1994) Pollen tube growth is coupled to the extracellular calcium ion flux and the intracellular calcium gradient: Effect of BAPTA-type buffers and hypertonic media. Plant Cell 6:1815–1828

Plyushch TA, Willemse MTM, Franssen-Verheijen, Reinders MC (1995) Structural aspects of in vitro pollen tube growth and micropylar penetration in Gasteria verrucosa (Mill.) H. Duval and Lilium longiflorum Thumb. Protoplasma 187:13–21

Rathore KS, Cork RJ, Robinson KR (1991) A cytoplasmic gradient of Ca^{2+} is correlated with the growth of Lily pollen tubes. Develop Biol 148:612–619

Steer, MW, Steer, JL (1989) Pollen tube tip growth. New Phytol 111:323–358

Südhof TC (1995) The synaptic vesicle cycle: a cascade of protein-protein interactions. Nature 375: 645–653

Taylor LP, Hepler PKH (1997) Pollen germination and tube growth. Ann Rev Plant Physiol Plant Mol Biol 48:461–491

Tiwari SC, Polito VS (1988) Organization of the cytoskeleton in pollen tubes of *Pyrus communis*: a study employing conventional and freeze-substituion electron microscopy, immunofluorescence and rhodamin-phaloidin. Protoplasma 147:100–112

Trewavas AJ, Malhó R (1997) Signal perception and transduction: the origin of the phenotype. Plant Cell 9:1181–1195

Tsien RW, Tsien RY (1990) Calcium channels, stores and oscillations. Annu Rev Cell Biol 6:715–760

Weisenseel MH, Jaffe LF (1976) The major growth current through lily pollen tubes enters as K^+ and leaves as H^+. Planta 133:1–7

Weisenseel MH, Kicherer RM (1981) Ionic currents as control mechanism in cytomorphogenesis. In: Kermeyer O (ed) Cytomorphogenesis in Plants, vol 8. Springer-Verlag, Berlin, Heidelberg, New York, pp 379–399

Weisenseel MH, Nucitelli R, Jaffe LF (1975) Large electrical currents traverse growing pollen tubes. J Cell Biol 66:556–567

West BJ (1985) An essay on the importance of being non-linear. In: Levine S (ed) Lectures Notes in Biomathematics. Vol 62. Springer-Verlag, Berlin, Heidelberg

Yang X-C, Sachs F (1989) Block of stretch-activated ion channels in *Xenopus* Oocytes by gadolinium and calcium ions. Science 243:1068–1071

CHAPTER 23

Fertilization and Zygotic Embryo Development *in Vitro*

E. Kranz · J. Kumlehn · T. Dresselhaus

Centre for Applied Plant Molecular Biology, AMP II, University of Hamburg, Ohnhorststr. 18, D-22609 Hamburg, Germany
e-mail: ekranz@botanik.uni-hamburg.de
telephone: +49-40-8 22 82-2 27
fax: +49-40-8 22 82-2 29

Abstract. Techniques have been developed recently to isolate *in vivo* fertilized zygotes, egg and sperm cells and to fuse gametes of higher plants *in vitro*. These procedures include the careful handling and micromanipulation of single cells and the use of equipment with devices allowing isolation, selection, transfer and fusion of individual cells under microscopical observation.

The products from *in vitro* fusions of pairs of sperm and egg cells and also zygotes, isolated after *in vivo* fertilization, can develop into embryos and fertile plants. With the experimental access, early developmental processes can now be studied immediately after fertilization with living material and independently from the maternal tissue. It is expected that these techniques will be used widely for developmental studies, e.g. to isolate and to elucidate the function of gamete-specific and fertilization-induced genes, to study gametic interaction and hybridization. Advances in this field have been reviewed recently by Kranz and Dresselhaus (1996) and Rougier et al. (1996). Here we consider recent results obtained from isolated *in vitro* and *in vivo* fertilized egg cells and give prospects for future work.

23.1 Fertilization

In higher plants one sperm cell fuses with the egg cell to generate a zygote that develops into an embryo, the other one fuses with the central cell to develop into the endosperm. This double fertilization process occurs in the embryo sac which is usually deeply embedded in the nucellar tissue of the ovule. Not only this tissue, but also other cells of the embryo sac: the synergids and the antipodal cells, have to be removed to gain access to egg and central cells and to fuse the latter with a sperm cell *in vitro*. The sperm cells have also to be isolated from pollen grains or tubes. The isolated gametic protoplasts can be fused by electrical pulses, by polyethylene glycol (PEG) or in a calcium containing medium. All these procedures presuppose experience in micromanipulations.

23.1.1 Isolation and Selection of Gametes

Mechanical isolation procedures determine the amount of isolated cells from the embryo sac. Egg cells have been isolated in reasonable amounts for experimentation from some monocot species, such as maize (Kranz et al. 1991a, Faure et al. 1994), barley (Holm et al. 1994), wheat (Holm et al. 1994; Kovács et al. 1994) and ryegrass (Van der Maas et al. 1993), and recently from the dicot plants tobacco (Tian and Russell 1997a) and rape seed (Katoh et al. 1997). In some species, as for example in maize, a short treatment of nucellar tissue with a mixture of low concentrated cell wall-degrading enzymes is necessary to obtain reasonable amounts of the finally mechanically isolated cells of the embryo sac.

Sperm cell isolation methods have been developed for a wide range of higher plants, reviewed by Theunis et al. (1991). Sperm cell isolation is performed mainly by the grinding or squashing of mature pollen or tubes or by bursting of the grains via an osmotic shock. With the access to isolated gametes defined studies now make it possible to elucidate surface molecules and investigate adhesion and putative recognition events.

23.1.2 Fusion of Gametes

With the involvement of individual cell fusion in gamete research and in investigations of fertilization and of early embryogenesis, *in vitro* methods with isolated gametes became a versatile technique (Table 23.1). It is now possible to follow the developmental transitions from the egg to the embryo and finally to the plant under defined experimental conditions *in vitro*.

Gametic pairs can be fused in microdroplets of fusion medium covered with mineral oil on a coverslip under microscopical observation. Egg-sperm cell fusions and fusions of sperm cells with other protoplasts were performed electrically in maize (Kranz et al. 1991a, 1995; Kranz and Lörz 1993; Sauter et al. 1997) and wheat (Kovács et al. 1995). Additionally in maize, egg-sperm cell fusions were achieved in calcium-containing media (Kranz and Lörz 1994; Faure et al. 1994; Digonnet et al. 1997). Gametes can also fuse spontaneously, without any fusogenic agent, because they are protoplasts. In addition, they can be fused using other fusion methods commonly applied in somatic protoplast fusion experiments, for example by polyethylene glycol (PEG) (Tian and Russell 1997b). PEG-mediated cell fusion was also performed using protoplasts of egg and central cells, of synergids, generative and somatic cells (Sun et al. 1995).

It is a peculiarity of sperm cell protoplasts that they can occasionally sustain their spindle-like shape after isolation. These fuse much more efficiently than round ones (Kranz et al. 1995). In somatic cell hybridization, exclusively round protoplasts are isolated and used for fusion. *In vitro* egg-sperm fusion occurs fast, and, depending on the turgor of the gametes, it generally happens within one second.

The electrical fusion is induced by short DC-pulses on one of the two microelectrodes that are fixed to a support under the condensor of an inverted microscope. In maize, 20–56 maize zygotes were produced electrically per experimentator a day. The use of media with suitable osmolalities during the isolation and fusion procedures is important for such efficiency of cell fusion (Kranz 1997). An isolated sperm can be fused very efficiently with other sperm protoplasts or an egg cell or somatic protoplast. Foreign cytoplasms can be easily transferred by the sperm or by an additional somatic cytoplast into the egg. The fate of chloroplasts and mitochondria can then be followed in the developing *in vitro* zygote (Kranz et al. 1991b; Faure et al. 1993).

It has been observed that two egg protoplasts can fuse in mannitol solutions with or without calcium. Also, fusion of two sperms and of a sperm and a mesophyll protoplast can take place in calcium containing solutions (Kranz and Lörz 1994; Faure et al. 1994; Kranz et al. 1995; Zhang et al. 1997). It is well known that calcium can promote membrane fusion as described in early reports on somatic protoplast fusion (for example, Keller and Melchers 1973; Grimes and Boss 1985). *In vitro* fusion methods might be used to elucidate the conditions which promote membrane fusion after gamete adhesion and possible recognition has occurred.

Table 23.1. *In vitro* fusion of gametes and fusion combinations with non-gametic cells of the embryo sac and with somatic cells to study fertilization and developmental processes

Method *Species*, Protoplasts	Topic of investigation	Reference
Electrofusion		
Maize + Maize		
Sperm + Egg	Fusion, zygote development	Kranz et al. 1991a
Sperm + Sperm	Fusion	Kranz et al. I991b
Sperm + Central Cell	Fusion	Kranz et al. 1991b
Sperm + Synergid	Fusion	Kranz et al. I991b
Sperm + Egg + Mesophyll	Fusion	Kranz et al. I991b
Sperm + Egg	Karyogamy	Faure et al. 1993
Sperm + Egg	Embryogenesis, plant regeneration	Kranz and Lörz 1993
Sperm + Egg	Early cytological events, karyogamy	Tirlapur et al. 1995
Sperm + Egg	Early cytological events, zygote development	Kranz et al. 1995
Sperm + Egg	cDNA library, gene isolation	Dresselhaus et al. 1996
Sperm + Egg	Cell cycle	Sauter et al. 1997
Egg + Egg	Cell- and nuclear fusion	Kranz et al. 1995
Egg + Suspension Cell	Fusion, fusion products develop	Kranz et al. 1995
Barley + Maize		
Sperm + Egg	Hybridization, fusion products develop	Kranz et al. 1995
Coix + Maize		
Sperm + Egg	Hybridization, fusion products develop	Kranz et al. 1995
Sorghum + Maize		
Sperm + Egg	Hybridization, fusion products develop	Kranz et al. 1995
Wheat + Maize		
Sperm + Egg	Hybridization, fusion products develop	Kranz et al. 1995
Wheat + Wheat		
Sperm + Egg	Fusion, zygote development	Kovacs et al. 1995
Calcium Fusion		
Maize + Maize		
Sperm + Egg	Adhesion, fusion, zygote formation	Faure et al. 1994
Sperm + Egg	Adhesion, fusion, zygote development	Kranz and Lörz 1994
Sperm + Egg	Free cytosolic Ca^{2+}, zygote development	Digonnet et al. 1997
Sperm + Sperm	Fusion	Faure et al. 1994
Sperm + Mesophyll	Fusion	Faure et al. 1994
Egg + Egg	Fusion	Kranz et al. 1995
Sperm + Sperm	Fusion	Zhang et al. 1997
Polyethyleneglycol (PEG) Fusion		
Tobacco + Tobacco		
Sperm + Egg	Fusion, zygote formation	Tian and Russell 1997b
Sperm + Sperm	Fusion	Tian and Russell 1997b
Egg + Generative Cell	Fusion	Sun et al. 1995
Egg + Synergid	Fusion	Sun et al. 1995
Egg + Central Cell	Fusion	Sun et al. 1995
Egg + Central Cell	Fusion	Tian and Russell 1997b
Synergid + Synergid	Fusion	Tian and Russell 1997b
Central- + Central Cell	Fusion	Tian and Russell 1997b

23.2 Zygotic Embryo Development

Herein zygotes are named *in vitro* zygotes, when they are produced via fusion of isolated egg and sperm protoplasts *in vitro*. *In vivo* zygotes are zygotes, that are isolated after *in vivo* fertilization.

Zygotic embryogenesis and plant regeneration are possible from *in vitro* and *in vivo* fertilized isolated egg cells. Up to now this was exclusively achieved in cereals, including maize, barley and wheat. Due to the limited amount of cells, individual handling and culturing of these zygotes are necessary. Zygotes can be cultured in microdroplets for short-time experiments, for example to study physiological and cytological processes at and shortly after fertilization. However, to follow early and further development, the zygotes have to be cultured using single-cell culture systems together with feeder cells. Developmental frequencies can be obtained which are far better than in any other culture system based upon isolated somatic plant cells. Moreover, isolated zygotes can be implanted into a cultivated ovule for embryo development. In these studies, embryogenesis mimics the development *in vivo* to various degrees.

23.2.1 *In Vitro* Zygotes

In vitro zygotes are metabolically highly active. In maize, cell wall material is newly formed as early as 30 sec after egg-sperm *in vitro* fusion (Kranz et al. 1995). Also, karyogamy happens quickly; it has been reported as early as 35 min, generally within 60 min after *in vitro* cell fusion (Faure et al. 1993; Tirlapur et al. 1995). Cell divisions in *in vitro* zygotes were described in maize (Kranz et al. 1991a, 1995; Kranz and Lörz 1993, 1994; Digonnet et al. 1997; Sauter et al. 1997) and wheat (Kovács et al. 1995). The first division of *in vitro* maize zygotes was observed 29 h, but generally 42–46 h after egg-sperm fusion (Kranz et al. 1995). The first division of maize zygotes can occur *in planta* 32 hap (hours after pollination) (about 16 h after karyogamy; Mól et al. 1994). As *in vivo*, the *in vitro* fertilized maize egg cell divides asymmetrically (Fig. 23.1a).

In maize, 85 % of the *in vitro* zygotes reproducibly and genotype independently developed into multicellular structures (Kranz et al. 1995). Also, sustained growth of *in vitro* zygotes was achieved. This was possible because of the development of a highly efficient microculture system. *In vitro* zygotes were cultured on a semipermeable membrane of inserts and cocultivated with actively growing maize suspension cells. In this way, large embryos were obtained, which were able to regenerate via zygotic embryogenesis into plants (Kranz and Lörz 1993).

These *in vitro* systems can contribute to the understanding of the role of polarity and asymmetric cleavage. *In vitro* zygotes show a polar distribution of the cell organelles. The main cytoplasm, including plastids and mitochondria, surround the zygotic nucleus (Tirlapur et al. 1995). This cluster of organelles is located at the cell periphery. During *in vitro* zygote development, the organelles become more densely located around the nucleus than in the isolated egg cell (Kranz et al. 1995). This uneven distribution of the cell organelles in the *in vitro* maize zygotes can also be observed in fusion products of maize egg and sperm cells of barley, wheat, *Coix* and *Sorghum*, resulting in an asymmetrical first cell division (Kranz et al. 1995). The egg cell is the predominant cell in heterologous egg-sperm fusions and determines the plane of the first cell division. This was observed in the fusion combinations maize egg + maize sperm, maize egg + wheat sperm, wheat egg + wheat sperm and wheat egg + maize sperm (Fig. 23.1; E. Kranz and J. Kumlehn, unpublished results). When a maize egg is fused with a wheat sperm cell, the first division in the fusion product is unequal as in *in vitro* zygotes of maize (Fig. 23.1a and b). However, wheat *in vitro* zygotes do not show a pronounced asymmetrical first cell division (Kovács et al. 1995; Fig. 23.1d). This can also be observed in the fusion products of a wheat egg and a maize sperm cell (Fig. 23.1c). Early development of fusion products occurs after heterologous gametic fusions with more related species. However, fusion products of a maize egg and a *Brassica* sperm cell do not survive. This indicates a zygotic barrier of incompatibility (Kranz et al. 1995).

Zygote formation *in vitro* allows studies under defined conditions, as for example of cytoskeleton changes and nuclear movement. Also, studies of signalling processes during egg activation are feasible (Dudits 1995; Tirlapur et al. 1995). In the *in vitro* fertilized maize egg, a transient elevation of free cytosolic Ca^{2+} was observed (Digonnet et al. 1997). Studies like this will facilitate investigations of the role of calcium in early processes of zygote development.

In vitro zygotes are able to self-organize and to develop into embryos in a predictable way without endosperm or any maternal tissues. Direct zygotic embryogenesis from *in vitro* maize zygotes is possible *in vitro*, and fertile hybrid plants were obtained (Kranz and Lörz 1993). Embryogenesis mimics the *in vivo* situation as, for example, indicated by the unequal first cell division and the formation of transition-phase embryos. These oblong embryos typically consist of a meristematic region and a suspensor. Maize transition phase embryos strongly resemble embryos derived from cultured wheat *in vivo* zygotes (Kumlehn et al. 1997b). A coleoptile and subsequently a plantlet are formed surrounded by scutellum-like tissue. Secondary embryogenesis can occur, especially when the growth-promoting substance 2,4-D is used in the regeneration media, resulting in more than one plantlet. Rapidly, that is 100 days after *in vitro* fertilization, a phenotypically normal and fertile hybrid plant can arise from a single embryo derived from an *in vitro* zygote.

In vitro zygotes divide and form multicellular structures with a frequency of around 85 %. The frequency of formation of large embryos, capable of further differentiation, can be high when rapidly growing feeder aggregates are used, consisting of maize suspension cells which are rich in cytoplasm. For example, nine globular embryos,

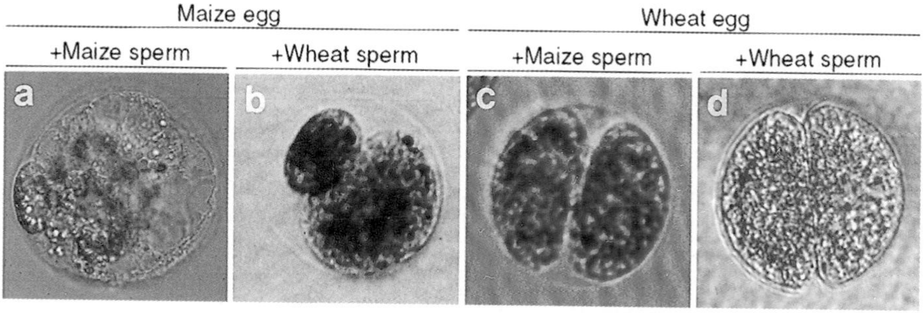

Fig. 23.1a–d. First cell division of *in vitro* zygotes and of fusion products after gametic homologous and heterologous *in vitro* fusions. **a** Pronounced unequal cell division after fusion of maize egg and maize sperm. **b** Pronounced unequal cell division after fusion of maize egg and wheat sperm. **c** Rather more equal cell division after fusion of wheat egg and maize sperm. **d** Rather more equal cell division after fusion of wheat egg and wheat sperm

including six transition-phase embryos, were obtained from twenty eight *in vitro* maize zygotes.

Unlike the case when using simple growth-promoting media, special maturation media have to be developed to obtain embryo maturation *in vitro*.

23.2.2
In Vivo Zygotes

The isolation procedures for zygotes are usually similar to those for unfertilized egg cells of the same species. As a first step, it is necessary to isolate the ovules (Holm et al. 1994) or a part of the ovules containing either the embryo sac (Leduc et al. 1996) or at least the egg apparatus (Kumlehn et al. 1997a,b). In barley and wheat this means removing one or both integumental layers. In this way, the *in vivo* zygotes can be recognized easier through an inverted microcope during subsequent cell dissection. Cell isolation itself is performed using fine-tipped glass needles or forceps. Since the maize embryo sac is more tightly embedded in the nucellar tissue compared to small-grain cereals, it is necessary to pre-dissect the embryo sac and to treat them with cell wall digesting enzymes. Freshly isolated maize zygotes are therefore disturbed in their process of cell wall formation which has to be re-initiated during subsequent culture.

It is not possible to isolate as many *in vivo* zygotes as required for a suitable population density in the subsequent culturing steps. Therefore, feeder systems are ultimately needed to ensure further zygotic development. Sporophytically induced micropore cultures, co-cultivated with isolated barley zygotes, have an excellent potential to promote *in vitro* zygotic development (Holm et al. 1994). This finding was later confirmed for isolated maize (Leduc et al. 1996) and wheat (Kumlehn et al. 1997b) *in vivo* zygotes which also underwent embryonic development when fed with barley microspore cultures. To keep the embryonic structures derived from *in vivo* zygotes and microspores separated, the *in vivo* zygotes can be cultured in insert vessels with a semipermeable membrane at the bottom which are submersed in the liquid medium of the culture dish containing the feeder aggregates. An alternative approach was pursued in wheat: *in vivo* zygotes were implanted into parthenocarpically induced ovules cultivated *in vitro* (Kumlehn et al. 1997a). Different genotypes were used as donor material for *in vivo* zygote isolation and parthenocarpic induction of ovule development, respectively. Embryo development and plant regeneration were proven to be derived exclusively from the implanted *in vivo* zygotes, since all regenerants were identified as being the same genotype as the *in vivo* zygotes.

Although media of the different protocols vary quite widely in formulation, they are comparable in their high complexity, particularly concerning the organic ingredients (Table 23.2). Good results were obtained without the use of exogenously applied auxins in barley (Holm et al. 1994) and wheat (Kumlehn et al. 1997b) *in vivo* zygote cultures. In maize, supplementation with 1 or 2 $mg \cdot l^{-1}$ of 2,4-D was necessary for the development of the isolated zygotes into globular structures (Leduc et al. 1996).

In contrast to most other cell and tissue culture applications, the *in vitro* development of isolated zygotes turned out to be largely genotype-independent, as is also the case for *in vitro* zygotes. This might be attributed to the zygote's natural fate to undergo embryogenesis and to form a new plant. Furthermore, it was shown in barley and wheat that almost every cultivated *in vivo* zygote can be triggered to develop into an embryo and finally into a plant.

Table 23.2. Survey of published methods for culture of *in vivo* zygotes

	Barley[a]	Maize[b]	Wheat	
			Method 1[c]	Method 2[d]
Harvest of pistils	During the entire zygote stage	24 hours after pollination	3–9 hours after pollination	3–6 hours after pollination
Solution for cell isolation	Culture medium Kao 90	Culture medium Kao 90	0.5 M Mannitol	0.55 M Mannitol
Enzymatic digestion of nucellus tissue	No	Yes	No	No
Pre-dissected tissue	Ovules	Embryo sacs with nucellus	Ovule tips	Ovule tips
Efficiency of viable zygote isolation (% of dissected pistils)	~ 75%	No information	~ 75%	~ 75%
Culture media	Kao90	Kao90 + 2,4-D	ModifiedOC, Modified Kao	N6Z
Feeder system	Co-culture with sporophytically induced microspores of barley	Co-culture with sporophytically induced microspores of barley	Zygote implantation into cultured wheat or barley ovules	Co-culture with sporophytically induced microspores of barley
Frequency of first cell division (% of cultivated zygotes)	No information	~ 28%	No information	~ 96%
Symmetry of first cell division	No information	Asymmetrical	No information	Rather symmetrical
Formation of multicellular structures (% of cultured zygotes)	~ 75%	~ 16%	~ 17%	~ 94%
Embryonic differentiation	Single or multiple embryo-like structures	Secondary somatic embryogenesis after callus formation	Dorsiventral embryo formation	Dorsiventral embryo formation; partially multiple embryos
Plantregeneration (% of cultivated zygotes)	~ 50%	No information	~ 15%	~ 93%

[a] (Holm et al. 1994); [b] (Leduc et al. 1996); [c] (Kumlehn et al. 1997a); [d] (Kumlehn et al. 1997b)

Cultures of *in vivo* zygotes provide a highly synchronized system of individual cells undergoing embryogenesis. Development of cultured *in vivo* zygotes can be structurally and temporarily homologous to that *in planta* (Kumlehn et al. 1997b).

Formation of cell wall could neither be detected with light microscopy nor with transmission electron microscopy in freshly isolated barley zygotes (Holm et al. 1994). Newly formed cell wall when stained with calcofluor white was observed in cultivated maize *in vivo* zygotes (Leduc et al. 1996),

Using quantitative measurements of nuclear DNA contents, Mogensen and Holm (1995) investigated the first cell cycle of isolated barley zygotes. Their results can be summarized as follows: (1) both egg and, by inference, sperm cells contained the 1C level of DNA, (2) within 1 hap, the male nucleus had reached the female one, (3) nuclear fusion was completed within 5 to 6 hap resulting in the 2C DNA level, (4) zygotic S-phase was initiated around 9 to 12 hap and (5) the 4C DNA level was between 22 to 29 hap. In a similar set of experiments using *in vivo* maize zygotes, Mogensen et al. (1995) showed that S phase was typically underway by 3 hours after fertilization.

An alternative approach to investigate cytological events is being pursued in wheat by focusing on nucleolar kinetics through observation of individually cultivated *in vivo* zygotes (J. Kumlehn, H. Lörz and E. Kranz, unpublished results). As reviewed by Risueno and Testillano (1994), the phases of different ribosomal activity and gene expression during the cell cycle are reflected by the appearance and morphological organization of nucleoli. The unfertilized wheat egg cell as well as the early zygote are characterized by one prominent nucleolus. The emergence of a second nucleolus after around 6 hap indicates that nuclear fusion is likely to have been completed in cultivated wheat *in vivo* zygotes. Between 9 to 12 hap, the large nucleolus of the zygote has attained highest activity which is displayed by a large central nucleolar vacuole typical for G2 phase. Coupled with the gradually decreased nucleolar vacuolation, transcription might be subsequently reduced until around the 17th hap at which all nucleoli of a zygote disappear within a few minutes. Disappearence of the nucleoli indicates initiation of mitotic prophase. Re-formation of nucleoli might occur during telophase. In the cultivated wheat zygotes, the latter event was observed around 20 hap. Cytokinesis was typically completed by 23 hap.

In studies with *in vivo* zygote culture in barley (Holm *et al.* 1994) and maize (Leduc *et al.* 1996), plant regeneration was reported to occur upon differentiation of zygote-derived proembryogenic masses into embryo-like structures, secondary somatic embryos and/or multiple shoots. In wheat, two different culture methods were established which both lead to direct zygotic embryo development including formation of scutellum, plumula, coleoptile, epiblast, coleorhiza and suspensor, that is, all typical structures for wheat embryo differentiation *in planta* (Kumlehn et al. 1997a, b).

23.3 Gene Expression after *In Vitro* Fertilization

Zygotic gene activation (ZGA) is the critical event that governs the transition from maternal to embryonic control of development. In some animal systems it has been demonstrated that abundant maternal transcripts are stockpiled in the unfertilized egg cell and expression of the embryo' s own genome is not required for the first cell division. In mammals, the onset of ZGA has been shown to occur between the two-cell (mouse, hamster) to sixteen-cell stage (sheep, cow, for review see Schultz 1993). In other animal species like *Caenorhabditis elegans* the first embryonically transcribed RNAs were not detected until the three-four cell stage (Seydoux et al. 1996), or even much later at the midblastula transition stage (MBT) in *Xenopus* and zebra fish, when the embryo already consists of thousands of cells (Newport and Kirschner 1982; Zamir et al. 1997). In *Drosophila*, the zygotic genome is not transcribed before cell cycle eleven, when the embryo is a syncytium containing a few thousand nuclei. Transcription of most genes does not begin until cycle fourteen, at the beginning of the cellularization stage (Edgar and Schubinger 1986; Orr-Weaver 1994). For review see also Fig. 23.2.

To investigate the ZGA in a higher plant species and to further analyse gene expression during fertilization and very early embryogenesis, representative cDNA libraries

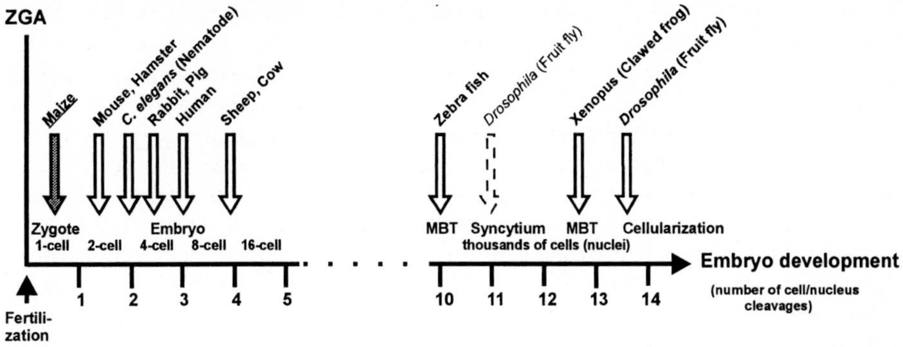

Fig. 23.2. Onset of zygotic gene activation (ZGA). The switch from maternal to embryonic control of gene activation during embryo development is indicated comparing several animal species, human and maize. See text for further explanations

were generated from unfertilized egg cells (Dresselhaus et al. 1994) and zygotes 18 h after *in vitro* fertilization (Dresselhaus et al. 1996) of maize. These libraries were analysed using differential screening methods and gene specific probes. More than 50 different transcripts have been identified whose amounts are up- or downregulated after *in vitro* fertilization (T. Dresselhaus, S. Cordts, S. Heuer; unpublished results). Some of them are newly induced 18 h after *in vitro* fertilization (see also Table 23.3), long before the unequal first cell division of the maize *in vitro* zygote. As mentioned above, generally the first zygotic cell cleavage occurrs between 42–46 h after gametic fusion. The identification of at least two transcripts that are newly induced and others that are strongly upregulated after fertilization indicates that the switch from maternal to embryonic control of development, occurrs in higher plants (at least in maize) shortly after fertilization at the zygotic stage, that is, at a much earlier developmental stage than in all investigated animal systems.

Among the isolated cDNAs there are two which encode proteins involved in RNA stability and processing (see Table 23.3). The expression of both genes is upregulated, indicating already an increased transcription ratio after fertilization. Transcripts encoding proteins involved in translation show a diverse expression pattern: transcripts of an initiation factor, an elongation factor and an acidic ribosomal protein (eIF-5A, EF-TU, RP-P0, respectively, see Table 23.3) are stockpiled in the unfertilized egg cell and their amounts slightly or strongly decrease in the zygote. Gene expression of two further ribosomal proteins (RP-S21A, RP-L39) is strongly induced, although notable expression is also found in the unfertilized egg cell. Transcripts needed for DNA replication are only partially stored in the egg cell. A strong upregulation is found for the expression of a gene encoding a protein needed for the initiation of DNA replication (a member of the MCM2/3/5 family of DNA binding proteins), while the expression of a gene encoding an endonuclease needed for DNA repair (FEN) is newly induced upon fertilization. From the expression pattern of the latter genes we assume that the zygote 18 h after *in vitro* fertilization, is at the beginning or already in the S-phase of the first cell cycle.

Further cDNAs have been identified (Table 23.3), encoding, e.g., calreticulin whose expression is clearly correlated to cell division (Dresselhaus et al. 1996) and which is probably needed as a chaperone to modify other proteins in the lumen of the endoplasmic reticulum (Dedhar 1994). These proteins might themselves be needed in the zygote e.g. for generating the cell plate during cytokinesis of the first cell division. The expression of a ubiquitin carrier protein gene (UBC7) is strongly downregulated after fertilization. This protein might be involved in a proteasome complex in the degradation of specific proteins that are no longer needed after fertilization. Finally, the amounts of three other transcripts that code for proteins involved in energy metabolism, glyceraldehyde-3-phoshate dehydrogenase (GAPDH), phosphoglycerate kinase (PGK) and aspartate aminotransferase (AAT), strongly decrease. During glycolysis GAPDH generates $NADH+H^+$ and PGK generates ATP. AAT generates $NADH+H^+$ for ATP synthesis during the malate-aspartate shuttle (for review see Voet and Voet 1995). The expression of these genes might be downregulated, because glucose is the carbon-source in the culture medium for the *in vitro* zygotes and high amounts of ATP can therefore be synthesized during these biochemical pathways. A high ATP level negatively regulates the described pathways (Voet and Voet 1995). Nevertheless, whether gene regulation of these genes is the same in *in vitro* and *in vivo* zygotes remains to be determined.

Recently, a novel RT-PCR method was described, allowing gene expression studies with single plants cells (Richert et al. 1996). This method permits investigation of the time course of gene expression from known genes and can now be used to study the expression of the above described genes in more detail before and after *in vitro* fertilization. The expression of known cell-cycle-regulators of maize was investigated already in

Table 23.3. Some of the genes whose expression is induced (+, + +) or repressed (−, − −) 18 h after *in vitro* fertilization code for proteins with homology to known proteins involved in translation, DNA replication and other processes

Function	Protein	Gene expression[a]	Acc. No.[b]
RNA stability and processing			
RNase L inhibitor	RLI	+[c]	[e]
Splicing factor	SF2	+	[e]
Translation			
Initiation factor	eIF-5A	−	y07920[e]
Elongation factor	EF-TU	−	[e]
Acidic ribosomal phospoprotein	RP-P0	−	y079S9[e]
Ribosomal protein	RP-S21A	+ +	X98656[e]
Ribosomal protein	RP-L39	+ +	X95458[e]
DNA replication			
Initiation factor	MCM2/3/5	+ +	[e]
FLAP-endonuclease	FEN	+ +[c]	[e]
Protein modification and degradation			
Calreticulin	CR	+	X89813[d]
Ubiquitin carrier protein	UBC7	−	AJ002959[e]
Energy metabolism			
Phosphoglyceratekinase	PGK	−	[e]
Glyceraldehyde-3-P dehydrogenase	GAPDH		[e]
Aspartate aminotransferase	AAT	−	[e]

[a] + or − (up or downregulated after *in vitro* fertilization, respectively); + + or − − (strongly up or downregulated after *in vitro* fertilization, respectively)
[b] (accession number in EMBL, GenBank and DDBJ Nucleotide Sequence Databases)
[c] (newly induced after *in vitro* fertilization)
[d] (Dresselhaus et al. 1996)
[e] (T. Dresselhaus, S. Cordts, S. Heuer and A. de Vries; unpublished results)

more detail using this method (Sauter et al. 1997). It was shown that cdc2 is constitutively expressed before fertilization and during zygote development. Cyclin genes display a more diverse pattern of gene expression: the mitotic B-type cyclin Zeama;*CycB2*;1 is newly expressed 24 h after *in vitro* fertilization. The expression of a second B-type cyclin Zeama;*CycB1*;2 oscillated: transcripts are detected between 12–14 h, 24–26 h and around 36 h after *in vitro* fertilization. The A-type cyclin Zeama;*CycA1*;1 is first completely downregulated and then newly induced 17 h after *in vitro* fertilization. All known maize A- and B-type cyclins (Renaudin et al. 1994, 1996) are probably mitotic cyclins and expressed in G2-phase of the cell cycle (Jacobs 1995). Cyclin gene expression in animals is different: cycB cyclins are expressed during G2-phase of the cell cycle, while animal cycA cyclins are expressed during S-phase of the cell cycle (Renaudin et al. 1996). The expression pattern of maize cyclin genes after *in vitro* fertilization further supports the assumption that the zygotic cell cycle is different from a somatic cell cycle and that the *in vitro* zygote is already in S-phase 18 h after *in vitro* fertilization, the time point when the cDNA library from zygotes was generated.

Up to now, all molecular analyses of zygote development have been made at the transcript level. It is possible, that some transcripts are not translated into proteins and even as proteins, they might not be necessarily active, because they might be post-translationally modified, which is e.g. the case for the cell-cycle regulator cdc2 (Jacobs 1995).

23.4 Conclusions and Prospects

These *in vitro* studies clearly demonstrate that zygotes are capable of self-organization in culture, apart from maternal tissue. They can do this comparably well to the *in vivo* situation, resulting in fertile plants. Also, egg cells without fertilization can be activated to initiate development, and further studies will show whether they develop in a comparable manner as zygotes undergo embryogenesis or not.

With the experimental access to single cells, comparative biochemical and molecular studies are now performed with sperms, eggs and zygotes to gain insight into early developmental processes. The mechanisms that initiate post fertilization development are not known. Further, there is neither information on the biochemistry of membrane fusion nor on the electrophysiology in higher plant gametes. The electrical conditions applied might not be as artificial as they seem. High currents were measured in plant cells. These could well be present in the embryo sac to induce fusion of gametic membranes *in vivo*. The access to isolated angiosperm gametes and the possibility of handling them allow detailed studies of surface biology of generative cells. One of the open questions to be answered is: whether fertilization in angiosperms occurs in a non-gametic specific manner as somatic protoplasts fuse or whether specific receptors are involved in this process.

Clearly, *in vitro* fertilization techniques can be used for cell hybridization to overcome interspecific barriers. With experimental access to interspecific gametic fusion, zygote formation and early embryogenesis, the processes that initiate hybrid formation can be dissected. The culture of *in vivo* zygotes of hybrids might help avoid postfertilization barriers. Further studies will reveal to what extent these techniques can be used to produce new hybrid plants.

Isolated egg cells are obtained only in relatively low amounts, which limits the experimental design. Also, plant regeneration from single zygotes is laborious. Sustained growth and therefore also differentiation of *in vitro* and *in vivo* zygotes decisively depend more on the quality of the feeder cells than on the origin of the cell suspension (for example excised zygotic embryo- or microspore-derived cell aggregates). The technique of implantation of zygotes into ovules is an alternative method, avoiding feeder cells. However, it would be very useful to develop a single cell culture technique which includes chemically defined conditions but avoids the use of feeder cells. Moreover, it should make it possible to follow the zygote development under continuous microscopical observation and sustained growth, efficient embryogenesis and plant regeneration.

A first step has now been taken in a higher plant system to study the fertilization process and very early embryo development also at the molecular level. This is only the beginning. A much clearer picture of the regulation of these important developmental stages can be expected in the near future from the analysis of specific genes (from the different cells of the embryo sac, the zygote, the two-cell embryo etc.), developmental regulators like MADS-box and Homeo-box genes and mutants disturbed in megagametogenesis, fertilization, zygote and early embryo development.

Acknowledgements
We thank Dr. O. da Costa e Silva for critical reading of the manuscript. The financial support by the Deutsche Forschungsgemeinschaft grant no. Kr 1256/1-4, the Körber Foundation, Hamburg, Germany and the European Commission (grants BIO4-960390 and BIO4-960275) is acknowledged.

References

Dedhar S (1994) Novel functions of calreticulin: interaction with integrins and modulation of gene expression. Trends Biochem Sci 19:269–271
Digonnet C, Aldon D, Leduc N, Dumas C, Rougier M (1997) First evidence of a calcium transient in flowering plants at fertilization. Development 124:2867–2874
Dresselhaus T, Hagel C, Lörz H, Kranz E (1996) Isolation of a full-size cDNA encoding calreticulin from a PCR-library of *in vitro* zygotes of maize. Plant Mol Biol 31:23–34
Dresselhaus T, Lörz H, Kranz E (1994) Representative cDNA libraries from few plant cells. Plant J 5: 605–610
Dudits D, Györgyey J, Bögre L, Bakó L (1995) Molecular biology of somatic embryogenesis. In: Thorpe TA (ed) In vitro Embryogenesis in Plants. Kluwer Dordrecht, pp 267–308
Edgar BA, Schubiger G (1986) Parameters controlling transcriptional activation during early *Drosophila* development. Cell 44:871–877
Faure J-E, Digonnet C, Dumas, C (1994) An *in vitro* system for adhesion and fusion of maize gametes. Science 263:1598–1600
Faure J-E, Mogensen HL, Dumas C, Lörz H, Kranz E (1993) Karyogamy after electrofusion of single egg and sperm cell protoplasts from maize: Cytological evidence and time course. Plant Cell 5:747–755
Grimes HD, Boss WF (1985) Intracellular calcium and calmodulin involvement in protoplast fusion. Plant Physiol 79:253–258
Holm PB, Knudsen S, Mouritzen P, Negri D, Olsen FL, Roué C (1994) Regeneration of fertile barley plants from mechanically isolated protoplasts of the fertilized egg cell. Plant Cell 6:531–543
Jacobs TW (1995) Cell cycle control. Annu Rev Plant Physiol Plant Mol Biol 46:317–339
Katoh N, Lörz H, Kranz E (1997) Isolation of egg cells of rape seed (*Brassica napus* L.). Zygote 5:31–33
Keller WA, Melchers G (1973) The effect of high pH and calcium on tobacco leaf protoplast fusion. Z Naturforsch 28c:737–741
Kovács M, Barnabás B, Kranz E (1994) The isolation of viable egg cells of wheat (*Triticum aestivum* L.). Sex Plant Reprod 7:311–312
Kovács M, Barnabás B, Kranz E (1995) Electro-fused isolated wheat (*Triticum aestivum* L.) gametes develop into multicellular structures. Plant Cell Rep 15:178–180
Kranz E (1997) *In vitro* fertilization with isolated single gametes. In: Hall, R (ed) Plant cell culture protocols. Humana Press, Totowa, in press
Kranz E, Bautor J, Lörz H (1991a) *In vitro* fertilization of single, isolated gametes of maize mediated by electrofusion. Sex Plant Reprod 4:12–16

Kranz E, Bautor J, Lörz H (1991b) Electrofusion-mediated transmission of cytoplasmic organelles through the *in vitro* fertilization process, fusion of sperm cells with synergids and central cells, and cell reconstitution in maize. Sex Plant Reprod 4:17–21
Kranz E, Dresselhaus T (1996) *In vitro* fertilization with isolated higher plant gametes. Trends Plant Sci 1:82–89
Kranz E, Lörz H (1993) *In vitro* fertilization with isolated, single gametes results in zygotic embryogenesis and fertile maize plants. Plant Cell 5:739–746
Kranz E, Lörz H (1994) *In vitro* fertilisation of maize by single egg and sperm cell protoplast fusion mediated by high calcium and high pH. Zygote 2:125–128
Kranz E, von Wiegen P, Lörz H (1995) Early cytological events after induction of cell division in egg cells and zygote development following *in vitro* fertilization with angiosperm gametes. Plant J 8:9–23
Kumlehn J, Brettschneider R, Lörz H, Kranz E (1997a) Zygote implantation to cultured ovules leads to direct embryogenesis and plant regeneration of wheat. Plant J: in press
Kumlehn J, Lörz H, Kranz E (1997b) Differentiation of isolated wheat zygotes into embryos and normal plants. Submitted
Leduc N, Matthys-Rochon E, Rougier M, Mogensen L, Holm P, Magnard J-L, Dumas, C (1996) Isolated maize zygotes mimic *in vivo* embryonic development and express microinjected genes when cultured *in vitro*. Dev Biol 177:190–203
Mogensen HL, Holm PB (1995) Dynamics of nuclear DNA quantities during zygote development in barley. Plant Cell 7:487–494
Mogensen HL, Leduc N, Matthys-Rochon, E, Dumas C (1995) Nuclear DNA amounts in the egg and zygote of maize (*Zea mays* L.). Planta 197:641–645
Mól R, Matthys-Rochon E, Dumas C (1994) The kinetics of cytological events during double fertilization in *Zea mays* L. Plant J 5:197–206
Newport J, Kirschner M (1982) A major developmental transition in early *Xenopus* embryos: II. Control of the onset of transcription. Cell 30:687–696
Orr-Weaver TL (1994) Developmental modification of the *Drosophila* cell cycle. Trends Genet 10: 321–327
Renaudin J-P, Colasanti J, Rime H, Yuan Z, Sundaresan V (1994) Cloning of four cyclins from maize indicates that higher plants have three structurally distinct groups of mitotic cyclins. Proc Natl Acad Sci USA 91:7375–7379
Renaudin J-P, Doonan JH, Freeman D, Hashimoto J, Hirt H, Inzé D, Jacobs T, Kouchi H, Rouzé P, Sauter M, Savoure A, Sorrell DA, Sundaresan V, Murray AH (1996) Plant cyclins: a unified nomenclature for plant A-, B- and D-type cyclins based on sequence organization. Plant Mol Biol 32:1003–1018
Richert J, Kranz E, Lörz H, Dresselhaus T (1996) A reverse transcriptase polymerase chain reaction assay for gene expression studies at the single cell level. Plant Sci 114:93–99
Risueno MC, Testillano PS (1994) Cytochemistry and immunocytochemistry of nucleolar chromatin in plants. Micron 25:331–360
Rougier M, Antoine AF, Aldon D, Dumas C (1996) New lights in early steps of *in vitro* fertilization in plants. Sex Plant Reprod 9:324–329
Sauter M, von Wiegen P, Lörz H, Kranz E (1997) Cell cycle regulatory genes from maize are differentially controlled during fertilization and first embryonic cell division. Sex Plant Reprod: in press
Schultz RM (1993) Regulation of zygotic gene activation in the mouse. BioEssays 15:531–538
Seydoux G, Mello CC, Pettitt J, Wood WB, Priess JR, Fire A (1996) Repression of gene expression in the embryonic germ lineage of *C. elegans*. Nature 382:713–716
Sun M-X, Yang H-Y, Zhou C, Koop H-U (1995) Single-pair fusion of various combinations between female gametoplasts and other protoplasts in *Nicotiana tabacum*. Acta Bot Sin 37:1–6
Theunis CH, Pierson ES, Cresti, M (1991) Isolation of male and female gametes in higher plants. Sex Plant Reprod 4:145–154
Tian HQ, Russell SD (1997a) Micromanipulation of male and female gametes of *Nicotiana tabacum*: I. Isolation of gametes. Plant Cell Rep 16:555–560
Tian HQ, Russell SD (1997b) Micromanipulation of male and female gametes of *Nicotiana tabacum*: II. Preliminary attempts for *in vitro* fertilization and egg cell culture. Plant Cell Rep 16:657–661
Tirlapur U, Kranz E, Cresti M (1995) Characterisation of isolated egg cells, *in vitro* fusion products and zygotes of *Zea mays* L. using the technique of image analysis and confocal laser scanning microscopy. Zygote 3:57–64
Van der Maas HM, Zaal, MACM, de Jong ER, Krens FA, Van Went JL (1993) Isolation of viable egg cells of perennial ryegrass (*Lolium perenne* L.). Protoplasma 173:86–89
Voet D, Voet JG (1995) Biochemistry (2nd edition). John Wiley and Sons, Inc, New York
Zamir E, Kam Z, Yarden A (1997) Transcription-dependent induction of G1 phase during the zebra fish midblastula transition. Mol Cell Biol 17:529–536
Zhang G, Liu D, Cass DD (1997) Calcium-induced sperm fusion in *Zea mays* L. Sex Plant Reprod 10: 74–82

CHAPTER 24

MADS Box Genes Controlling Ovule and Seed Development in Petunia

L. Colombo · G. C. Angenent

Department of Developmental Biology, DLO-Centre for Plant Breeding
and Reproduction Research (CPRO-DLO); P.O. Box 16, 6700 AA Wageningen, The Netherlands

24.1 Introduction

The ovule contains the female gametes and gives rise to the seed upon fertilisation. Recently, regulatory genes involved in ovule development have been isolated from Arabidopsis and petunia. Here, we will review the current state of our knowledge about petunia ovule ontogeny and concentrate on members of the MADS box gene family controlling the development of this important reproductive organ. Furthermore, we will elaborate on the hypothesis that ovules are separate floral organs based on molecular and paleobotanical evidences.

24.2 Ovule Development

In petunia the ovules originate from the surface of the placenta and arise at first, as a dense group of meristematic cells (Figure 24.1A). These ovule primordia protrude producing a funiculus, which connects the ovule with the placenta and the nucellus containing the megasporocyte (Figure 24.1B). Figure 24.2 shows a schematic and microscopic presentation of ovule development in Petunia. The megasporocyte undergoes two meiotic divisions (megasporogenesis) followed by three mitotic divisions (megagametogenesis) giving rise to the mature embryo sac. During megasporogenesis, a single integument develops from the base of the nucellus (Figure 24.1C) and covers the mature embryo sac completely leaving a small opening, the micropyle (Figure 24.1D). Through the micropyle the pollen tube enters the embryo sac in which the two sperm nuclei are released.

In many Angiosperm species including petunia, the embryo sac develops according to the Polygonum type (Willemse and van Went, 1984). In this type of embryo sac the two meiotic divisions result in four linearly arranged megaspores. Three of them degenerate while a single megaspore survives. The megaspores at the dyad stage (Figure 24.1E) and at the tetrad stage (Figure 24.1F) are surrounded by callose, which has been suggested as playing an important role in the degeneration of the three megaspores (Rodkiewicsz, 1970). A strong indication for this is that the callose disappears around the megaspore that is located at the chalazal part of the ovule from which the embryo sac will originate, whereas the callose remains deposited around the three degenerating megaspores (Figure 24.1F). This callose wall may physically prevent the import of nutrients required for the completion of megasporogenesis and consequently causes the starvation of three megaspores (Rodkiewicsz, 1970). The functional megaspore under-

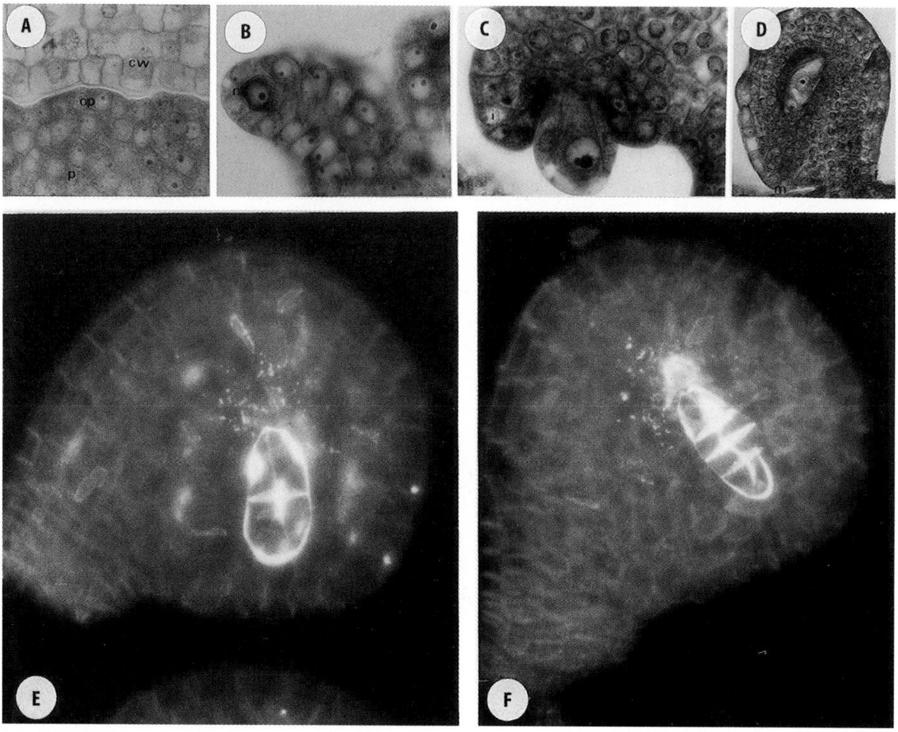

Fig. 24.1. Light micrographs of developing Petunia wild type ovules. **(A)** Close-up of ovule primordia. **(B)** Developing ovule. **(C)** Ovule with a pre-meiotic megasporocyte within the nucellus. **(D)** Mature ovule with embryo sac. **(E), (F)** The megasporocyte divides in four megaspores after two cycles of meiotic divisions. During megasporogenesis the dyad **(E)** and tetrad **(F)** are surrounded by a callose wall as can be visualized by aniline blue staining. Abbreviations used are: op, ovule primordia; cw, carpel wall; p, placenta; f, funiculus; n, nucellus; m, micropyle; i, integument

goes three mitotic divisions producing an eight nucleated and seven celled embryo sac. Three cells, the antipodal cells, are localised at the chalazal part of the ovule. Their function is unclear and soon after completion of the embryo sac, these cells degenerate. The large bi-nucleated central cell gives rise to the endosperm after fusion with one of the male gametes. At the micropilar side of the embryo sac the egg apparatus is formed consisting of two synergids and the egg cell. The egg cell develops into the embryo after fertilisation.

Fig. 24.2. The ABC model according to Coen and Meyerowitz (1991)

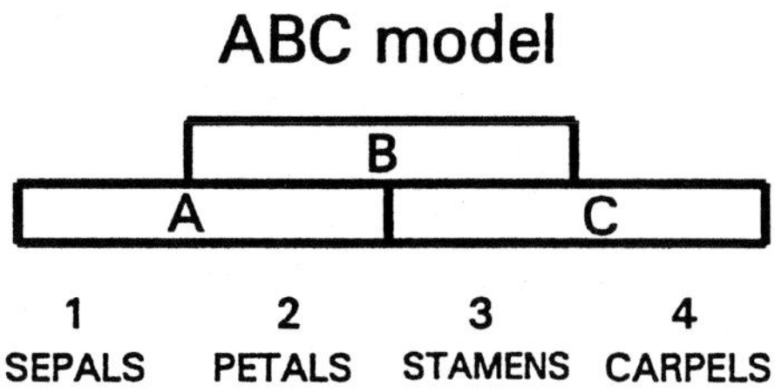

24.3 MADS Box Genes and the ABC Model

During the last decade, it has been shown that transcription factors belonging to the MADS box family play an essential role in various developmental processes (for reviews see Riechmann and Meyerowitz, 1997; Colombo et al., 1997a). The first MADS box identified were of yeast (Passamore et al., 1988) and human (Norman et al., 1988) origin. Since then a large number of MADS box genes have been isolated from plants (Theissen et al., 1995). The MADS box proteins have a highly conserved N-terminal domain of 56 amino acids in common which has been demonstrated to be the DNA binding domain. This conserved domain is the MADS box, which is an abbreviation of the first four isolated genes belonging to this transcription factor family: *MCM* (Passamore et al., 1988); *AGAMOUS* (Yanofsky et al., 1990); *DEFICIENS* (Sommer et al., 1990) and *SRF* (Norman et al., 1988). Plant MADS box proteins have a second conserved domain of about 70 amino acid residues, namely the K box, which shares similarity to the keratin coiled-coil domain (Ma et al., 1991). The K box has the potential to form an amphipatic α-helical structure which is involved in the dimerisation of MADS box transcription factors (Schwarz Sommer et al., 1992).

MADS box genes have important functions in various steps of flower development. For instance, the *Arabidopsis* MADS box genes *APETALA1* (*AP1*) (Bowman et al., 1993, Gustafson-Brown et al.,1994) and *CAULIFLOWER* (*CAL*) (Kempin et al., 1995) control the transition from inflorescence to flower meristem and several other MADS box genes such as *AGAMOUS* (*AG*) (Yanofsky., 1990), *APETALA3* (*AP3*) (Jack et al., 1992), *PISTILLATA* (*PI*) (Goto and Meyerowitz, 1994), and *AP1* control the determination of floral organ identity. Mutations in these MADS box genes cause homeotic transformation of floral organs. Based on the analysis of *Arabidopsis* and *Antirrhinum* homeotic mutants a model has been proposed describing floral organ formation (Figure 24.2), (Coen and Meyerowitz, 1991). According to this model floral organ identity is determined by the action and interaction of three classes of homeotic genes, A, B, and C, each active in two adjacent whorls. The genes of class A control sepal development and are also required, in combination with class B genes, to determine petal identity. Expression of class B and C genes together results in stamen development, while class C genes are responsible for carpel formation. The analysis of flower homeotic mutants in a number of diverse species has shown that the ABC model has a general validity for all the Angiosperm species analysed up till now. In petunia, *Floral Binding Protein 1 (FBP1)* (Angenent et al., 1992) and p*MADS1* (van der Krol et al., 1993) facilitate the B-function, whereas *FBP6* (Angenent et al., 1994) and p*MADS3* (Tushimoto et al., 1993; Kater et al., 1998) are regarded as the C-function genes (for reviews see: Van der Krol et al., 1993, Colombo et al., 1997a).

24.4 MADS Box Genes Controlling Ovule Development in Petunia

Several members of the MADS box gene family have been isolated from petunia (Colombo et al., 1997a), from which two of them, *FBP7* and *FBP11* are specifically expressed in ovules. The FBP7 and FBP11 proteins are very homologous, they share about 90% of their amino acid sequence (Angenent et al., 1995)

In situ hybridisation experiments performed using *FBP11* antisense RNA as probe revealed that *FBP11* starts to be expressed very early during pistil development. At first

FBP11 transcripts are detectable in the meristem that emerges from the recepticle in-between the two carpel primordia. This meristem is the progenitor of the placenta and ovules. Later, *FBP11* expression is restricted to the area of the placenta where the ovule primordia arise. When ovule primordia become apparent the expression of *FBP11* is detectable in these primordia only. The tissue of the ovule in which *FBP11* is highly expressed is the endothelium, the inner cell layer of the integument. No expression was observed in either nucellus or embryo sac. The expression pattern of *FBP7* is similar to that of *FBP11* (Angenent et al., 1995).

Simultaneous inhibition of *FBP7* and *FBP11* gene expression using an *FBP11* cosuppression construct, resulted in homeotic transformation of the ovules in style-stigma structures (Figures 24.3A and B) (Angenent et al., 1995). Pollen grains germinate on the stigmatic tissue and pollen tubes are able to grow along the stylar structures (Figure 4A). This mutant phenotype indicates that *FBP7* and *FBP11* are required for determining ovule identity. In addition, these genes are also sufficient to induce ovule formation. This was demonstrated by analysing transgenic petunia plants in which either *FBP7* or *FBP11* were ectopically expressed using the constitutive CaMV 35S promoter (Colombo et al., 1995). In these transgenic plants ovules were formed on the inner side of the sepals (Figure 24.4B) and on the tube of the petals. The presence of ovule on the sepals coincides with the homeotic transformation of the inner epidermis of the sepal into placenta like tissue (Figure 24.4C), (Colombo et al., 1995). No placenta-like cells were observed surrounding the ovules on the petals indicating that ovules can develop ectopically without any change in epidermal cell identity (Colombo et al., 1995).

These results have shown that *FBP7* and *FBP11* are required to determine ovule identity, moreover they are able to induce ovule formation on other floral tissues when ec-

Fig. 24.3. Scanning electron microscopy of ovary wild type and ovary of *FBP7/FBP11* cosuppression plants. **(A)** Ovary of a wild type plant. **(B)** Ovary of *FBP7/FBP11* cosuppression plant

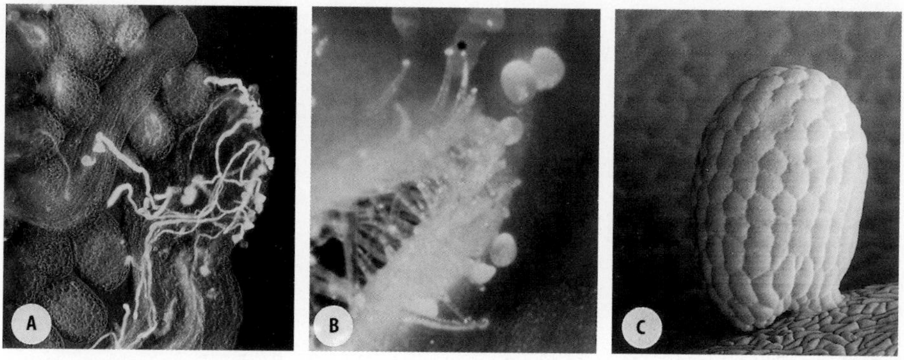

Fig. 24.4. (A) Transgenic plants in which *FBP7* and *FBP11* are down regulated: pollen tubes growth on style-sigma structures which replaced the ovules. **(B)** FBP11 ectopic expression plant with ovule-like structures on sepals. **(C)** FBP11 ectopic expression plant, ovule-like structure on a sepal: the inner epidermis of the sepal is transformed into placenta-lke tissue

topically expressed. Therefore they represent a new class of MADS box genes determining ovule identity (Angenent et al., 1995; Colombo et al., 1995; Angenent and Colombo 1996).

In the ABC model proposed by Coen and Meyerowitz in 1991, the ovule is considered as a part of the pistil, of which the development is controlled by the action of the C-type MADS box genes. Based on the results obtained with *FBP7* and *FBP11*, an extension of the ABC model into ABCD model was proposed (Colombo et al., 1995). In this extension the D genes represent a new class of MADS box genes required for ovule identity determination (Figure 24.5). *FBP7* and *FBP11* facilitate this function in petunia.

24.5 The Ovule: A Separate Floral Organ

In Angiosperms the ovules are formed inside the pistil. In many species, the ontogeny of the ovules is tightly linked with the development of the carpels. For instance, Arabidopsis ovules develop on the edges of the carpels. In contrast, the carpel primordia in petu-

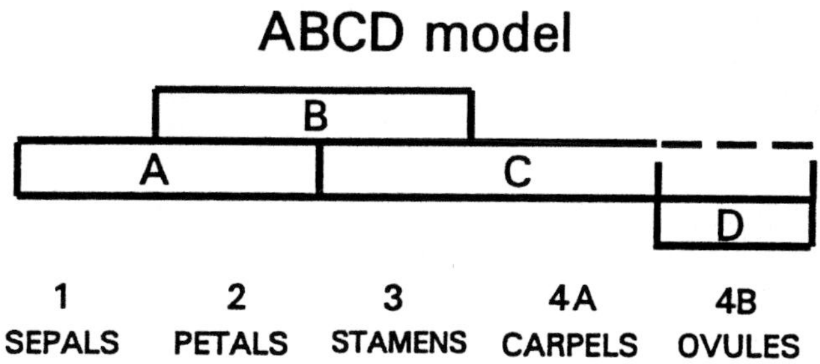

Fig. 24.5. The ABCD model according to Colombo et al, 1995. New extended model in which the D gene function determines the identity of ovules. The involvement of C type genes is still unclear and is indicated by a dashed line

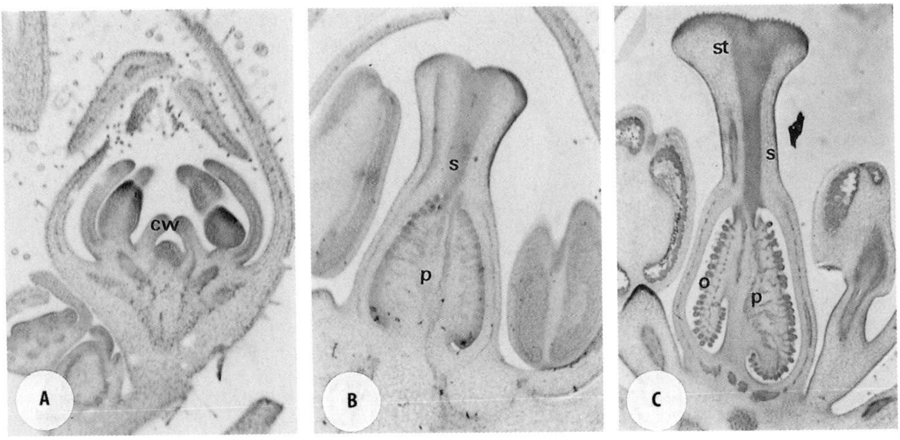

Fig. 24.6. Light micrographs of developing Petunia wild type pistils. **(A)** Developing gynoecium. Carpel primordia surround the placenta primordium (stage 3, Angenent et al, 1995). **(B)** Developing gynoecium (stage 6, 7, Angenent et al., 1995) Ovule primordia and transmitting tissue are formed. **(C)** Pistil, papillae are apparent on top of the stigma (stage10, Angenent et al., 1995). Abbreviation: cw, carpel wall, s, style, p, placenta, st, stigma, o, ovule

nia are completely separated from the central meristem that later give rise to the placenta bearing the ovules (Figure 24.6A). At later stages the carpel primordia elongate and fuse at the top with the emerging placenta. Subsequently, ovule primordia arise from the edge of the placenta as shown in Figure 24.6B. Although the ovules develop within the ovary (Figure 24.6C), the ontogeny described above combined with fossil records supports the hypothesis that the ovule can be considered as an independent flower organ (Angenent and Colombo., 1996; Baker et al., 1997). The fossil records suggest that the first megasporangia were formed at the tip of leaf-like structures hundreds of millions of years before any carpel or ovary structure evolved (Stewart, 1983). Furthermore, gymnosperms lack carpels and free ovules develop on a sporophyll indicating that carpel development is not necessarily required for ovule formation (Gifford and Foster 1989). In some Angiosperms, such as petunia, this independence is still visible during carpel and ovule ontogeny.

24.6 Other MADS Box Genes Expressed in the Ovule

Besides *FBP7* and *FBP11* (Figure 7A) also *FBP6* (Angenent et al., 1994) and p*MADS3* (Tsuchimoto et al., 1993) are expressed during ovule development (Figure 24.7B and C). *FBP6* and p*MADS3* are most likely class C homeotic genes involved in stamen and carpel development (Tsuchimoto et al., 1993, Kater et al., 1998), however their role in ovule development is still unclear. Nevertheless, at least in petunia class D gene expression is always accompanied by C-type gene expression. For instance, in transgenic petunia plants overexpressing *FBP7* and *FBP11,* also *FBP6* and p*MADS3* expression is induced ectopically (Colombo et al., 1995). This suggests that both type of genes are required for ovule formation.

It has been shown that C-type genes can affect the identity of ovules. Ectopic expression of C class genes of *Brassica napus* (*BAG1*) (Mandel et al., 1992)., and of Arabidop-

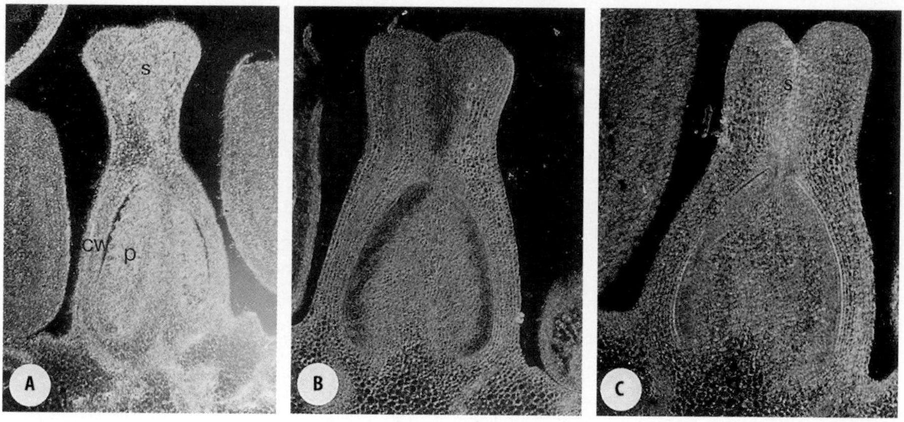

Fig. 24.7. Expression of *FBP11*, *FBP6* and *pMADS3* genes during ovule development. Longitudinal sections were hybridized with digoxigenin-labeled antisense *FBP11*, *FBP6* and *pMADS3* RNA. All sections were viewed using dark-field microscopy. **(A)** *FBP11* expression. **(B)** FBP6 expression. **(C)** pMADS3 expression. Bars = 0.5 mm. cw, carpel wall; o, ovule; pl, placenta

sis (*AG*) (Ray et al., 1994) in tobacco and Arabidopsis, respectively, resulted in the conversion of ovules into stigma-style structures reminiscent of the structures observed in *FBP7*/ *FBP11* mutants (Angenent et al., 1995). This suggests that during ovule development, the balance between the expression of D-and C-type gene is crucial. High expression levels of C genes compared to D genes result in carpel formation, while relatively high expression of D genes, as in wild type, leads to ovule formation.

In Arabidopsis several others MADS box genes have been found to be expressed in ovules: *AGL1* (Flanagan., 1996), *AGL5* (Savidge et al., 1995), *AGL11* (Rounsley et al., 1995), and *AGL13* (Rounsley et al., 1995). The role of these MADS box genes in ovule development is still unclear.

24.7 *FBP7* and *FBP11* are Required for Correct Seed Development

Fertilisation of the egg cell gives rise to the zygote that subsequently develops into an embryo, and the fusion between the two central nuclei and a sperm cell nucleus leads to the formation of triploid endosperm. During seed development the integument undergoes morphological changes and subsequently becomes the seed coat. Besides a role in determining ovule identity, *FBP7* and *FBP11* have an important function during seed development (Colombo et al., 1997). Down-regulation of *FBP7* and *FBP11* in a weak co-suppression mutant resulted in seeds without or drastically reduced endosperm (Figures 24.8A and B), (Colombo et al., 1997). This defect in endosperm development is a secondary effect due to the degeneration of the inner cell layer of the seed coat which originates from the endothelium of the ovule. The degeneration of this cell layer (Figures 24.8C and D) occurs when the embryo is in the globular stage and affects the transport of nutrients from the mother plant to the endosperm and embryo (Colombo et al., 1997b) Cell proliferation, starch and protein biosynthesis in the endosperm largely depend on this nutrient flow (Murray, 1984, 1987). Despite the defect in endosperm devel-

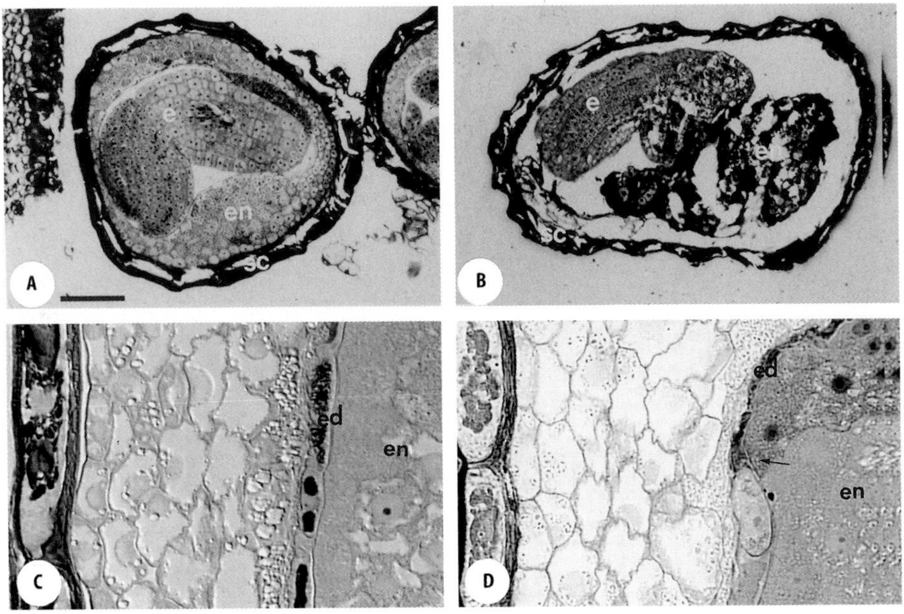

Fig. 24.8. Light micrographs of developing Petunia seed in wild type and cosopression plants. **(A)** Section of a mature wild-type seed with well developed embryo and endosperm. **(B)** Section of a mature seed of a plant in which *FBP7* and *FBP11* have been down-regulated by cosuppression; the embryo has reached the final stage of developmental process but the endosperm is degenerated. **(C)** Wild type seed 13 DAP. **(D)** Seed of the trangenic plant in which *FBP7* and *FBP11* have been down-regulated by cosuppression. Some of the endothelium cell start to degenerate and the endosperm cells bordering the crushed endothelium (see arrow) cells start to degenerate. Abbreviations; e, embryo; sc, seed coat; ed, endothellium

opment, many seeds contain a mature and viable embryo (Figure 24.8B) suggesting that, in petunia, endosperm is not required during later stages of embryo development (Colombo et al., 1997b).

24.8 Conclusions and Perspectives

Members of the MADS box gene family have important regulatory functions during ovule and seed development. In petunia, they specify ovule identity and are required for the maintainance of the endothelium cell layer in the seed coat. This cell layer connects the maternal tissue with the paternal tissue in the seed. The relationship and interdependence between these two types of tissues of different origin is poorly understood. The petunia mutants described in this review will enable us to gain a better understanding of the interactions between the tissues in a seed.

In addition to these MADS box genes, several other regulatory genes involved in ovule development have been identified, mainly from Arabidopsis (for reviews see: Angenent and Colombo, 1996; Baker et al., 1997) The challenge for the future is to understand the interactions between all these genes and the target genes that are controlled by these regulatory genes. This may enable us to control female gametophyte development and fertility in the future. Knowledge about the key factors that control the transition from ovule to seed tissue may lead to the development of new breeding tools with a great impact on plant biotechnology

References

Angenent GC, Busscher M, Franken J, Mol JNM, Van Tunen AJ (1992) Differential expression of two MADS box genes in wild-type and mutant Petunia flowers. Plant Cell 4:983–993

Angenent GC, Franken J, Busscher M, Weiss D, van Tunen AJ (1994) Co-suppression of the Petunia homeotic gene *FBP2* affects the identity of the generative meristem. Plant J 5:33–44

Angenent GC, Franken J, Busscher M, van Dijken A, van Went J, Dons H, van Tunen AJ (1995) A novel class of MADS box genes is involved in ovule development in Petunia. The Plant Cell 7:1569–1582

Angenent GC, Colombo L (1996) Molecular control of ovule development Trends in Plant science 1: 228–232

Baker SC, Robinson-Beer K, Villanueva JM, Gaiser JC, Gasser CS (1997) Interaction among genes regulating ovule development in Arabidopsis thailana. Genetics 145:1109–1124

Bowman JL, Smith DR, Meyerowitz EM (1993) Genetic interaction among floral homeotic genes of Arabidopsis. Development 122 1–20

Coen ES, Meyerowitz (1991) The war of the whorls: Genetic interactions controlling flower development. Nature 353:31–37

Colombo L, Franken J, Koetje E, van Went J, Dons HJM, Angenent GC, Van Tunen AJ (1995) The Petunia MADS box gene *FBP11* determines ovule identity. The Plant Cell 7:1859–1868

Colombo L, Van Tunen AJ, Dons HJM, Angenent GC (1997a) Molecular Control of Flower Development in *Petunia hybrida*. Petunia hybrida Advances in Botanical Research 26:229–250

Colombo L, Franken J,van der Krol AR, Wittich PE, Dons HJM, Angenent GC (1997b) Downregulation of ovule specific MADS box genes from Petunia results in maternally defects in seed development. Plant Cell 9:703–715

Flanagan CA, Hu Y, Ma H (1996) Specific expression of the AGL1 MADS box gene suggests regulatory functions in Arabidopsis gynoecium and ovule development. Plant journal 10:343–353

Gifford EM, Foster AS (1989) Morphology and Evolution of vascular plants. WH Freeman, New York

Goto K, Meyerowitz EM (1994) Function and regulation of the Arabidopsis floral homeotic gene *PISTILLATA*. Genes and Devel 8:1548–1560

Gustafson-Brown CB, Savidge B, Yanofsky MF (1994) Regulation of the Arabidopsis floral homeotic gene *APETALA1*. Cell 76:131–143

Jack T, Fox GL, Meyerowitz EM (1994) Arabidopsis homeotic gene *apetala3* ectopic expression: Transcriptional and post transcriptional regulation determine floral organ identity. Cell 76:703–716

Kay R, Chan A, Daly M, McPherson J (1987) Duplication of CaMV 35S promoter sequences creates a strong enhancer for plant genes. Science 236:1299–1302

Kater MM, Colombo L, Franken J, Busscher M, Masiero S, Lookeren-Champagne M, Angenent GC (1998) Multiple AGAMOUS homologs from cucumber and petunia differ in their ability to induce reproductive organ fate. The Plant Cell 10:171–182

Kempin SA, Savadge B, Yanofsky MF (1995) Molecular basis of the *cauliflower* phenotype in Arabidopsis. Science 267:522–525

Lopez MA, Larkins BA (1993) Endosperm origin, development and function. The Plant Cell 5: 1383–1399

Ma H, Yanofsky MF, Meyerowitz EM (1991) *AGL1-AGL6*, an *Arabidopsis* gene family with similarity to floral homeotic and transcription factor genes. Genes Dev 5:484–495

Mandel MA, Bowman JL, Kempin SA, Ma H, Meyerowitz EM, Yanofsky MF (1992) Manipulation of flower structure in transgenic tobacco. Cell 71:133–143

Modrusan Z, Reiser L, Feldmann KA, Fischer RL, Haughn GW (1994) Homeotic transformation of ovules into carpel-like structures in Arabidopsis. Plant Cell 6:333–349

Murray DR, (1985) Seed physiology. 2 vols. Academic Press, London

Murray DR, (1987) Nutritive role of the seed coat in developing legume seeds. American Journal of Botany 74:1122–1137

Norman C, Runswick M, Pollock R, Tresiman R (1988) Isolation and properties of cDNA clones encoding SRF, a transcriptional factor that bind to the *c-fos* serum response element. Cell 55:989–1003

Passamore S, Maine GT, Elble R, Christ C, Tye BK (1988) A *Saccaronmyces cerevisiae* protein involved in plasmid maintenance is necessary to mating of MAT cell. J Mol Biol 204:593–606

Ray A, Robinson-Beers K, Ray S, Baker SC, Lang JD Preuss D, Milligan SB, Gasser CS (1994) *Arabidopsis* floral homeotic gene BELL (*BEL1*) controls ovule development through negative regulation of AGAMOUS gene (*AG*). Proc Natl Acad Sci USA 91:5761–5765

Reiser L, Modrusan Z, Margossian L, Samach A, Ohad N, Haughn GW, Fischer RL (1995) The *BELL1* gene encode a homeodomain protein involved in pattern formation in the Arabidopsis ovule primordium. Cell 83:735–742

Riechmann JL, Meyerowitz EM (1997) MADS Domain Proteins in Plant Development. Biol Chem 378: 1079–1101

Rodkiewicz B (1970) Callose in cell walls during megasporogenesis in angiosperms. Planta 93:39–47

Rounsley SD, Ditta GS, Yanofsky MF (1995) diverse roles for MADS genes in Arabidopsis Development. The Plant Cell 7:1259–1269

Savidge B, Rounsley SD, Yanofsky MF (1995) temporal relationship between the transcription of two Arabidopsis MADS box genes and the floral organ identity genes. The Plant Cell 7:721–733

Schwarz-Sommer, Z, Hue I, Huijser P, Flor PJ, Hansen R, Tetens F, Lonning WE, Saedler H and Sommer H (1992) Characterization of the Anthirrinum floral homeotic MADS box gene DEFICIENS: evidence for a DNA binding and autoregulation of its persistent expression throughout flower development. EMBO J 11:251–263

Sommer H, Beltran JP, Huijser P, Pape P, Lonning WE, Saedler H, Schwarz-Sommer Z (1990) Deficiens, a homeotic gene involved in the control of flower morphogenesis in Anthirrinum majus: the protein shows homology to transcription factors. EMBO J 9:605–613

Theißen G, Saedler (1995) MADS box genes in plant ontogeny and phylogeny: Haeckel's "biogenetic law" revisited. Current opinion in Gen & Dev 5:628–639

Tsuchimoto S, van der Krol AR, Chua N-H (1993) Ectopic expression of *pMADS3* in transgenic Petunia phenocopies the Petunia *blind* mutant. Plant Cell 5:843–853

van der Krol AR, Brunelle A, Tsuchimoto S, Chua NH (1993) Functional analysis of Petunia MADS box gene *pMADS1*. Genes & Development 7:1214–1228

van der Krol AR, Chua NH (1993) Flower development in Petunia. The Plant Cell 5:1195–1203

Willemse MTM, van Went JL (1984) The female gametophyte. In: Johri BM (ed) Embryology of angiosperms. Springer-Verlag, Berlin, pp 159–196

Yanofsky MF, Ma H, Bowman JL, Drews GN, Feldmann KA, Meyerowitz EM (1990) The protein encoded by the Arabidopsis homeotic gene *AGAMOUS* resemble transcriptional factors. Nature 346:35–39

CHAPTER 25

Domains of Gene Expression in Developing Endosperm

H.-A. Becker · G. Hueros · M. Maitz · S. Varotto · A. Serna · R. D. Thompson

Max-Planck-Institut für Züchtungsforschung, Carl-von-Linné Weg 10, D-50829 Köln, Germany
e-mail: thompson@mpiz-koeln.mpg.de
telephone: +49221-5062440
fax: +49221-5062413

Abstract. Cereal endosperm provides directly or *via* livestock feed a major component of human nutrition. Using molecular probes, gene expression within the developing maize endosperm can be shown to be organized into a series of spatial domains which belie its apparent uniformity. Current knowledge of the physiological and genetic controls of gene expression in these domains is sketchy. Domain boundaries can be influenced by both hormonal and nutritional factors. Transcription factors so far identified operate generally within one endosperm domain, although the factor(s) interacting with one class of conserved promoter sequences, termed collectively the RY-repeats, for example, may act throughout the endosperm. Understanding how domains are established and maintained may help attempts to improve grain quality and yield.

25.1 Origin of Endosperm Tissue

The evolutionary origin of the Angiosperm endosperm reflects its role as a nurse tissue for the embryo. In more primitive plants such as *Gnetales* female gametophytic tissue in the form of the nucellus carries out this function. The substitution of the endosperm in this role has a number of possible advantages, one being a heterotic effect on the endosperm phenotype which may be subject to selection, and the second, the opportunity of tight coupling of embryo and nurse tissue development. It is thought that the endosperm may have arisen from a supernumerary embryo, multiembryonic fertilizations being common in basal seed plants such as *Ephedra* (Friedman, 1994). In support of this hypothesis are numerous parallels between embryo and endosperm development. Basal seed plants develop via a multinucleate (coenocytic) embryo phase; the angiosperm endosperm also develops coenocytically up to 3 dap (days after pollination). Although embryo and endosperm have thus a very similar genetic origin, the embryo is normally diploid, being derived from one generative (sperm tube) nucleus fused to the nucleus of the egg cell, whereas endosperm is a triploid tissue originating from fusion of $2 \times n$ central cell nuclei with the other generative nucleus of the pollen tube. This, at first sight small difference, has immense consequences for the two cell lineages. Whereas the embryo develops primordia for the basic vegetative plant structures, the endosperm is relatively uniform, with little obvious differentiation in discrete cell types. In contrast to the embryo, endosperm is referred to as an "end-cell" system i.e., endosperm cells have lost the totipotency typical of plant cells. Endosperm cells cannot be persuaded in culture to undergo organogenesis, and furthermore, endosperm cell cultures usually change their gene expression profile markedly from that of the starting tissue. One possibility is that endosperm cells may be subject to a form of programmed cell death, and the cells that survive in culture may have lost an endosperm-specific programming component. In contrast, the endosperm of dicotyledonous plants does in some cases possess viable cells after seed maturation (see, for example, Leubner-Metzger et al., 1995. Further differences in dicot endosperm development to that of cereals will be expanded more fully in Section 5).

25.2 Stages in Cereal Endosperm Development

Endosperm development is characterized by a rapid proliferation of cells highly specialized for nutrient assimilation and storage. In response to positional cues from the surrounding embryo sac, the final form of the organ is arrived at and a number of distinct domains of gene expression are established in the endosperm, within the first few days after pollination (DAP). Development can be divided into four stages (Olsen et al., 1992, Figure 25.1).

25.2.1 Coenocytic Phase

The triploid endosperm nucleus undergoes a series of rapid mitotic divisions over the first 72 h to produce, in the absence of cell wall formation, a coenocyte of up to 1000 nu-

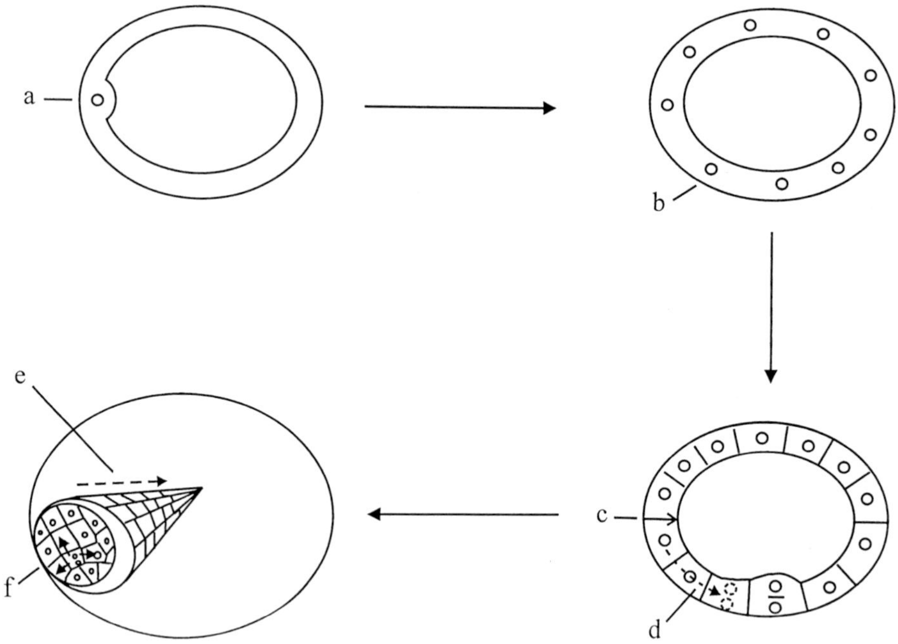

Fig. 25.1. Scheme of cereal endosperm development. **(a)** Primary endosperm nucleus located at the micropylar end of the embryo sac. **(b)** Free nuclear divisions resulting in a syncytium surrounding a large central vacuole. Spacing of nuclei determined by microtubular arrays. **(c)** Cellularization arising from wall ingrowths to form anticlinal cell walls. **(d)** First periclinal cell divisions to give the (external) aleurone initial and (internal) starchy endosperm initial. **(e, f)** Fully cellularized endosperm consisting of files of cells of increasing age towards the centre, and (in maize), a single aleurone cell layer at the periphery

clei, suspended in a thin layer of cytoplasm over the walls of the embryo sac, surrounding a large central vacuole.

25.2.2
Cellularization Phase

Up to this point, the direction of nuclear divisions has been essentially at random, but because of the existence of a large central vacuole, the resulting nuclei are located in a single layer at the periphery. There then follows a process of cell wall formation, directed by the development of inter-nuclear lamellae with their origin as pegs on the embryo sac wall (Figure 25.1c). Nuclei thus separated from one another again divide, this time in a more orderly manner, to give anticlinal cell plates dividing an inner layer of nuclei, which will give rise to the central or starchy endosperm initials, from an outer layer of nuclei, which form the aleurone initials (Figure 25.1d). This article will use the term endosperm to cover aleurone, basal and central, or starchy, endosperm cells

25.2.3
Differentiation Phase

During the differentiation phase, cell divisions continue, and this process, accompanied by cell elongation, proceeds until the endosperm cavity is filled. Aleurone initials divide anticlinally to give typically a single-cell outer layer to the endosperm cavity. Starchy endosperm initials divide predominantly periclinally to give files of cells of increasing age towards the center of the endosperm (Figure 25.1e, f). A deviation from this relationship is seen in the basal endosperm, adjacent to the pedicel, where the most basal cell layer is morphologically quite distinct from the rest of the aleurone. Instead of the typical vesicle-rich appearance of aleurone cytoplasm, the basal cells are characterized by extensively modified cell walls (Davis et al., 1990). A series of cell wall lamellae project from the basal face of the cell, and to a lesser extent from the side walls. Adjacent central endosperm cells up to three or four layers into the endosperm also have similar modified basal cell walls. The greatly increased surface area of the associated plasma membrane adapts these cells for their function in solute transfer.

The clonal origin of endosperm cells, which results from cell divisions during the differentiation phase, can be readily analysed using *Wx*-mutable alleles. The *Wx* gene encodes a starch granule-bound glucosyl transferase responsible for amylose synthesis. *Wx* and *wx* cells can be distinguished by KI/I_2 staining; the amylose in *Wx* cells stains blue-black, and *wx* cells, containing amylopectin, stain a much fainter lilac. An unstable transposable element insertion in the *Wx* gene is capable of excising and thus restoring gene activity to that cell lineage. Such events are seen as dark-staining sectors or patches in the endosperm (McClintock, 1978). The origins of sectors frequently lie close to the centre of the endosperm and reflect reversion events very early in endosperm development. Such sectors confirm the cell lineage from the centre towards the outermost layer. The basal transfer region is compressed by the embryo in the mature seed, so tracing such a lineage is more difficult. However, similar sector origins are also seen.

25.2.4
Maturation Phase

The beginning of the maturation phase (ca. 10–12 DAP) is characterized by a rapid change in cytokinin/auxin ratio (Lur and Setter, 1993), due to concurrent decrease/increase in concentrations of the two phytohormones in the endosperm. This shift has been proposed to trigger the starch and storage protein accumulation which begins during this period. Cell divisions continue with reduced frequency up to around 19 DAP, with the latest divisions taking place in the subaleurone layer. A further characteristic of this phase of endosperm development is the occurrence of endoreduplication in the starchy endosperm cells. This process of DNA replication in the absence of mitosis or cytokinesis results in large nuclei having DNA contents of 16C or higher. Endoreduplication is correlated with the phosphorylation of a maize Rb homologue (Grafi et al., 1996). Endoreduplication has been suggested as a mechanism to favour efficient synthesis of storage proteins, by increasing the supply of DNA template for transcription, but the process appears to involve the whole genome, and the storage protein genes do not appear to be preferentially replicated during this phase.

A further developmental feature of the maturation phase is dehydration. At 25–30 DAP the relative water content of the endosperm begins to decrease. In contrast to the gene expression programme operating in the developing embryo and aleurone, the central endosperm cells of cereals do not accumulate gene products associated with desiccation tolerance, and the central endosperm cells are inviable on seed germination. Endosperm seed reserves are mobilized by enzymes either activated from proenzymes deposited during kernel development or synthesized in the aleurone and the filial scutellum during imbibition. In addition to storage products, all endosperm cells synthesize a battery of defense-related proteins, for example thionins, protease inhibitors and ribosome-inactivating proteins, which are assumed to protect against pathogen ingress or insect predation during maturation and/or germination. In general, these proteins are encoded by small gene families, and individual endosperm cell-types (aleurone, basal cell, starchy endosperm cell) each express a unique spectrum of the corresponding gene products.

25.3 Mutants Affecting Endosperm Development

Mutants affecting endosperm development are an invaluable tool for dissecting the steps taking place during differentiation. The large number of defective endosperm mutants which have been isolated can be divided into four classes: opaque-class, starch synthesis, viable defective endosperm, and defective kernel (dek) mutants.

Some mutants fit into more than one class. The phenotypes of *Opaque* and starch synthesis mutants are usually restricted to the maturation phase, after 10 DAP. These mutants have been relatively well-characterized, because of their potential agronomic interest, whereas the defective mutant types which often result from lesions in earlier developmental stages, have received less attention. All classes include large numbers of mutants with similar phenotypes but having otherwise unrelated genetic lesions.

25.3.1 Opaque-Class Mutants

Opaque-mutants possess a kernel which transmits light poorly, in contrast to the translucent wild-type appearance. This is due to defects in either the formation or the maturation of protein bodies. The first opaque mutant to be characterized in detail, *opaque-2 (o2)*, was shown to have an increase in kernel lysine content over that of wild-type lines. This is attributable to the reduction in the lysine-poor alpha-zein storage proteins, and an increase in the absolute as well as the relative, amounts of lysine-rich endosperm proteins. In subsequent searches for further mutants with increased kernel lysine contents, an initial screening of opaque kernel types was carried out. Among the mutants identified, both *opaque-7* and *Floury-2* were also shown to have increased lysine in the kernel, although this is not the case for all opaque mutants. The molecular cause of the opaque lesion is known for two loci; *O2* encodes a bZIP transcription factor (25.4.2), and *Fl2* corresponds to a defective α-zein whose accumulation interferes with protein body assembly. *o6* is a lethal, proline-requiring mutation, and *o6* embryos can be rescued by various amino acid supplements, although it seems not to be a simple auxotroph (Manzocchi et al., 1986).

25.3.2 Starch Synthesis Mutants

A detailed description of the functions of this class of genes falls outside the scope of this article and is discussed elsewhere (see for example Smith et al., 1997). The major storage product in the cereal endosperm, starch is deposited in modified plastids, the amyloplasts. At least five starch synthases are known, and the process also involves pyrophosphorylases, and starch branching enzyme. Maize kernel mutants have been critical in identifying the key enzymes.

25.3.3 Viable Defective Endosperm Mutants

This is a very heterogeneous class of mutants with reduced endosperm development to varying degrees, which does not affect viability of the seed (Soave and Salamini, 1984). The mutants have for the most part received little attention from molecular biologists, and the loci affected have not been identified or isolated. During investigations into the mode of action of the *deB* series of mutants, endogenous auxin levels were tested, and in a few cases, dramatic differences to wild-type were discovered. *deB18*, for example, has less than 10% of the WT IAA concentration in the endosperm (Torti et al., 1986). The reduced rate of grain filling exhibited by the *deB18* mutant could be rectified by external application of an IAA analogue, NAA, in lanolin. As the vegetative growth and IAA content of the plant is near-normal, it appears to be affected in a seed-specific auxin biosynthetic pathway. Recent work suggests the *deB* mutants are nearly all non-allelic, i.e., define a large class of endosperm-specific genes (F. Salamini, pers. comm.). The *miniature, (Mn),* loci also belong in this class of mutants. *mn-1* to *-4* mutants have severely reduced kernel filling, resulting from a reduced endosperm size, whereas the maternal pericarp, and embryo development and germination are normal. *mn-1* has been shown to cosegregate with *Incw-2*, an endosperm-specific cell wall-located invertase, which is expressed specifically in the endosperm transfer layer (Cheng et al., 1996). The genes affected by the other *mn* loci are unknown, but they may also be candidates for transfer cell-specific defects.

25.3.4 Defective Kernel (dek) Mutants

Neuffer and Sheridan (1980) isolated a large class of defective endosperm mutants from a maize population derived from ethyl methane sulphonate mutagenesis. Most of these mutants are embryo-lethals or embryo-defectives (Sheridan and Neuffer, 1980, Sheridan and Clark, 1987). A fraction of the defective embryo class can be rescued by amino acid supplement of *in vitro* embryo and seedling culture, suggesting a non-seed-specific metabolic defect. Many of these mutants have been placed on the genetic map of maize, and 63 mutants of this class have also been isolated from *Mutator* stocks (Scanlon et al., 1994) and mapped to 14 new loci. The effect on embryo development of the genetic lesions has been analysed, and an *emb-* class, has been identified, which are defective at different stages of embryo development, but have normal endosperm development (Clark and Sheridan, 1991). Although in some *emb-* mutants the embryo may be arrested after a few cell divisions only or be completely lacking, endosperm development is unaffected, indicating the maize embryo does not play an essential role in regulating endosperm development.

25.4 Regulation of Gene Expression in the Endosperm

25.4.1 Domain Organization

The development of the endosperm can be summarized as the establishment of a number of multicellular domains of gene expression, and this process can be monitored by using marker genes specific for individual domains. A number of distinct cell-types arise during endosperm development, of which the most obvious are the aleurone and the central or starchy endosperm cell. The aleurone cells serve as a source of hydrolases, which mobilize starch and storage protein reserves of the inert central endosperm cells during germination. The aleurone enzymes are partly activated by cleavage of vacuolar proenzymes, but also synthesized *de novo* on imbibition. Compatible with their ability to survive desiccation, aleurone cells also accumulate a spectrum of ABA-inducible drought-associated proteins. The central endosperm cells accumulate starch granules and storage protein bodies to form a tightly packed matrix. With the advent of *in*

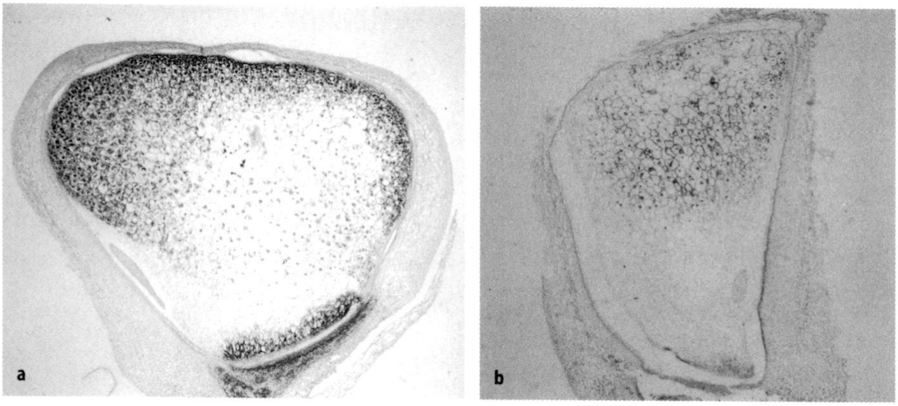

Fig. 25.2. Endosperm gene expression domains revealed by *in situ* hybridizations with 22kDa α-zein and *Betl1* cDNA probes. 15 DAP maize wild-type **(a)** or *dek30/dek30* mutant **(b)** kernels were sectioned as described (Hueros et al., 1995), and hybridized with digoxigenin-labelled probes. 22 kDa α-zein sequence was detected with Fast Blue B, and the BETL-1 probe was detected with Fast Red TR

situ hybridization techniques, further "domains" of gene expression have been revealed, which supplement those apparent from cytomorphological differences. One such domain includes the basal transfer cells. Genes expressed here show a gradient of expression from the pedicel-proximal cells toward the central endosperm (Hueros et al., 1995, Figure 25.2). A further domain of gene expression, surrounding the suspensor and developing embryo, was recently revealed by the *in situ* hybridization pattern of *Zmesr*, an early endosperm-expressed cDNA of unknown function (Opsahl Ferstad et al., 1997).

25.4.2
Central Endosperm Domain

The best characterized domain in terms of genetic regulation is that represented by the starchy endosperm cells, located in the crown of the kernel. Genes expressed here are typically most active in the outer, sub-aleurone cells, with a gradient of declining transcript concentration towards the center and towards the basal end of the kernel, as shown by *in situ* hybridization for the regulation of a storage protein probe (alpha-zein), and PPDK in the central endosperm domain (Gallusci et al., 1996, Figure 25.2). This spatial distribution of gene expression within the endosperm can be modified, as for example, is the case in genetic backgrounds containing modifier genes suppressing the *o2* phenotype (Or et al., 1993). The pattern may also be compromised in defective kernel mutants such as *dek30*, shown in Figure 25.2. In this case, a reduced rate of cell division gives rise to a thinner kernel compared to wild-type, but most notable is the absence of a gradient of zein mRNA concentration, and the reduced size of the expression domain, which may indicate a link between cell division and continued zein gene expression during the maturation phase.

A comparison of storage protein promoter sequences has revealed the presence of a short conserved motif, called variously the -300, prolamin or endosperm box (Forde et al., 1985). The endosperm box was found to be necessary for conferring endosperm-specific expression in tobacco. *In vivo* footprinting provided evidence for the binding of at least two factors to separate motifs in the endosperm box sequence (Hammond-Kosack et al., 1993). One factor bound at 10 DAP to the prolamin or endosperm motif, TGTAAAG, (see Table 25.1), whereas a second motif in the endosperm box, termed GCN4-like due to sequence similarity with GCN4 binding sites, was first protected by a protein from 14 DAP onwards (see Table 25.1 for a summary of factors and their binding sites). A component of the GCN4-motif binding complex in maize appears to be the *Opaque-2 (O2)* gene product. The *o2* mutant fails to accumulate a major zein class, the 22 kD α-zeins. In addition, a series of other more minor proteins are absent or reduced in amount. *O2* encodes a transcription factor of the bZIP family. The factor possesses the ability to activate 22 kD α-zein and b-32 promoters in tobacco protoplasts, suggesting it is required to activate gene expression during the endosperm maturation phase (Lohmer et al., 1991).

Recently, a gene encoding the prolamin box binding factor, PBF, has also been isolated. This protein interacts directly with O2 protein *in vitro*, and binds to prolamin promoters at the endosperm motif sites, adjacent to the O2 target sites in the endosperm box (Vicente-Carbajosa et al. 1997). As the motif to which PBF binds is found on most promoters expressed in the maturation phase of endosperm development, but O2 only regulates a small fraction of these genes, it is probable that PBF has a more universal regulating activity in the endosperm than O2 does. It is likely that in order to carry out these additional functions, PBF will be shown to interact with other transcription factor

Table 25.1. *Cis*-elements and the corresponding *trans*-acting factors implicated in conferring cell-type specificity in endosperm cells

Site of action/factor	Target sequence	Reference
Aleurone-specific		
C1	C/CTAACG/TG	Bodeau and Walbot 1996
R	CACGTG	
VP1	CGTCCATGCAT	Kao et al., 1996
?	CAA/N_{2-9}/TGG	Leah et al., 1994
?	"RY" (CATGCATG)	Dickinson et al., 1988
Central Endosperm		
Opaque-2	TCCACGTAGA GATGAC/TG/ATGG/A	Schmidt et al .,1990 Lohmer et al., 1991
Prolamin binding factor	TGTAAAG	Vicente-Carbajosa et al., 1997
Basal Layer		
?	"RY" repeats	Guo and Thompson, in prep.

partners. This hypothesis may be testable if a PBF mutant can be identified, or if one can be generated by a transgenic approach. *In situ* hybridization of *O2* (Dolfini et al. 1992), and Northern blot hybridizations with PBF (Thompson, unpublished) indicate that both transcripts are restricted to the crown endosperm cells, and therefore do not appear to be involved in regulation of gene expression in the aleurone and basal transfer cells. Both O2 and PBF are first detectable after 10 DAP, and their expression must itself be controlled by factors conferring endosperm-specificity on the cell-type.

25.4.3 Regulation in the Aleurone Domain

The expression of anthocyanin pigments in maize aleurone has enabled the visual screening of a series of mutations in structural and regulatory genes controlling anthocyanin biosynthesis. More recently, cis-elements responsible for aleurone-specific expression of a lipid transfer protein have been identified in transgenic rice (Kalla et al., 1994) and in barley, an aleurone-specific enhancer sequence was identified by particle gun bombardment of immature aleurone layers. The enhancer identified in barley contains a conserved sequence resembling the binding site of HMG-domain-containing proteins from mammals such as TCF-1 (Leah et al., 1994). The anthocyanin-regulatory loci identified, *C1* and *R1*, are representatives of the Myb- and bHLH transcription factor families respectively. The two factors interact on the same target promoters and genetic evidence suggests they heterodimerize *in vitro*, although only C1 binds to DNA directly. The binding sites through which *C1* and *R* act on the *Bronze-2 (Bz2)* locus have been mapped (Bodeau and Walbot, 1996, Fig. 25.4). They consist of clusters of two interspersed, functionally redundant repeats (C1-motif: TAACTG, R-motif: CACGTG). The role of *R* in transcriptional activation is not understood; it does not appear to possess its own activation domain. One possibility is that it may serve to link *C1* with another factor which stabilizes the DNA-binding of the *C1*-complex. A further *R*-like gene, *In1*, (*Intensifier*), probably corresponding to a gene duplication of the *R*-locus, appears to repress anthocyanin production, possibly by competing with *R1* protein in nucleoprotein complex formation (Burr et al., 1996). A further regulatory gene, *Vp1*, exhibits pleiotropic effects as it is involved in both the regulation of anthocyanin biosynthesis, and in the acquisition of seed dormancy, which is an ABA-dependent response. Molecular analysis of *Vp1* protein shows it also to be a transcription factor of a novel sequence type, which interacts with DNA as part of a multicomponent complex. *Vp1* activates the anthocyanin-regulatory gene *C1* at the Sph-box (Hattori et al., 1992), an RY-motif containing sequence. The RY-motif, consisting of alternating pyrimidine and purine bases (in this case CATGCATG) is also implicated in the regulation of seed-specific expression in cotyledons of the dicotyledonous species *P. vulgaris* (Bobb et al.,1997) where this site is bound by a bZIP factor, PvALF. Interestingly, the *C1* promoter can only be activated by VP1 in transient expression using seed-derived cells, implying the need for a further, seed-specific factor for the Sph-box-mediated interaction.

In order for aleurone colour to develop, exposure to light is also needed, and *C1*-activation is also light-dependent. This aspect of *C1* promoter regulation is mediated by a further *cis*-element, distinct from the *VP1*-responsive site (Kao et al., 1996). Aleurone differentiation also requires the product of the *Crinkly-4 (Cr4)* locus, a trans-membrane kinase (Becraft et al., 1996). *cr4* mutants exhibit alterations in leaf epidermis morphology as well as patchy aleurone development, implying common signalling processes for the differentiation of both cell-types, although the precise role of the *Cr4* gene product is unknown.

Comparison of the primary structures and target binding sites of the aleurone-specific transcription factors so far identified with those of the central endosperm regulators O2 and PBF does not reveal obvious homologies which might have indicated a common evolutionary origin. A possible link may be represented by the factor(s) hypothesized to bind to the RY-element present in all aleurone-specific promoters so far analysed. By analogy to PvALF, this sequence may also be bound by a bZIP factor.

25.4.4
Regulation in the Basal Transfer Domain

The regulation of gene expression in basal transfer cells is at an early stage of investigation, but the information so far available suggests that controls distinct from those in aleurone cells or central endosperm cells are operating from an early stage of endosperm development. The genes so far isolated, termed BETL-1 to -4 (BETL=Basal Endosperm Transfer Layer), all encode novel, highly expressed small polypeptides (8–10 kDa), three of which have sequence homology to putative defense-related genes (Hueros and Thompson, unpublished). The modified cell wall structure typical of BETL cells is seen, in successively reduced extent, up to four cell layers into the central endosperm; this gradient of expression is mirrored by a BETL-1 mRNA gradient, as revealed by *in situ* hybridizations. This expression pattern may reflect a solute gradient along the pedicel-endosperm crown axis. A zone around the developing embryo and suspensor, corresponding to the site of expression of *Zmesr* (Opsahl Ferstad et al., 1997), does not express the BETL genes so far isolated.

In contrast to the BETL-type domain of expression, crown cell-specific transcripts typically show a gradient of expression in the opposite direction, with maximal expression in the sub-aleurone layer, declining towards the pedicel (Fig. 25.2). Such gradients must be interpreted cautiously, however, due to the increasing age of the cells nearer the center of endosperm. So far, the substance(s) responsible for establishing these expression patterns have not been identified, but *in vitro* culture experiments suggest determination of the transfer cell phenotype has already taken place by 5 DAP. Basal cell gene expression is seen in immature kernels after isolation at 5 DAP and incubation for 5 days on media approximating to the composition of phloem sap. If the crown cell region of the kernel alone is cultured under the same conditions, no BETL-gene expression is seen, suggesting the domain cell-type commitment may have taken place earlier than 5 DAP and that it cannot be revoked by simple changes in solute supply. Possibly the transfer cell/aleurone cell-type divergence is established as early as the first anticlinal cell division of the endosperm initials, as the antipodal cells and other structural components of the surrounding nucellus could confer asymmetry on the embryo sac at this stage.

Two mechanisms might account for the development of the BETL-domain. In the first model, a modification of basal aleurone cells and adjacent central endosperm cells induced by the same factor(s), derived from the pedicel, takes place (Figure 25.3). In the second version, a modification of the basal aleurone layer alone occurs initially, followed by transmission of modifying factor(s) (possibly mRNAs?) from these cells to the adjacent endosperm cells (Figure 25.3).

A striking feature of gene expression in the basal endosperm transfer layer (BETL) is its transient nature. Analysis of four BETL-specific cDNA clones indicated almost identical expression patterns, with appearance of mRNA at 8 DAP, peaking at 14–16 DAP and declining to zero by 25 DAP, in contrast to the steady accumulation of storage protein (α-zein) mRNA between 15 and 35 DAP or the accumulation of hydrolases such as α- and β-amylases as proenzymes, and the presumed desiccation-protective LEA (Late Embryogenesis Abundant) proteins in the aleurone, during late kernel development (25–35 DAP). Whereas the aleurone is metabolically active in the mature kernel on imbibition, the transfer cells, which are crushed between the embryo and the black or closing layer of the pedicel, do not appear to be. Studies on transfer cell-defective mutants (Miller and Chourey, 1992, Cheng et al., 1996, Maitz and Thompson, unpublished) indicate that a misfunctioning transfer layer can also affect pedicel development, with reduced formation of chalazal cell walls. In addition to effects on grain filling, therefore, the transfer layer may contribute indirectly to the control of imbibition and germination of the kernel.

25.5 Manipulation of Endosperm Gene Expression *In Vitro*

Long-term endosperm *in vitro* culture produces cell lines which are significantly altered in expression profiles from the starting tissue. An alternative is to culture endosperm,

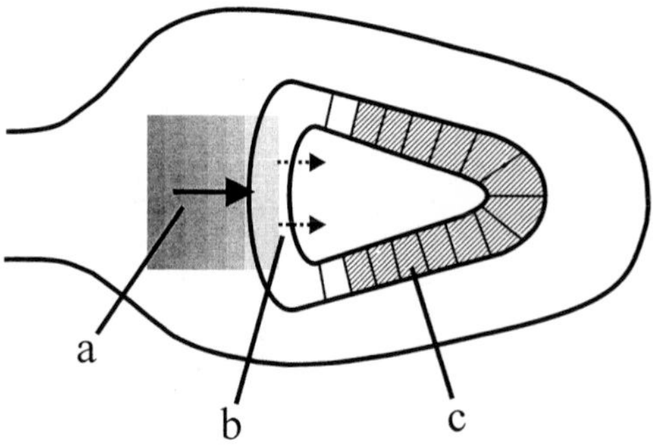

Fig. 25.3. Models to account for origin of transfer cell phenotype in basal endosperm cells. The differentiation of the transfer cell layer is suggested to arise in two stages: (i) perception by the basal aleurone cell layer of a signal originating in the pedicel; (ii) induction of transfer cell modifications in adjacent central endosperm cells either indirectly via a signalling mechanism, or by mRNA transport. (*a*) pedicel, (*b*) basal aleurone cell layer, (*c*) crown aleurone cells

immature kernels or cobs, for short periods of up to 15 days. Transfer layer gene expression can be studied by starting cultures at 5 DAP or less, before transfer cell markers are detectable, and storage product accumulation in the maturation phase can be investigated using kernels cultured from 10–14 DAP for 5 or more days. Under these conditions, grain filling can be compared to that of *in vivo* grown seeds, and the effects of individual media components can be studied. One application of this technique has been to analyse the kernel's response to nitrogen input in the form of amino acids. Endosperm reacts to increased amino acid availability with increased storage protein accumulation (Singletary and Below, 1989). This response involves changes in mRNA accumulation, and zein mRNA concentration increases sharply (Figure 25.4). In contrast, neither the level of *O2* mRNA nor that of the protein increases in response to nitrogen input. An unexpected finding was that 22 kD zein mRNA, which is under the control of the *O2* transcription factor, increased in response to amino acid input even in an *o2m(r)* mutant line, which completely lacks *O2* transcript. This *O2*-independent effect is attributed to the induction or activation of another factor which can substitute for O2 protein (Balconi et al., 1993, Müller et al., 1997).

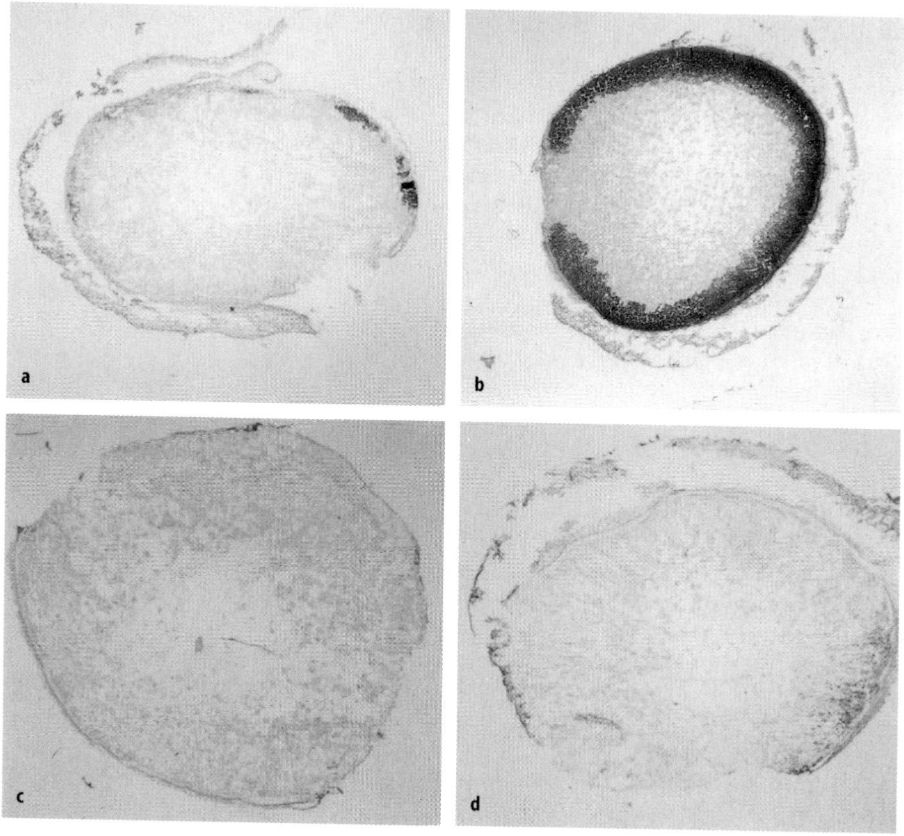

Fig. 25.4a–d. *In vitro* cultured maize kernels synthesize 22 kDa α-zein storage protein mRNA in response to media amino acid composition. *In situ* hybridization of DIG-labelled 22 kDa α-zein probe to 15 DAP wild-type kernels **(a and b)**, and to *o2m(r)* mutant kernels **(c and d)**. Kernels were cultured on low amino acid medium **(a and c)**, or high amino acid medium **(b and d)**, as described (Müller et al., 1997)

How do endosperm cells perceive the increase in amino acid supply and transduce this into a >10-fold increase in zein mRNA concentration? The finding that *O2*-independent 22 kd zein mRNA accumulation could be stimulated even at low amino acid concentrations by the addition of Methyl Jasmonate or ABA, indicates that phytohormones may mediate this response. Taken together, these results point to the existence of a 'nitrogen-responsive' factor, which is induced or activated by a signal transduction pathway susceptible to changes in either amino acid or in phytohormone concentrations. To what extent this reflects the regulation of zein synthesis in the whole plant remains to be established.

25.6 Endosperm Development and Function in *Arabidopsis*

In comparison to maize, knowledge about the endosperm in dicotyledons, taking as our example *Arabidopsis thaliana* L., is more limited. In contrast to maize, the *Arabidopsis* endosperm is a transient tissue, which is largely resorbed during late seed development, starting after 5–6 dap, drawing into question its role as a storage organ in this plant. In *Arabidopsis*,as in most dicotyledons, the cotyledons assume a comparable storage function to that played by the cereal endosperm. Possible functions for *Arabidopsis* endosperm are nutrient supply to the developing embryo, hormone supply, osmotic protection and mechanical support for the developing embryo (Mansfield, 1994). Nourishing of the developing embryo might require low-molecular-mass readily transferable metabolites instead of the production of long term storage products, as in the cereals. The high density of dictyosomes and rough endoplasmic reticulum in the chalazal chamber indicates that extensive protein synthesis does occur, however, and these features are reminiscent of the inter-lamellar cytoplasm of cereal transfer cells. Genes associated elsewhere with storage protein accumulation such as Calreticulin, a molecular chaperone, are also expressed in the developing *Arabidopsis* endosperm (Nelson et. al., 1997).

As seen for maize, the *Arabidopsis* endosperm also exhibits a coenocytic phase showing free nuclear divisions of the primary endosperm nucleus following the typical angiosperm double fertilization. Cellularization starts at the micropylar pole at the time of cotyledon initiation, and finishes at the chalazal end of the embryo sac. Cellularization is complete by the time the embryo starts growing in the endosperm (Mansfield and Briarty, 1990a, b; Mansfield, 1994). The polarity of cellularization corresponds to the polarity of embryo sac and ovule. This polarity is reflected furthermore in the subcellular structures present in the developing tissue, for example the high density of dictyosomes and rough ER observed in the chalazal domain, prior to cellularization. The maturation phase of endosperm development as observed in maize is absent, only a single cell layer of endosperm, (also referred to as the aleurone layer), remaining in mature seeds.

The regulation of *Arabidopsis* endosperm development shows a strict dependence on several factors. First the diploid central cell originated by fusion of two gametophytic nuclei in female gametophyte development needs a further haploid genome for division. Possible regulators of *Arabidopsis* endosperm development have been revealed by the isolation of *fis* (fertilization independent seed) and *fie* (fertilization independent endosperm) mutants (Chaudhury et al., 1997, Ohad et al., 1996). These allow fertilization-independent proliferation of the diploid central cell nucleus, and subsequent endosperm development, without influencing the egg cell. The uncoupling of endosperm and zygote development is also a feature of apomixis (Kultonow, 1993).

A number of embryo defective mutants, such as *raspberry 1* and *2* provide evidence of a regulatory role on endosperm development being played by the embryo itself (Yadegari et al, 1994). These mutants which fail to undergo the embryo transition to heart stage. In addition, an enlarged suspensor is seen, which resembles a "twin" embryo. In contrast to the wild type endosperm being located at the periphery of the embryo sac, the endosperm cells in this mutant seem to be concentrated around the suspensor, and endosperm development is maintained in a prolonged proliferative stage. A role for maternal factors influencing dicotyledonous endosperm development has been provided by the analysis of petunia plants downregulated in ovule-specific MADS box genes (Colombo et al., 1997).

The *Arabidopsis* endosperm may also act itself as a "regulator" in influencing the development of surrounding tissues, a similar role to that revealed for maize endosperm by the *mn1* mutant. For example, in *fie* mutants, silique extension and seed coat differentiation take place when endosperm develops in the embryo sac in the absence of fertilization. Here the cue is probably hormonal, as mutations of the *Spindly* locus also show parthogenetic silique development without fertilization, and the *spindly* mutant displays an altered gibberellin response (Jacobsen and Olszewski, 1993). The domain concept of endosperm gene expression may help in understanding differences in endosperm development between monocotyledons and dicotyledons, once the factors governing establishment of domains are identified.

References

Balconi C, Rizzi E, Motto M, Salamini F, Thompson RD (1993) The accumulation of zein polypeptides and zein mRNA is modulated by nitrogen supply. Plant J 3:325–334

Becraft PW, Stinard PS, McCarty DR (1996) Crinkly4: A TNFR-like receptor kinase involved in maize epidermal differentiation. Science 273:1406–1409

Bobb AJ, Chern MS, Bustos MM (1997) Conserved RY-repeats mediate transactivation of seed-specific promoters by the developmental regulator *PvALF.* Nucleic Acids Research 25:641–647

Bodeau JP, Walbot V (1996) Structure and regulation of the maize *Bronze2* promoter. SO – Plant Molecular Biology 32:599–609

Burr FA, Burr B, Scheffler BE, Blewitt M, Wienand U, Matz EC (1996)The maize repressor-like gene intensifier1 shares homology with the *r1/b1* multigene family of transcription factors and exhibits missplicing. Plant Cell 8:1249–1259

Cameron JW (1947) Chemico-genetic bases for the reserve carbohydrates in maize. Genetics 32: 459–485

Chaudhury AM, Ming L, Miller C, Craig S, Dennis ES, Peacock WJ (1997) Fertilization-independent seed development in Arabidopsis thaliana. Proc Natl Acad Sci USA 94(8):4223–4228

Cheng W-H, Talierco EW, Chourey PS (1996) The *Miniature 1* seed locus of maize encodes a cel wall invertase required for normal development of endosperm and maternal cells in the pedicel. Plant Cell 8:971–983

Clark JK, Sheridan WF (1991) Isolation and characterization of 51 embryo-specific mutations of maize. Plant Cell 3:935–952

Coleman CE, Clore AM, Ranch JP, Higgins R, Lopes MA, Larkins BA (1997) Expression of a mutant alpha-zein creates the floury2 phenotype in transgenic maize. Proceedings of the National Academy of Sciences of the United States of America 94:7094–7097

Colombo L, Franken J, Van der Krol AR, Wittich PE, Dons HJM, Angenent GC (1997) Downregulation of Ovule-Specific MADS Box Genes from Petunia Results in Maternally Controlled Defects in Seed Development Plant Cell, Vol 9:703–715

Davis RW, Smith JD, Cobb BG (1990) A light and electron microscopic investigation of the transfer cell region of maize caryopses. Can J Bot 68:471–479

Dickinson CD, Evans PR, Nielsen NC (1988) RY repeats are conserved in the 5′flanking regions of legume seed protein genes. Nucl Acids Res 16:371

Dolfini SF, Landoni M, Tonelli C, Bernard L, Viotti A (1992) Spatial regulation in the expression of structural and regulatory storage-protein genes in *Zea-mays* endosperm. Developmental Genetics 13 (4): 264–276

Forde BG, Heyworth A, Pywell J, Kreis M (1985) Nucleotide sequence of a B1 hordein geneand the identification of possible upstream regulatory sequences in endosperm storage protein genes from barley, wheat and maize. Nucl Acids Res 13:7327–7339

Friedman WE (1994) The evolution of embryogeny in seed plants and the developmental origin and early history of endosperm. Am J Bot 81:1468–1486

Gallusci P, Varotto S, Matsuoko M, Maddaloni M, Thompson RD (1996) Regulation of cytosolic pyruvate, orthophosphate dikinase expression in developing maize endosperm. Plant Molecular Biology 31:45–55

Goff SA, Cone KC, Chandler VL (1992) Functional analysis of the transcriptional activator encoded by the maize b gene evidence for a direct functional interaction between two classes of regulatory proteins. Genes & Development 6:864–875

Grafi G, Burnett RJ, Helentjaris T, Larkins BA, Decaprio JA, Sellers WR, Kaelin WG Jr (1996) A maize cDNA encoding a member of the retinoblastoma protein family Involvement in endoreduplication. Proc Natl Acad Sci (US) 93:8962–8967

Hammond-Kosack MCU, Holdsworth MJ, Bevan MW (1993) *In vivo* footprinting of a low molecular weight glutenin gene (*LMWG-1D1*) in wheat endosperm. EMBO J 12:545–554

Hattori T, Vasil IK, Rosenkrans L, Hannah LC, McCarty DR, Vasil IK (1992) The *Viviparous-1* gene and abscisic acid activate the *C1* regulatory gene for anthocyanin biosynthesis during seed maturation in maize. Genes Dev 6:609–618

Hueros G, Varotto S, Salamini F, Thompson RD (1995) Molecular characterization of BET1, a gene expressed in the endosperm transfer cells of maize. Plant Cell 7:747–757

Jacobsen SE, Olszewski NE (1993) Mutantions at the SPINDLY locus of Arabidopsis alter gibberellin signal transduction. Plant Cell 5:887–896

Kalla R, Shimamoto K, Potter R, Nielsen PS, Linnestad C, Olsen O-A (1994) The promoter of the barley aleurone-specific gene encoding a putative 7 kDa lipid transfer protein confers aleurone cell-specific expression in transgenic rice. Plant Journal 6:849–860

Kao CY, Cocciolone SM, Vasil IK, Mccarty DR (1996) Localization and interaction of the cis-Acting elements for abscisic acid, *Viviparous1*, and light activation of the *C1* gene of maize. Plant Cell 8:1171–1179

Kultonow AM (1993) Apomixis: Embryo sacs and embryos formed without meiosis or fertilization in ovules. Plant Cell 5:1425–1437

Leah R, Skriver K, Knudsen S, Ruud-Hansen J, Raikhel NV, Mundy J (1994) Identification of an enhancer/silencer sequence directing the aleurone-specific expression of a barley chitinase gene. Plant J 6:579–589

Leubner-Metzger G, Frundt C, Vogeli Lange R, Meins F Jr (1995) Class I beta-1,3-glucanases in the endosperm of tobacco during germination. Plant Physiology (Rockville) 109:751–759

Lohmer S, Maddaloni M, Motto M, Di Fonzo N, Hartings H, Salamini F, Thompson RD (1991) The maize regulatory locus *Opaque-2* encodes a DNA-binding protein which activates the transcription of the b-32 gene. EMBO J 10:617–624

Lur HS, Setter TL (1993) Endosperm development of maize defective kernel *dek* mutants: auxin and cytokinin levels. Annals of Botany (London) 72 (1):1–6

Mansfield SG, Briarty LG (1990a) Development of the free-cellular Endosperm in Arabidopsis thaliana (L.). Arab Inf Serv 27:53–64

Mansfield SG, Briarty LG (1990b) Endosperm cellularization in Arabidopsis thaliana (L.). Arab Inf Serv 27:65–72

Mansfield SG, Briarty LG (1994) Arabidopsis: an atlas of morphology and development 367–373. pub Springer Verlag

Manzocchi L, Tonelli C, Gavazzi G, Di Fonzo N, Soave C (1986) Genetic relationship between *o6* and *pro-1* mutants in maize. Theor Appl Genet 72:778–781

McClintock (1978) Development of the maize endosperm as revealed by clones. Cold Spring Harbor Symp Quant Biol 217–237

Miller ME, Chourey PS (1992) The maize invertase-deficient *miniature-1* seed mutation is associated with aberrant pedicel and endosperm development. Plant Cell 4:297–305

Müller M, Dues G, Balconi C, Salamini F, Thompson RD (1997) Nitrogen and hormonal responsiveness of the 22kDa α-zein and b-32 genes is displayed in the absence of the transcriptional regulator *Opaque-2*. Plant J 12:281–291

Nelson DE, Glaunsinger B, Bohnert HJ (1997) Abundant Accumulation of the Calcium-Binding Molecular Chaperone Calreticulin in Specific Floral Tissues of *Arabidopsis thaliana*. Plant Physiol 114:29–37

Neuffer MG, Sheridan WF (1980) Defective kernel mutants of maize. I Genetic and lethality studies. Genetics 95:929–944

Ohad N, Margossian L, Hsu Y-C, Williams C, Repetti P, Fischer RL (1996) A mutation that allows endosperm development without fertilization. Proc Nat Acad Sci 93:5319–5324

Olsen O, Potter R, Kalla R (1992) Histo-differentiation and molecular biology of developing cereal endosperm. Seed Science res 2:117–131

Opsahl Ferstad HG, Le Deunff E, Dumas C, Rogowsky PM (1997) *ZmEsr*, a novel endosperm-specific gene expressed in a restricted region around the maize embryo. Plant Journal 12:235–246

Or E, Boyer SK, Larkins BA (1993) *Opaque2* modifiers act post-transcriptionally and in a polar manner on gamma-zein gene expression in maize endosperm. Plant Cell 5 (11):1599–1609

Sainz MB, Grotewold E, Chandler VL (1997) Evidence for direct activation of an anthocyanin promoter by the maize *C1* protein and comparison of DNA binding by related Myb domain proteins. Plant Cell 9:611–625

Scanlon MJ, Stinard PS, James MG, Myers AM, Robertson DS (1994) Genetic analysis of 63 mutations affecting maize kernel development isolated from *Mutator* stocks. Genetics 136:281–294

Schmidt RJ, Burr FA, Aukerman MJ, Burr B (1990) Maize regulatory gene *Opaque-2* encodes a protein with a "leucine zipper" motif that binds to zein DNA. Proc Natl Acad Sci USA 87: 46–50

Sheridan WF, Neuffer MG (1980) Defective kernel mutants of maize. II Morphological and embryo culture studies. Genetics 95:945–960

Sheridan WF, Clark JK (1987) Maize embryogeny a promising experimental system. Trends in Genetics 3:3–6

Singletary GW, Below FE (1989) Growth and composition of maize kernels cultured *in vitro* with varying supplies of carbon and nitrogen. Plant Physiol 89:341–346

Smith AM, Denyer K, Martin C (1997) The synthesis of the starch granule. In: Jones RL (ed) Annual Review of Plant Physiology and Plant Molecular Biology, Vol 48. x+819p 67–87. Annual Reviews Inc: Palo Alto,California, USA, ISBN 0-8243-0648-1

Soave C, Salamini F (1984) The role of structural and regulatory genes in the development of maize (*Zea mays*) endosperm. Developmental Genetics 5:1–26

Torti G, Manzocchi L, Salamini F (1986) Free and bound IAA is low in the endosperm of the maize mutant *defective endosperm-B18*.Theoretical & Applied Genetics 72:602–605

Vicente-Carbajosa J, Moose SP, Parsons RL, Schmidt RJ (1997) A maize zinc-finger protein binds the prolamin box in zein gene promoters and interacts with the basic leucine zipper transcriptional activator Opaque-2. Proc Natl Acad Sci USA 94:7685–7690

Yadegari R, de Paiva GR, Laux T, Koltunow AM, Apuya N, Zimmermann JL, Fischer RL, Harada JJ, Goldberg RB (1994) Cell Differentiation and Morphogenesis Are Uncoupled in Arabidopsis *raspberry* Embryos Plant Cell 6:1713–1729

Advances on the Study of Sexual Reproduction in the Cork-Tree (*Quercus suber L.*), Chestnut (*Castanea sativa Mill.*) and in Rosaceae (*Apple and Almond*)

J. A. Feijó[1] · A. C. Certal[1] · L. Boavida[1] · I. Van Nerum[2] · T. Valdiviesso[3] · M. M. Oliveira[1]
W. Broothaerts[2]

[1] Dep. Biologia Vegetal, Faculdade de Ciências da Universidade de Lisboa, Campo Grande, C2, 1700 Lisboa, Portugal
[2] Fruitteeltcentrum, Katholieke Universiteit Leuven, W. de Croylaan 42, 3001 Heverlee, Belgium
[3] Estação de Fruticultura Vieira Natividade, Alcobaça, Portugal
e-mail: jose.feijo@fc.ul.pt telephone: +351.1.7500069 fax: +351.1.7500048

26.1 Introduction

Research in sexual plant reproduction has been boosted by the concentration of means and minds around central issues, such as incompatibility, microspore development or, to a less extent, the progamic phase of fertilisation. These areas are now pivotal since they involve basic biological questions still unanswered and which became recently accessible due to powerful genetic and molecular techniques. While these combined efforts resulted in the outcome of major findings that projected the field to a place not easy to by-pass even in the general context of Biology, this was only possible because of the existence of a few model species, particularly adapted to the experimental needs of the molecular weaponry. Most notably, *Arabidopsis* now plays a central role, since it possesses most of the development paths that characterise the majority of Angiosperms (with the exception of secondary growth). While this approach is not questionable, since it permits a faster advance in models and concepts, the reductionism underlying the focus on a few model species may create a void of knowledge on the natural variability and diversity of the cellular and molecular mechanisms. Most likely, the majority of the flowering species will never find their place as models species since they cannot compete as regards easiness to grow, maintain and analyse. Yet, an historical perspective shows beyond doubt that some of these species may bear variations to the basic models which may open new avenues and question established paradigms. Such was the case of *Arabidopsis*, for instance, often neglected and rejected for funding because of its condition of weed (Goldberg 1996). Another example is now emerging on the incompatibility field, where a major critical effort was made to categorise the field (de Nettancourt 1977, Heslop-Harrison 1975) and triggered two decades of fecund research. One of the major results was the outcome of a molecular model for the gametophytic incompatibility in *Nicotiana alata* (McClure et al. 1990, short review on Dickinson 1994). Originally thought of as the molecular template for all gametophytic systems, research on other species is now challenging this model as universal, and new hypotheses are developing (Franklin et al. 1995).

The lack of basic knowledge in sexual reproduction becomes even more evident in trees. They are difficult, when not impossible, to grow in controlled conditions and seasonal flowering is not easy to overcome. This creates a strong experimental constraint and the necessity to develop most of the experimental work in just a few months (or weeks!). Many of these species, however, have strong economic value based on fruit production, a direct result of sexual reproduction. It's generally accepted that a better knowledge of the mechanisms of pollination and fertilisation may result in better productivity (Pesson and Louveaux 1984). Yet the generalisation of the knowledge described in the model species is not, always, straightforward, and demands an analysis of the specific sexual cycle. Firm grounds on the structural and chronological aspects of fertilisation should be established before genetic and molecular approaches can be applied.

The demand for knowledge of the sexual reproduction of important economic trees and the general interest in the diversity of specific cellular and molecular mechanisms that may arise from the study of these species, established the grounds for a number of projects we undertook in the last years. The advances reported in this chapter refer to four of the species under study, where the gathered data already resulted in interesting patterns. They necessarily reveal approaches at different levels, since there is a large heterogeneity of background information about these species. Also the objectives are somehow different since, although fruit production is usually the major issue to be addressed, in some species, genetic diversity and the way to monitor and control it may constitute an important objective, especially at the population level.

26.2 The Cork-Tree (*Quercus suber*)

Quercus suber is one of the native forest trees of higher economic interest in Portugal, mainly for the production of cork used in a variety of industrial products. Present in large areas of agroforestry management, there has been a decline in cork-oak populations due to ageing, debilitation and the occurrence of free hybridisation, which produces a great variability of undesirable phenotypes. The increasing demand for good cork and the declining area of the species calls for improvement programs in the fields of breeding, biodiversity and conservation of genetic resources (Varela and Eriksson 1995).

The cork-oak has a complex reproductive behaviour, and the knowledge of the species is very scarce. Statistical data on oaks shows an evident fluctuation in acorn production from year to year and among individual trees. These variations are dependent not only on the success of pollination and fertilisation but also on the susceptibility to many environmental conditions, parasites and weather cues at the time of flowering and fruiting (Sharp and Chisman 1962; Sharp and Sprague 1967; Wolgast and Stout 1977; Farmer 1981; Freeman 1981; Feret et al. 1982). Understanding the reproductive biology of the species in order to make accurate predictions about patterns of acorn production, will be useful in both seed orchard management and increasing mast production in natural or planted stands (Farmer 1981; Sork and Bramble 1993). However little attention was paid to pollen, pollination mechanisms, pollen-stigma interactions, pollen tube growth and their important role in the reproductive success of cork-oak. The first studies, mainly concerning flowering and fruiting started about 50 years ago (Natividade 1950, Corti 1955) little more has been done ever since. It is thus not surprising, that the major source of information about the sexual reproduction in the genus *Quercus* come from studies on other species. The first detailed study presented the ontogeny of the staminate and pistillate flowers, detailing sporogenesis, gametogenesis, and early embryonic development in *Quercus velutina* (Conrad 1900). Langdon (1939) described the formation of the flower and dome of the various groups comprising the Amentiferae, including red oak (*Quercus rubra*). Embryo sac development in *Quercus robur* was described as being monosporic by Hjelmqvist (1953). This author also presented evidence that several megaspore mother cells may occur within one nucellus, revealing the presence of a *caecum* as a chalazal extension of the embryo sac. Turkel et al.(1955) reported the ontogeny of the staminate and pistillate flowers of *Quercus alba* and followed the development of the pistillate flowers through the completion of the mature embryo sac. Corti (1959) described the reproductive structures in *Quercus ilex*, the development of the embryo sac and few phases of the subsequent embryo development. Stairs (1964) presented the floral structure, phenology, microsporogenesis in *Quercus alba* and *Quercus ilicifolia* and described embryogenesis in *Quercus coccinea*. Mogensen (1965) studied the very early stages of embryo to the mature acorns in *Quercus alba* and *Quercus velutina*. Brown and Mogensen (1972) described important aspects on the structure and development of the ovule and young embryo in *Quercus gambelii*. Based on ultrastructure and histochemical studies, Mogensen (1973) suggested a translocation route of metabolites to the egg and developing embryo. In a later report (Mogensen 1975) the occurrence and abundance of four different types of abortive ovules in *Quercus* was also described. Merkle et al. (1980) provided a record of the development of the floral primordia of *Quercus alba*, especially of the pistillate structures.

While these earlier studies established the grounds for much of what is known in terms of structure, some recent reports began to explore the new research techniques available for the comprehension of the reproductive process. Because successful pollination and fertilisation are critical for seed production and fruit development, information on pollen tube growth can be useful for interpreting the variation in pistillate flower survival and acorn production. Cecich (1997) recently described the way of the growing pollen tube from the time pollen landed on the stigmas to the occurrence of fertilisation in three species of *Quercus* (*Q. alba, Q. velutina, Q. rubra*). Recently Yacine and Bouras (1997) investigated the phenotypic responses of the self-incompatibility system, and were to demonstrate that in self-pollination the pollen tube growth was slower and ovules stopped the development, which decreased seed production. If self pollen reaches the stigma first or simultaneously this will result inflowers abortion.

26.2.1 Flowering Biology

Flowering structure in *Quercus suber* has been reported in various studies (Corti 1955, Natividade 1950). Phenological phases of the male and female flowers were reported as an essential step for future research, especially aiming at controlled pollination (Varela and Valdiviesso 1996). *Quercus suber* is a monoecious specie, with two distinct types of flowers. However the occurrence of hermaphroditic flowers is known to occur, perhaps as a result of unusual environmental conditions, as in gambel oak (Tucker et al. 1980). The flowering season happens from April to the end of May but other flushes may occur during the year. Oaks seems to have the potential to produce perfect flowers several months after the normal spring period. Floral primordia differentiate very early, just a few months after spring flowering and are structurally mature before the onset of dormancy in October; Merkle et al. (1980) described a similar event in *Q. alba* (1980). This could provide the oak with an alternative opportunity for sexual reproduction under an environment where spring flowering fails. However, the impact of the delayed flowering on acorn productivity will depend on the timing of the event with respect to the length of the growing season, and with the reproductive effort favouring one or the other gender. For this reason, the cork-oak is recognised as having a sub-continuous flowering. A strong variation on gender allocation for flowering may be seen in relation to the period of flowering, between populations and between trees of the same stand. For instance, in 1997, a strong flush of pistillate flowers was observed in July, with simultaneous occurrence of hermaphroditic flowers in the same axis of the female spike, but no male flowers developed. In October one more flush of flowering occurred with less intensity, but only male catkins were present.

26.2.1.1
Pistillate Flowers

The pistillate flower appears in spikes with three to five flowers in the axils of the leaves produced in the current year. The earlier stages of differentiation are difficult to recognise since they are very similar to the vegetative buds. The female flower is partially enclosed by a dome of imbricate scales. Usually three styles emerge from the dome and become yellow when receptive (Fig. 26.1). The stigmas are of the "dry" type, without papillate cells, but presenting receptive cells in the surface concentrated in distinct ridges (Heslop-Harrison and Shivanna 1977). The receptive period may last 5–6 days, after which the receptive cells become necrotic and the stigmas acquire a brownish colour.

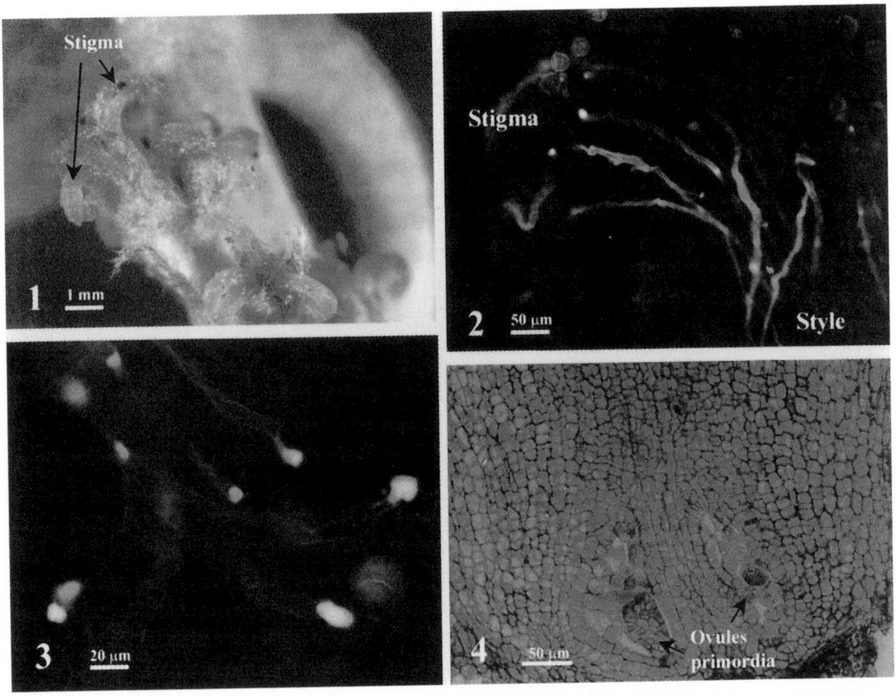

Figs. 26.1 to 26.4. Some aspects of the sexual reproduction of the cork-oak (*Quercus suber*). **Fig. 26.1.** General aspect of the receptive stigma. **Fig. 26.2.** Pollen germination and penetration in the style as visualised by fluorescence microscopy after aniline blue staining. Grains hydrate and germinate in the dry stigma surface and progress into the solid style through intercellular spaces, regularly forming callose plugs. **Fig. 26.3.** Some aspects of growth in self-pollination. Typical figures of incompatibility are formed, with branched tubes and callose plugs at the tip. **Fig. 26.4.** Ovary section after pollination. While pollen tubes already traversed the style, they are arrested at the entrance of the ovary since ovules are still in the first steps of their development. Only more than one month later will pollen tubes progress into the ovary and fertilise the ovules

The ovary is inferior consisting of three carpels with two loci. From the six ovules, however, only one will develop into seed and the fruit is monospermic (see Fig. 26.5).

26.2.1.2 *Staminate Flowers*

Male flowers are grouped in catkins presenting a perianth with 4-6 tepals with an equal or double number of anthers. The catkins develop in the base of the shoots of the current year or in the shoots of the previous year. At the time of anthesis the binuclear pollen grains disperse by wind currents (anemophilous pollination). There is a lag period between the anthesis of the male flowers and the receptivity of female flowers in the same tree: the dehiscent anthers shed the pollen generally one week before the female spikes start to appear. This phenomenon may also play a role in the self-incompatibility known in *Quercus* (Hagman 1975).

26.2.2 Pollen Viability and Germination

The importance of pollen viability in breeding projects and studies on pollination usually involve routine determinations of *in vitro* pollen germination. Controlled pollination may be used to produce seeds of improved or selected strains in reforestation programs and may be a way of increasing seed or fruit production. In most plants pollen quality is measured by the germination *in vitro*. The maintenance of pollen germination capacity depends on the conditions of storage. A protocol of pollen storage was developed to assure the primary needs for application on controlled pollination in the study of the reproductive mechanism on cork-oak. Collections of pollen were made from dehiscing anthers and a sample of each batch was used to determine the viability by the fluorochromatic reaction (Heslop-Harrison and Heslop-Harrison 1970). Cryopreservation was achieved by freezing pollen in liquid nitrogen and subsequent storage at $-180°C$. This method offered a very successful way of preserving the quality of pollen for several months. The germination capacity was measured by *in vitro* tests of pollen during storage. A germination medium was optimised for cork-oak, since the culture mediums described for other species were not effective regarding germination rates and pollen tube growth. Rates of pollen tube growth similar to *in vivo* were obtained (approximately 0,2 $\mu m \cdot min^{-1}$).

Anemophilous species produce pollen in great abundance, which permits large collections. Long-term preservation of the collected pollen was effective. Pollen cryopreservation is an area of considerable importance for the conservation of genetic resources since it is a viable source of germplasm (Bajaj 1987). The establishment of cork-oak pollen banks will be of great importance in future breeding projects.

26.2.3 Pollen-Stigma Interaction

Oak species have relatively small pollen grains (Solomon 1983). Cork-oak pollen grain is tricolpate in shape and in the hydrated state it measures about 32 µm. The flowering season starts in April and extends until the end of May. The anthers start to shed the pollen about one week before flowers are fully receptive. After landing in the stigmas, pollen adhesion, hydration and germination must occur. Since no secretion is present at the stigma surface, hydration is slow and controlled. The pollen tube emerges and penetrates the epidermis of the stigmas passing through intercellular spaces and growing along the solid transmitting tissue (Fig. 26.2). The transmitting tissue extends from the tip of the stigma to the loci of the ovary coating the ovary locus. The transmitting tissue is surrounded by vascular and cortical tissue. In controlled pollination, 24 hours after pollination pollen tubes started to penetrate the stigma surface. In the transmitting tissue pollen tubes vary in width and number of callose plugs. Callose plugs usually form in the pollen tube all along the style length. Most of the pollen grains detached from the surface of the stigma without affecting the pollen tube growth. The growth rate is slow and pollen tubes reach the zone above the stylar junction only ten to twelve days after pollination. Until the end of May the pollen tubes observed never grew under this zone. Furthermore, the transmitting tissue below this zone was not really differentiated and no ovules were present at this time (Fig. 26.4). To ensure good seed set, excess pollen relative to the number of ovules is necessary to allow for pollen tube competition (Cruden 1976). Many workers suggest that gametophytic selection may be an important component of natural selection through pollen competition (Mulcahy 1979; Spira et al. 1992). A great number of pollen tubes were observed in the transmitting tissue, but only few of them reached the base of the styles. Apparently some mechanism of pollen tubes competition may be present. Some pollen tubes arrested growth with the deposition of callose in the tip. Oaks are recognised to be self-incompatible species with gametophytic control, particularly due to morphological evidence of the flower and pollen grain (Hangman 1975, Yacine and Bouras 1997). In oaks self-incompatibility occurs with inhibition inside the style, observed by the slowing or arresting of the tubes. Sometimes callose depositions on the tips were also observed (Fig. 26.3).

From the end of May to mid June ovules are still developing and no fertilisation occurs (Fig. 26.4). This happens as the pollen tube apparently arrests or slows down growth until full maturation of the ovule. Only then do pollen tubes reach the ovary. Ovules are individualised by incomplete *septi* that do not reach the top of the cavity and allow any pollen tube to fertilise any ovule. Under the fusion of the styles the *copitum* permits the convergence of the pollen tubes before going into the locules. One or two pollen tubes were observed entering each micropyle, but not all ovules were fertilised. The ovary *locus* is covered by secretory cells and hairs.

26.2.4
Fertilisation

At the time of pollination the ovary is rarely mature and ovules are just beginning to develop. By the end of April, tissues start to differentiate, with the consequent formation of the ovary cavity internally coated with hairs. There, a mass of sporogeneous cells emerges from the base of the cavity and gives rise to the placenta and the future ovule. The integuments started to differentiate in May, but still no pollen tubes were observed inside the ovary by this time. From mid June to early July ovule's development is completed and fertilisation may then occur. Pollen tubes were observed entering the micropyle in early July. Generally only one ovule proceeds development, presumably the first one to be fertilised. The other ovules will degenerate, giving rise to a monospermic seed that will be mature by the end of November. Mogensen (1975) described this phenomena in *Quercus gambelii* but this author assumed that the abortive ovules may also have been fertilised. Based on this assumption, a system of four different types of abortive ovules was described: ovule abortion due (1) to zygote or embryo failure, (2) to the absence of an embryo sac, (3) to an empty embryo sac or (4) to the lack of fertilisation. In this view all fertilised ovules had the potential to develop normal seeds, but the first one to be fertilised would suppress the growth of the others by mechanisms still unknown. The fact remains, though, that the studies reported so far do not elucidate clearly whether it is just a temporal matter of fertilisation, or whether some more sophisticated mechanism of competition may be present.

The first fertilised ovule enlarges rapidly and the embryo starts to differentiate near the micropyle. By this time the entire ovary is coated inside by nuclear endosperm. From mid July to the end of November, there is a rapid increase in volume of the fruit and abscission starts in October-November.

26.2.5 Fruit Maturation

The genus *Quercus* comprises two subgenera, which differ in the time required for fruit maturation. The *Lepidobalanus* group (white oaks) requires one season for seed maturation, while *Erythrobalanus* (red or black oaks) requires two seasons. In the second group, pollen germinates on the stigma surface in spring, but fertilisation and seed maturation are delayed until the second growing season after pollination. *Quercus suber* is included in the *Lepidobalanus* subgenera, but several reports verify two reproductive strategies, one annual and one biennial cycle on cork-oak (Corti 1955, Natividade 1950, Elena-Rossello et al. 1993).

However, there is still some discussion about the provenience of the fruits, which has not been clarified until now. Corti (1955) states in his work that the different types of fruits are the result of the same flowering period but are products of fertilisation in different seasons. On the other hand (Natividade 1950) three different types of fruits, with physiological differences, could be the result of different flowering seasons. Studies are being made to clarify this situation. Corti's hypothesis would depend upon the arrest of the active pollen tube growth somewhere inside the carpel until the next season, when fertilisation would occur. Natividade's hypothesis would depend on atypical flowering seasons that would produce fruits with different properties. However, as mentioned above, during this atypical flowering, only one flower gender is generally formed. Therefore, if no male flowers are formed and seeds are set, some sort of apomyxis must occur. The clear demonstration of apomictic fruits remains, however, to be proved.

26.3 The Chestnut (*Castanea sativa* Mill.)

Chestnuts (*Castanea sativa* Mill) have a long history of cultivation in all Mediterranean countries. It has represented the principal food and income source of local populations for centuries. Chestnut culture still retains significant social and economic importance namely in Central and North Portugal, due not only to its tradition and wide dispersion but also to the superior quality of its fruit, well recognised on the European market.

The new market technological imperatives and diversification of chestnut fruit use have pointed out the urgency of the initiation of actions aiming at fruit production with good quality and calibre in order to increase the production value of the existing orchards and to program the implantation of new ones. Since the major economical output of chestnut is fruit production, this objective implies a better knowledge of the chestnut sexual reproduction process concerning the aspects of floral morphology and

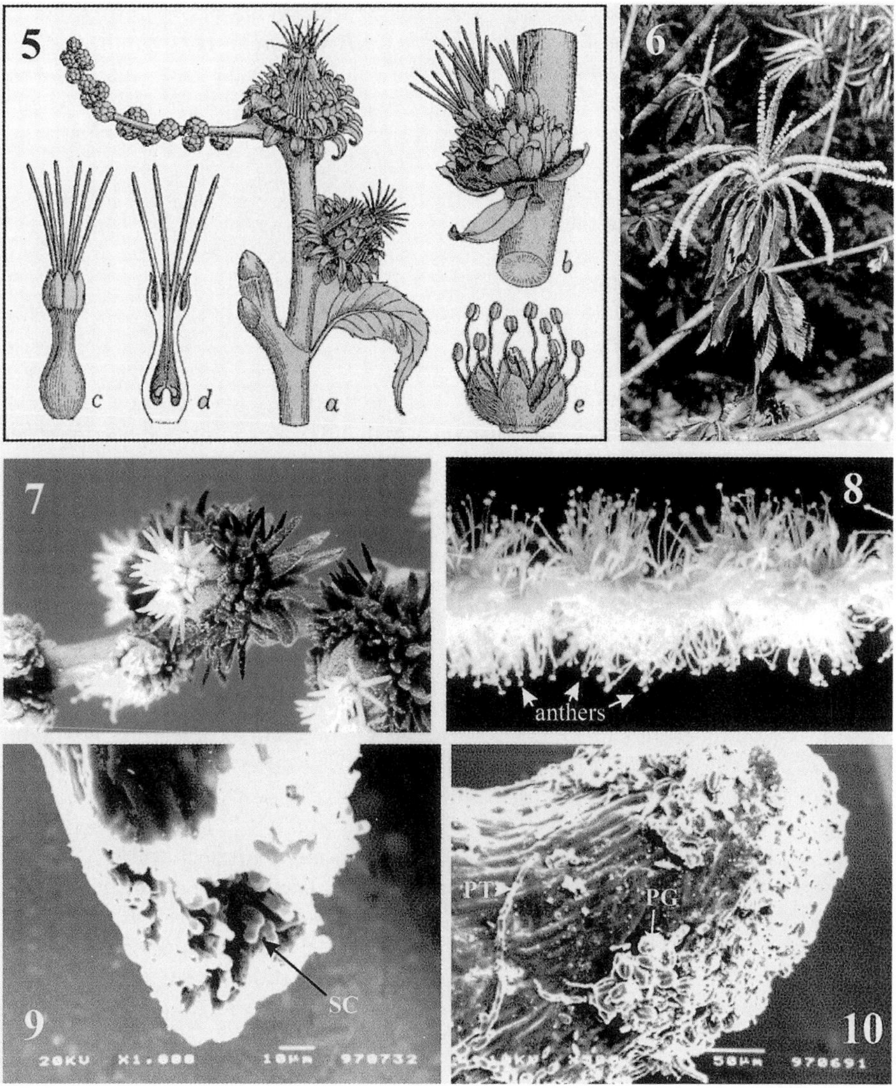

Figs. 26.5 to 26.10. Some aspects of sexual reproduction in chestnut (*Castanea sativa*). **Fig. 26.5.** Anatomical description of the different flowers and flower parts: **a** bisexual catkin with two female inflorescences; **b** female inflorescence; **c** female flower; the five spike-like structures that extend from the ovary are the styles, but were erroneously considered as the stigma surface (see text for details § 3.2); **d** longitudinal section of female flower; **e** male flower (adapted from Piccioli, 1922). **Fig. 26.6.** Male inflorescence of longistaminate variety. **Fig. 26.7.** Female flower with typical spike like styles; the stigma is located at its extreme and opens only when receptive. **Fig. 26.8.** Electron scanning micrograph showing receptive stigma after exudate removal; the secretory cells layering the inner surface of the style are clearly visible. **Fig. 26.8.** Aspect of a receptive stigma, with conspicuous secretion; many pollen grains have adhered and some germinated; a great number of fungi hyphae are also usually seen growing in the lipid rich secretion

fertilisation. Remarkably, despite the economical importance of chestnut cultivation in all of Southern Mediterranean Europe, practically nothing is known about the process of sexual reproduction and the published reports, besides being scarce, are based on very ancient studies, carried out when many of the currently standard analysis methods were not available.

26.3.1 Flower Morphology and Function

Chestnuts are monoecious and flower development occurs on the current year's growth. Two types of inflorescence are found: the unisexual staminate catkins, which are located on the lower parts of the shoot and the bisexual catkins towards the terminal end of the shoots (see Fig. 26.5). Staminate flowers are spirally arranged along the axis of the catkin in clusters of three to seven (Fig. 26.6). Pistillate inflorescences appear alone or in clusters of two or three at the base of bisexual catkins (Fig. 72.6.).

26.3.1.1 *Male Flowers*

Staminate flowers are spirally arranged along the axis of an inflorescence known as catkin (Fig. 26.8). The shoots with catkins flush in buds on the apex of the year shoot (Shad et al. 1952). Catkins are yellow to green during vegetative development, yellow at the beginning of flowering and brown by the end of this period. Catkin length and bearing differ within and among species and cultivars. In C. sativa cultivars catkins measure between 10 to 15 cm with straight bearing and 25 to 40 cm with following bearing in most catkins of *C. sativa*, *C. mollissima* and cultivars of *C. crenata*. In *C. sativa* and *C. crenata* catkin flowers are grouped in round and prominent clusters, flatter and longer in *C. mollissima* (Solignat et al. 1952). Generally each flower has four to six tepals adherents in the lower region. During flowering, flower perianth is yellow, with 8 to 12 stamens (Camus 1929, Fenaroli 1945). In some chestnuts there are no stamens, but silk white filaments are present in the central part of the flower (Solignat and Chapa 1975). This occurrence gives rise to the existence of two groups of chestnut flowers: the astaminate (flowers without stamens and pollen) and staminates (flowers with stamens) (Morettini 1949, Solignat et al. 1952, Solignat and Chapa 1975, Valdiviesso et al. 1993). Filaments of staminate young flowers are packed inside and after blooming they can present different lengths depending on the cultivar. According to the filaments length, 3 classes of staminated flowers were reported (Morettini 1949, Solignat et al. 1952, Solignat 1973, Solignat and Chapa 1975): (1) the astaminate (without stamens and pollen, trees are male sterile) and the staminate (with stamens, but with variable filament length) which includes (2) the brachistaminate (filaments with 1 to 3 mm; pollen is rare in anthers), (3) mezostaminate (filaments with 3 to 5 mm; small quantities of pollen) and (4) longistaminate (filaments with 5 to 7 mm; abundant quantities of pollen). In this classification system, astaminate and brachistaminate chestnuts are morphologically sterile. Thus, for fruit set, these three types of chestnut must be pollinated with abundant pollen from longistaminate cultivars (Solignat et al. 1952, Solignat 1973) or, with less efficiency, by mezostaminate cultivars (Soylu and Ayfer 1993).

In brachistaminate we frequently found pollen grains which had not matured. At bloom it is common to find mature pollen grains, tetrads, and pollen mother cells all in the same anther (Valdiviesso et al. 1993).

26.3.1.2
Female Flowers

Female flowers are found in bisexual catkins known as androgenic (Fig. 26.7). They appear in clusters at the base of androgenic catkins, with male inflorescences at the upper portion (Camus 1929, Coutinho 1936, Fenaroli 1945, Solignat et al. 1952). The number of inflorescences typically ranges between 1 to 5 in each catkin (Solignat, 1973) but only two or three are fertile (Solignat et al. 1952) and all the others abort (Solignat and Chapa, 1975). The inflorescence position in the shoot is a characteristic of the species. The androgenic catkins appear in an upper position than the unisexual in C. sativa (Solignat et al. 1952, Solignat 1973, Breisch 1995).

Each female inflorescence usually has 3 flowers, one terminal and two secondary lateral flowers (Camus 1929, Fenaroli 1945, Breisch 1995, Nienstaedt 1955, Bergamini 1975, Valdiviesso et al. 1993) usually inside a cluster which fissure in 2 or 4 valves (Solignat and Chapa 1975, Jaynes 1978, Coutinho 1936, Guerreiro 1948). Female flowers possess various viable ovules with ability to give rise to fruits with various seeds. In France the number of seed marks is used to define the difference between the regular "chestnut" (fruit with various seeds) and the "marron" (fruit with only one seed). In the same tree these two types of fruits are usually found (Solignat 1973). Camus (1929) was the first author to describe a central stigmatic channel, covered with a specific kind of epidermis which develops a conspicuous quantity of protection hairs. This stigmatic channel is inserted immediately bellow the fusion of a number of "spikes" which were wrongly considered as the stigmatic surface (see § 3.2.). The placenta and the internal part of the ovary also present the same type of hairs. The ovary has 5 to 7 loci with two ovules each. After fertilisation the endosperm is composed of a small number of cellular layers, which during embryo development are absorbed together with the nucellus.

26.3.1.3
Hermaphroditic Flowers

The presence of hermaphroditic flowers was first noted by Guerreiro (1948). Kientzler (1959), studying the Annecy region, detected the presence of hermaphrodite flowers in all androgenic catkins. In each flower he observed stamens fused at the base of stigmata and covered by 6 bracts. In the same flower stamens were morphologically identical to those in the male flowers. The anthers were large and without apertures in dehiscence, but some studies in young female inflorescences confirmed the presence of fertile stamens with dehiscence anthers bearing pollen. The available information on this subject is rare but Solignat et al. (1952) and Solignat (1973) briefly reported the presence of rare hermaphroditic flowers and their limited viability. We could confirm (Valdiviesso et al. 1993) the presence of anthers in female flowers in some cultivars, usually located at the base of the stigma, under the bracts. The pollen quantity of hermaphrodite flowers is variable, and in some cultivars anthers do not complete their development and pollen is not present (ex. Verdeal, CSA Ve1). In Amarelal (CSA am) and Rebordã (CSA re), the percentage of germinated pollen in hermaphroditic flowers is higher than in male flowers. We further showed that hermaphroditic flowers, when isolated, presented fruit set in some cultivars.

26.3.1.4
Pollen and Pollination

Natural colour and morphology are two characteristics useful to the recognition of pollen grains (Failla and Marro 1993). These features have been well characterised in chestnuts by several authors (Solignat et al. 1952, Breviglieri 1955, Solignat and Chapa 1975, Sedgley and Griffin 1989). Solignat and Chapa (1975) assayed the ability of pollen grains germination in a sucrose medium (10 to 15%) and could show that the viability rates are variable depending on the cultivars. Furthermore, these authors observed that at the time of natural pollination, atmospheric mean temperatures were of only 16 to 18°C. In these conditions pollen germination is low or non-existent in all *in vitro* assay so far described.

To confirm this finding we developed *in vitro* germination assays. Pollen was extracted by dropping the male catkin on a petri dish and scraping it into a vial with a razor blade. This crude method has limitations since chestnut pollen is difficult to extract because it cakes easily and therefore does not fall freely from the anthers. Furthermore, male catkins produce large amounts of nectar. This tendency to cake, together with the presence of nectar in the extracted pollen, was undoubtedly one reason for the frequent occurrence of fungal contamination. Best results after 24 h were obtained with a modified F-medium (Feijó, 1994) at pH 7.0 and 20% of sucrose. While these conditions are still far from ideal we could obtain the longest pollen tubes so far described (about 5 times the diameter of the pollen grain). Hydrated pollen grains measure between 15 and 20 µm and the diameter of pollen tubes is about 3 and 3,5 µm. These values, as well as the very slow growth rates measured, are compatible with the *in vivo* rates.

We could confirm that both pollen viability and storage time are relatively short. If collected at room temperature the viability decreases by 50% in four days and 90% if it's done with some humidity. However partial drying for 24 h after collection, followed by storage at 4°C at low humidity, allowed effective pollination one month after collection.

26.3.2
Receptivity and Fertilisation in the Pistillate Flowers

The receptivity of female flowers was investigated in six different cultivars. This issue was brought about by the necessity of determining beyond any doubt the correct functional anatomy of the pistil. In the former literature there was remarkable confusion over the description and location of the stigma, and the actual landing and germinating surface for pollen was attributed to different parts of the flower. Namely Camus (1929) considered the stigma to be the whole surface of the spike-like hard structures that are inserted on the top of the ovary. Pollen, after landing and germinating on its surface, would then penetrate directly the small stylar channel and the ovary.

To resolve this open issue we collected female inflorescences in different stages of development of the pistil and followed them in electron scanning and fluorescence microscopy. The aim of this approach was to establish the precise site of pollen tube penetration and the period of full receptivity. We could determine that only the terminal zone of the style corresponds to the stigmatic region (Figs. 26.9 and 26.10). At full receptivity it opens and this aperture is layered with secretory cells (Fig. 26.9). This secretion is important in the process of pollen adhesion and germination (Fig. 26.10). Histochemical characterisation of this secretion revealed the presence of some proteins and polysaccharides, but mostly a conspicuous presence of lipids, which distinctly covered all the stigma surface and layered the inner channel of the style along the tube growth path.

At the beginning of June most cultivars had receptive stigmata. During this period great quantities of pollen grains adherent and hydrated were found at the stigma surface, some developing distinct tubes (Fig. 26.10). The receptivity period is restricted to a few days, and is synchronised within the flower: on one hand the style enlarges to form the inner channel only upon opening and secretion; secondly only one style at a time becomes receptive. If it is successfully pollinated, then the other styles of that flower will arrest the final development. If not, then the next becomes receptive and so on.

Fluorescence microscopy after aniline blue staining revealed the progress of pollen tubes inside the style and the formation of clearly distinguishable callose plugs. They reach the ovule after 10 to 20 days depending on the pollen quality. During this period, the final steps of ovule development occur.

Careful examination of the ovary and the ovary tract revealed that only a few pollen tubes grow up to this point. Furthermore, only one ovule seems to be fertilised in the majority of cases. However, as in the oaks (see discussion above § 26.2.4) it was not possible to discriminate whether this is due to a single fertilisation or to the dominance of the first fertilised ovule.

26.4 Self-Incompatibility in the Rosaceae

Nearly all deciduous fruit species need fertilisation for fruit set (Stösser et al. 1996). Because of mechanisms of self-incompatibility, this process implies, in many cases, the transfer of compatible pollen from a polleniser tree to a fruit producing tree. In the Rosaceae family, to which most of the commercial fruit trees belong, the available data suggests a system of self-incompatibility very similar to that of Solanaceae, although both systems may have evolved independently. Such may be the case for almond and apple.

26.4.1 The Self-Incompatibility System in Apple: What Does it Look Like?

Apples are eaten worldwide. They are produced on small or large trees, which are vegetatively propagated. There are not as many different varieties which have largely recognised commercial attractiveness. e.g. there are but a few varieties with a storage ability allowing delivery to the market almost all the yearround and this feature is very important nowadays. Orchards in Western Europe, therefore, have mostly only one main variety grown and this variety produces the quality fruits that provide the growers' income and the consumers' satisfaction. However, apple trees, like many other plant species, are self-incompatible and, therefore, polliniser trees are grown between the main variety trees, hence ensuring successful cross-fertilisation. As breeding programmes generally do not select for good pollination skills (large amounts of viable pollen produced, long period of pollen shed, bee-attractiveness, cross-compatibility with other varieties, etc.), most commercial varieties are inappropriate as pollenisers. As a result, the varieties ideal for cross-pollination have often much less commercial value.

How does the self-incompatibility system in apple work? Why are the self-pollen tubes inhibited as they grow through the style, while the cross-pollen tubes are unaffected? How does the pistil discriminate between self- and cross-pollen, what is the mechanism underlying the growth arrest, and how are recognition and inhibition linked with each other? Evolution allocated a gametophytic SI system to the Rosaceae family to which apple belongs, but is it based on the RNase model, discovered in Solanaceous species (McClure et al. 1990), or on the *Papaver rhoeas* model (Foote et al. 1994), or on the system that is operating in the grasses (Li et al. 1995)? The Rosaceae family is unrelated to either of the plant families mentioned and the elucidation of the SI mechanism in this family could also imply a better understanding of the origin of SI systems in plants. These kinds of questions and thoughts were the ones that we asked ourselves at the beginning of the current decade. We decided to attempt to answer them using a molecular-genetics approach.

From numerous early studies on gametophytic SI we know that the system is regulated by products of a single polymorphic S-locus. One of the genes that reside at this locus is the S-gene, which encodes the pistil discriminant of the SI reaction. In species like tobacco and petunia, the encoded proteins have ribonuclease activity and are called the S-RNases (Lee et al. 1994). Each S-RNase, being the gene product of an allele of the S-gene, is directed against pollen bearing the corresponding S-specificity. In apple, we found out that similar S-RNases are present in the pistil (Broothaerts et al. 1991) and

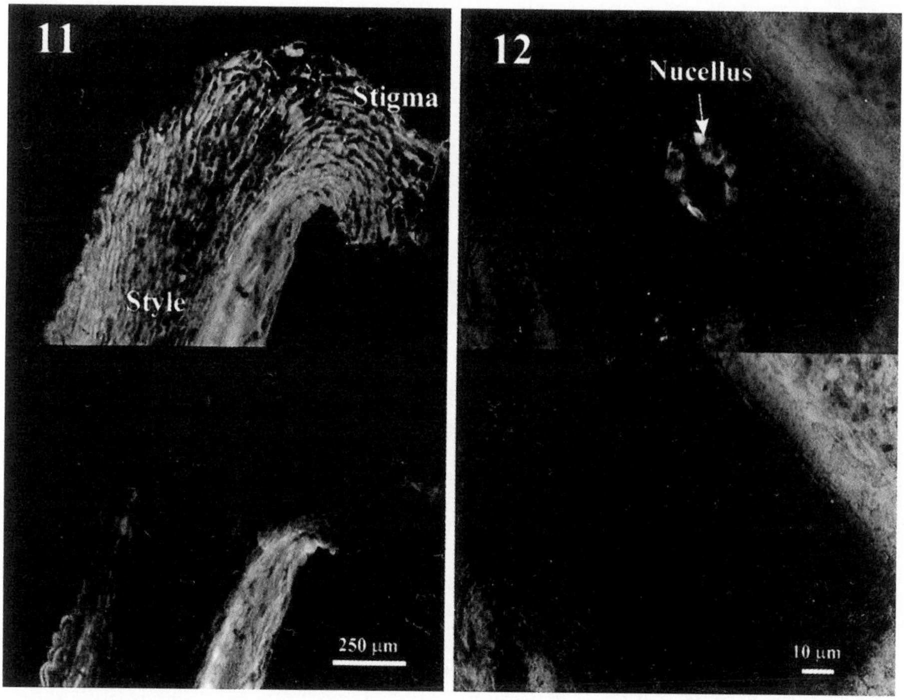

Figs. 26.11 and 26.12. Double excitation confocal images of apple style **(Fig. 26.11)** and ovule cross-section **(Fig. 26.12)** after immuno-labelling with FITC. The blue excited channel (top half) shows the labelling and some auto-fluorescence. The red excited channel (bottom half) shows only the auto-fluorescence. **Fig. 26.11.** Style of a "Golden" cultivar labelled with an antibody raised against a specific peptide encoded by the S3-allele. Labelling of the cell walls is evident at the stigma and in the transmitting tissue along the style. Auto-fluorescence is confined to the cortical vascular bundles and epidermis. **Fig. 26.12.** Cross section of a "Golden" ovule imaged at the nucellus level after labelling with the same S3 antibody. A ring of cells defining the nucellus is clearly labelled. Furthermore, the labelling is intracellular, in contrast with the labelling of the transmitting tissue cells

this was later confirmed for several rosaceous fruit tree species (Sassa et al. 1992, 1994, Boscovic et al 1997, Tao et al. 1997). Complementary DNA for the corresponding S-alleles in apple have been cloned since and at the moment the coding sequences for 10 different S-alleles are known (Broothaerts et al. 1995, Janssens et al. 1995, Sassa et al. 1996, Verdoodt et al. in press). Amazingly, only 25% of amino acid residues are shared between S-RNases in the Solanaceae and the Rosaceae. Nevertheless, all of them have the conserved regions that are characteristic for the T2/S class of RNases and, additionally, a few other residues, occurring dispersed in the sequence, are perfectly conserved between all known S-RNases (Richman et al. 1997).

At this point we know that rosaceous species express S-RNases in their pistils. From Northern blot analysis, we also know that the S-transcripts are only found in the pistil, and not in leaves, pollen or other flower parts (Broothaerts et al. 1995). It is, therefore, very likely that the cytotoxic model for the SI reaction, developed for Solanaceous species, also holds true for the Rosaceae. In this model, the S-RNases degrade the RNA inside incompatible pollen tubes and arrest their further growth. At which level in the pistil does the recognition between S-RNase and incompatible pollen occur? To study the precise distribution of the S-RNases in the apple pistil, we took an immuno-histochemistry approach. One of the antibodies that were used in these studies was raised against a synthetic peptide that is specific for the S3-RNase, while the other antibody is directed towards a region that is conserved in all apple S-RNases, but absent from S-RNases in other plant families (the antibodies were isolated from chicken eggs after immunisation of chicken and this was done by H. Kokko, University of Kuopio, Finland). Fresh pistils were collected from the varieties "Elstar" and "Golden Delicious", bearing the S3-allele, and as a negative control for S3-specificity of the reaction, "Braeburn" and "Prima" were used which don't contain the S3-allele. Cryosections were made through the apple pistil, and incubated with one of the primary antibodies, followed by the secondary goat anti-chicken antibody. In the initial experiments we used rhodamine as the labelling component of the secondary antibody. However, we found out that it was very difficult to distinguish between the auto-fluorescence of the tissue and the fluorescence of the rhodamine-label. Therefore, we replaced the secondary antibody with a FITC-labelled goat anti-chicken antibody (Sigma), which yielded much better results. In both, when viewed under fluorescence microscopy and by CSLM, the clear labelling of the transmitting tissue of the stigma and style was apparent. No specific fluorescence in the green wavelength, typical for FITC, was seen in the cortex tissue, although the yellowish auto-fluorescence of this tissue was high. The intensity of the signal seen in the transmitting tissue was the same at the stigma level as it was in the upper and lower style. When viewed at a higher magnification, it was clear that the labelling was mainly extracellular and both in the cell wall and in the intercellular spaces. This was particularly evident from transverse sections of the style. The signal observed with the S3-specific antibody applied to cv. "Elstar" and "Golden Delicious" styles was completely absent in sections from cv. "Braeburn" and "Prima" styles, providing strong evidence that the antibody is binding to the S3-RNase-specific epitope to which it was raised against. When applied to sections through the ovary, a ring of cells, putatively corresponding to the nucellus, also showed fluorescence. We are currently studying these ovary signals further, as they may signify specific additional SI barriers in the apple ovary. Moreover, what is very intriguing about the highly labelled cells of the nucellus, is that they have their labelling intracellularly only. Why is the protein not secreted to the exterior of the cells? And further, being an RNase, where is it stored in the cell in order to ensure the preservation of the protein synthesis apparatus? What is the function of this protein in these cells? Using the S3 specific antibody, intracellular labelling was also found in the nucellus of the cv. "Golden Delicious" (which has the S3 allele) and in "Braeburn" and "Prima" (which do not have the S3 allele). When using the antibody for the conserved region (present in all apple S-alleles) we could still find labelling on the nucellus cells. These results mean that we' are probably detecting some component which is different from the S3-RNase, but containing the same epitopes, allowing binding of the antibodies. Could this unknown protein be an S-like RNase without implications on the SI system? From other studies we already know that S-like RNases present great homology with S-RNases, and further experiments are now being carried out in order to elucidate these unanswered questions.

What we know at this point is that S-RNases an present along the path of pollen tube growth through the style, exactly as we expected. The proteins are constitutively present in the transmitting tissue, and as far as we know from other species, their expression is in no way affected by pollination. What then happens upon pollination? Do S-RNases bind to the pollen tube wall? Are they internalised? Studies in tobacco have provided evidence for the presence of S-RNases in the pollen tube cytoplasm (Gray et al. 1991). But what is the difference in reaction between compatible and incompatible pollen tubes? What is the nature of the pollen component to which the S-RNase specifically binds, hence providing the basis for the recognition of incompatible pollen? Considering the large diversity in S-RNase primary sequences between Solanaceae and Rosaceae, it is obvious that the same level of diversity will characterise the pollen S-locus product in both plant families. Or is this low level of similarity indicative of a completely different mechanism of pollen recognition, hence different proteins being involved? It is these kind of questions that need to be addressed in future studies of self-incompatibility in apple.

26.4.2
The Self-Incompatibility in Almond

The substitution of seed-derived almond orchards for orchards of clonal origin has revealed problems related to pollination and fruit production. These problems were carefully studied, in the forties, by Almeida (1945) who performed controlled pollination in several thousands of flowers of 21 almond varieties. The results of these experiments revealed that 10 of these varieties were 100% self-incompatible, while 9 showed 99% of SI (varying between 97.8% and 99.6%) and two were partially self-compatible ("José Dias" and "Duro Italiano") with 34% of fruit production in self-pollination experiments (Almeida 1945). The self-incompatibility system of almond is not well known and almost all the studies developed until now were made at the crossing/selection level (Almeida 1945, Grasselly 1985, Jraidi and Nefzi 1988, Socías i Company, 1989, Kester et al. 1994).

SI has a large impact on the profitability of an orchard where, to ensure fruit set, it is generally recommended to have 1 line of pollenizer trees between each 3 lines of producers. However, the use in the field of 2 or more varieties has not proved to be enough to overcome pollination problems due to partial or defective pollination. For successful fruit formation it is necessary to use cross-compatible varieties, with simultaneous flowering periods, and to have efficient pollen transport between trees. Because wind pollination is not sufficient, the use of beehives is recommended, but the temperatures during almond flowering period may be too low for bees. The use of late flowering cultivars is a strategy to escape low temperatures at bloom. Contrary to what happens with other fruit species, partial pollination in almond is not counterbalanced by an increase in weight and size of the fruits on the tree. Moreover, in Prunoideae, fruit formation through parthenocarpy or estenospermocarpy does not occur as it does in Pomoideae (i.e. the case of pear, Stösser et al. 1996). As a consequence, the total and efficient pollination of the flowers in an almond tree is a goal to achieve because it is related to the number of kernels (Duval and Grasselly 1994). In the two last congresses of GREMPA ("Groupe de Recherches et d'Etudes Méditerranéenes pour le Pistachier et l'Amandier"), almond self-compatibility has been widely dealt with. Self-compatibility (i.e.: self-fertility – SF) limits the problems of pollination, making it more regular, avoiding the problem of flower synchronisation during pollination and, moreover, permitting only one date of harvest with the standardisation of the associated agricultural and industrial processes. Since reduced pollination does not increase the size of the fruits on a tree, it is not expected that overcropping may reduce fruit size. However, for "Marcona", a Spanish SI cultivar, it was found that occasionally, after the 5th year of production, strong fertility may result in alternation of production (Grasselly and Crossa-Raynaud 1980).

Breeding self-compatible cultivars has thus become a priority in most almond breeding programmes. In fact, since the work of Almeida (1945) other self-fertile varieties were found by various researchers, particularly in Italy in the region of Apulia (i.e.: "Tuono", "Filippo Ceo" and "Genco", among others) (Jaouani 1973, Grasselly and Olivier 1976, Godini 1977). Moreover, since the 1940s, in California, the incorporation of self-fertility (*SfSf*) has been attempted in almond, mostly through crosses with the self-compatible peach. Selection of self-compatibility was difficult but, in some cases, SF cultivars could be obtained (Kester 1970, Socías i Company et al. 1976, Socías and Felipe 1988). Some of these cultivars, however, were either not highly self-compatible or the tree was difficult to harvest (Kester 1994). The same hybridisation and selection strategy was conducted by numerous authors searching for high quality self-fertile varieties (Dicenta and Garcia 1993, Garcia et al. 1994, Grasselly and Crossa-Raynaud 1980, Jraidi and Nefzi 1988, Grassely 1994, Kester 1994, Kester et al. 1994). From 1970 to 1984, SF almond genitors, "Tuono" mainly, have been used in breeding programs to transfer the *Sf* character. In Europe, through a GREMPA network, some self-compatible almond cultivars could be studied. These cultivars presented very low productivity, intermediate productivity, reduced vigour or increased percentage of double kernels (which is a major drawback for the use of whole almonds). For the best SF cultivar selected in France, the simultaneous growth in the orchard of inter-compatible cultivars has been recommended to ensure high pollination (Duval and Grasselly 1994). Several studies have shown that self-fertility is a dominant allele of the S series (Socías i Company and Felipe 1977, Socías and Felipe 1988, Socías 1990, Dicenta 1991, Dicenta and Garcia 1993, Grasselly 1994). In the majority of crosses with a SF genitor, 50% of progenies are SF, which provides an effective method to obtain new SF cultivars. Additionally, the study of the molecular mechanism of SI can also give rise to new SF cultivars through the introduction of antisense copies of the S-alleles by genetic transformation, as already done in Petunia (Lee et al. 1994) and apple (unpublished) with the advantage of maintaining the good qualities of the selected cultivars.

In recent investigations, we verified that proteins with similar dimensions to S-proteins could be detected in styles of cherry and almond using an antibody raised against one of the conserved active site domains of S-proteins (Broothaerts et al., unpublished results). More recently, four different stylar S-RNases from almond SI cultivars were identified and characterised (Tao et al. 1997). Moreover, an almond pistil cDNA encoding a putative S-allele was identified, cloned, sequenced and shown to be an S-like

RNase gene (van Nerum et al. in preparation). S-like genes are known to exist in Solanaceae (Lee et al. 1992) and Rosaceae (Broothaerts, unpublished results) and, although they share great sequence similarity with the S-alleles, they are not related to the SI system.

26.5 Conclusions

While some of the projects presented are at various degrees of development, a number of interesting questions could be raised from the study of the sexual reproduction in these species. What determines pollen tube competition? What controls pollen tube arrest, and synchronisation of ovule development? What determines ovule domination or competition? Are different growth rates or paths of progression in the style reflected in the cellular and molecular characteristics of the pollen tubes? Are they signalised towards the mycropyle using the same spatial clues? What is the molecular basis of the incompatibility which seems to be present at different levels in these species? These and other questions may not only help to unravel new cellular and molecular mechanisms, but, even in the described state of knowledge, provide answers and suggest improvements to breeders. While the difficulties underlying this experimental approach may not be compared with those of model species we hope that multi-disciplinary studies and the comparison of different models will be of added value and point out issues that cannot be envisaged on a more reductionist approach around a limited number of model species.

Acknowledgements
Parts of this paper were financed by projects Praxis/3/3.2/Hort/2136/95 and Praxis/3/3.2/Flor/2100/95. AC Certal is granted by PRAXIS 4/4.1/BIC/3474, and LC Boavida by PRAXIS 4/4.1/BIC/3476. We thank Carolina Varela for field coordination on the cork-oak project and E. Leopoldo Ferreira for field coordination on the almond project.

References

Almeida CRM (1945) Acerca da improdutividade na amendoeira. Anais Instituto Superior Agronomia 15:1–184
Bacilieri R, Roussel G, Ducousso A (1993) Hybridization and mating system in a mixed stand of sessile and pedunculate oak. Ann Sci For 50 (Suppl 1):122s–127s
Bajaj YPS (1987) Cryopreservation of pollen and pollen embryos, and the establishment of pollen banks. Int Rev Cyt 107:421–435
Bergamini A (1975) Osservazioni sulla morfologia fiorale di alcune cultivar di Castagno. Riv Ortoflorofrut It 59:103–108
Boscovic R, Russell K, Tobutt KR (1997) Inheritance of stylar ribonucleases in cherry progenies, and reassignment of incompatibility allelels to two incompatibility groups. Euphytica 95:221–228
Breisch H (1995) Châtaignes et Marrons: Monographie. Ctifl. Paris
Bretaudeau J (1964) Atlas d'Arboriculture Fruitière. JB Baillière et fils. Paris, pp 182–197
Breviglieri N (1955) Ricerche sulla disseminazione e sulla germinazione del polline nel Castagno. La Ricerca Scientifica (suppl) 2:5–25
Broothaerts W, De Bondt A, Cammue BPA, Broekaert WF (1991) Genetic evidence for the overall involvement of ribonucleases in gametophytic self-incompatibility. Abstracts 3rd Int Congr Plant Mol Biol, Tucson, USA, p 544
Broothaerts W, Janssens GA, Proost P, Broekaert WF (1995) cDNA cloning and molecular analysis of two self-incompatibility alleles from apple. Plant Mol Biol 27:499–511
Brown RC, Mogensen HL (1972) Late ovule and early embryo development in Quercus gambelli. Am J Bot 59:311–316

Camus A (1929) Les Chataigniers: Monographie des Genres *Castanea* e *Castanopsis*. Encyclopédie Économique de Sylviculture vol III. Paul Lechevalier, Paris, pp 8 a 75
Cecich RA (1997) Pollen tube growth in *Quercus*. For Sci 43:140–146
Conrad A (1900) A contribution to the life history in *Quercus* Bot Gaz 29:408–418
Corti R (1955) Richerche sul ciclo reprodutivo di specie del genere *Quercus* della flora italiana. II. Contributo alla biologia ed alla sistematica di *Quercus suber* L. in particulare delle forme a sivilupo biannale della ghianda. Ann Accad Ital Sci For 4:55–136
Corti R (1959) Richerche sul ciclo reprodutivo di specie del genere *Quercus* della flora italiana. IV. Osservazioni sulla embriologia e sul ciclo riproduttivo in *Quercus ilex* L. Ann Accad Ital Sci For 8:19–42
Coutinho AXP (1936) Esboço de uma flora lenhosa Portuguesa. Serviços Florestais Portugueses. Lisboa, pp 60–62
de Nettancourt D (1977) Incompatibility in Angiosperms. Monographs Theor Appl Gen vol 3. Springer-Verlag, Berlin, Heidelberg, NY
Cruden RW (1977) Pollen-ovule rayios: a conservative indicator of breeding systems in flowering plants. Evolution 31:32–46
Dickinson HG (1994) Simply a social disease? Nature 367:517–518
Dicenta F (1991) Mejona genetica del Almendro por intervarietales, pereneia de characteres φ selecion. PhD thesis, Univ Barcelona, p 313
Dicenta F, García JE (1993) Inheritance of self-compatibility in almond. Heredity 70:313–317
Duval H, Grasselly Ch (1994) Behaviour of some self-fertile almond selections in the South-East of France. Acta Hortic 373:69–74
Elena-Rossello JA, de Rio JM, Valdecantos JLG, Santamaria IG (1993) Ecological aspects of the floral phenology of the cork-oak (*Q. suber* L.): Why do annual and biennial biotypes appear? Ann Sci For 50 (Suppl 1):114s–121s
Failla O, Marro M (1993) Metodi e finalità nella raccolta e conservazione del polline e tecniche di impollinazione. Riv Frutticoltura 4:87–91
Farmer RE (1981)Variation in seed yeld of white oak. For Sci 27:377–380
Feijó JA (1994) Contribuição oara o estudo da Biologia da Reprodução Sexual em Ophrys lutea Cav. PhD Thesis, Univ Lisboa, Lisboa
Fenaroli L (1945) Il Castagno: Trattati di Agricoltura. Ramo Ed Agric, Roma
Feret PP, Kreh RE, Merkle SA, Oderwald RG (1982)Flower abundance, premature acorn abscission, and acorn production in Quercus alba L Bot Gaz 143:216–218
Foote HCC, Ride JP, Franklin-Tong VE, Walker EA, Lawrence MJ, Franklin FC (1994) Cloning and expression of a distinctive class of self-incompatibility (S) gene from *Papaver rhoeas* L. Proc Natl Acad Sci USA 91:2265–2269
Franklin FCH, Lawrence MJ, Franklin-Tong VE (1995) Cell and Molecular Biology of self-incompatibility in flowering plants. Int Rev Cytol 158:1–64
Franklin WM (1959) Staining and observing pollen tubes in the style by means of fluorescence. Stain Tech 34:125–127
Freeman DCE, McArthur ED, Harper KT Blauer AC (1981) Influence of environment on the floral sex ratio of monoecious plants. Evoution 35:194–197
Garcia JE, Egea L, Berenguer T, Dicenta F, Egea J (1994) News about the almond breeding program in CEBAS-CSIC. Murcia (Spain). Acta Hort 373:65–68
Godini A (1977) Contributo alla connoscenza della cultivar di Mandorlo (*Prunus amygdalus*, Batsch.) della Puglia: 2) Un quadriennio di ricerche sull'autocompatibilità. 3° Colloques GREMPA, Bari, pp 150–159
Goldberg RB (1996) To grant or not. The Plant Cell 8:346–347
Grasselly CH (1985) Avancement du programme "auto-compatibilité chez l'amandier". Options Méditerr Serie Études 1:39–41
Grasselly CH (1994) Forty years of almond and its rootstocks breeding. Acta Hortic 373:29–33
Grasselly Ch, Olivier G (1976) Mise en évidence de quelques types autocompatibles parmi les cultivars d'amandier (*Prunus amygdalus* Batsch.) de la population des Pouilles. Ann Amél Pl 26:107–113
Grasselly G, Crossa-Raynaud P (1980) "L'amandier". Techniques Agricoles et Productions Méditerranéennes, vol 2. G.-P. Maisonneuve et Larose, Paris
Graves AH, Nienstaedt H (1953) Chestnut Breeding. 44th Ann Rep North Nut Growers Assoc, pp 136–144
Gray JE, McClure BA, Bönig I, Anderson MA, Clarke AE (1991) Action of the style product of the self-incompatibility gene of *Nicotiana alata* (S-RNase) on *in vitro* grown pollen tubes. Plant Cell 3: 271–283
Guerreiro MG (1948) Alguns estudos no género *Castanea*. Pub Serv Flor Port 15:10–15
Hagman M (1975) Incompatibility in forest trees. Proc R Soc Lond B 188:313–326
Heslop-Harrison J (1975) Incompatibility and the pollen-stigma interaction. Ann Rev Plant Physiol 26: 403–425
Heslop-Harrison J, Heslop-Harrison Y (1970) Evaluation of pollen viability by enzimatically induced fluorescence; intracellular hydrolysis of fluorescein diacetate. Stain Tech 45:115–120

Heslop-Harrison Y, Shivanna KR (1977) The receptive surface of the angiosperm stigma. Ann Bot 41:1233–1258

Hjemqvist H (1953) The embryo sac development of Q. robur L. Phytomorphology 4:377–384

Janssens GA, Goderis IG, Broekaert WF, Broothaerts W (1995) A molecular method for S-allele identification in apple based on allele-specific PCR. Theor Appl Genet 91:691–698

Jaouani A (1973) Autocompatibilité de l'amandier "Mazzetto". Rapp Activites INRAT, p 20

Jaynes RA (1975) Chestnuts. In: Janick J, Moore JN (eds) Advances in Fruit breeding. Purdue Univ Press. West Lafayette, Ind, USA, pp 490–503

Jaynes RA (1978) Selecting and Breeding Blight Resistant Chestnut Trees. In: Janick J, Moore JN (eds) Advances in Fruit Breeding. Purdue Univ Press, West Lafayette, Ind, USA, pp 350–375

Jraidi B, Nefzi A (1988) Transmission de l'autocompatibilité chez l'amandier. 7 Colloque GREMPA. Reus, June 1987, pp 47–57

Kester DE (1970) Transfer of self-fertility from peach (*Prunus persica* L.) to almond (*Prunus amygdalus* Batsch.). Western Sect Amer Soc Hort Sci Berkeley (Abstr)

Kester DE (1994) Almond cultivars and breeding programs in California. Acta Hortic 373:13–28

Kester DE, Gradziel TM, Micke WC (1994) Identifying pollen incompatibility groups in California almond cultivars. J Amer Soc Hort Sci 119:106–109

Kientzler L (1959) Cas d'Hermaphodrisme chez le Châtaignier. Bull Soc Bot France 106:211–212

Langdon LM (1939) Ontogenic and antomical studies of the flower and fruit of the Fagaceae and Jungladaceae. Bot Gaz 101:301–327

Lee H, Singh A, Kao T-H (1992) RNase X2, a pistil-specific ribonuclease from *Petunia inflata*, shares sequence similarity with solanaceous S proteins. Mol Biol 20:1131–1141

Lee H, Huang S, Kao T-H (1994) S-proteins control rejection of incompatible pollen in *Petunia inflata*. Nature 367:560–563

Li X, Nield J, Hayman D, Langridge P (1995) Thioredoxin activity in the C terminus of *Phalaris* S protein. Plant J 8:133–138

McClure BA, Gray JE, Anderson MA, Clarke AE (1990) Self-incompatibility in *Nicotiana alata* involves degradation of pollen rRNA. Nature 347:757–760

McKay JW (1972) Pollination of Chestnut by Honey Bees. 63th Ann Rep North Nut Growers Assoc, August 20–23/72, pp 83–86

Merkle SA, Feret PP, Croxdale JG, Sharik TL (1980) Development of floral primordia in white oak. For Sci 26:238–250

Mogensen HL (1965) A contribution to the anatomical development of the acorn in *Quercus* L. Iowa State Jour Sci 40:221–255

Mogensen HL (1973) Some histochemical, ultrastructural, and nutritional aspects of the ovule of *Quercus gambelli*. Amer J Bot 60:48–54

Mogensen HL (1975) Ovule abortion in *Quercus* (Fagaceae). Amer J Bot 62:160–165

Morettini A (1949) Biologia Fiorale del Castagno. Italia Agricola 12:264–274

Mulcahy DL (1974) Adaptative significance of gametic competition. In: Linskens HF (ed) Fertilization in higher plants. North-Holand Pub Co, Amsterdam–Oxford, pp 27–30

Natividade J (1950) Subericultura. DGSF, Lisboa

Nienstaedt H (1956) Receptivity of the Pistillate Flowers and Pollen Germination Tests in Genus Castanea. Z Forstgenetik Forstpflzüchtg 5:40–45

Pesson P, Louveaux L (1984) Pollinisation et productions végetales. INRA, Paris

Piccioli L (1922) Monografia del castagno. Ledoga SA, Milano, pp 5–40

Richman AD, Broothaerts W, Kohn JR (1997) Self-incompatibility RNases from three plant families: homology or convergence? Am J Bot 84:912–917

Santos ASA, Pereira JAS (1993) Geographic and economic expression of the Portuguese Chestnut fruit production. In: Antognozzi E (ed) Proc Int Cong Chestnut Univ Peruggia, Peruggia, pp 69–72

Sassa H, Hirano H, Ikehashi H (1992) Self-incompatibility-related RNases in styles of Japanese pear (*Pyrus serotina* Rehd.) Plant Cell Physiol 33:811–814

Sassa H, Mase N, Hirano H, Ikehashi H (1994) Identification of self-incompatibility-related glycoproteins in styles of apple (*Malus x domestica*) Theor Appl Genet 89:201–205

Sassa H, Nishio T, Kowyama Y, Hirano H, Koba T, Ikehashi H (1996) Self-incompatibility (*S*) alleles of the Rosaceae encode members of a distinct class of the T2/S ribonuclease superfamily. Mol Gen Genet 250:547–557

Schad C, Solignat G, Grente J, Venot P (1952) Recherches sur le Chataignier a la St. de Brive. Ann INRA 3:369–458

Sedgley M, Griffin AR (1989) Sexual Reproduction of Tree Crops. Academic Press, London

Sharp WM, Chisman HH (1961) Flowering and fruiting in the white oaks. I. Staminate flowering through pollen dispersal. Ecology 42:365–372

Sharp WM, Sprague VG (1967) Flowering and fruiting in the white oaks. Pistillate flowering, acorn development, weather, and yelds. Ecology 48:243–251

Socías i Company R (1989) Estado actual y perspectivas de la mejora genetica del almendro para la autocompatibilidad. Infecion Téc Econ Agra 85:3–22

Socías i Company R, Felipe AJ (1977) Ereditabilitá dell'autocompatibilitá nel mandorlo. 3 Colloque GREMPA. Bari, October 1977, 221–223
Socías i Company R, Kester DE, Bradley MV (1976) Effects of temperature and genotype on pollen tube growth of some self-incompatible and self-compatible almond cultivars. J Amer Soc Hort Sci 101: 490–493
Socías R (1990) Ou sommes nos dans l'amélioration génétique pour l'autocompatibilité. 8^{éme} colloque du GREMPA, 27–48
Socias R, Felipe A (1988) Self compatibility in Almond, transmission and recent advances in breeding. Acta Hort 224: 307–317
Solignat G (1973) Un Renouveau de la Chataigneraie Fruitière. – INRA-BollTech Inf Centre Rech Agr Bordeaux 280: 1–15
Solignat G, Chapa J (1975) La Biologie Florale du Châtaignier. – Invuflec, Paris, pp 5–31
Solignat G, Chapa J (1975) Principales espèces de Châtaignier. Invuflec, Paris, pp 15–35
Solignat G, Grente J, Schad C, Venot P (1952) Recherches sur le chaitaignier a la St. de brive. INRA, Paris, pp 2–89
Solomon AM (1983) Pollen morphology and plant taxonomy of white oaks in eastern North America. Amer J Bot 70: 481–494
Sork VL, Bramble JE (1993) Prediction of acorn crops in three species of North American oaks: *Quercus alba, Q. rubra and Q. velutina*. Ann Sci For 50 (Suppl 1): 128s–136s
Soylu A, Ayfer M (1993) Floral Biology and Fruit Set of some Chestnut Cultivars (*C. sativa Mill.*). In: Antognozzi E (ed) Proc Int Cong Chestnut Univ Peruggia, Peruggia, pp 125–130
Spira TP, Snow AA, Whigham DF, Leak J (1992) Flower visitation, pollen deposition, and pollen-tube competition in *Hibiscus moscheutos* (Malvaceae). Am J Bot 79: 428–433
Stairs GR (1964) Microsporogenesis and embryogenesis in *Quercus*. Bot Gaz 125: 115–121
Stösser R, Hartmann W, Anvari SF (1996) General aspects of pollination of pome and stone fruit. Acta Hort 423: 15–22
Tao R, Yamane H, Sassa H, Mori H, Gradziel TM, Dandekar AM, Sugiura A (1997) Identification of stylar RNases associated with gametophytic self-incompatibility in almond (*Prunus dulcis*) Plant Cell Physiol 38: 304–311
Tucker,JM, Neilson RP, Wullstein LH (1980) Hermaphroditic flowering in gambel oak. Amer J Bot 67: 1265–1267
Turkel, HS, Rebuck AL, Grove Jr AR (1955) Floral morphology of white oak. Trends Farm Prod Bull 24: 594–597
Valdiviesso T, Abreu CP, Medeira C (1993) Contribution for the study of the Chestnut Floral Biology. In: Antognozzi E (ed) Proc Int Cong Chestnut Univ Peruggia, Peruggia, pp 95–97
Varela MC, Ericksson MG (1995) Multipurpose gene conservation in *Quercus suber* – a Portuguese example. Silvae Genetica 44: 28–37
Varela MC, Valdiviesso T (1996) Phenological phases of *Quercus suber* L. flowering. For Gen 3: 93–102
Verdoodt L, Van Haute A, De Witte K, Keulemans J, Broothaerts W (1998) Use of the multi-allelic self-incompatibility gene in apple to assess homozygocy in shoots obtained through haploid induction. Theor Appl Genet (in press)
Wolgast LJ, Stout BB (1977) The effects of relative humidity at the time of flowering on fruit set in bear oak (*Quercus ilicifolia*). Amer J Bot 64: 159–160
Yacine A, Bouras F (1997) Self- and cross-pollination effects on pollen tube growth and seed set in holm-oak *Quercus ilex* L. (Fagaceae). Ann Sci For 54: 447–462

CHAPTER 27

Pollen as Food for Humans and Animals and as Medicine

H. F. Linskens

Dipartimento di Biologia Ambientale, Università degli Studi di Siena, Via P. A. Mattioli 4, 53100 Siena, Italy

The uses of pollen are normally considered to be limited to feed for bees. Like many other solitary and social insects, honey bees collect and store pollen. However, the ubiquity, chemical composition and nutritional value of pollen make it suitable for human consumption, especially as a food supplement.

27.1 Pollen Everywhere

A surprising amount of pollen is produced by flowering plants. Pollen grains are so small that they usually pass unobserved, except during so-called "pollen rain", when enormous quantities of pollen are shed in a short time and settle visibly on pools and other surfaces in a yellowish layer. The amount of pollen settling out varies with geographical position, site, vegetation, and time of year and has been estimated to range from 20 $mg/m^2/day$ to 128 $g/m^2/year$ (Linskens 1992).

Different species of plants have different mechanisms of shedding pollen. Dehiscence of the anther, in which pollen is produced, depends on temperature, humidity and light conditions (Linskens and Cresti 1988).

Pollen is a ubiquitous particle fraction of aeroplankton, and is transported over long distances. It can be found in the remotest parts of the earth, the Arctic (Linskens 1996 a), Antarctica (Linskens et al. 1993) and Greenland (Linskens 1995), where no anemophilous plants (wind pollinators) grow, or indeed no plants at all. There is apparently one exception: no pollen has been observed under the canopy of the tropical rainforests (Linskens 1996c). The use of moss cushions as natural pollen traps confirmed a total absence of anemophilous plants in these environments.

Only a small fraction of the pollen released reaches the female gametophyte and performs its primary task: that of fertilisation. In some species, a second part of the airborne pollen, participates in the pollination process by selectively attracting and rewarding pollinators. Most of the airborne pollen, however, is scattered and seems to be wasted.

Airborne pollen determined an influx of minerals and organic molecules in certain areas, providing the starting material for primary colonisation. It is therefore of great ecological importance, especially for the recycling of organic matter in an ecosystem (Linskens 1992).

A new problem, related to the presence of pollen everywhere, is the cultivation of genetically modified plants which produce genetically modified pollen grains. Genes inserted into crop plants are much more likely to escape into the wild than was previously thought. Environmentalists have argued that genes importing resistance to herbicides or drought could be transmitted from crops to weeds, via pollen (Coghlan 1995). The consequences are still difficult to estimate.

27.2 Chemical Composition of Pollen

The importance of pollen as food and feed is evident from its chemical composition. Although honey has been consumed by man since prehistoric times the significance and possibilities of pollen were not recognised until quite recently. During the Second World War American scientists, considered whether so-called bee-bread, the pollen gathered by bees and stored in their hives, could be a significant reserve of high grade food. It was calculated that bees collected about 80.000 t of pollen per year in the US, an amount similar to the total annual yield of honey (Vivino and Palmer 1944).

Table 27.1. Nutrients, electrolytes and vitamins in pollen and some foods. 1 oz. equals 28,5 grams. Source: Lauks Testing Laboratories Inc., Seattle 4, Washington

	Pollen mg/oz.	Beef meat mg/oz.	Wheat bread mg/oz.	Milk mg/oz.
Proteins	4.000	4.400	2.700	1.030
Lipids	400	8.800	900	1.160
Carbohydrates	22.000	0,7	13.600	1.440
Calcium	82	2,6	17	34
Phosphorus	122	48	100	28
Iron	0,45	0,66	0,7	0,022
Vitamine A	27	0	0	0,011
Vitamine B 1	1	0,024	0,008	0,011
Vitamine B 2	0,4	0,03	0,04	0,044
Nicotinic acid	5,2	1,2	1	0,033
Ascorbic acid	1,2	0	0	0,3
Kcal	125	97	74	20

The chemical composition of pollen is highly variable (Stanley and Linskens 1985) and depends on the analytical methods used, the seasons of harvesting, the plant species, and whether the pollen was gathered by bees or collected by hand directly from the flower (Herbert and Shimanuki 1978). Table 27.1 lists the components of mixed pollen and other foods (beef, bread, milk). The protein, carbohydrate, calcium and phosphorus contents of pollen are higher than those of meat, bread, milk and legumes. Vitamins are generally much higher than in other foods. The ascorbic acid content of pollen is similar to that of fresh lettuce, endive, cooked potatoes and tinned tomatoes. Its calorie content makes pollen a concentrated, high energy food.

The chemical composition of pollen suggests that it is a food suitable for humans and livestock.

27.3 Pollen as Food for Humans

The importance of pollen as provision can be evaluated by clinical and experimental studies of its nutritive value in a quantitative way. Depending on the plant species the protein content of pollen varies between 16 and 30%, the carbohydrate content between 10 and 40%, and the rate of lipids was estimated between 3 and 10%, ashes between 1 and 9%. Carbohydrates and lipids of pollen can contribute to man's food intake, as well as the proteins. The daily demand of protein of a mature person of about 70 kg. body weight is between 20 and 26,5 grams of completely absorbed and transformed protein. This demand can be met by taking 90 grams of pollen per day, without having to look for an alternative source of protein.

Few experimental studies have addressed the question of digestibility. In Siena, *in vivo* and *in vitro* studies have been conducted with hazelnut (*Corylus avellana* L.) pollen (Franchi 1987). The *in vitro* tests involved prolonged contact with diluted aqueous solutions of sodium hydroxide and or hydrochloric acid containing surfactants or with digestive enzyme mixtures (Franchi 1987). For the *in vivo* experiments, mice were fed a suspension of pollen in water by gastric intubation. Both types of experiment gave identical results: digestion of pollen was found to be time dependent. Substances located in and on the external pollen surface were reached by the enzymes and digested. Digestion of the interior started in the pore region and proceeded inwards.

An interesting task for the future will be to explore the digestibility of the pollen of other species by these and other methods.

27.3.1 Pollen as a Food Supplement

Pollen was part of the normal diet of American Indians (Linskens 1996b). It was consumed as bread, mash and cookies, sometimes mixed with corn flour. It was also stored as provisions for periods of famine. Warriors used pollen as refreshment on long journeys and carried it in special bags (Fig. 27.1).

In western civilisation pollen is only consumed indirectly as a contaminant (e.g. of honey) and as a so-called "health" food. Finnish athletes take pollen as a supplement during training. Popular literature (see Linskens and Jorde 1997) suggests that pollen is a "miracle food" good for appetite, slimming, indigestion, neurasthenia, brain damage and growth deficit. Apparently the high content of vitamins in pollen mixture has a supplementary effect in human nutrition.

Fig. 27.1. "Medicine bag" of an Appalachian Indian, made from deer skin, containing hoddentin (pollen). From the 9th Annual Report of the Bureau of Ethnology to the Secretary of the Smithsonian Institute 1892

27.3.2
Ways of Consuming Pollen

Pollen has been eaten as food since prehistoric times. Coproliths (faecal stones) found in America date back to 1400 – 200 B.C.; they indicate that pollen was collected and eaten by the ancient Indians. American Indians made extensive use of pollen as food, especially in times of hardship. The fine yellow powder of corn and cattail was used as flour or mixed with wheat flour to make bread (Peterson 1978), or mush. It was also eaten raw and used for thickening soups (Tomikel 1976) and coloring rice dishes.

The main indirect way of eating pollen is with honey. Pollen is a normal biological contaminant of honey (Stanley and Linskens 1985). The amount ranges from 20.000 to more than 100.000 pollen grains per 10g of honey. Pollen spectra of honey samples can be used to identify the origin of the honey or to discover adulteration.

Another way in which pollen is eaten is the consumption of edible flowers or flower buds (Clifton 1984), which of course contain pollen. The flower buds of the caper plant (*Caparis edulis* L.), i.e. capers, contain anthers full of pollen.

Cooking with florets is becoming more and more fashionable. This is another way in which pollen is consumed indirectly with food. Many spices, such as saffron and its substitute, *Calendula* florets, also known as pot marigold (*Calendula officinalis* L.) are contamined with pollen. A popular vegetable in Italy is baby zucchini with their florets rich in pollen, still attached, and zucchini flowers themselves. Water and garden cress (*Nasturtium officinale* R.Br., and *Lepidium sativum* L., resp.) are other sources of pollen.

27.3.3
Pollen Inside the Human Body

In the digestive tract, pollen takes up water, increases in size and is enzymatically activated. The intine is emitted and a structure reminiscent of a germination tube is formed (Linskens and Mulleneers 1967). Substances attached to the outer wall are removed. The pollen wall bursts open and the contents are released for digestion. Gastrointestinal acids and enzymes cause hydolysis, breaking down polymers so they can be absorbed. Exine does not break down the human gastrointestinal tract, which is why exine is found in coproliths (Reinhard et al. 1991).

However, a small fraction of pollen grains reaches the blood stream from the gastrointestinal tract, by a process known as persorption (Volkheimer 1972) observed in dogs, rabbits and man. When volunteers ate 100–150 grams of pollen, 6.000–10.000 pollen grains were persorbed into the blood stream (Jorde and Linskens 1974). Surprisingly, the exine of the persorbed pollen showed signs of breakdown.

27.4
Pollen as Medicament

Pollen has been officially recognised as medicine by the German Federal Board of Health. Crude pollen, pollen preparations and pollen mixed with honey is a tonic for general debilitation and lack of appetite. Pollen has also been successfully given for chronic prostatitis. In urology, an extract of rye pollen, called "Cernitol", is used as an anticongestive. Experiments suggest that the liposoluble fraction of pollen contains molecules that inhibit the biosynthesis of prostaglandins and leukotrienes. In benign prostatic hyperplasia, pollen is an anticongestive and antinflammatory agent and has

Table 27.2. Use of pollen as a food additive against high altitude sickness. The oxygen partial pressure was equivalent to that at an altitude of 12.000 m. Experimental animals: mice, weighing 20 + 2 g. Pollen supplementation began 3 days before exposure to low pressure. Daily dose 0.05 g pollen per animal. Source: From Peng et al. 1990

	No of tested animals	Surviving animals %
Without pollen feeding control	73	4.1
Typha sp.	83	51.8
Zea mays L.	35	40
Vicia faba L.	35	40
Helianthus annuus L.	73	31.5
Fagopyrum esculentum Moench	74	29.7
Gossypium sp.	35	28.6
Sapium sebiferum (L.) Roxb.	84	71.4

relaxant and antiproliferative effects (Loschen and Ebeling 1991). A dose of 250 mg pollen twice daily is given for gastric ulcers. Pollen is an adjuvant during radiotherapy for gynecological carcinomas and is claimed to have immunogenetic effects.

Experiments in China (Peng et al. 1990) have demonstrated the efficacy of pollen for high altitude sickness (Table 27.2). Various species of animals, previously dosed with pollen, showed increased survival rate at an altitude of 12.000 m with respect to animals who were not given pollen. After the experiments with rats and mice, tests on humans were also carried out. In two different years it was demonstrated that subjects, who were given pollen for 3–7 days prior to going to an altitude of 5.000 m showed either no symptoms of low pressure or fewer symptoms than controls who did not eat pollen. The conclusion was that the consumption of pollen increases adaptability to high altitudes and low oxygen tension.

27.4.1 Pollen Allergy and Desensitisation

In spring and summer the air is loaded with pollen. About 20% of people in developed countries are allergic to pollen which lodges in the moist tissue lining the eyes, nose and airways. Pollen acts as an allergen carrier (Jorde and Linskens 1978), releasing chemicals (allergens) in minute quantities, but a few pollen grains are sufficient to induce hay fever. Allergic people are sensitive to these chemicals, which are located in the exine. Starch grains inside pollen form the reserve material for the early stages of pollen germination, and proteins are coated in it which also have allergenic properties. Inhalable particles, containing allergen molecules are released apparently into the atmosphere in two different ways: osmotic rupture of pollen grains in rain drops (observed for grass pollen), or partial germination of pollen settling on leaves (observed for birch pollen). "Pollen rain" is a source of secondary emission of allergens into the atmosphere (Sophioglu et al. 1992; Schäppi et al. 1997).

The pollen allergy is highly specific. People in central Europe are usually allergic to rye and rye-grass, Scandinavians to birch pollen, and Americans to ragweed pollen. Recent identification of the rye-grass gene that codes the main allergen in rye-grass pollen has raised hopes for a vaccine that will block the allergic response at the outset.

Desensitisation is a current therapy for pollen allergy. The patient is inoculated with low doses of pollen extract from the plant to which he or she is sensitive. In this way, the immuno system learns not to react to the allergen.

Deliberate or accidental ingestion of pollen by an allergic person has complex implications. Patients with an allergic disposition should therefore be analyzed before taking pollen. Although pollen is rich in nutrients, it should be prescribed with caution because of its allergenic potential.

27.5 Pollen as Feed for Animals

Although the therapeutic value of pollen for man is still being discussed, its nutritional value for animals has never been disputed. Apart from bees, there are many other insect species – even predatory mites – which make use of pollen as an alternative food. For Phytoseiidae, i.e. herbivorous mites, its nutritional value depends on the kind of pollen consumed (Baier and Karg 1992). Stored pollen is known as "bee bread" and has been transformed by biochemical processes which result in an enrichment in lactic acid. This conserves the pollen grains in the comb, though its germination capacity is lost.

Some neotropic ants collect pollen by licking wind-transported pollen from leaves. This pollen is stored in the foregut and regurgitated as feed for nestmates. Indigestible pollen membranes are disposed of outside the nest (Baroni Urbani and de Andrade 1997).

Twenty-three families of beetles are known to feed on pollen. Primitive butterflies utilise pollen by chewing it. Beetles crush it in their jaws. In Collembola the digestion of pollen grains can be observed through the semi-transparent exoskeleton.

Experiments have demonstrated that domestic animals, like piglets, calves and chickens, react positively to pollen supplements. Pollen has been found in the digestive tract of flower visiting animals, such as bats. Pollen is evidently also a protein source for these higher animals (Faegri and van der Pijl 1979).

References

Baier B, Karg W (1992) Untersuchungen zur Biologie, Ökologie und Effektivität oligophager Raubmilben. Mitt Biol Bundesanstalt Land- und Forstwirtsch, Berlin-Dahlem, Vol 281

Baroni Urbani C, de Andrade ML (1997) Pollen eating, storing, and spitting by ants. Naturwissenschaften 84:256–258

Clifton C (1984) Edible flowers. Mc Graw Hill, New York

Coghlan A (1995) Far-flung pollen raises spectre of superweed. New Scientist 148:10

Faegri K, van der Pijl L (1979) The principles of pollination ecology. Third edition, Pergamon Press, Oxford

Franchi GG (1987) Researches on pollen digestibility. Atti Società di Scienze Naturali Memorie Ser 94: 43–52

Herbert EW, Shimamuki M (1978) Chemical composition and nutritive value of bee-collected and bee-stored pollen. Apidiologie 9:33–40

Jorde W, Linskens HF (1974) Zur Persorption von Pollen und Sporen durch die intakte Darmschleimhaut. Acta Allergologica 29:165–175

Jorde W, Linskens HF (1978) Pollen als Allergenträger. Allergologie 1:7–10

Linskens HF (1992) Mature pollen and its impact on plant and man. In: Cresti M, Tiezzi A (eds) Plant Sexual Reproduction. Springer, Berlin, Heidelberg, New York, pp 203–217

Linskens HF (1995) Pollen deposition in mosses from northwest Greenland. Proc Kon Ned Akad Wet, Amsterdam 98:45–53

Linskens HF (1996a) Pollen deposition in arctic mosses from Severnya Zemlya archipelago. Frans Joseph Land and from Novaya Zemlya. Proc Kon Ned Akad Wet, Amsterdam 99:71–84

Linskens HF (1996b) Indianer und Pollen. In: B. Wolters, Agave bis Zaubernuß, Heilpflanzen der Indianer Nord- und Mittelmerikas. Urs Freund, Greifenberg, pp 222–227

Linskens HF (1996c) No airborne pollen within tropical rain forest. Proc Kon Ned Akad Wet, Amsterdam 99:175–180

Linskens HF, Cresti M (1988) The effect of temperature, humidity, and light on the dehiscence of tobacco anthers. Proc Kon Ned Akad Wet, Amsterdam 91:369–375
Linskens HF, Jorde W (1997) Pollen as food and medicine – a review. Econ Bot 51:78–87
Linskens HF, Mulleneers JML (1967) Formation of "Instant Pollen Tubes". Acta Bot Neerl 16:132–140
Linskens HF, Bargagli R, Cresti M, Focardi S (1993) Entrapment of long-distance transported pollen grains by various moss species in coastal Victoria Land, Antarctica. Polar Biol 13:81–87
Loschen G, Ebeling L (1991) Hemmung der Arachnidonsäure-Kaskade durch einen Extrakt aus Roggenpollen. Arzneimittel-Forschung 41:162–167
Peng Hong-fu, Xue Zheng-sheng, Miao Fang, Pan Dung-pi, Liu Zi-ming, Liu Zhong-wen, Liu Shau-rong, Tao Shung-xing (1990) The effect of pollen enhancing tolerance to hypoxia and promoting adaptation to highlands (In Chinese). Chung Hua I Hsung Tsa Chin (J Chinese Medicine) 70:77–81
Peterson W (1978) A field guide to edible wild plants in Eastern and Central Nord America. Mifflin, Boston
Reinhard KJ, Hamilton DL, Hevly RH (1991) Use of pollen concentration in palaeopharmacology, coprolite evidence in medical plants. J Ethnobiol 11:117–132
Schäppi GF, Taylor PE, Staff IA, Suphioglu C, Knox RG (1997) Source of Bet v1 loaded inhalable particles from birch revealed. Sex Plant Reprod 10:315–323
Sophiogly C, Singh MB, Taylor P, Bellomo R, Holmes P, Puy R, Knox RB (1992) Mechanism of grass-pollen-induced asthma. The Lancet 339:569–572
Stanley RG, Linskens HF (1985) Pollen – Biologie Biochemie Gewinnung und Verwendung. Urs Freund, Greifenberg
Tomikel J (1976) Edible wild plants of Eastern United States of Canada. Alleghany Press, California/Pennsylvania
Vivino AE, Palmer LS (1944) The chemical composition and nutritional value of pollen collected by bees. Arch Biochem Biophys 4:129–136
Volkheimer G (1972) Persorption. Gasteroenterologie und Stoffwechsel. Band 2, Thieme, Stuttgart

CHAPTER 28

Field Release of Transgenic Virus Tolerant Tomatoes

S. Valanzuolo* · M. M. Monti · M. Colombo · G. Cassani

* Tecnogen S.C.p.A., Loc. La Fagianeria, 81015 Piana di Monte Verna (CE), Italy
e-mail: valan@tecnogen.it
telephone: 39+8 23 61 22 10
fax: 39+8 23 61 22 30

Abstract. Tomato production in Southern Italy has fallen in the last ten years owing to the outbreak of different pathogenic agents. Cucumber Mosaic Virus (CMV) particularly affected the high quality variety named San Marzano, mainly utilized by the canning industry, causing severe yield losses. Up to now, no natural form of resistance from tomato wild relatives has been successfully introduced in cultivated lines. Therefore, biotechnological approach represents an effective alternative to classical breeding.

Two San Marzano lines were transformed with a cDNA copy of a benign CMV satellite RNA. First evidence of virus tolerance in transformants was obtained during greenhouse experiments. Parental and hybrid transformants were subsequently tested for two years in open field, in the CMV epidemic area, in order to evaluate both the agronomic performance and the environmental impact of transgenic plants. The transgenic satellite potential spread was evaluated by sequencing different satellite RNAs found on tomato and other crop plants surrounding the release field.

28.1 Introduction

The cultivated tomato (*Lycopersicon esculentum*) has become, during the 20th century, one of the most widely consumed horticultural crops.

The annual production, in 1996, exceeded 88 million MT. In the US, tomato per capita consumption during the last sixty years has more than tripled, while potato consumption, for example, has fallen in the same period by 20% (Rick 1978).

This large acceptance is mainly due to the taste and versatility of the berry, while tomato does not rank particularly high in nutritional value: broccoli, spinach, peas, cauliflowers, artichokes, asparagus, carrots and even potatoes have higher relative concentrations of vitamins and mineral salts; but, by virtue of consumed volume, tomato contributes primarily to the American dietary intakes of vitamins A and C as well as essential minerals (preceding orange, lettuce and potato) (Tigchelaar 1986).

European Community in 1996 has produced 14.9 million MT of tomato, and Italy is the leading producing European country with about one half of the total amount (*source*: FAO Agricultural Database).

The yield, in terms of MT/ha, increased in Italy by 14% between 1984 and 1995, but, in the same period, the tomato gross production has fallen by 20% because of the relevant decrease (−29%) in terms of tomato productive surfaces (*source*: ISMEA 1996).

One of the reasons inducing Italian farmers to give up the tomato culture has been represented by the strong Cucumber Mosaic Virus (CMV) epidemic that, expecially in Southern Italy, has greatly reduced the tomato production over the past ten years. Some high quality varieties, such as San Marzano, mainly utilized by the canning industry, were dramatically hit (Gallitelli et al.1988a; Kaper et al. 1990; Crescenzi et al.1993).

Viruses are main loss biotic factors affecting tomato production, like insects, fungi and bacteria too. CMV is a cucumovirus with a multipartite genome composed by three genomic RNAs and a fourth subgenomic RNA which acts as a messenger-RNA for coat protein synthesis (Douine et al. 1979; Kaper and Waterworth 1981; Palukaitis et al. 1992). A small single stranded RNA molecule, known as "satellite RNA", is sometimes present in CMV inocula: satellites are capable of modulating disease symptoms caused by their helper virus (probably by competing with viral RNAs for the replicative machinery), either by attenuating or exacerbating the disease expression (Kaper et al. 1976; Kaper and Waterworth 1977; Kaper et al. 1981; Kaper 1993). They depend on a helper virus for their replication and share no significant sequence homology with viral RNAs nor with plant RNAs (Murant and Mayo 1982; Francki 1985).

Satellite RNAs can be classified into two groups, benign or necrogenic, according to their ability to produce symptoms on tomato (Devic et al. 1990).

28.2 CMV Resistance: The Biotechnological Approach

Classical breeding methods, based on crossing and backcrossing programs, have been widely utilized within the past 50 years in order to obtain new tomato cultivars improved in fruit quality, colour of berries, growth habit for machine-harvest adaptability, stress resistance (Tigcheelar 1986). This has been made possible by the natural genetic variability of tomato germplasm, considered as a whole of nine species within the genus *Lycopersicon.* (Rick 1976; Rick 1979; Tigcheelar 1986)

In some cases, anyway, classical breeding proved insufficient to get good results in terms of improvement. CMV resistance, for example, is a goal that can hardly be carried

Tab. 28.1. Tomato biotechnology: some important applications

Genetically Modified Organisms	Transgene	Modified trait	Tested by
FLAVR SAVR	Polygalacturonase antisense RNA	Shelf-life extension	Calgene
High viscosity	Pectinesterase antisense RNA	High pectin content	ICI, Calgene, Monsanto, Campbell
Virus resistance	Viral coat protein	CMV resistance	Monsanto
Virus resistance	Viral satellite RNA	CMV resistance	AGC, Univ. Of Maryland
Insect resistance	Bacillus thurigiensis endotoxin	Coleopters and Lepidopters grubs resistance	Ciba Geigy, PGS, Monsanto
Herbicide tolerance	Mutant bacterial gene encoding an EPSP synthase	Low sensitivity to Glyphosate	Calgene, Monsanto, Agracetus
Malesterility	Ribonuclease under anther specific promoter	Selfpollination is prevented	PGS
Slow ripening	Ethylen forming enzyme antisense RNA	Ethylen endogenous formation is prevented	ICI

out by crossing methods (Mc Garvey and Kaper 1993), since no source of resistance genes has been definitely detected in tomato cultivated varieties and in wild relatives, while the attempts to transfer supposed tolerance traits from *Lycopersicon hirsutum* to *Lycopersicon esculentum* have not produced any final result so far.

In these cases, the biotechnological approach could prove critical to reaching the goal (Gasser and Fraley 1989).

In Table 28.1, some significant results reached by biotechnological applications in tomato breeding have been reported (Beck and Ulrich 1993)

Different methods based on genetic transformation of tomato plants have been described in order to induce virus resistance or tolerance in tomato (Baulcombe 1994)

The most common ones involve:

a) the transfer in tomato plants of a gene coding for a viral coat protein (Powell-Abel et al. 1986; Klee et al. 1987; Nelson et al. 1988; Cuozzo et al. 1988);
b) the transfer in tomato plants of a gene coding for a viral satellite RNA (Baulcombe et al. 1986; Harrison et al. 1987; Mc Garvey et al. 1990; Tien and Wu 1991; Saito et al. 1992).

28.3 Tecnogen Experiment With Satellite RNA

28.3.1 The Laboratory Steps

San Marzano tomato lines have been transformed, via *Agrobacterium tumefaciens*, with a chimeric gene whose transcript is a satellite RNA named CL14-3. This is classified as a benign satellite, structurally distant from the necrogenic ones since its sequence (340 bases) shows 12 nucleotide differences in the necrogenic box (bases 277→305) and 7 out of 50 at the 3′ end.

The satellite RNA CL 14-3 was cloned as cDNA and inserted into the binary vector pBI 121.1 (Jefferson et al. 1987) after the removal of Hind III-Eco RI GUS cassette. The resulting plasmid, named pSat, (Fig. 28.1) contains, between *Agrobacterium tumefaciens* right and left borders, two transcriptional cassettes expressing a chimeric form of the satellite and the enzyme neomycin phosphotransferase II (npt II) under the control of plant promoters. This plasmid was used to transform the *Agrobacterium* strain LBA 4404 by direct transformation.

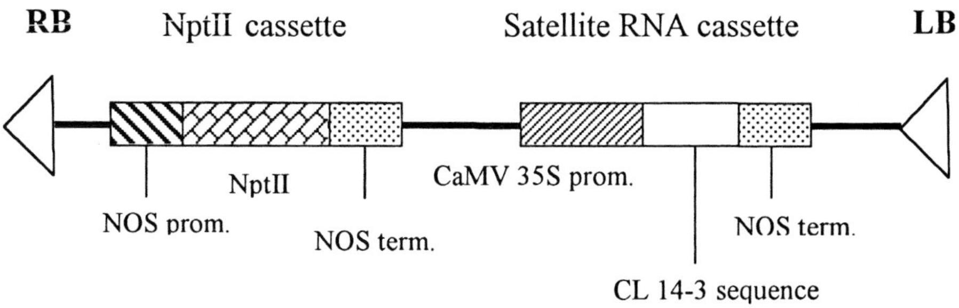

Fig. 28.1. The T-DNA of pSat

In vitro grown cotyledons of several different San Marzano genotypes were chosen as starting explants to induce tomato regenerants. Regeneration and rooting of transformants occurred within 6–8 weeks on a medium containing kanamycine as a selectable marker: the transformation overall efficiency was higher than 2%. (Valanzuolo et al. 1994)

All transformants were checked by Southern blot in order to detect the integration pattern of the foreign gene in tomato genome. The analysis of R0 transformed plants showed three different integration patterns of the chimeric gene. These patterns were conserved in all the progenies we analysed: each progeny obtained from a single regenerant was considered as a family. The segregation data of four families, analysed by the χ^2 test, related to a Mendelian single dominant gene pattern; these four families have been monitored for the presence of the transgene by PCR up to R3 generation, and homozygous plants were screened. (Valanzuolo et al. 1994)

The transgenic satellite RNA was detected, as mature form (338 bases), by Northern blot of total RNA from plants infected with a CMV strain lacking satellite RNA. The basal level of transcription of foreign gene is low and the primary transcript can hardly be detected on standard blots. The amount of transcript is anyway sufficient to act as a template for the viral RNA polymerase associated with the helper strain, resulting in the production of relevant amounts of satellite RNA in transgenic plants. (Valanzuolo et al. 1994)

The transformed plants were artificially challenged in greenhouse, using the mentioned CMV satellite-free strain, to preliminarily test their tolerance to CMV: no virus induced symptoms were observed on the transformants, while alterations in leaf shape ("fern leaf") appeared on controls.

Homozygous transgenic plants were subsequently tested, using infected plants as source of inoculum, under aphid proof net. CMV infection was constantly monitored and associated to final yield of the tested plants. Results are shown in Fig. 28.2 (average yield per plant).

28.3.2 The Field Tests

28.3.2.1 *Materials and Methods*

As the following step, about 800 homozygous plants and F1 hybrids (obtained using transformants as male parentals) were tested for two years in different San Marzano to-

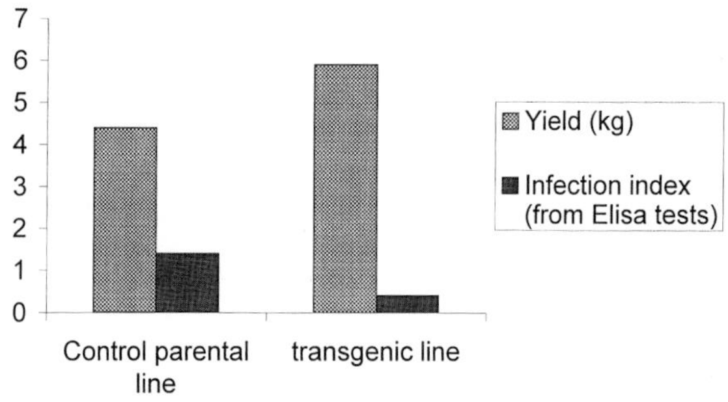

Fig. 28.2. Preliminary greenhouse tests for CMV resistance

mato production areas during CMV epidemic, in order to investigate both their virus tolerance and potential environmental impact. This has been the first Italian deliberated release in open field of satellite RNA transformed plants.

A design of the experimental field is reported in Fig. 28.3.

The CMV tolerance attitude of transgenic parental lines and F1 hybrids was evaluated in terms of yield, comparing their agronomical performances with control plants. ELISA tests were periodically performed during the experiment in order to detect the presence of CMV and other main viruses (AMV – Alfalfa Mosaic Virus; TSWV-Tomato Spotted Wilt Virus; PVY – Potato Virus Y; TMV – Tobacco Mosaic Virus) on the released plants.

The environmental impact of all transgenic plants was evaluated in terms of satellite spread (with related potential effects) on tomato and other horticultural crops (zucchini, melon, pepper: *host plants*). Spreading of satellite RNA from transgenic plants was monitored by analysing satRNA populations isolated from neighbouring plants and control tomato. Two samplings of green material were performed, in June and July. Total RNA was isolated, retrotranscribed and amplified by the use of two primers complementary to the conserved ends of CMV satellite RNAs. The amplified products were purified by agarose gel electrophoresis and submitted to DNA sequencing by the dideoxy chain terminators technique.

Different satellite RNA sequences were analysed and compared, by the use of the program ALIGN (part of DNASTAR software) based on the Wilbur & Lipman alignment method, for:

- percentage of dissimilarity with the transgenic satellite
- number of bases and presence of diagnostic bases in the necrogenic box

Fig. 28.3. Field test design

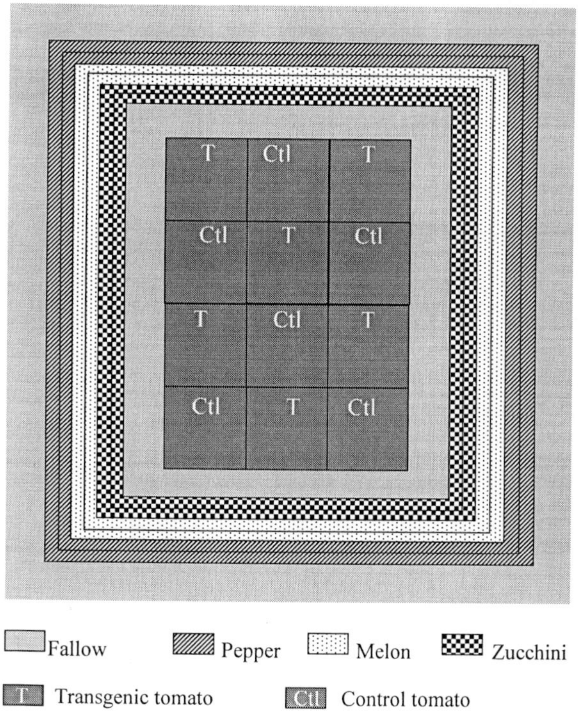

28.3.2.2
Results

Yield data related to the second field test year is reported in Fig. 28.4. Transgenic tomato plants, both parentals and hybrids, gave higher percentages of marketable production, while no relevant differences were observed as total production.

No differences were remarked as regards the main physical and chemical characters influencing the fruit quality (size, shape, pH, colour, optical residue, dry matter, viscosity, citric acid, soluble sugars, potassium and calcium content) comparing control plants and transformants.

The population of satRNAs found in the field looks like a mixture of different sequences associated to several CMV strains. About 60 different sequences were compared and analysed, resulting in a dendrogram in which three distinct branches are evident. One of them is composed by 5 satRNAs recognized as necrogenic on the basis of specific sequence features in the region known as "necrogenic box" (Kaper *et al.* 1988) on the 3' side of the molecule. The remaining two branches are composed of satRNAs recognizable, on the same structural basis, as benign variants. One of them comprises the transgenic CL14-3 satellite (found 11 times on different host plants) and other closely related sequences (different in no more than 2–5 bases). The other one is composed of benign variants more markedly distant (10–30 base differences).

Predominant symptoms detected on these plants include mosaic, dwarfism and fern leaf. On the other hand, leaf and fruit necrosis appear on plants harboring the necrogenic variants.

28.4
Discussion

According to these results, we are allowed to say that expression in tomato plants of a gene coding for a satellite viral RNA is a safe method resulting in the attenuation of symptoms induced by CMV attack.

Our data shows that satellite RNA is capable of spreading from transgenic plants but this phenomenon does not correlate with the appearance of new symptoms on host plants. Furthermore, the presence of producing satRNA transgenic plants does not seem to significantly alter the natural variability of satRNA populations.

Finally, the satellite mediated resistance to CMV, in spite of its effectiveness, could prove to be inadequate in assuring total protection of tomato culture, because of the

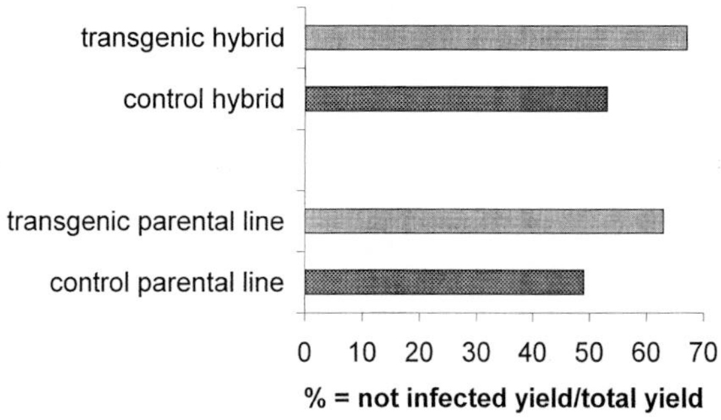

Fig. 28.4. Controls vs. transformants in open field: marketable yield

critical influence in field of many yield decreasing factors, such as different viruses (AMV – Alfalfa Mosaic Virus; TSWV-Tomato Spotted Wilt Virus) and fungal pathogens (mainly Tomato cork root: *Pyrenochaeta lycopersici*) detected during our experimental release. Therefore, the satellite CMV resistance must be considered as an essential starting point in obtaining new tomato varieties carrying multiple resistances.

References

Baulcombe DC (1994) Novel strategies for engineering virus resistance in plants. Current Opinion in Biotechnology 5:117–124

Baulcombe DC, Saunders GR, Bevan MW, Mayo MA, Harrison BD (1986) Expression of biologically active viral satellite RNA from the nuclear genome of transformed plants. Nature 321:446–450

Beck CI and Ulrich T (1993) Le biotecnologie nell'industria alimentare. BIOTEC vol 8 n 6:27–30

Crescenzi A, Barbarossa L, Gallitelli D, Martelli GP (1993) Cucumber Mosaic Cucumovirus populations in Italy under natural epidemic conditions and after a satellite mediated protection test. Plant Disease 77:28–33

Cuozzo M, O'Connell MK, Kaniewski W, Fang RX, Chua NH, Tumer NE (1988) Viral protection in transgenic tobacco plants expressing the cucumber mosaic virus coat protein or its antisense RNA. Bio/Technology 6:549–557

Devic M, Jaegle M, Baulcombe D (1990) Cucumber Mosaic Virus satellite RNA (strain Y): analysis of sequences which affect systemic necrosis on tomato. Journal of General Virology, 71:1443–1449

Douine L, Quiot JB, Marchoux G, Archange P. Ann Phytopatol 11:439–475

FAO Agricultural Database. Website http//apps.fao.org/

Francki RIB (1985) Plant virus satellites. Ann Rev of Microb 39:151–174

Gallitelli D, Di Franco A, Crescenzi A, Vovlas C, Ragozzino A (1988) Una grave virosi del pomodoro in Italia meridionale. L'Informatore Agrario 30:67–70

Gasser CS and Fraley RT (1989) Genetically engineered plants for crop improvement. Science 244: 1293–1299

Harrison BD, Mayo MA, Baulcombe DC (1987) Virus resistance in transgenic plants that express cucumber mosaic virus satellite RNA. Nature 328:799–802

ISMEA (1996) Filiera Ortofrutta

Jefferson RA, Kavanagh TA, Bevan MW (1987) GUS fusions: β-glucuronidase as a sensitive and versatile gene fusion marker in higher plants. Embo J 6:3901–3907

Kaper JM (1993) Satellite-mediated symptom modulation: an emerging technology for the biological control of viral crop disease. Microb Releases 2:1–9

Kaper JM and Waterworth HE (1977) Cucumber-mosaic-virus-associated RNA 5:casual agent for tomato necrosis. Science 196:429–431

Kaper JM and Waterworth HE (1981) In: Kurstak E (editor) Handbook of Plant Virus Infections and Comparative Diagnosis. Amsterdam Elsevier/North Holland pp 257–332

Kaper JM, Gallitelli D, Tousignant ME (1990) Research in Virology. Virology 141:81–95

Kaper JM, Tousignant ME, Lot H (1976) A low-molecular-weight replicating RNA associated with a divided genome plant virus: defective or satellite RNA? Biochem Biophys Res Commun 72:1237–1243

Kaper JM, Tousignant ME, Steen MT (1988) Cucumber mosaic Virus-associated RNA 5. – XI. Comparison of 14 CARNA-5 variants relates ability to induce tomato necrosis to a conserved nucleotide sequence. Virology 163:284–292

Kaper JM, Tousignant ME, Thompson SM (1981) Cucumber mosaic virus associated RNA 5. VIII. Identification and Partial chracterization of a CARNA 5 incapable of inducing tomato necrosis. Virology 114:526–533

Klee H, Horsch R, Rogers S (1987) *Agrobacterium*-mediated plant transformation and its further applications to plant biology. Ann Rev Plant Physiol 38:467–486

Mc Garvey PB and Kaper JM (1993) Transgenic plants for conferring virus tolerance: Satellite approach. In: Transgenic plants, vol I. Academic Press, New York, pp 277–296

Mc Garvey PB, Kaper JM, Avila-Rincon MJ, Pena L, Diaz-Ruiz JR (1990) Transformed tomato plants express a satellite RNA of cucumber mosaic virus and produce lethal necrosis upon infection with viral RNA. Biochem Biophysic Res Commun 170:548–555

Murant AF, Mayo MA (1982) Satellites of plant viruses. Annu Rev Phytopatol 20:49–70

Nelson RS, McCormick SM, Delannay X, Dube P, Layton J, Anderson EJ, Kaniewska M, Proksch RK, Horsch RB, Rogers SG, Fraley RT, Beachy RN (1988) Virus tolerance, plant growth, and field performance of transgenic tomato plants expressing coat protein from tobacco mosaic virus. Bio/Technology 6:403–409

Palukaitis P, Roossinck MJ, Dietzgen RG, Francki RIB (1992) Cucumber Mosaic Virus. Adv In Vir Res 41:281–347

Powell-Abel P, Nelson RS, De B, Hoffmann N, Rogers SG, Fraley RT, Beachy RN (1986) Delay of disease development in transgenic plants that express the tobacco mosaic virus coat protein gene. Science 232:738–743
Rick C (1976) Tomato (family Solanaceae). In: Simmonds N (ed) Evolution of crop plants. Longman Publications, New York, pp 268–273
Rick CM (1978) The tomato. Scientific American 239:76–87
Rick CM (1979) Biosystematic studies in *Lycopersicon* and closely related species of *Solanum*. In: Hawkes JG, Lester RN, Skelding AD (eds) The Biology and the Taxonomy of the *Solanaceae*. Academic Press, New York, pp 667–678
Saito Y, Komari T, Masuta C, Hayashi Y, Kumashiro T, Takanami Y (1992) Cucumber mosaic virus-tolerant transgenic tomato plants expressing a satellite RNA. Theor Appl Genet 83:679–683
Tien P, Wu G (1991) Satellite RNA for the biocontrol of plant disease. In: Advances in virus research. Academic Press, New York 39:321–339
Tigcheelar EC (1986) Tomato Breeding. Breeding Vegetable Crops. AVI Publishing Co. 135–171
Valanzuolo S, Catello S, Colombo M, Dani M, Monti MM, Uncini L, Petrone P, Spigno P (1994) Cucumber Mosaic Virus resistance in transgenic tomatoes. Acta Horticulturae 376: 377–386

CHAPTER 29

Grapevine Biotechnology Coming on to the Scene

R. Vignani · M. Scali · M. Cresti

Dipartimento di Biologia Ambientale, Università degli Studi di Siena, Via P. A. Mattioli, 4
53100 SIENA, Italia
e-mail: vignani@unisi.it
telephone: 577-29 88 56
fax: 577-29 88 60

29.1 Introduction

The ancientness of viticulture and the mythology of wine have given the grapevine a specially favorable and beneficial position among cultivated fruit plants.

It is thought that cultivation of the grapevine (*Vitis vinifera L.*) began during the Neolithic era (6000–5000 BC) in the region known as Transcaucasia corresponding to the eastern shores of the Black Sea. However archeological finds of grape seeds indicate that either the grapevine or at least its progenitor, *Vitis sylvestris*, was already distributed throughout much of Europe during the Atlantic and Sub-Boreal paleoclimatic periods (7500–2500 years ago). Quite recently archeological grape seeds, dating according to radiocarbon method, between 4350 and 3950 years, were discovered in Spain (Walker 1985).

It is likely that the discovery of wine occurred accidentally. Prehistoric people probably just ate the berries of the wild vines climbing on the forest trees. With the development of village settlements and spreading of the habit of storing grape berries, naturally hosting fermenting microbial populations, it is possible that wine became the unexpected product of such a procedure. That was probably the beginning of viticulture (grape growing) and oenology (the art and science of making wine). Even though historically, grapevines occurred as far North as Belgium and grape growing and wine-making traditions extended from Transcaucasia to Asia Minor (especially Egypt and Babylon) and as far East as China, viticulture stands as a major crop almost exclusively in countries sitting along the Mediterranean basin, due to the fact that *Vitis vinifera L.* is a temperate-climate species. Traditionally, the innovation in viticulture is based on the introduction and selection of new, agronomically favorable genotypes primarily done at the level of husbandry. This is probably a general feature which distinguishes broad-acre agriculture and horticulture consisting of fruit-growing crop species.

Grapevines have been perpetuated for centuries by vegetative propagation and especially in those countries where the wine industry represents a great part of the national economy based on agriculture, like Italy or France, regional differences in terms of type of cultivars spreading, nomenclature and use of synonyms which are linked to the diffusion of local dialects or linguistic tradition, make the world of viticulture very complex and highly enriched by traditional-popular inputs and relatively recalcitrant to the introduction of technological or biotechnological innovations.

In viticulture, especially in European, Mediterranean countries, the main response to biological constraints or economic change has been to manipulate the existing traditional cultivars by applying progressively higher inputs of husbandry. These include standard husbandry procedures like rootstocks maintenance, pruning and training, or chemical-based husbandry based on the use of fertilizers or pesticides to increase health and quality of the vines, through mechanization and postharvest technology (winemaking). Grapevines whose fruit is not only employed for wine production since there are also table grapes and grapes grown for drying, have several technical constraints which make breeding and vineyard management of this crop particularly difficult. Grapevines are highly heterozygous outcrossers and they do not breed true from seeds. Moreover most of the characters of economical importance which make a good cultivar are polygenic. The traditional cultivars are characterized by complex gene combinations which, under a biological profile, are vegetatively, randomly propagated.

Strong pressures for change in methods of production are being applied to viticulture by economic and social forces. Some national economies require a modern, intensive viticulture involving a significant energy input and high costs. Chemical crop protection is not only expensive but an increasingly stringent environmental protection legislation, causes a general disquiet to consumers, concerned about public health and quality of food. All of these factors point to the conclusion that chemical controls must be replaced by genetic resistance possibly reducing the level of husbandry and increasing the reproducibility of performance of each genotype through technology or biotechnological aids.

In recent years, plant biotechnology through new scientific acquisitions, has made impressive progress opening up new frontiers for many crop species having a crucial economic importance.

Recombinant DNA technology, including gene cloning, DNA sequencing, hybridization technologies established in the 1970s, and more recent methodologies such as Polymerase Chain Reaction (PCR) or modern application to cell culturing, have been rapidly applied to the molecular biology of plants in order to establish biotechnological improvements to several crop species, including fruit-growing ones. Even if for grapevines many biotechnological approaches are still to be set and established due to the various technical difficulties which characterize this species, still biotechnology is beginning to be a useful tool, which may help to ameliorate viticulture and winemaking industrial standards. In the present chapter some of the more common biotechnological applications to grapevines are described, from cell culture for the production of secondary metabolites, to DNA technologies applied to varietal genotyping.

This work was written in the belief that a better knowledge of the grapevine biology will ease the difficult task of reconciling the opposing forces of innovation and tradition.

29.2 Biotechnology for a Better Understanding of Grapevine Biology

Grape products contribute significantly to the world agricultural economy, especially in the Mediterranean area. The possibility of increasing the quality of the wines or of using grapevines either as a tool or target for biotechnological applications, necessarily passes through a better understanding of the biology and physiology of the plant.

Molecular biology techniques are not only the fundamental tools for most biotechnological applications, but are also currently used for a better comprehension of the complex phenomena participating in the grapevine biology and physiology. Only after the acquisition of a basal knowledge of some crucial biological events characterizing the grapevine life cycle, like fruit set and ripening, natural mechanism of response to pathogens attack, mechanisms of adaptation to environmental stresses, will it be possible either to select within natural or man-introduced population or to control via biotechnological methods the phenotypic features contributing to making a good cultivar.

However, under a strict technical point of view it is to be emphasized that the high content of vines in secondary metabolites like tannins, phenolics, acids and carbohydrates, make *Vitis vinifera* a difficult plant to work with for molecular or biochemical studies.

In many cases conventional methods applied in molecular biology for basic laboratory procedures like nucleic acid extraction (DNA or RNA) which are essential to obtain information on the genome characteristics (mapping, gene structure, gene expression, gene cloning and sequencing and so forth) need to be revised and corrected when used in grapevine. To cite a few examples we will mention the modification of RNA extraction methods introduced by Loulakakis (1996), the DNA extraction method reported in Sensi (1996) which revises the method described by Mulchay (1993) or the genomic DNA isolation procedure described by Kim (1997).

Despite the technical difficulties faced when working with grapevines, the great diffusion and expansion of molecular biology have allowed researchers to dissect and understand some of the crucial biochemical pathways which characterize the physiology of this plant (berry ripening) and influence the response to external stimuli (environmental stress and pathogen attacks). In particular we would like to mention at least two general examples of how the basic research in molecular biology has eventually helped in applied biotechnology in grapevine:

(i) the molecular characterization and expression pattern controlling the flavonoid and anthocyanin biosynthesis pathway and (ii) that of a group of genes responsible to environmental stimuli, including the alcohol dehydrogenase gene or the gene(s) controlling the resveratrol synthesis (Stilbene synthase).

(i) Flavonoids are phenolic compounds present in the plant tissues as secondary metabolites which are largely known among plants, as being involved in many plant functions such as plant-pathogen interaction, UV protection and pollen tube growth. Flavonoids are also precursors of anthocyanins, which form the main pigments in flowers and fruits, where they play as much a role in the pollination process as insects or animal attractants. The color of red and black grapes results from the accumulation of anthocyanins in the berry skin. The quantity and quality of color in grape berries at harvest are crucial factors that influence significantly the wine quality. Each variety of grapevine is characterized by a typical anthocyanins profile and several cultivars like the so called "Colorino" or colored accessions, widely grown in the Tuscany region for the production of the Chianti wine, are selected for their properties of conferring to the wine an intense, deep red tonality. Recently, analysis of the expression of flavonoid and anthocyanin pathway genes, and their molecular characterization during berry ripening have been studied in detail (Sparvoli et al. 1994; Boss et al. 1996 (a); Boss et al. 1996 (b)). These findings prove that the onset of anthocyanin synthesis in ripening grape berry skins (typically found only in red grapevines) seems to be influenced by a set of regulatory genes according to a pattern, which appears to differ from the anthocyanins biosynthetic pathway observed in maize, petunia and snapdragon. This kind of molecular data makes it possible to monitor and eventually control one of the more characteristic phases of the development of grapes, like the character of the color of the berry, which is strictly correlated to a commercial and economically valuable trait of the wine.

(ii) Alcohol dehydrogenases (ADH; EC 1.1.1.1) catalyzing the conversion of aldehyde to the relative primary alcohol, are known in plants to participate in environmental or stress stimuli, including anaerobiosis, UV light, chemicals exposure (Llewellyn et al. 1987; Longhurst et al. 1994). Different isoforms of ADH have already been described as being distinctive for each variety (Wolfe 1976). Recently, Sarni-Manchado et al. (1997) obtained the complete sequence of one alcohol dehydrogenase gene (GV-Adh 1) expressed during ripening of grapevine berry. The comparison of the Adh gene from grapevine with other Adh genes from both monocts and dicots, performed using their nucleic acid sequence and the deduced amino acid sequences, allowed a phylogenetic analysis showing that the grapevine gene is more strictly related to the Rosaceae than to the Solanaceae. This study appears to be essential for further inquiries into the molecular control of berry ripening in grapevine. A comparable study as concerns the possible biotechnological implications, has led to the molecular characterization of the stilbene synthase gene (STS) also called resveratrol synthase in grapevine (Sparvoli et al. 1994) which permitted a further clarification of some crucial steps in the mechanism of inducibility of this gene by UV irradiation and exposure to ozone or pathogen attack. In particular it has been shown that the ozone-responsive promoter region of the grapevine Vst 1 gene differs from the pathogen-responsive promoter region (Schubert et al. 1997). The Vst 1 promoter might be used for biotechnological applications related to the enhancement of pathogens resistance, when using "tailor-made" constructs excluding the "ozone-responsive element". The stilbene synthase gene from grapevine has been successfully integrated into protoplasts genome of a commercially important japonica rice (*Oryza sativa L.*) cultivar ("Nipponbare"), where it is able to confer an enhanced resistance to rice blast caused by the fungus *Pyricularia oryzae* (Stark-Lorenzen et al. 1997). Resveratrol present predominantly in red wines, is also considered to be important as a biologically active product able to inhibit the production of oxidized low-density lipoproteins (Frankel et al. 1993; Meyer et al. 1997) thus preventing the atherogenesis in man and blocking the aggregation of platelets, involved in the progression of atherosclerosis and myocardial infarction and stroke. A great degree of variation in resveratrol or related compounds has been observed among wines obtained according to the oenological practices used, or derived from grapevines grown in different climatic areas, or from different cultivars, or from grapes exposed to different fungal attacks (Goldberg et al. 1996). Thus, the comprehension of the molecular mechanisms regulating the production of this family of compounds, certainly important for the grapevine, and possibly also to human health, appears to be of great importance for a further employment of biotechnological applications in the direction of controlling grapevine berry ripening.

29.3 Grapevine Cell Cultures as a Tool for Genetic Manipulation and for the Production of Secondary Metabolites

Even if speaking about biotechnology is commonly interpreted as DNA technology i.e. the ability to isolate, modify and transfer genetic molecules, according to a broader definition the term biotechnology should also include at least antibody production, as well as molecular diagnostics and cell or tissue culture.

Tissue culture (micropropagation) or cell culture in grapevine are used for different purposes, and widespread biotechniques are beginning to substitute the more traditional cultural practices.

Grapevine cell suspension cultures have been recently employed to understand the mechanism of action at cellular level of a systemic fungicide, the fosethyl-Al (aluminum tris[ethyl phosponate]), commonly used to defeat diseases caused by oomycetes (Derks et al. 1989). Elucidation of the molecular mechanism of action of this compound and the individuation of intracellular or metabolic targets, could be of strategic importance in the control of the host-parasite interaction. It has been demonstrated that Ca^{2+} and Mg^{2+} ions counteract the reduction by fosethyl-Al of peroxidase activity from suspension-cultured grapevine cells, and that the probable mechanism of action on peroxidase is carried out by an apoplastic Ca^{2+}/Mg^{2+} displacement (Lopez-Serrano et al. 1997).

The production of haploid or homozygous *Vitis* plants is of crucial importance under the biotechnological and commercial profile, since it may provide a new insight into conventional breeding and genetic transformation technology. A recent study (Sefc et al. 1997) enriched previous experiences in the field of anther culture, demonstrating that it is possible to obtain embryoid structures out of *Vitis* microspores, which showed well defined epidermal layers. Even if neither calli nor embryoid were able to regenerate plants, the study could be considered a fundamental acquisition in the field of haploid tissue regeneration out of *Vitis* spp. microspores. Other studies employing tissue or cell grapevine culture set up medium conditions which promote in vitro rooting (Roubelakis et al. 1991) and "explored" the effect of genotype on somatic embryogenesis and plant regeneration of several *Vitis* spp. (Mozsár et al. 1996).

Plant cell culture offers interesting possibilities for the large scale production of secondary metabolites, which may be employed by the pharmaceutical and food chemistry industries. Yet, in particular for *Vitis* spp., it seems necessary to acquire a better knowledge of the effect of artificial in vitro conditions used in tissue cultures on the metabolic pathways of isolated plant cells as well as appearing crucial to understanding the feasibility of realizing *Vitis vinifera* cell bioreactor cultures, which could promote the employment of this techniques on a large scale. Recent findings showed how for *Vitis vinifera* cell culture biomass production is not completely correlated with cell division, since the latter ceased while the former production increased further (Pépin et al. 1995). These results strongly suggest the necessity of measuring cell as well as biomass concentration to better characterize cell growth conditions and maximize their production performances. According to the study of Decendit et al. (1996) suspension cultures of *Vitis vinifera* in a stirred fermenter showed characteristics of growth and polyphenol metabolism similar to that found in shake flasks. These results seem promising for the use of grapevine cell cultures in bioreactors.

29.4 Biotechnology for Genetic Improvement and Maintenance of the Grapevine Germoplasm

Grapevine cultivars represent highly subtle gene combinations which can be disrupted by the sexual process. The probability that recombining genes in a hybrid can recreate the essential characters of a traditional cultivar is very low.

Wine grapes, which represent 80% of viticultural production, are bred for quality rather than for yield and a consequence of this is that grapevine cultivars of established characteristics have been perpetuated by vegetative propagation. This is a mode of propagation in which the genetic apparatus of the mother plant is reproduced with high fidelity in the offspring.

Innovation in vine-growing is made difficult by several factors in the interaction between genotype and environment. Some cultivars adapt to a wide range of environments but others perform well only in a particular environment (Branas et al 1980).

Another disincentive to genetic innovation in vine-growing is the conservatism of producers and consumers. The good quality in table wine is based on a restricted spectrum of flavors, aromas and other gustatory attributes. The wines with unfamiliar flavors are generally not well accepted. To be successful a new wine cultivar must be very similar, or perhaps indistinguishable from the traditional cultivar that it is designed to replace. Therefore modern wine production is based primarily on traditional cultivars.

Clonal selection, which has become the most widely used procedure for improvement of wine grapes, has been designed as a "high tech" agricultural practice allowing the perpetuation of traditional cultivars. All members of a family asexually propagated from a mother vine are clones and in theory they may be considered genetically identical, but the existence of variation within clonal cultivars of wine grapes is firmly established. The genetic differences among the traditional varieties or among clones arising within a cultivar, can be exploited to obtain better genotypes, selecting those which seem to have the better characteristics (Becker 1977). Nevertheless, good progress has been made in breeding new cultivars of wine grapes by both intraspecific hybridization (crossing between *vinifera* cultivars) and interspecific hybridization. Table and raisin grapes are not subjected to the same constraints as wine grapes regarding tradition and consumer acceptance, and breeding programs based on intraspecific hybridization have produced several new cultivars. Despite the strong pressure exerted by tradition a few notable successes have also been obtained with wine grapes. In both California and Australia new cultivars of wine grapes, which are well adapted to dry hot, irrigated conditions have been produced (Antcliff 1975). Besides, the advent of tissue culture and genetic engineering, and the application of these technologies to crop improvement, had much significance for viticulture.

Hopefully, in the near future molecular technologies will be used for grapevine genetic improvement, as for instance the insertion of foreign genes into the genomes of traditional cultivars in order to make them resistant against diseases, weed killers and insects, without altering the genes concerning any of their other characteristics, including those influencing wine quality. Of special interest as milestone studies because of their tremendous biotechnological innovation impact, we would like to mention the conferring of resistance to virus disease by expression of plant antibody in *Nicotiana benthamiana* (Tavladoraki et al. 1993), the incorporation of viral coat protein genes in *Nicotiana tabacum* and *Cucumis melo* (Maiti et al.1997) or the expression of virus satellite RNA in *Nicotiana tabacum* (Reavy et al.1992).

The genetic improvement of woody plants, such as the vine, has great commercial importance and there are already interesting scientific perspectives as concerns the possibility of transforming vines belonging to the group commonly used as rootstocks. The first positive results in the production of transgenic vines, have been obtained with the introduction of exogenous genes into *Vitis rupestris S.* genome by Martinelli et al. (1994). Besides, in respect to the selection with markers associated with the quantitative characters, Lodhi et al. (1995) supplied the genetic linkage map for the genus *Vitis,* which can be used to tag valuable phenotypical traits associated to different loci present in each species.

Nevertheless the grapevine is still a recalcitrant material for genotype improvement by genetic engineering since it is a perennial plant with a long juvenile period showing a high degree of heterozygosis, features, which contrast to the main characters showed by other horticultural crops such as tomato, apple, maize, potato for which genetically transformed plants are already commercially widespread (Richard 1992).

29.4.1
Molecular Technology and Germoplasm Biodiversity

The plant ability to adapt to new environmental conditions depends on the genetic variability existing in the population and gives the species the ability to survive the adversity and the environmental stress. Each breed or variety or cultivar present in a particular habitat constitutes a genetic patrimony, which at any time can become useful in participating in the formation of new genetic combinations produced by the evolution of vegetable and animal productions, as a result of environment adaptation.

If the maintenance of genetic variability (biodiversity) in a species is considered fundamental for the preservation of germoplasm resources, the term biodiversity is also generally recognized as being important because the physiological differences among species provide various sources of food and medicine for man.

In the course of this century many grapevine varieties risked disappearing as a result of some profound changes in the vine-growing structures and orientations.

The extinction of cultivars and biotypes concerns the grapevine as well as all the other arboreal and herbaceous species, in Italy and in the world, causing a restriction of the genetic variability (Pisani 1986).

Among the strategies used to oppose the biodiversity loss, the constitution of characterized grapevine germoplasm collections became of some importance as a genetic reserve of the representative variability of a population.

Recently, biochemical and molecular techniques have been combined with ampelography and ampelometry for evaluating and conserving plant biodiversity.

The biochemical methodologies have been widely employed to resolve taxonomic problems and for the characterization of grapevine rootstocks. These methodologies include the analysis of the secondary metabolites present in grapevines (Di Stefano 1996), and the use of isozyme markers (Royo et al. 1997; Walker et al. 1995).

Some molecular techniques have been used for mapping and tagging traits of interest in the plant genome and to detect diversity at the DNA level. They can provide, at species level, some information to help define the distinctiveness of a species and can also be used to resolve the problems concerning the hybridization and the polyploidy. At the population level, the molecular markers can help in determining how many different genetic classes are present, the genetic similarities among them, how much diversity is present in those classes and their evolutionary relationships. (Karp et al. 1997).

Molecular marker techniques can be grouped into different categories according to the methodology adopted. Basically, genotyping tests may be divided in PCR- and non PCR-based ones, which can use specifically designed primers for a known sequence or arbitrary primers for unknown sequences.

The choice of a molecular marker rather than another for genome analysis depends on different factors such as the level of polymorphism, the technical and economical considerations, the aim of the research, the genetics and biology of the studied species (Vendramin et al. 1996).

For example, to distinguish between closely related genotypes, a variable number of tandem repeats fingerprint (VNTR), amplified fragment length polymorphisms (AFLP), and all arbitrary primed approaches (as RAPD), can be used because they produce multilocus profiles.

Besides, the DNA markers can provide information useful in determining how populations of given species are distributed, how genetically distinct different populations are, how much genetic variation is present in/and between populations.

Recently, Powell et al. (1995) showed the high efficiency of the use of microsatellites localized inside the plastidial genome to study the genetic variability between seven natural populations of *Pinus leucodermis*. The plastidial genome is normally uniparental so that the analysis of the plastidial microsatellites polymorphism can be useful to study the gene flow in/and among populations.

Recently, the topic of biodiversity also interested cultivated species, including *Vitis vinifera*. Especially for ancient and geographically dispersed cultivars, it is known that many phenotypical variants are commonly propagated. The maintenance of this variability is thought to be crucial for the adaptability of this species to different habitats. Several studies have focused on the topic of grapevines biodiversity, and the degree of phenotypical variability occurring in existing vineyards (Scienza 1993; Bogoni et al. 1993; Qu et al. 1996).

29.4.1.1
Molecular Technologies for Grapevines Genotype Identification

The characterization of a DNA sample for individual identity according to its sequence information, often referred to as DNA fingerprint, has been used for several kinds of analysis, like genome linkage mapping, identity testing, determination of family relationship and genetic variation, population and pedigree analysis (Caetano-Anollès et al. 1991). The identification of different grapevine varieties and clones is very important especially in consequence of the ever more intense exchange of the propagation material between wine producing and consumer countries (Stavrakakis et al. 1983).

The detection of a polymorphic DNA locus characterized by a number of variable-length restriction fragments, termed restriction fragment length polymorphism (RFLP), was originally the most diffused technique used for the grapevines genotype characterization.

Many studies used the RFLP test for grapevine varietal identification because of its high level of consistency and its reliability in assessing genetic similarity among grapevine cultivars (Bowers end Meredith 1996).

A preliminary study on *Vitis vinifera* "Sangiovese" by RFLP analysis showed that a biotype supposed to be a "Sangiovese" showed distinct patterns with respect to the other "Sangiovese" clones analyzed, supporting the idea that it either did not belong to the "Sangiovese" cultivar or that it could be the result of an accumulation of somatic mutation within this variety (Vignani et al. 1994). These findings supported by further analysis, (Vignani et al. 1996) gave new emphasis to the open discussion related to the existence either of polyclonal cultivars (Scienza, 1993) or to the more traditional and strict definition of cultivar as being composed of identical genotypes (Pisani, 1986).

Nevertheless in the last years new approaches to DNA fingerprinting have been set-up. They exploit the enzymatic amplification of specific DNA sequences using the PCR (Erlich et al. 1991) and can be subdivided into two groups, based on the criterion adopted for the primer design:

1) DNA amplification using arbitrary primers (Random Amplified Polymorphic DNA (RAPD));
2) DNA amplification using primers that are specific for known sequences (Amplification of Microsatellites also called Variable Number of Tandem Repeats (VNTR), Amplified Fragment Length Polymorphism (AFLP), Inverse Sequence Tagged Repeats (ISTR), Sequence Characterized Amplified Region (SCAR)).

RAPD analysis has been largely used because of the facility and rapidity of its application.

The early studies concerning the use of this technique in detecting genetic differences among closely related organisms have been employed by Welsh et al. (1990) and by Williams et al. (1990), while the RAPD application in the grapevine genotype analysis started to be experimented upon a few years later (Collins and Symons 1993; Newbury et al. 1993). The RAPD has been adopted successively to identify molecular genetic markers associated with seedlessness in table grapes (Striem et al. 1996) and Qu et al. (1996) showed the RAPD markers utility for investigating the genetic diversity between two morphologically distinct types of grapes, belonging to the subgenera *Euvitis* and *Muscadina* in the genus *Vitis.*

Despite the large diffusion of this technique, same workers have described possible difficulties using RAPD due to data reproducibility and interpretation especially for woody plant species having complex genomes (Mulchay et al. 1995; Xu et al. 1995).

The analysis of microsatellites, where the primers amplify simple sequence repeats of the genome, guaranties a higher degree of consistency. The primers are designed on the grounds of the information sequence of the flanking regions, which is normally univocal and genus characteristic (Cipriani et al. 1994). Microsatellites are ubiquitous, short tandemly arranged di- or trinucleotides repetitions, interspersed in the genome, highly variable in length and therefore ideal for polymorphisms detection. Another characteristic is that these markers are codominant, different from RAPD which is dominant or null, allowing to distinguish heterozygous genotypes.

The microsatellites markers proved to be more suitable for the characterization of near-isogenic grapevine genotypes, than other molecular tests (Cipriani et al. 1994). Besides, the high polymorphism of microsatellites (Bowers et al. 1996) enables the individualization of alleles differing even for a single nucleotide (Thomas et al. 1994). Recently, DNA polymorphism at seven microsatellite loci among 12 clones of *Vitis vinifera* "Sangiovese" was employed in order to assess their genetic uniformity (Vignani et al. 1996). The same kind of analysis was carried out for six microsatellite loci among 5 clones of "Fortana", showing that molecular tests based upon microsatellite DNA analysis can objectively separate different genotypes (Silvestroni et al. 1997) and possibly contribute to determining the historical origin or probable parentage of known, traditional cultivars (Bowers and Meredith 1997).

Among the advantages of the microsatellite markers technology, great interest is growing in the possibility, originally suggested by Thomas et al. (1994), of linking this kind of analysis to a suitable automatic system for the DNA analysis connected to an electronic data-base in such a way as to make it easier for the creation of data banks on grapevine genotyping and to favour exchanges among different laboratories. In relation to this, an international consortium stipulated in collaboration with several research groups, for the characterization of new microsatellite loci in *Vitis vinifera* is being activated, and it should lead in the near future to an increase in the accuracy of grapevine genotyping (Meredith, personal communication).

Even if the microsatellite markers are useful for the grapevine genotypes identifications, new analytical techniques have been developed in order to compare the controversial and difficult aspects of clonal differentiation and/or identification.

The Sequence Characterized Amplified Region (SCAR), which represents an "evolution" of the RAPD test, has been used for the identification of grape rootstocks by Xu et al. (1996).

The AFLP and ISTR, introduced as new tools for genetic analysis (Vos et al. 1995; Rohde 1996), were used for the characterization of grapevine genetic biodiversity. Both technologies proved to be powerful tools in the characterization of intraspecific variation among cultivars of *Vitis vinifera* (Sensi et al. 1996). These DNA marker technologies are more sensitive than RFLP and generally more reproducible and reliable than RAPD, especially for woody plants. With reference to microsatellite DNA amplification, which allows allele polymorphism analysis at one specific locus, AFLP and ISTR provide a general overview of the genome structure based on multiple loci/single allele polymorphism analysis.

Due to their specificity in separating cultivar genotypes, AFLP and ISTR together with other DNA marker techniques currently available, will hopefully routinely complement the more traditional ampelometric analysis of grapevine.

Encouraging results have been obtained in terms of the possible application of AFLP and ISTR methods to the difficult subject of clonal distinction and identification in *Vitis vinifera* (Sensi et al. 1996).

29.5 Conclusions and Future Perspectives

The availability of effective molecular methods for the grapevine genome analysis, which can complement the more traditional ampelometric analysis, could favour the grapevine varietal identification.

As reported above in the last few years same modern techniques for the DNA analysis have been widely diffused. The perspectives of the use of same molecular markers for the identification of species and varieties allow a methodological objective study of the plant, realized in a controlled environment and at any moment of the plant's life, to be carried out.

The molecular analysis is already applied in medicine for resolving law and fatherhood problems. It is expected, in the future, to use molecular markers in the nursery and plant propagation industry as well, in order to supply a genetic guaranty. The DNA fingerprinting should be a genetic guaranty added to the traditional certification system used for the propagation material.

The molecular techniques used for the genome identification are subjected to continuous development becoming more and more sophisticated. **Very interesting** on this subject is the recently developed DNA chip technology and its applications to gene discovery and expression, detection of mutations or polymorphisms and mapping. Two variants of the chip exist. On one hand the DNA is immobilized to a solid surface such as glass and exposed to a set of labeled probes either separately or in a mixture; on the other an array of oligonucleotide probes is synthesized either in situ or by conventional synthesis followed by on-chip immobilization. The array is exposed to labeled sample DNA, hybridized, and complementary sequence is determined.

In the medical field DNA microchips are already available for genetic studies concerning the expression profile, the polymorphism analysis and diagnostic, and the novel genes identification.

There are also some good perspectives for the use of DNA chips in plant genetics. cDNA from *Arabidopsis thaliana* was robotically printed on glass microscope slide coated with poly-L-lysine and denatured. Fluorescently-labeled probes were labeled with fluorescein or lissamine and hybridized to the array under stringent conditions. Hybridization was measured with a laser scanner and displayed as a pseudocolor display of differential expression (Ramsay 1998). Shalon et al. (1996) extended this technology by producing microarrays of genomic DNA from λ clones of *Saccharomyces cerevisiae.*

It is desirable that, in the future, the DNA chip technology will be used for the grapevine genome analysis to make the genotype identification process faster.

Although modern biotechnology tends to adopt more and more sensible techniques able to detect even low degrees of DNA polymorphism, the main aim of the use of molecular markers for the vine-growing industry must be the clonal differentiation and/or identification. In relation to this, it seems to be of extreme importance that the scientific community working on grapevine genotyping will focus its attention on the mapping of molecular markers that are characteristic for the identification of valuable phenotypical traits.

The clonal identification is very important especially considering that the producers need and want better grapes possibly as specialized biotypes within traditional cultivars. Thus, the economic viability of producers depends on the continuous use of grape varieties for which there is an established market demand. Therefore the strategy of producers till now, has been to exploit spontaneous variation within the old cultivars through clonal selection, but it seems likely that in the near future biotechnological tools will become essential for grapevine selection programs.

References

Antcliff AJ (1975) Four new grapes varieties released for testing. F Aust Inst Agric Sci 41:262–264

Becker H (1977) Methods and results of clonal selection in viticulture. Acta Hort 75:111–112

Bogoni M, Reina A, Valenti L, Scienza A (1993) Valutazione della variabilità intravarietale attraverso procedure di pressione selettiva debole. Vignevite 12:25–30

Boss PK, Davies C, Robinson SP (1996a) Expression of anthocyanin biosynthesis pathway in red and white grapes. Plant Mol Biol 32:565–569

Boss PK, Davies C, Robinson SP (1996b) Analysis of the expression of anthocyanin pathway genes in developing *Vitis vinifera* L. cv Shiraz grape berries and the implications for pathway regulation. Plant Physiol 111:1059–1066

Bowers JE, Meredith CP (1997) The parentage of a classic wine grape, Cabernet Sauvignon. Nature Genetics 16:84–87

Bowers J E, Meredith C P (1996) Genetic similarities among wine grape cultivars revealed by restriction fragment-length polymorphism (RFLP) analysis. J Amer Soc Hort Sci 121(4):620–624

Bowers JE, Dangl GS, Vignani R, Meredith CP (1996) Isolation and characterization of new polymorphic simple sequence repeat loci in grape (*Vitis vinifera*). Genome 39:628–633

Branas J (1980) Sol, vigne, qualitè des vins. Progr Agric Vitic 24:529–532

Caetano-Anollès G, Bassam B J, Gresshoff P M (1991) DNA amplification fingerprint: a strategy for genome analysis. Plant Molecular Biology Reporter 9(4):294–307

Cipriani G, Frazza G, Peterlunger E, Testolin R (1994) Grapevine fingerprinting using microsatellite repeats. Vitis 33:211–215

Collins GG, Symons RH (1993) Polymorphism in grapevine DNA detected by the RAPD PCR technique. Plant Mol Biol Rep 11:105–111

Decendit A, Ramawat KG, Waffo P, Deffieux G, Badoc A, Merillon JM (1996) Anthocyanins, catechins, condensed tannins and piceid production in *Vitis vinifera* cell bioreactor cultures. Biotechnology Letters 18(6):659–662

Derks W, Creasy LL (1989) Influence of fosethyl-Al on phytoalexin accumulation in the Plasmopara viticola-grapevine interaction. Physiol Mol Plant Pathol 34:203–213

Di Stefano R (1996) Metodi biochimici nella caratterizzazione varietale. Riv Vitic Enol 1:51–56

Erlich HA, Gelfand D, Sninsky JJ (1991) Recent advances in the polymerase chain reaction. Science 252:1643–1651

Frankel EN, Waterhouse AL, Kinsella JE (1993) Inhibition of human LDL oxidation by resveratrol. Lancet 314:1103–1104

Goldberg DM, NG E, Karumanchiri A, Diamandis EP, Soleas GJ (1996) Resveratrol glucosides are important components of commercial wines. Am J Enol Vitic 47(4):415–420

Karp A, Edwards KJ, Bruford M, Funk S, Vosman B, Morgante M, Seberg O, Kremer A, Boursot P, Arctander P, Tautz D, Hewitt GM (1997) Molecular technologies for biodiversity evaluation: opportunities and challenges. Nature Biotechnologies 15:625–628
Kim CS, Lee CH, Shin JS, Chung YS, Hyung NI (1997) A simple and rapid method for isolation of high quality genomic DNA from fruit trees and conifers using PVP. Nucleic Acid Res 25:1085–1086
Llewellyn DJ, Finnegan EJ, Ellis JG, Dennis ES, Peacock WJ (1987) Structure and expression of an alcohol dehydrogenase 1 gene from *Pisum sativum* (cv Green-feast). J Mol Biol 195:115–123
Lodhi MA, Daly MJ, Ye GN, Weeden NF, Reisch BI (1995) A molecular marker based linkage map of *Vitis*. Genome 38:786–794
Longhurst T, Lee R, Hinde R, Brady C, Speirs J (1994) Structure of the tomato Adh2 pseudogenes, and a study of Adh2 gene expression in fruit. Plant Mol Biol 26:1073–1084
Lopez-Serrano M, Ferrer MA, Pedreño MA, Barceló AR (1997) Ca^{2+} and Mg^{2+} ions counteract the reduction by fosethyl-Al (aluminum tris[ethyl phosponate]) of peroxidase activity from suspension-cultured grapevine cell. Plant Cell, Tissue and Organ Cult 47:207–212
Loulakakis KA, Roubelakis-Angelakis KA, Kanellis AK (1996) Isolation of functional RNA from grapevine tissues poor in nucleic acid content. Am J Enol Vitic 47:181–185
Maiti IB, Hunt AG (1997) Genetically Engineered Protection of Plants Against Potyviruses. In: Gresshoff PM (ed) Technology Transfer of Plant Biotechnology. Handbook A CRC Series in Current Topics in Plant Molecular Biology, pp 51–65
Martinelli L, Mandolino G (1994) Genetic transformation and regeneration of transgenic plants of grapevine (*Vitis rupestris* S). Theor Appl Genet 88:621–628
Meyer AS, Yi O-S, Pearson DA, Waterhouse AL, Frankel EN (1997) Inhibition of human low-density lipoprotein oxidation in relation to composition of phenolic antioxidants in grapes (*Vitis vinifera*). J Agric Food Chem 45:1638–1643
Mozsar J, Viczian O (1996) Genotype effect on somatic embryogenesis and plant regeneration of *Vitis* spp. Vitis 35(4):155–157
Mulcahy DL, Cresti M, Linskens HF, Intrieri C, Silvestroni O, Vignani R, Pancaldi M (1995) DNA fingerprinting of italian grape varieties: a test of reliability in RAPDs. Adv Hort Sci 9:185–187
Mulcahy DL, Cresti M, Sansavini S, Douglas GC, Linskens HF, Bergamini-Mulchay G, Vignani R, Pancaldi M (1993) The use of random amplified polymorphic DNAs to fingerprint apple genotypes. Sci Hortic 54:89–96
Newbury HJ, Ford Lloyd BV (1993) The use of RAPD for assessing variation in plants. Plant Growth Regulation 12:43–51
Pepin MF, Archambault J, Chavarie C, Cormier F (1995) Growth kinetics of *Vitis vinifera* cell suspension cultures: I. shake flask cultures. Biotechnology and Bioengineering 47:131–138
Pisani P L (1986) Primi risultati di ricerche sul patrimonio varietale viticolo della Toscana. L'Enotecnico 1001–1005
Powell W et al (1995) Polymorphic simple sequence repeat regions in chloroplast genomes: application to the population genetics of pines. Proc Natl Acad Sci 92:7759–7763
Qu X, Lu J, Lamikanra O (1996) Genetic diversity in Muscadine and America Bunch grape based on Randomly Amplified Polymorphic DNA (RAPD) Analysis. J Amer Soc Hort Sci 121(6):1020–1023
Ramsay G (1998) DNA chips: State-of-the art. Reviev. Nature Biotecnology 16:40–44
Reavy B, Mayo MA (1992) Genetic Engineering of Virus Resistance. In: Gatehouse AMR, Hilder VA, Boulter D (eds) Plant Genetic Manipulation for Crop Protection. Handbook of Biotechnology in Agriculture Series 7:183–214
Richard J A, Barfoot C, Barfoot P D (1992) The Development of Genetically Modified Varieties of Agricultural Crops by the Seeds Industry. In: Gatehouse AMR, Hilder VA, Boulter D (eds) Plant Genetic Manipulation for Crop Protection. Handbook of Biotechnology in Agriculture Series 7:45–73
Rohde W (1996) Inverse sequence-tagged repeats (ISTR) analysis: a novel and universal PCR-based technique for genome analysis in the plant and animal kingdom. J Genet & Breed 50 (3):249–261
Roubelakis KA, Zivanovitc SB (1991) A new culture medium for in vitro Rhizogenesis of grapevine (*Vitis* spp.) genotypes. Hort Science 26(12):1551–1553
Royo J B, Cabello F, Miranda S, Gorgorcena Y, Gonzalez J, Moreno S, Itoiz R, Ortiz J M (1997) The use of isoenzyme in characterization of grapevine (*Vitis vinfera L.).* Influence of the environment and time of sampling. Scientia Horticulturae 69:145–155
Sarni-Manchado P, Verriès C, Tesnière C (1997) Molecular characterization and structural analysis of one alcohol dehydrogenase gene (GV-Adh 1) expressed during ripening of grapevine (*Vitis Vinifera* L.) berry. Plant Sci 125:177–187
Scienza A (1993) Vigneti policlonali e valorizzazione della diversità dei vini. Vignevini 12:23–24
Sefc KM, Ruckenbauer P, Regner F (1997) Embryogenesis in microspore culture of *Vitis* subspecies. Vitis 36:15–20
Sensi E, Vignani R, Rohde W, Biricolti S (1996) Characterization of genetic biodiversity with *Vitis vinifera* L. Sangiovese and Colorino genotypes by AFLP and ISTR DNA marker technology. Vitis 35(4):183–188

Shalon D, Smith JS, Brown PO (1996) A DNA microarray system for analyzing complex DNA samples using two color fluorescent probe hybridization. Genome Res 6:639–645
Shubert R, Fisher R, Hain R, Schreier PH, Bahnweg G, Ernst D, Sandermann HJr (1997) An ozone-responsive region of the grapevine resveratrol synthase promoter differs from the basal pathogen-responsive sequence. Plant Mol Biol 34:417–426
Silvestroni O, Di Pietro D, Intrieri C, Vignani R, Filippetti I, Del Casino C, Scali M, Cresti M (1997) Detection of genetic diversity among clones of cv. Fortana (*Vitis vinifera* L) by microsatellite DNA polymorphism analysis. Vitis 36(3):147–150
Sparvoli F, Martin C, Scienza A, Gavazzi G, Tonelli C (1994) Cloning and molecular analysis of structural genes involved in flavonoid and stilbene biosynthesis in grape (*Vitis vinifera* L.). Plant Mol Biol 24:743–755
Stark-Lorenzen P, Nelke B, Hänßler G, Mühlbach HP, Thomzik JE (1997) Transfer of a grapevine stilbene synthase gene to rice (*Oryza sativa* L.). Plant Cell Rep 16:668–673
Stavrakakis M, Loukas M (1983) The between- and within-grape cultivar genetic variation. Sci Hort 19:321–334
Striem MJ, Ben-Hayyim G, Spiegel-Roy P (1996) Identifying molecular genetic markers associated with seedlessness in grape. J Amer Soc Hort Sci 121(5):758–763
Tavladoraki P, Benvenuto E, Trinca S, De Martinis D, Cattaneo A, Galeffi P (1993) Transgenic plants expressing a functional single-chain Fv antibody are specifically protected from virus attack. Nature 366:469–472
Thomas MR, Cain P, Scott NS (1994) DNA typing of grapevine: a universal method and database for describing cultivars and evaluating genetic relatedness. Plant Molecular Biology 25:939–949
Vendramin GG, Lelli L (1996) Marcatori molecolari nel miglioramento genetico delle piante: i microsatelliti. BioTec 4:49–56
Vignani R, Bowers JE, Meredith CP (1996) Microsatellite DNA polymorphism analysis of clones of *Vitis vinifera* "Sangiovese". Scientia Horticulturae 65:163–169
Vignani R, Silvestroni O, Intrieri C, Meredith CP, Cresti M (1994) Studio preliminare mediante RFLP su cloni di "Sangiovese" e di "Montepulciano". In: Università degli Studi di Bologna (eds) Tecnologie avanzate per l'identificazione varietale ed il controllo genetico sanitario nel vivaismo fruttiviticolo. Agro Bio-Frut 1994, pp 97–102
Vos P, Hogers R, Bleeker M, Reijans M, Van De Lee T, Hornes A, Frijters A, Pot J, Peleman J, Kuiper M, Zabeau M (1995) AFLP: a new technique for DNA fingerprint. Nucleic Acids Res 23:4407–4414
Walker M A, Liu L (1995) The use of isozymes to identify 60 grapevine rootstoks (*Vitis* spp.). Am J Enol Vitic 46(3):299–305
Walker MJ (1985) 5000 años de viticulture en Espagna. Riv Arqueol 6:44–47
Welsh J, Clelland M (1990) Fingerprint genomes using PCR with arbitrary primers. Nuc Acids Res 18:7213–7218
Williams JGK, Kubelik AR, Livak KJ, Rafalsky JA, Tingey SV (1990) DNA polymorphisms amplified by arbitrary primers are useful as genetic markers. Nuc Acids Res 18:6531–6535
Wolfe WH (1976) Identification of grape varieties by isoenzyme banding pattern. Am J Enol Vitic 27:68–73
Xu H, Bakalinsky TA (1996) Identification of grape (*Vitis)* rootstocks using sequence characterized amplified region DNA markers. Hort Science 31(2):267–268
Xu H, Wilson DJ, Arulsekar S, Bakalinsky TA (1995) Sequence specific Polymerase Chain Reaction markers derived from randomly amplified polymorphic DNA markers for fingerprinting grape (*Vitis*) rootstocks. J Amer Soc Hort Sci 120(5):714–720

Subject Index

Color Plates

Fig. 5.1. Key stages of male meiosis in pollen mother cells of a wheat-rye translocation line T5AS·5RL analysed by light microscopy (scale bar = 15 µm in **a–d** and 10 µm in **e–g**). **a**) Silver staining of the synaptonemal complex (thin dark brown lines) in a surface-spread meiocyte at late zygotene showing almost complete SC formation. The nucleolus remnants stain darkly (right side). **b–g**) Fluorescent *in situ* hybridization using total genomic rye DNA as a probe to localize the rye chromosome arms (in red or, **d**, green) in cells from the interphase before meiotic prophase (**b**) through leptotene (**c** showing DAPI staining of DNA, and **d** with *in situ* hybridization), zygotene (**e**), pachytene (**f**) and to metaphase I (**g**, where the red rye arm is seen pairing in a ring bivalent with wheat chromosomes stained blue with DAPI)

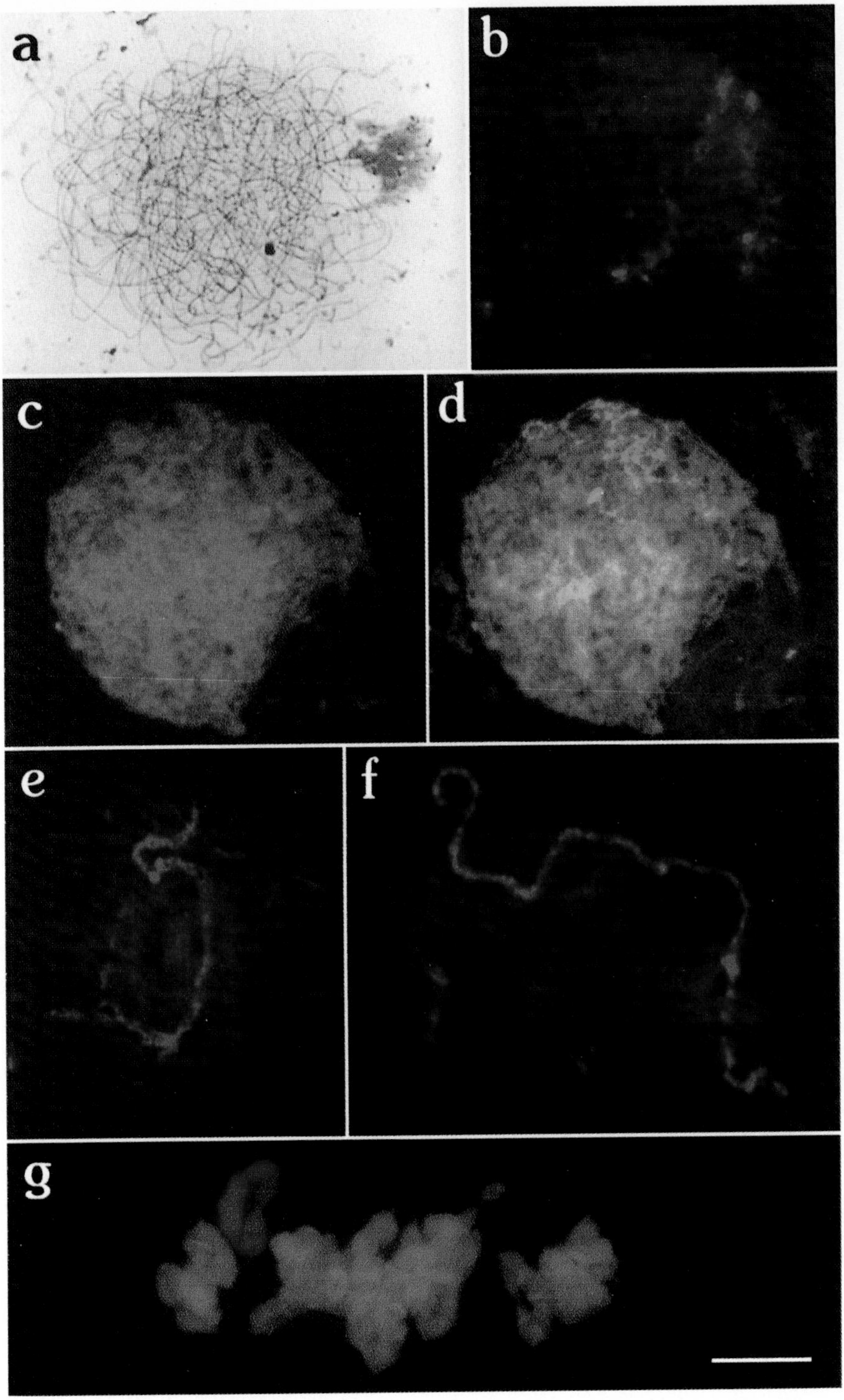
a
b
c
d
e
f
g

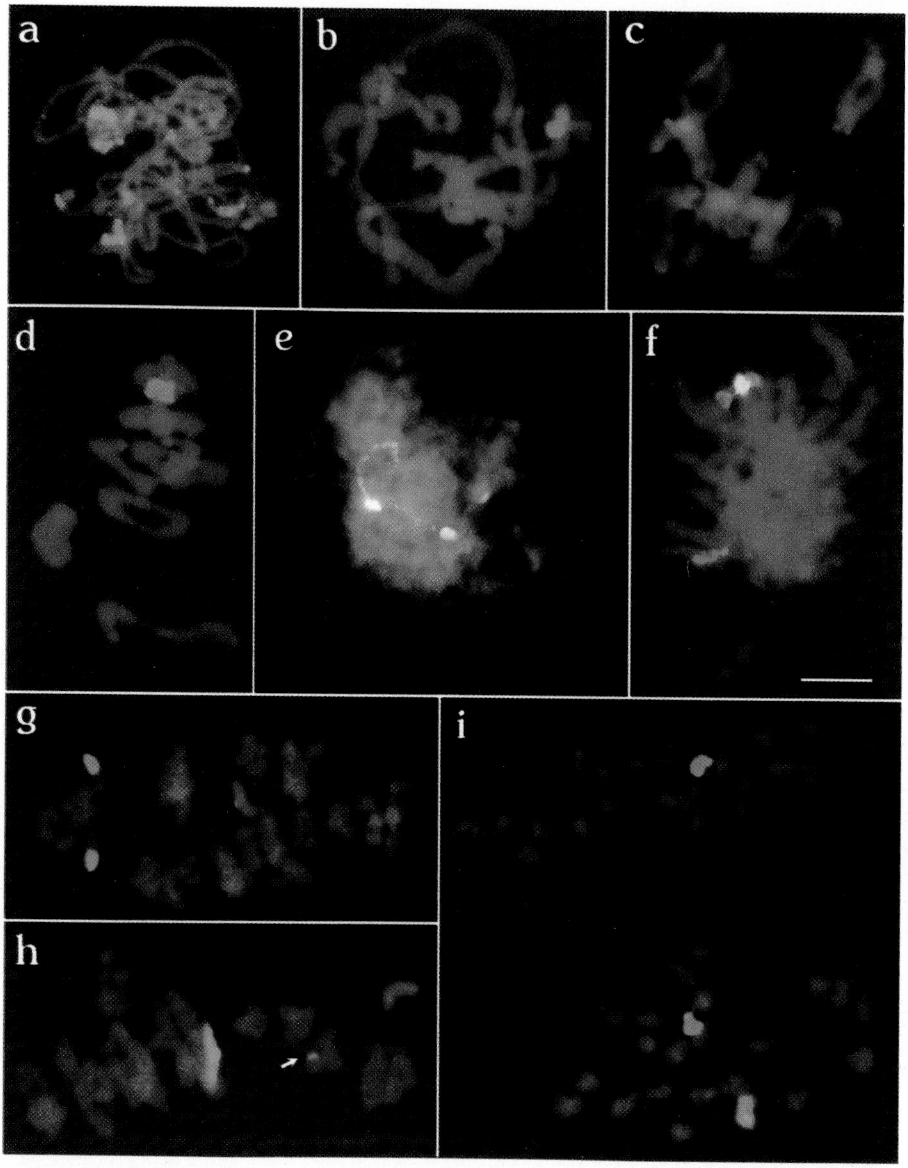

Fig. 5.2. Multi-target fluorescent *in situ* hybridization micrographs showing the localization of two DNA probes (green and red) on the DAPI stained (blue) chromosomes. (Scale bar = 8 μm in **a–f**; 10 μm in **g–i**). **a–d**) Stages of the first meiotic division in rye from pachytene (**a**), diplotene (**b**), diakinesis (**c**), and metaphase I (**d**) with the 18S–25S rDNA labelled in red (**a, c**) or green (**b, d**) and the tandemly repeated sequence pSc119.2 (green) seen in (**a**). (**e, f**) *In situ* hybridization of rye genomic DNA (red) to a wheat cultivar including a 1BL.1RS translocation at (**e**) pachytene and (**f**) diplotene; the rye chromosome arm carrying a 18S–25S rDNA site (green) can be followed through both nuclei. (**g, h**) Metaphase I and (**i**) anaphase I in a wheat line with a 1BL.1RS translocation (probed green with genomic rye DNA) and an additional *Thinopyron bessarabicum* chromosome (orange). The wheat chromosomes form 20 rod and ring bivalents (brown colour from cross-hybridization of both probes), while the wheat-rye bivalent is a rod in (**g**) with no chiasma in the rye arm and a ring in (**h**). The segregation of the wheat-rye bivalent is clear in the anaphase I (**i**). The *T. bessarabicum* chromosome forms an unpaired univalent (**g**) that segregates to one pole at anaphase I (**i**). A *de novo* translocation has occurred in the metaphase I in (**h**) between the *T. bessarabicum* and a wheat chromosome, indicating that inter-genomic recombination is a possible, even if exceptionally rare, event

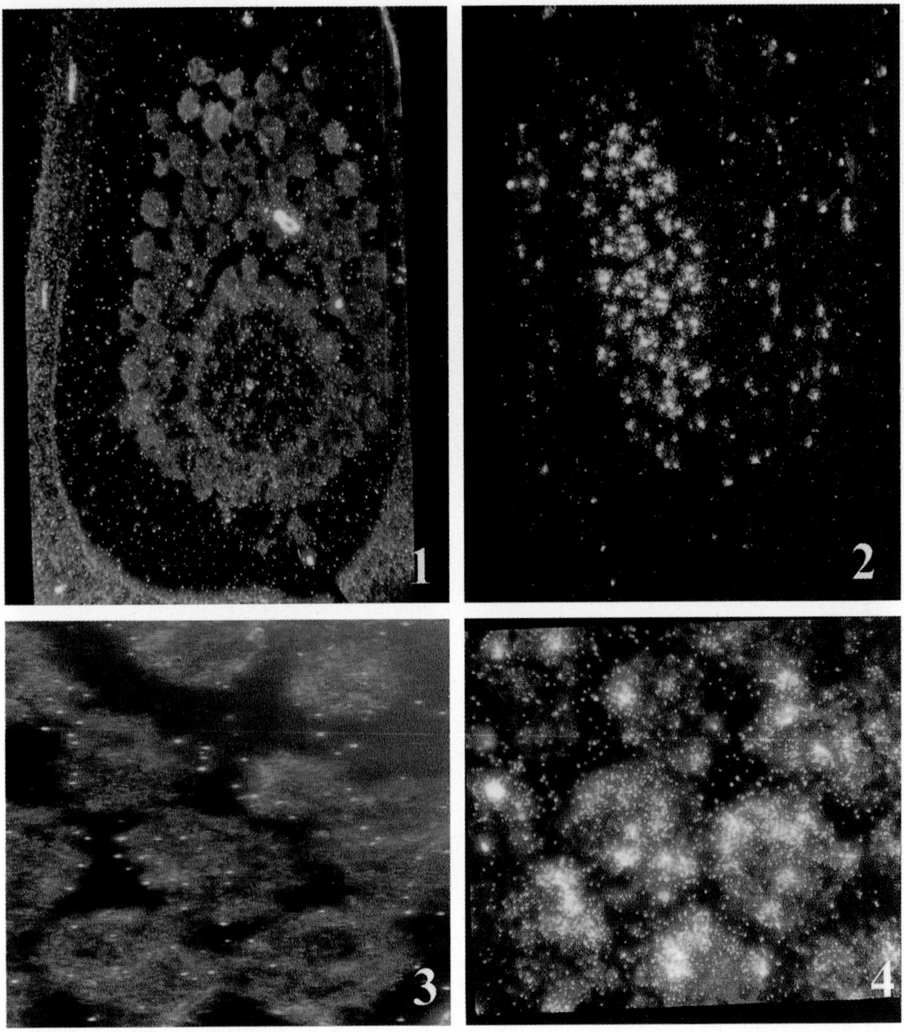

Fig. 8.5. Localisation of *Pet cyc1* mRNA in longitudinal sections of stage 4 ovaries. (*1*) and (*3*) in situ hybridisation of a *Pet cyc1* sense probe. (*2*) and (*4*) in situ hybridisation of a *Pet cyc1* antisense probe. White coloration represents regions containing RNA/RNA hybrids

Tetrad from WILD TYPE **Diad from SPO13**

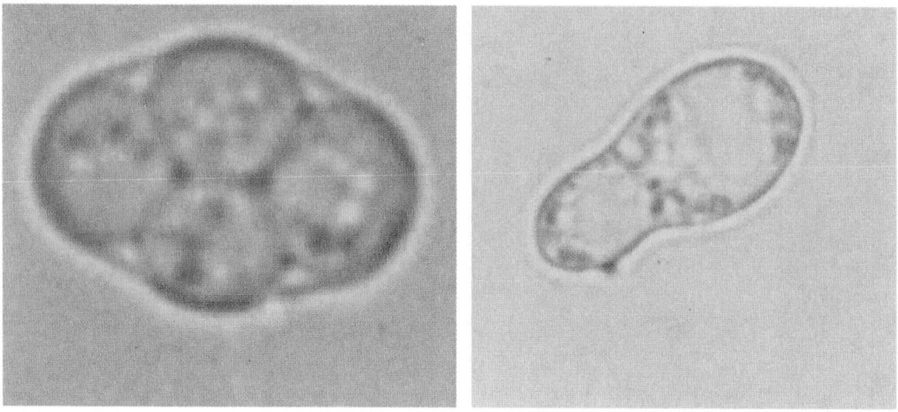

Fig. 8.6. Tetrads of haploid spores resulting from wild type yeast meiosis. Two diploid spores resulting from the yeast *spo13-1* mutant meiosis

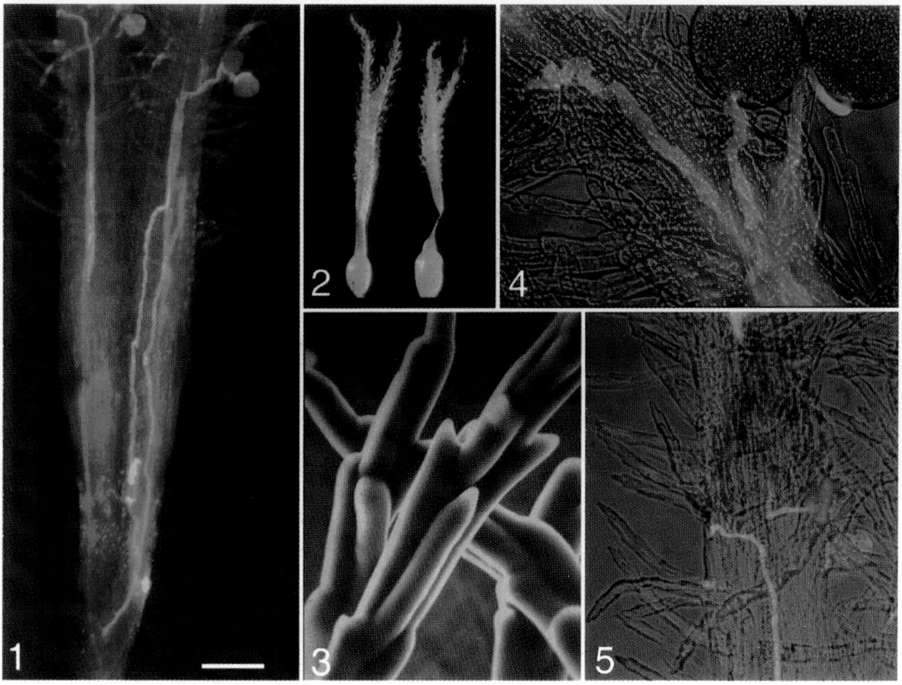

Fig. 11.1. A stigma of the grass, pearl millet, after pollination with compatible millet pollen and staining with decolourized aniline blue before observation with an epifluorescence microscope. The callose (a beta-1-4 linked glucan) in the pollen tube walls fluoresces bright blue with this stain, while chlorophyll, cell walls and other molecules show weak red and green fluorescence. Pollen tubes are seen to penetrate the stigma and grow down towards the ovule in the pollen tube transmitting tract. There is a restriction in the number which pass the abscission zone (see also Fig. 11.2), where calcium oxalate crystals are visible as green dots. Scale bar = 0.75 mm

Fig. 11.2. Ovules, styles and stigmas of pearl millet of the same age. That on the right was pollinated 72hr before collection while the other (left) remained unpollinated. After pollination, a sharp constriction forms in the style above the ovule, and the stigma trichomes have started to collapse. When disturbed, the stigma will abscise at the constriction

Fig. 11.3. Electron micrograph of fresh, uncoated stigma papillae of millet showing the receptive trichomes and the dry, plumose form characteristic of the grasses

Figs 11.4 and 11.5. Light micrographs of pearl millet stigmas pollinated by hand with fresh *Zea mays* pollen. Staining as fig. 11.1 but some red transmitted illumination added. In contrast to Fig. 11.1 (pollinated with millet), the grains do not adhere well to the stigma trichomes and the pollen tubes (yellow staining) are not guided correctly to follow the orientation of the (relatively much smaller) cells of the stigma. In fig. 11.5, two pollen tubes make sharp bends into the transmitting tract, but one turns to grow away from the ovule towards the stigma tip

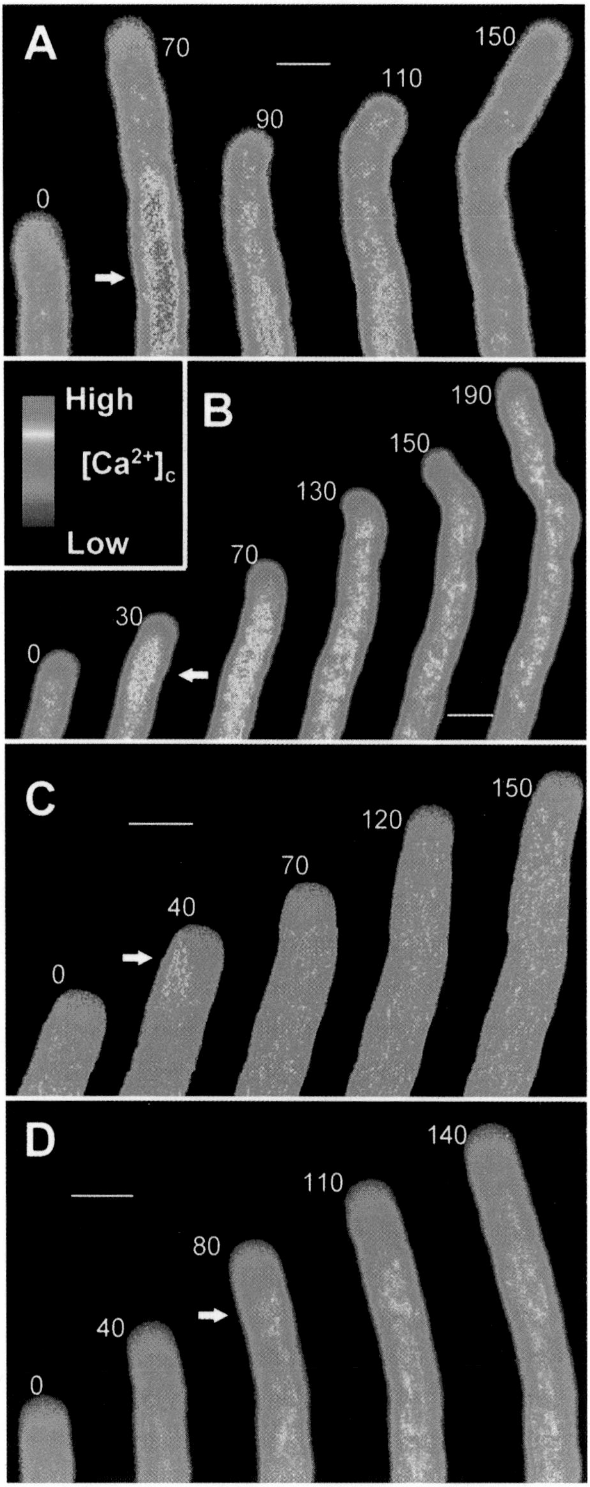

Fig. 18.4. Effect of UV photolysis of loaded caged Ins(1,4,5)P_3 in different regions of a growing Pollen Tube on the distribution of $[Ca^{2+}]_c$ and orientation. The times (sec) at which images were taken are shown adjacent to the tip of the growing tube. Images are displayed with the colour coding shown next. Bar = 10 μm. (**A**) Confocal time course series of images where caged Ins(1,4,5)P_3 was flash photolyzed in the nuclear zone (arrow) at ≈ 55 sec. Random reorientation of the tube growth axis occurs after the elevation in $[Ca^{2+}]_c$ reaches the tip. (**B**) Photolysis of caged Ins(1,4,5)P_3 in the sub apical zone at ≈ 25 sec (arrow). A transient increase in $[Ca^{2+}]_c$ and random reorientation were observed. (**C**) Photolysis of caged Ins(1,4,5)P_3 in the left hemisphere of the apical zone at ≈ 35 sec (arrow). A slight and short lasting reorientation of the tube growth axis follows the localized elevation in $[Ca^{2+}]_c$ in the left side of the tube apex. (**D**) Photolysis of caged Ins(1,4,5)P_3 was flash photolyzed in the sub apical zone at ≈ 65 sec (arrow) of a pollen tube loaded with heparin. Heparin inhibited almost completely the increase in $[Ca^{2+}]_c$ and no reorientation was observed

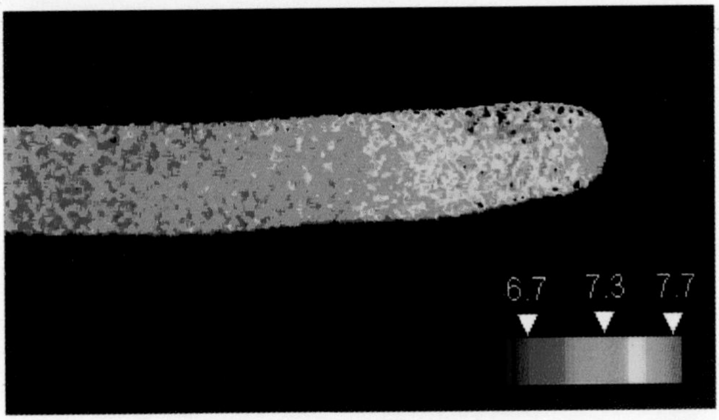

Fig. 22.3. Typical aspect of the cytosolic pH in a growing pollen tube of *Lilium*. Imaging was done using the ratio probe BCECF-Dextran (PM 10.000) and widefield microscopy with a high-sensitivity cooled CCD (for methods check Feijó et al. 1998). An alkaline band seems to divide a slightly acidic tip from a neutral tube. This alkaline band was generally associated with the so called clear cap, and could be correlated with localised effluxes of protons, detected with the extracellular vibrating probe (see. Figure 21.7 in Shipley and Feijó, this volume). This alkaline band is also spatially correlated to the zone where vectorial streaming of vesicles and organelles is disrupted, putatively by disorganisation of the actin bundles (Miller et al. 1996)

Printing: Mercedesdruck, Berlin
Binding: Buchbinderei Lüderitz & Bauer, Berlin